Springer Series in Advanced Microelectronics

Volume 46

For further volumes:
http://www.springer.com/series/4076

The Springer Series in Advanced Microelectronics provides systematic information on all the topics relevant for the design, processing, and manufacturing of microelectronic devices. The books, each prepared by leading researchers or engineers in their fields, cover the basic and advanced aspects of topics such as wafer processing, materials, device design, device technologies, circuit design, VLSI implementation, and subsystem technology. The series forms a bridge between physics and engineering and the volumes will appeal to practicing engineers as well as research scientists.

Xingcun Colin Tong

Advanced Materials for Integrated Optical Waveguides

Springer

Xingcun Colin Tong, Ph.D.
Laird
1751 Wilkening Ct.
Schaumburg, IL 60173, USA

ISSN 1437-0387
ISBN 978-3-319-01549-1 ISBN 978-3-319-01550-7 (eBook)
DOI 10.1007/978-3-319-01550-7
Springer Cham Heidelberg New York Dordrecht London

Library of Congress Control Number: 2013947926

Printed on acid-free paper

Springer is part of Springer Science+Business Media (www.springer.com)

This book is dedicated to my wife Dali,
our daughter Lingbo, and our sons William
and Alan. Their love fully filled my heart
during the long hours of work on this book.

Preface

With the rapid advance of integrated optics, the importance of optical waveguides, which are the fundamental elements of optical integrated circuits, has been widely recognized. For example, the further success of broadband communications in optical networking, metro/access communications, and computing will rely on advances in optical interconnects, optical components, such as splitters, combiners, multiplexers and demultiplexers, optical switches/modulators, tunable filters, variable optical attenuators, amplifiers, and integrated optical circuits that are based on optical waveguides. The most widespread optical waveguide has been the optical fiber, which provides the huge bandwidth required to carry all of the information between tele-/datacommunication centers. After having moved from long haul backbones to metropolitan area networks and local area networks, integrated waveguides are and will become increasingly more important for rack-to-rack, board-to-board, chip-to-chip, and component-to-component within a single chip or on-chip interconnect.

Integrated optics presents a potentially low cost and higher performance alternative to electronics in optical communication systems. A dramatic scaling down in feature sizes has mainly driven semiconductor productivity and performance increases at exponential rates in the past decades, as smaller devices on lager wafers led to larger yield, lower cost, and faster circuits. However, with approach of Giga-scale integration, the scaling-down in sizes has almost reached a physical limitation due to its negative impact on the resistance and inductance of metal interconnects with current copper-trace base technology. As a result, electronic interconnects have become the primary limit on high speed/high frequency circuits, causing significant propagation delays for clock signals, overheating, information latency, and electromagnetic interference. The optical interconnect is therefore an increasingly attractive alternative, which can provide much greater bandwidth, lower power consumption, decreased interconnect delays, resistance to electromagnetic interference, and reduced crosstalk. Optical interconnects in which light can be generated, guided, modulated, amplified, and detected need to be integrated with standard electronic circuits to combine information processing capabilities of electronics with data handling of photonics. However to be widely adopted, these covering technologies must provide significant performance breakthrough with a cost-effective engineering. In other words, an optical interconnect could replace an electrical interconnect when it has higher performance at lower cost and strong manufacturability in high volume. This can be done through integrated optics with optimal waveguide materials selection, and improved component design and integration.

Integrated waveguide optics now represents a truly multidisciplinary field of science and engineering, which encompasses not only a rather broad range of research topics but also a diverse variety of well-rooted industrial applications. As the technology of waveguide optical devices and systems has matured, it has penetrated a number of markets by providing performance-competitive integrated optic and fiber-optic device solutions. This growth, in turn, requires new developments in modeling, further advances in material science, and innovations in integration platforms. In addition, the processing and fabrication of these new devices must be optimized in conjunction with the development of accurate and precise characterization and new testing techniques. In response to these critical needs, there have been many investigations and revolutionary advances in optical waveguide materials and fabrication technologies for integrated optics that promise integrated and cost-effective optical interconnect solutions. As a result, a great many papers, articles, and presentations have been published on the development of advanced materials for optical waveguides. However, few comprehensive discussions devoted to the buildup of a fundamental system from optical waveguide materials have been available to students, scientists, and engineers.

To meet this need, this book aims to introduce in a comprehensive manner advanced materials and fabrication technologies, characterization methodology, design and simulation guidelines, and perspectives on likely future trends and challenges in integrated optical waveguides for information technology and data communications. Chapter 1 provides an outline of waveguiding theory, evolution of optical interconnects, waveguide component and integration technology, waveguide materials and fabrication techniques, and component simulation and design guideline. Chapter 2 gives an extensive review of assessment techniques and characterization methodology for advanced optical waveguide materials and components. Chapter 3 provides an overview of the state of the art of optoelectronic devices integrated with optical waveguides, including light sources, modulations, detectors, receivers, optical pathways, switches, and their integration technology. Chapters 4–11 are an in-depth introduction to the diverse and increasing number of advanced optical waveguide materials and fabrication techniques, including optical fibers, semiconductors, electro-optic materials, glasses, silicon-on-insulator technology, polymers, hollow waveguides, and metamaterials. Finally, Chapter 12 presents a perspective on anticipated future trends in advanced materials for integrated optical waveguides and their applications.

It is a great pleasure to acknowledge the help and support I have received from my colleagues who have contributed to my understanding of optical waveguides and integrated optics. I would like to express my sincere gratitude to Dr. David Packer, the series editors, and all other editing staff at Springer who helped to give the text its final polish.

Schaumburg, IL, USA Xingcun Colin Tong

Contents

Abbreviations

ABC	Activity-based costing
ABF	Air blown fiber
AD	Adsorption and desorption
aDSR	Asymmetric double split-ring resonators
AO	Acousto-optic
APD	Avalanche photodiodes
AR	Anti-reflection
ARDE	Aspect ratio dependent etching
ARROW	Anti-resonant reflecting optical waveguide
ASE	Amplified spontaneous emission
ATIR	Attenuated total internal reflection
ATR	Attenuated total reflection
AWG	Arrayed waveguide grating
BCB	Benzocyclobutene
BCL	Birefringence compensating layer
BD	Blu-ray Disc
BESOI	Bonded and etch-back silicon-on-insulator
BGA	Ball-grid-array
BIC	Boron interstitial cluster
BMI	Brain machine interface
BO	Bridging oxygen
BOX	Buried oxide
BPM	Beam propagation methods
BSI	Back-side illumination
BW	Backward
CAD	Computer aided design
CAPEX	Capital expenditure
CATV	Cable television
CBE	Chemical beam epitaxy
CCD	Charge-coupled device

CDR	Crash data retrieval
CL	Cathodoluminescence
CLBO	Cesium lithium borate
CML	Current mode logic
CMOS	Complementary metal–oxide–semiconductor
CMP	Chemical mechanical polishing
CMS	Core-multi-shell
CMT	Coupled mode theory
COC	Cyclo-olefin co-polymer
COP	Cyclic olefin polymer
CT	Crosstalk
CTE	Coefficient of thermal expansion
CVD	Chemical vapor deposition
CW	Continuous-wave
DBR	Distributed Bragg reflector
DCF	Dispersion compensating fiber
DFB	Distributed feedback
DFF	Dispersion flattened fiber
DFG	Difference frequency generation
DNA	Deoxyribonucleic acid
DNTT	Nucleotidylexotransferase
DOS	Density of optical states; digital optical switch
DSF	Dispersion shifted fiber
DSL	Digital subscriber line
DVD	Digital video disc
DWDM	Dense wavelength division multiplexing
DXRL	Deep X-ray lithography
EBL	Electron-beam lithography
EBW	Electron-beam writing
EDC	The electronic dispersion compensation
EDFA	Erbium doped fiber amplifier
EDWA	Erbium-doped semiconductor waveguide amplifier
EDWL	Er-doped waveguide laser
EIT	Electromagnetically induced transparency
EL	Electroluminescence
ELPCD	Electroless liquid phase chemical deposition
EMC	Electromagnetic compatibility
EME	Eigen mode expansion method
EMI	Electromagnetic interference
ENT	Ear, nose, and throat
EO	Electro-optic
EPON	Ethernet passive optical network
ESA	European Space Agency

EU	European Union
EXAFS	Extended X-ray absorption fine structure
FBT	Fused biconic taper
FCA	Free-carrier absorption
FCC	Face centered cubic
FCD	Free-carrier induced dispersion
FDTD	Finite difference time domain
FEM	Finite element method
FET	Field effect transistor
FHD	Flame hydrolysis deposition
FIB	Focused-ion beam
FOTS	Fiber optic transmission system
FP	Fabry-Perot
FPD	Flat panel display
FPGA	Field programmable gate array
FSL	Far-field superlens
FSO	Free-space optics
FSR	Free spectral range
FSS	Frequency selective surface
FTIR	Fourier transform infrared spectroscopy
FTS	Fourier transform spectrometer
FTTH	Fiber to the home
FWM	Four-wave mixing
GFP	Generic framing procedure
GI	Graded index
GLYMO	γ-glycidyloxipropyltrimethoxysilan
GOI	Germanium-on-insulator
GPTS	γ-glycidyloxypropyl trimethoxysilane
GPU	Graphic processing unit
GRIN-SCH	Graded index separate confinement heterostructure
GVD	Group-velocity dispersion
HBF	Hollow Bragg fiber
HBT	Heterostructure bipolar-transistor
HCG	Human chorionic gonadotropin
HCW	Hollow core waveguide
HD	High density
HDCVD	High density chemical vapor deposition
HDP	High density plasma
HGW	Hollow glass waveguide
HID	High intensity discharge
HMFG	Heavy metal fluoride glass
HMO	Heavy metal oxide
HP	Hard polymer

HPC	High performance computer
HPCF	Hard polymer clad fiber
HPW	Hollow polycarbonate waveguide
IC	Integrated circuit
ICP	Inductively coupled plasma
IFVD	Impurity-free vacancy disordering
IL	Interference lithography
IOR	Index of refraction
IPD	Integrated preamplifier detector
IR	Infrared
ISI	Inter-symbol interference
ITO	Indium tin oxide
ITOX	Internal oxidation
ITU	International Telecommunication Union
IVD	Inside vapor deposition
KTN	Potassium tantalite-niobate
KTP	Potassium titanyl phosphate
LADAR	Laser detection and ranging
LAN	Local area network
LASER	Light amplification by the stimulated emission of radiation
LC	Liquid-crystal
LCD	Liquid crystal display
LCW	Liquid core waveguide
LD	Laser diode
LED	Light-emitting diode
LFC	Liquid flow coating
LHM	Left-handed material
LIGA	Lithographie, galvanoformung, abformung (Lithography, electroplating, and molding)
LMA	Large mode area
LP	Linearly polarized
LPCVD	Low pressure chemical vapor deposition
LPE	Liquid phase epitaxy
LSP	Localized surface plasmon
LTA	Laser thermal annealing
MBE	Molecular beam epitaxy
MCM	Multi-chip module
MCVD	Modified chemical vapor deposition
MEMS	Micro-electro mechanical system
MIR	Mid-infrared
ML	Monolayer
MLS	M-line spectroscopy

MM	Metamaterial
MMF	Multimode fiber
MMI	Multimode interference
MO	Magneto-optic
MOCVD	Metal–organic chemical vapor deposition
MOF	Microstructed optical fiber
MOPA	Master oscillator power amplifier
MOSFET	Metal-oxide semiconductor field-effect transistor
MPI	Multiphoton ionization
MPTS	γ-methacryloxypropyl trimethoxysilane
MQW	Multiple quantum well
MRI	Magnetic resonance imaging
MSM	Metal–semiconductor-metal
MT	Mount
MTES	Methyltrietoxysilane
MWMLS	Multi-wavelength M-line spectroscopy
MWS	Metamaterial waveguide structure
MZI	Mach-Zehnder interferometer
NA	Numerical aperture
NASA	National Aeronautics and Space Administration
NBO	Nonbridging oxygen
NED	Nitridation enhanced diffusion
NGPON	Next Generation Passive Optical Network
NIL	Nanoimprint lithography
NIR	Near-infrared
NLO	Nonlinear optics
NMP	N-methylpyrrolidone
NMPM	Negative magnetic permeability material
NPV	Negative phase velocity
NRI	Negative refractive index
NRZ	Nonreturn-to-zero
NW	Nanowire
NZDSF	Non-zero dispersion shifted fiber
OADM	Optical add–drop multiplexer
OCDMA	Optical code division multiple access
OCT	Optical coherence tomography
OE	Electro-optical
OED	Oxidation enhanced diffusion
OEIC	Optoelectronic integrated circuit
OEO	Optoelectronic oscillators
OFDMA	orthogonal frequency division multiple access
OFNM	Optical fiber nanowires and microwires
OH	Hydroxide

OIC	Optical integrated circuit
OIF	Optical Internetworking Forum
OL	Optical lithography
OLED	Organic light-emitting diode
OPA	Optical parametric amplification
OPGW	Optical ground wire
OPO	Optical parametric oscillator
ORD	Oxidation retarded diffusion
ORMOSILS	Organic-modified silicates
OTDM	Optical time division multiplexing
OTDR	Optical time-domain reflectometer
OVD	Outside vapor deposition
PBG	Photonic-bandgap
PC	Polycarbonate
PCB	Printed circuit board
PCF	Photonic crystal fiber
PCOF	Primary coated optical fiber
PCVD	Plasma-activated chemical vapor deposition
PDE	Partial differential equation
PDL	Polarization dependent loss
PDMS	Polydimethylsiloxane
PDP	Plasma display panel
PDS	Photothermal deflection spectroscopy
PECVD	Plasma-enhanced chemical vapor deposition
PEI	Polyetherimide
PFCB	Perfluorocyclobutyl
PhC	Photonic crystals
PI	Polyimide
PIN	Semiconductor structure comprising of p-and n-type materials with an intrinsic material in between; addition of intrinsic layer changes properties of p-n junction
PL	Photoluminescence
PLC	Planar light wave circuit
PLD	Pulsed laser deposition
PMD	Polarization mode dispersion
PMMA	Poly (methyl methacrylate)
PN	A boundary or interface between two types of semiconductor material, p-type and n-type, inside a single crystal of semiconductor
POF	Plastic optical fiber
PON	Passive optical network
POP	Point-of-purchase
PS	Polystyrene
PU	Polyurethane
PVD	Physical vapor deposition

PVK	Poly-(N-vinyl carbazole)
PWE	Plane wave expansion
QD	Quantum dot
QPM	Quasi-phase-matching
QR	Quantum wire
QW	Quantum well
QWI	Quantum well intermixing
RBS	Rutherford back scattering
RC	Resistive-capacitive
RE	Rare earth
RECAP	Resonant-cavity photodetector
RF	Radio frequency
RFA	Raman fiber amplifier
RGB	Red, green, and blue
RHEED	Reflection high energy electron diffraction
RI	Relative intensity
RIE	Reactive ion etching
RIN	Relative intensity noise
RIU	Refractive-index unit
RNF	Refracted near-field
ROADM	Reconfigurable optical add–drop multiplexer
RoHS	Restriction of the use of certain hazardous substances
RTA	Rapid thermal annealing
RZ	Return-to-zero
SAW	Surface acoustical wave
SBC	Soleil-Babinet compensator
SBS	Stimulated Brillouin scattering
SCHG	Single-component halide glass
SCOF	Secondary coated optical fiber
SDN	Software defined network
SED	Silicidation enhanced diffusion
SERS	Surface-enhanced Raman spectroscopy
SESAM	Semiconductor saturable absorber mirror
SFP	Small form pluggable
SGOS	Silicon-germanium-on-silicon
SHG	Second harmonic generation
SHIP	Silicon hetero-interface photodetector
SIMOX	Separation by implanted oxygen
SLED	Surface emitting light-emitting diodes
SLM	Spatial light modulators
SMF	Single-mode fiber
SMIC	Submicroscopic interstitial clusters
SMT	Surface mount technology; source model technique
SNR	Signal-to-noise ratio

SOEI	Silicon-on-epitaxial-insulator
SOG	Spin-on glass
SOI	Silicon-on-insulator
SON	Silicon-on-nitride
SOS	Silicon-on-silicon; silicon-on-sapphire
SPASER	Surface plasmon amplification by stimulated emission of radiation
SPM	Self-phase modulation
SPP	Surface plasmon polariton
SPR	Surface plasmon resonance
SQW	Single quantum well
SRR	Split-ring resonator
SRS	Stimulated Raman scattering
SSM	Split-step method
STEM	Scanning transmission electron microscopy
TCK	Temperature coefficient of dielectric constant
TCO	Transparent conducting oxide
TDS	Time domain spectroscopy
TE	Transverse electric
TEA	Transversely-excited atmospheric-pressure
TEC	Thermal expansion coefficient; thermoelectric cooler
TED	Transient enhanced diffusion
TEM	Transverse electromagnetic
TEOS	Tetraethoxysilane
TER	Total external reflection
TIA	Transimpedance amplifier
TM	Transverse magnetic
TMM	Transfer matrix method
TNF	Trinitrofluorenone
TO	Thermo-optic
TPA	Two-photon absorption
TPP	Two-photon-photopolymerization
UHV	Ultrahigh vacuum
ULCF	Uniform-loss and cyclic-frequency
ULSI	Ultra-large scale integrated
UV	Ultraviolet
VAD	Vapor axial deposition
VCSEL	Vertical-cavity surface-emitting laser
VLS	Vapor–liquid-solid
VLSI	Very-large-scale integration
VOA	Variable optical attenuators
VPE	Vapor phase epitaxy
WAN	Wide area network
WBGS	Wide-band gap semiconductor

WDM	Wavelength-division multiplexing
WEEE	Waste electrical and electronic equipment
WGL	Waveguide-laser
XFP	10 Gigabit small form factor pluggable
XPM	Cross-phase modulation
YAG	Yttrium iron garnet

Chapter 1
Fundamentals and Design Guides for Optical Waveguides

Abstract Next-generation high-end data processing systems such as Internet switches or servers approach aggregate bandwidth in excess of 1 Tb/s. The task of providing hundreds of individual links as speeds in excess of 10 Gb/s over the link distances becomes increasingly difficult for conventional copper-based interconnect technology. Optical interconnects are foreseen as a potential solution to improve the performance of data transmission on chip, PCB, and system levels. They carry data signals as modulation of optical intensity, through an optical waveguide, thus replacing traditional electrical interconnects. Optical devices can overcome the bottleneck imposed by the limited bandwidth of electronic circuits in areas such as computing, data storage, or telecommunication networks. The basic element of any optical circuit is the optical waveguide, which permits to connect optically different devices. To build integrated optical circuits that substitute micro-electronic circuits, integrated optical waveguides with light confinement in a size of the order of the wavelength are mandatory. Optical waveguides can be classified according to their geometry (planar, strip, or fiber waveguides), mode structure (single-mode, multi-mode), refractive index distribution (step or gradient index) and material (glass, polymer, or semiconductor). They are designed as energy flow only along the waveguiding structure but not perpendicular to it, so radiation losses can be avoided. Usually, optical integrated waveguides rely on the principle of total internal reflection, using materials with low absorption loss. The waveguide cross section should be as small as possible to permit high-density integration, functionally linking devices or systems or implementation of complex functionalities, such as splitters/combiners, couplers, AWGs, and modulators. A wide range of materials can be used, with their corresponding advantages and drawbacks. Current commercial devices are mostly based on silicon/silica waveguides, III–V compounds, and lithium niobate waveguides. Silicon waveguides offers the possibility of mass-manufacturing and a high level of integration, which would result in cheaper chips. Novel materials such as photonic crystals can provide advantages to fulfill the requirements for high-density photonic integration. This chapter will review fundamentals and design

X.C. Tong, *Advanced Materials for Integrated Optical Waveguides*, Springer Series in Advanced Microelectronics 46, DOI 10.1007/978-3-319-01550-7_1,

guides of optical waveguides, including state-of-the-art and challenges, fundamental theory and design methodology, fabrication techniques, as well as materials selection for different level waveguide components and integration structures.

1.1 State-of-the-Art and Challenges

1.1.1 Rationale and Challenges for Optical Interconnects to Electronic Circuits

Advances in integrated electronic circuits have revolutionized the speed with which all kinds of computing and communications are performed. The continuing exponential reduction in feature sizes on electronic chips leads to ever larger numbers of faster devices at lower cost per device. This evolution is shifting the balance between devices and interconnection in digital processing systems; electrical interconnections do not scaled to keep up with the devices (Miller 2000). This may result in a situation in which the performance of the whole system becomes limited by the interconnect capacity, which is often called interconnection bottleneck. Electrical interconnects are facing plenty of challenges. The high-speed signals can be distorted due to dispersion, reflections and ringing, as well as attenuation and its variation with frequency. The high attenuation of high frequency signals results in a need to use high-power line-drivers, thus causing thermal management issues. Electromagnetic interference (EMI) causes noise and design constraints due to the need to fulfill electromagnetic compatibility (EMC) specifications. Moreover, the performance of parallel links is limited by the crosstalk due to coupling from neighboring traces and the signal skew caused by variations in the delay between different signal traces (Karppinen 2008). There are several possible approaches to such interconnection bottleneck problems, and likely all of these will be used to some degree. Architectures could be changed to minimize interconnection. Design approaches could put increasing emphasis on the interconnection layout. Signaling on wires could be significantly improved though the use of a variety of techniques, such as equalization. Optics is a different physical approach to interconnection that can in principle address most, if not all, of the problems encountered in electrical interconnections, such as scaling of the raw capacity, voltage isolation, timing accuracy, and overall ease of design (Miller 2000).

Optical interconnection refers to a kind of data transmission in which the data signal is transmitted as a modulation of optical carrier wave, i.e., light, through an optically transparent media, such as optical waveguide or air. Typically, the intensity of the light source, such as a laser diode (LD), is modulated and the transmitted optical signal is converted back to the electrical domain by the use of a photodetector. Optical interconnects have many advantages over conventional electrical interconnects, thus making it easier to achieve and maintain good signal integrity. These include extremely large information capacity (very broad bandwidth), low heat generation and transmission losses (lower signal dispersion, distortion, lower attenuation, lower skew and easier impedance matching of transmission lines), as well as immunity of the signal path to EMI. The latter enables reduced signal crosstalk, easier EMC design and the possibility for higher interconnection density in parallel links (Karppinen 2008). Furthermore, optical interconnects are light weight and smaller in size compared to electronic interconnects, and do not interact linearly when multiple wavelengths propagate in an optical medium, and thus allow parallel processing of different wavelengths (Ma et al. 2002).

The value of the aforementioned advantages certainly depends on the application and all other constraints involved in the design and implementation of the system. In order to ease transfer to a novel technology with the high bit-rate optical interconnects, the needed technological step should be made as small as possible. In other words, the novel technology has to be developed based on existing technology instead of developing a completely new one. For instance, on board-level interconnects, the technologies should be made compatible with the conventional circuit board processes and assembling practices. Moreover, there are at least two obvious prerequisites for the introduction of optical interconnects in commercial applications: the costs per interconnect capacity should be competitive with copper-based ones in mass-production, and the required PCB area or volume may not exceed that of equivalent copper solutions (Karppinen 2008).

Optical interconnects can be realized through free-space optics (FSO) or guided-wave optics. FSO refers to the transmission of modulated visible or infrared beams through the atmosphere to obtain broadband communications, which can function over distances of several kilometers as long as there is a clear line of sight between the source and the destination. Nevertheless, there are limitations to FSO due to rain, dust, snow, fog, or smog, which can invariably block the transmission path. Guided-wave optics, however, depends on the phenomenon of total internal reflection, which can confine light in the optical waveguide. Optical waveguides may be rectangular shapes, and planar or thin-film deposits used in integrated optical circuits, or a filament of waveguide material usually circular in the cross section used in fiber optics. Depends on the various possible patterns of propagating or standing electromagnetic fields, there are single-mode and multimode optical waveguides. Each mode is characterized by its frequency, effective refractive index, polarization, power distribution, electric field strength, and magnetic field strength (Ma et al. 2002).

Optical communications through fiber has long been the technology of choice for high-speed long-distance data links. Gradually, as the capacity requirements have increased, the optical links have emerged into shorter distance applications, such as fiber-to-the-home, local area network, and even into fiber-optic interconnects between boards and cabinets, eventually, optics would be used to interconnect integrated circuits on a board or even to be used in intra-chip interconnects. That is, electrical interconnects, which have dominated since the infancy of electronics, are likely to be replaced with optical interconnects in some cases (Karppinen 2008). Optical waveguiding was originally used with fiber optics that transmitted light through a glass pipe medium. With invention of lasers and the development of coherent optics, optical fiber for long-distance transmission and guiding structure with which to build optical components and connect them into optical circuits arose. These optical waveguides allowed the fabrication of planar components and their integration into planar optical circuits with many advantages over electronic circuits at all frequencies up to microwave range. To achieve this goal, rectangular dielectric waveguides, planar optical waveguides in the form of films and strips, as well as other planar components and circuitry were constructed, and resulted in the formation the integrated optics in 1960s. Since then, not only has the theory of wave propagation, excitation, and the conjugation of optical films and trips been advanced to encompass all aspects in the analysis and design of these guiding structures but also advanced materials as well as precise and reliable technologies have been developed to fabricate planar optical waveguides and integrated optical circuits (Ma et al. 2002).

A typical optical integrated circuit (OIC) has been made in the form of thin film and designed to have a certain function by integrating a laser diode as the source, functional components such as switches/modulators, interconnection waveguides, and photodiodes as detectors, on a single substrate. Through integration, a more compact, stable, and functional optical system can be built. In contrast to optical fiber system, most of integrated optical components are based on functional crystal materials. Therefore the OICs can not only confine but also modulate the guided waves. OICs are divided into three types from the standpoint of materials: monolithic OIC, hybrid OIC, and quasi-hybrid OIC. When three basic components (source, waveguide, and detector), are all integrated on a single substrate, the device is called a monolithic OIC. Compound semiconductors, such as GaAs and InP, can be the candidate substrates for this OIC. When the three components are made of three different materials, the device is called as a hybrid OIC. For example, in a hybrid optical IC the source, the waveguide and the detector are made of the compound semiconductor and the dielectric materials such as glass or $LiNbO_3$, and silicon (Si), respectively. A quasi-hybrid or quasi-monolithic OIC is an intermediate between the monolithic and hybrid ICs. For example, the source is made of a compound semiconductor and photodiodes are integrated on a Si substrate. Here, the waveguide layer is a dielectric film deposited on the thermally oxidized silicon substrate. The monolithic-type OIC is ideal as an OIC, but its implementation is very difficult at present. Comparably, the hybrid is relatively easy to fabricate, and has a great advantage that the most appropriate materials and processing techniques can be utilized although there is a challenge with assembling the three basic components. OICs are mainly constructed with single-mode waveguides whose widths are on the order of micrometers and in which a single-mode optical wave propagated. To fabricate an OIC with this micron structure, advanced thin film and micro-pattern processing techniques have been developed. As material variety increases and integration improves, more and more new waveguide fabrication techniques would become available (Li 2004).

Although it seems that optical waveguides are more related to short-distance-integrated optics, optical networking does mark a fundamental shift in the architecture of fiber communications. Early fiber-optic systems were simple long-distance transmission pipelines, carrying signals from point to point. Optical networks have

Optical interconnects	Illustrations	Distance	Number of lines	Use of optics
Internet, wide area net work		Multi-km	1	Since the 1980s and the early 1990s
Local area network		10 – 2000 m	1 - 10	Since the late 1990s
System level interconnects		30 – 100 m	~ 100	~ 2005
Cabinet level interconnects		1 – 30 m	~ 100	~ 2005
Backplane level interconnects		~ 1 m	~ 100 - 1000	~ 2010
Board level interconnects		0.1 – 0.3 m	~ 1000	2010 - 2015
Module level interconnects		5 – 100 mm	~ 10, 000	Probably after 2015
Chip level interconnects		0.1 – 10 mm	~ 100,000	Later, if ever

Fig. 1.1 Optical interconnects hierarchy: trends from long to short distance

been used to organize and process signals in optical form as well as transmit them optically, expanding the realm of light. The growth in communication networks has been accelerated particularly with the rapid and global spread of multimedia services including telephone, cable TV, digital video, data and internet capabilities. With the advent of optical dense wavelength division multiplexing (DWDM) in the quest for virtually unlimited bandwidth capacity, all-optical networks would be enabled using the existing optical fiber infrastructure. However, the continuing development of fiber-based communications networks to accommodate future demands will eventually depend on the availability of cheap, reliable, and robust integrated optical waveguide devices for routing, switching, and detection (Ma et al. 2002).

In addition, broadband communications are also being applied to ever-shortening distances, penetrating from metro-area to access and intra-computer interconnect networks. The current generation of computers is limited by the speed at which information can be transmitted between such electronic components as processors and memory chips. Under these conditions, bus traffic increases as the computing power of the microprocessor increases. Therefore, the limited bandwidth constitutes a major bottleneck to efficient communications at the board-to-board data interface. The development of optical interconnects that would replace the conventional passive backplane is desirable in order to achieve higher data throughput (Ma et al. 2002). Figure 1.1 shows a general optical interconnection hierarchy, trending from long to short distance. Based on the type of interconnected components, the international packaging level hierarchy is usually divided as chip level (intra-chip); multi-chip-module level (intra-MCM or chip-to-chip); board level (MCM-to-MCM or chip-to-chip); backplane level (board-to-board); cabinet level (rack-to-rack); and system level (cabinet-to-cabinet). The further success of broadband communications in optical networking, metro/access communications, and computing will rely on the advancement of optical interconnects, such as splitters, combiners, multiplexers, and demultiplexers, optical switches/modulators, tunable filters, variable optical attenuators, amplifiers, and integrated optical circuits that are based on optical waveguides. While the basic technologies for the design and production of many integrated optical waveguide devices are in place, a great many of waveguide materials have been developed. Today, glass optical fibers are routinely used for high-speed data transfer. Although these fibers provide a convenient means for carrying optical information over long distances, they are inconvenient for complex high-density circuitry. In additional to being fragile and vibration sensitive, glass fiber devices are difficult to fabricate especially when they have a high port count, and as a result are quite expensive. InP-based quaternary semiconductor materials are widely employed for waveguide devices, due mainly to their potential for integration with active devices such as lasers and photodetectors operating at around 1,550 nm. However, semiconductor processing remains complex and expensive, especially at the sales volumes presently experienced for optical components. By using manufacturing techniques closely related to those employed for silicon-integrated circuits, excellent optical components have already been demonstrated using silica-based planar light wave circuits (PLCs). Currently, optical switches and arrayed waveguide gratings (AWGs) for multiplexing and de-multiplexing multiple wavelengths in a DWDM system are the predominant applications of silica-based PLC technology. The cost issue, the high switching power needed in silica-based switching devices, the low wavelength tuning range, and the temperature dependence of the central wavelength of silica-based AWGs, however, are major problems with this technology. Among the candidate material systems, high expectations have been place on polymers as the materials choice for highly integrated optical components and circuits. State-of-the-art optical polymers are particularly attractive in integrated optical waveguide devices because they offer rapid processibility, cost-effectiveness, high yields, high performance, such as lower optical loss and smaller birefringence compared to silica, power-efficient thermal actuation due to a larger thermo-optic coefficient than in silica, and compactness owing to a large refractive index contrast. Furthermore, polymers provide an ideal platform for the incorporation of more complex material functionalities through selective doping or reaction, thereby enabling amplification and electro-optic effects to be achieved (Ma et al. 2002).

From the viewpoint of the packaging level, perhaps the most attractive optical solution proposed to overcome the interconnection bottleneck at the board and module level is the interconnection based on low-loss integrated waveguides. This stripline-like optical waveguide approach with polymer waveguides embedded into the board as additional optical layers has a potential to fulfill the compatibility requirements of the existing PCB technology. Such optical channels could also enable design techniques analogous to the ones used for wiring electrical signals on the board. In particular, PCBs made of conventional low-performance base materials, such as FR4 (Flame Retardant 4, the most common PCB base material), are seen as the most attractive substrates for the O/E-PCB technology. This would enable equipping the low cost, widely used technology basis with very high-speed interconnects. One of the biggest challenges hindering the introduction of optical interconnects at the board level is the need to implement a feasible and reliable optical coupling between the optoelectronic transmitter/receiver and the optical waveguide. High coupling efficiency is required due to the rather limited loss budget (typically around 10–15 dB) of the high bit-rate transceivers based on VCSEL (vertical-cavity surface-emitting laser) diodes. On the other hand, to achieve compatibility with conventional electronics manufacturing processes, all components (including optical ones) should be suitable for surface mounting or bare-die mounting procedures and all materials should withstand the reflow soldering temperatures (Karppinen 2008).

In addition, it would be very promising if dense optical interconnects could be integrated with silicon chips. This needs substantial technological work, and there are barriers, psychological, technological, and financial, to its introduction. Optics offers a great many advantages, both qualitative and quantitative, that, if it were introduced, it could substantially alter information processing systems, and perhaps could become as indispensable on and between chips as it is today in long-distance communications (Miller 2000).

1.1.2 Evolution of Optical Interconnects

As shown in Fig. 1.1, optical interconnects have been trended from long to short distance. At integrated optical circuit packaging level, since the electronic interconnects fail at short distances as speeds increase over time, demanding for short-range optical interconnects have been boomed from system level, backplane level, board level to chip level. Currently, backplanes are seen as the most attractive application for board-level optical interconnects. The existing and proposed optical backplane solutions are often classified into (Karppinen 2008): (a) fiber-based interconnects; (b) interconnects based on fiber laminates and/or shuffles; (c) board-integrated optical channels, which can be sub-categorized from overlaid optical pathways to board-embedded waveguides; and (d) free-space optical interconnects. It is the media and structure of the optical path in which the categories differ from each other. On the other hand, the optoelectronics devices and electronic circuitry of the transmitters and receivers used are essentially of a similar kind in all categories.

In addition, a broad range of novel optical technologies has also been investigated to support possible optical interconnects to silicon chips. Systems have been demonstrated using relatively conventional lenses for imaging whole

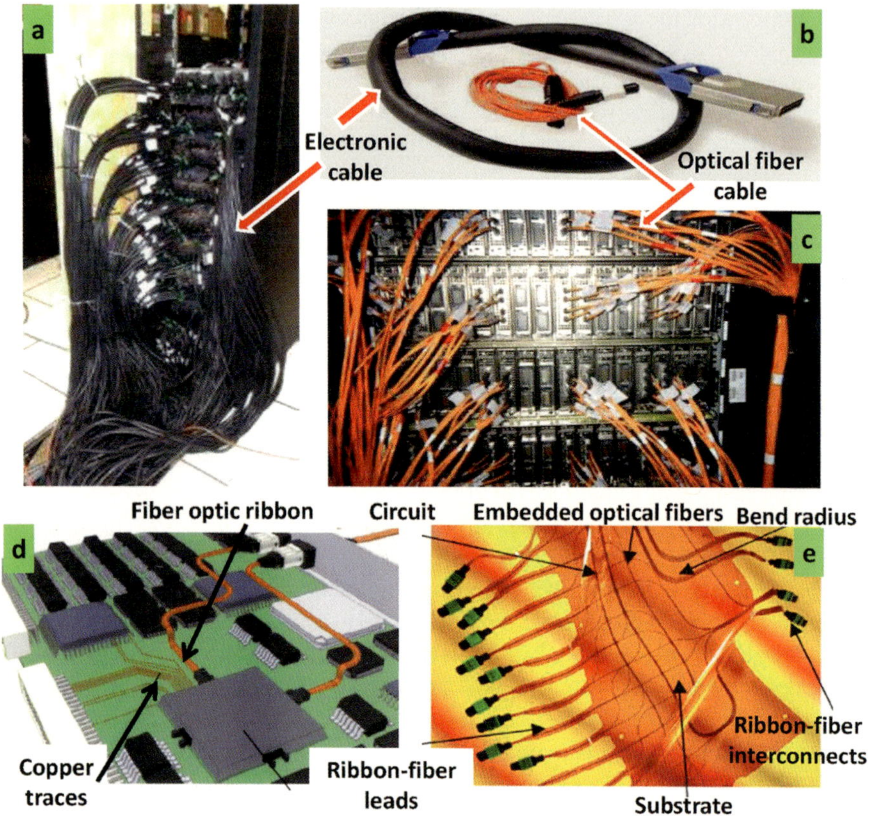

Fig. 1.2 Fiber-based optical interconnects: density advantage of optical fibers over copper cables (**a**–**c**); fiber-ribbon cables (**d**); and embedded optical fibers (**e**)

arrays of optical beams from one chip to another. Various micro-optical techniques, such as lens-let arrays and diffractive optics, as well as waveguides on silicon chips have been investigated though complete systems (Miller 2000).

1.1.2.1 Fiber-Based Optical Interconnects

Today, fiber-optic data links have been used inside electronic systems to interconnect between backplanes, racks, and cabinets. In fact, at longer distances than the backplane level, the optical alternative to conventional interconnects is to replace copper cables with fiber optics. In addition to high transmission capability and reduced EMI issues, the advantage of fiber is its lightness compared to bulky and heavy high-bandwidth electrical cables, as shown in Fig. 1.2a–c. Similar kinds of fiber-based links can also be used to interconnect daughter-boards in electronic equipment. That is, to form a kind of optical backplane with high bit-rate point-to-point links. Parallel optical links based on fiber-ribbon cables allow very high aggregate data rates with a relatively high integration level (Fig. 1.2d), thanks to multichannel transmitters and receivers. The highest integration level is achieved with transmitters based on VCSEL sources, mainly thanks to their low power consumption with high-speed direct modulation. Graded-index multimode fiber technology is favored because it allows high enough bandwidth for short-distance interconnects and the module technology is cheaper to implement than single-mode fibers due to the easier optical alignment. Using industry-standard fiber-ribbon cables and multi-fiber connectors, typically with a mechanical transfer ferrule, as well as VCSEL and PIN photodiode arrays, commercially feasible multichannel data links can be implemented with very high aggregate data rates. Typically, 12 parallel optical channels are used, and data links have been presented with channel bit-rates up to 10 Gb/s. Even higher channel counts are possible using 2D VCSEL arrays and 2D fiber-ribbon connectors. In optical backplane systems, flexible fiber laminates and optical shuffle components are utilized in order to handle several fibers and fiber ribbons in a narrow space, as well as to ease implementation of cross-connections. Polymer fibers allow a much smaller bending radius than silica fiber. Thus they have been proposed and demonstrated to be used as optical backplanes or intra-board interconnects. Furthermore, as shown in Fig. 1.2e, it is also possible to embed multimode silica fibers into an FR4 PCB to form intra-board optical interconnects (Kühner and Schneider 2007). Silica fibers provide negligible attenuation and high reliability; however, the cleaving of embedded fibers is difficult. Nevertheless, fiber-optic links are unattractive for board-level interconnects due to the fact that the packaging density is limited due to the fiber connectors/pigtails, the bend radius of the fibers is limited, and system assembly is not easily automated. In addition, fibers cannot exploit the planar fabrication technologies that are utilized to reduce packaging costs in electronics manufacturing. To conclude, higher integration density is probably needed before optical interconnects may become widely used in board-level and shorter interconnects (Karppinen 2008).

1.1.2.2 Optical Interconnects Overlaid on PCB

Various technologies have been studied to implement board-level interconnects using some kind of separate optical pathway component that is placed and used in parallel with the electronics PCB or perhaps overlaid on it (Fig. 1.3). The component can be an optical plate or sheet including beam steering refractive or diffractive elements. Another kind of pathway component widely proposed is a foil with integrated optical waveguides. An example is the polymer foil technology with integrated waveguides that was used in the overlaid configuration to demonstrate a fully assembled optical backplane system. A major benefit of a separate optical pathway component is that it does not need to be compatible with all the PCB manufacturing and assembly processes. For instance, materials that do not withstand lamination or soldering temperatures can be used (Karppinen 2008).

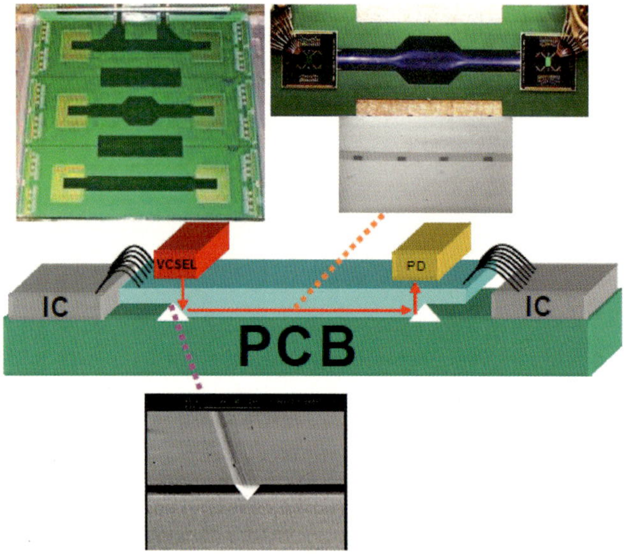

Fig. 1.3 Schematic of optical waveguides over PCB

1.1.2.3 Inter-chip Interconnects with Board-Embedded Waveguides

Several different approaches have been proposed to construct a chip-to-chip interconnect based on an optical waveguide integrated on/in the PCB. One of the earliest demonstrators was based on the simple structure with a board-mounted edge-emitting laser diode was butt-coupled to an optical ridge waveguide on top of the board, and the output from the waveguide was deflected towards the active surface of the photodiode with a laser-ablated mirror surface. Later on, most of the research interest has focused in the use of VCSEL sources and board-embedded waveguides (Fig. 1.4). In such cases, 45° deflection of the optical beam is often required both at the transmitter and receiver parts (Karppinen 2008).

In addition, rather than using a chip-on-board type mounting, it has been seen favorable to mount the optoelectronic transmitter and receiver components into a device package, possibly with the IC. In that case, it is often necessary to utilize micro-optical components, such as lenses, to enhance coupling efficiency. Coupling from board-to-board, such as from an embedded waveguide to another, allows the implementation of more complete optical interconnect systems. Optical waveguide layers can be processed on the PCB or laminated inside the board. The advantages of the latter, i.e., the embedded solution, are that the surface area of the board is completely left to the pads and components and the waveguide layers are shielded from ambient environment and mechanical stresses, thus improving reliability (Immonen et al. 2007). Moreover, multilayer optical waveguides would enable higher interconnect capacity and density as well as ease signal routing, especially if optical via are implemented (Karppinen 2008).

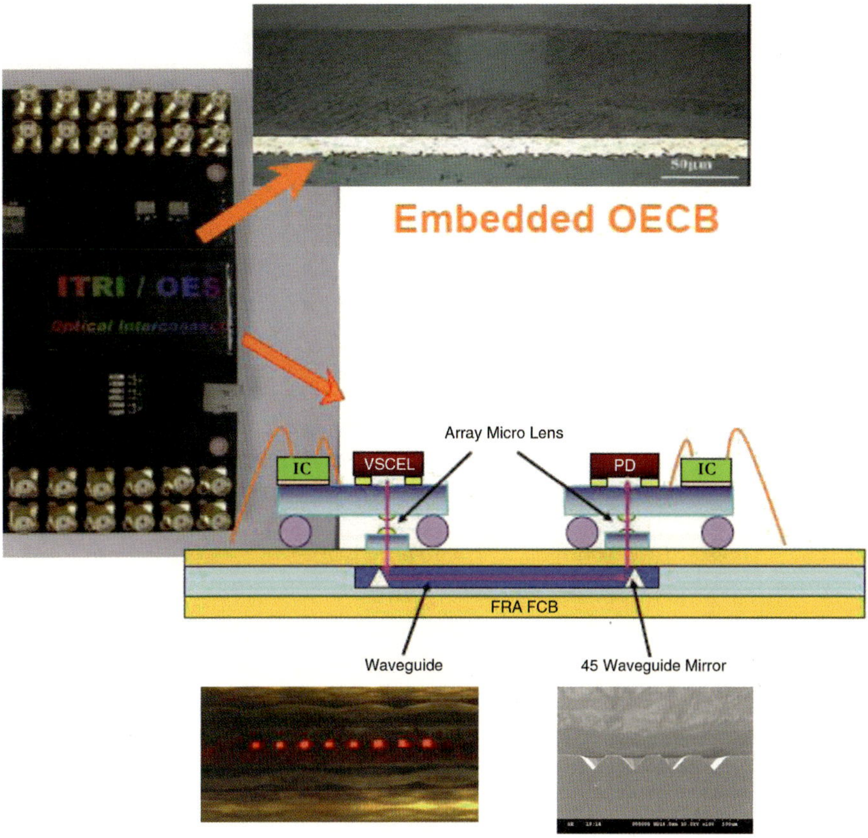

Fig. 1.4 Schematic of optical waveguides embedded in PCB

1.1.2.4 Free-Space Optoelectronic Interconnects

Free-space optical interconnect refers to the system where light is directed, or actually imaged, from the transmitter towards the receiver using optical components, such as lenses and mirrors, or there is just a transparent media between the transmitter and the receiver. This is in contradiction to the guided-wave optical transmission where the light is guided in a media based on total internal reflection phenomenon. In free-space interconnects, the light-guiding media between the transmitter and receiver can be air or another transparent material. A lot of research has been carried out on the free-space optical interconnects; mostly for very short distances, like intra-chip and intra-MCM (i.e., chip-to-chip) interconnects, but also for board-level and board-to-board applications, where ICs can be connected using massive parallel interconnects based on two-dimensional laser and detector arrays. There are major challenges hindering the feasibility of high bit-rate free-space optical links for board/backplane-level interconnection applications. First, it is extremely difficult to achieve the required optical alignment accuracy, especially when aiming to parallel interconnects with dense pitch. This is because, typically, the mechanics of the standard PCB technologies are not designed to provide alignment accuracies required by high bit-rate parallel optics. That is, the PCBs are not rigid and not forced to be straight on the racks. In addition, for the board-to-board application, the alignments between daughter-cards are not very accurate, and for the board-level application the alignment tolerances of the components and parts are only in the order of 50–200 μm, which is not high considering dense parallel interconnects. Secondly, the long-term reliability of a free-space optical link is questionable since the electronics are typically not in an isolated environment and, consequently, a cabinet eventually becomes dusty inside and in many applications the cabinet may be exposed to high humidity and large ambient temperature changes as well as mechanical stresses and vibrations. Consequently, very high bit-rate interconnects are difficult to implement without significant changes to the backplane systems (Karppinen 2008).

1.1.2.5 Optical Interconnects to Electronic Chips

The use of optics to make connections within and between chips could solve many of the problems experienced in current electrical systems. Many of the physical reasons for the use of optics are well understood and indicate many potential quantitative and qualitative benefits. Though there are, and will continue to be, electrical solutions that stretch the capabilities of electrical interconnects, optics is arguably the only physical solution available to solve the underlying problems of interconnects, and has the potential to continue to scale with future generations of silicon-integrated circuits. Optics may also solve the growing problems of system synchronization, allowing, for example, individual silicon chips to remain synchronous domains even as the chips grow. This synchronization advantage applies both to optical clock distribution and to optical interconnects themselves. Optics may reduce on-chip power dissipation in clock distribution, global on-chip interconnects, and off-chip interconnects. Optics also may relieve a broad range of design problems; for example, optics itself does not have to be redesigned as system speed is increased, optical interconnects provide voltage isolation, and problems of wave reflection, impedance matching, and pin inductance are essentially absent. The basic devices and laboratory demonstrations of the key elements of technology exist for dense optical interconnects to silicon ICs. No physical breakthrough is required to implement optical interconnects (though there are possibilities for revolutionary advances). However, substantial technological work remains (Miller 2000): (1) Optoelectronic devices require continued development to meet the yield, tolerance, and drive voltage requirements for practical systems with future generations of silicon CMOS. (2) Work will be required in the interface circuits between optics and electronics. Though there appears to be no current fundamental difficulty in making such circuits in CMOS, research is needed in circuits that (a) avoid issues such as crosstalk and susceptibility to digital noise, (b) have appropriately low power dissipation and latency, and (c) are tolerant to process variations. (3) The technology for integrating optoelectronics with silicon-integrated circuits can likely use hybrid approaches, such as solder bonding; such hybrid approaches require no modifications to the current process for fabricating silicon-integrated circuits except to add processes to fabricated silicon-integrated circuit wafers. Solder bump hybrid integration also simplifies design since the optoelectronics is then attached only to the top-level metal, allowing complete freedom of placement of electronic circuits underneath. Longer term approaches will likely be monolithic integration, though such integration is still under development. (4) Novel approaches will be required for the optics and the mechanics for such optical interconnections, though there are several opportunities for appropriate opto-mechanical technologies. (5) It will be important to research the systems and architectural benefits of optics for interconnect. Optics can likely enable kinds of architectures that are not well suited to electrical interconnect systems (e.g., architectures with many long connections, architectures with large "aspect ratios," architectures requiring synchronous operation over large domains), and can likely also allow continued use of current architectures that otherwise would have to be abandoned in the future because of the limitations of wired interconnects. There are also significant opportunities for optics in the somewhat simpler application of clock distribution. (6) The issue of cost is a major one for optical interconnects. A valid question is whether the ideas being researched could result in low enough cost if they were to be commercialized at reasonably large volume. There are, however, revolutionary technologies that might offer sufficient economies of scale and could be manufacturable at low cost. A fair assessment might be that any estimate of final cost is currently highly speculative, and achieving low cost will require both considerable ingenuity and the promise of volume markets.

1.1.3 Waveguide Component and Integration Technologies

1.1.3.1 Light Sources

A light-emitting device is one of the key components of any optical interconnect system.

The structure and characteristics of the emitter have a significant role in the design of the transmitter, especially affecting the choice of optical coupling and packaging scheme. To enable feasible integration of the optical interconnect into electronic system, the emitter has to fulfill several requirements, such as small size, low power consumption (high electrical-to-optical conversion efficiency, even >50 %), high modulation bandwidth (when a direct modulator is used), emission characteristics that enable efficient optical coupling, and low fabrication costs. The costs would preferably be in the order of (or less than) the costs of the other devices in the system. Also, the possibility to fabricate as one-dimensional or even two-dimensional emitter arrays is important in order to enable implementation of dense parallel interconnects and thus achieve high aggregate transmission rates.

Of the currently available light emitter technologies, only semiconductor devices fulfill the requirements of optical interconnection applications, such as very small size and high conversion efficiency. There are two types of semiconductor emitters: light-emitting diodes (LEDs) and laser diodes (LDs). They can also be classified as surface-emitting and edge-emitting devices, depending on how the emitted light exits from the die. VCSEL is commonly promoted as the most attractive optical source for the short-distance optical interconnections. It is a laser diode in which the gain medium is surrounded by Bragg mirrors in such a way that the laser emits in the direction perpendicular to the surface of the die. VCSELs have many advantageous properties that are especially valuable in short-distance interconnects, such as low threshold current, low drive current, low operation voltage, high direct modulation bandwidth, low-temperature sensitivity (as compared to edge-emitting lasers), circular output beam with rather small divergence (allowing easy coupling to waveguide), suitability for high-density integration as monolithic arrays, compatibility with wafer level testing and screening (reducing manufacturing costs). A particularly attractive choice is the 850-nm VCSEL array that has already become a mature technology in parallel fiber-optic links. This is due to the fact that 850 nm is the local attenuation minimum of a silica fiber and, thus, commonly used in short-reach fiber-optic links (Karppinen 2008).

LEDs have the advantage that they may be relatively easier to make than VCSELs and might avoid some of the problems of lasers, such as mode and polarization stability. They have disadvantages of limited speed of response and overall optical efficiency, which is limited by the spontaneous recombination rate of the charge carriers. Although edge-emitting LDs are very common in fiber-based data links, they are less attractive for board-level interconnects. This is because they have difficulties meeting the cost/performance requirements, especially as a multichannel source, due to the inherent way the device chips have to be processed, diced, cleaved, and assembled. In addition, the edge-emitters are not able to compete with VCSELs in power dissipation, which is a key property when aiming at dense-packaged interconnects. Moreover, the output beam of edge-emitting lasers is asymmetric with high divergence (30–50°) perpendicular to the PN-junction, which often complicates coupling into waveguides. An option technology to construct a high-speed source is to use an external optical modulator in combination with a continuous-wave-emitting laser diode or LED source. With this technology it would be possible to use a single emitter as a source of several transmission channels by dividing the power of the emitter among several modulators. The disadvantage in the use of an external modulator is that the integration and optical coupling structures become more complicated (Karppinen 2008).

Quantum-well modulators and VCSELs are strong candidates for viable output devices for dense optical interconnects to silicon. No major breakthrough is required for use of these devices, at least with hybrid integration. In the case of quantum-well modulators, the same devices can also be used effectively as input photodetectors, giving a potentially viable complete solution. In the case of VCSELs, different photodetectors are required, though there is no basic problem in making efficient photodetectors in III–V technology (Miller 2000).

1.1.3.2 Characteristics of VCSELs

VCSELs can be manufactured for several different wavelengths. Today, devices emitting at around 850 nm have been widely adopted as a standard in short-distance fiber-optic interconnects. The wavelength matches with the so-called first transmission window of the silica fibers. VCSELs operating at wavelengths below a micrometer are based on GaAs/AlGaAs-based process technology. Another VCSEL wavelength range which has gained interest within optical interconnect demonstrators is 980 nm. Since GaAs is transparent at 980 nm, it is possible to fabricate devices emitting from the rear of the device (i.e., through its substrate). VCSELs operating at a wavelength range of 1,200–1,600 nm have been developed used in single-mode fiber communications. It seemed challenging to produce VCSEL structures using the InP-based process technology required for that wavelength range, but today the devices are emerging for commercial use in some short-haul communications applications. Thanks to the small device size and the low power dissipation, dense arrays of VCSELs, both one-dimensional and two-dimensional, can be fabricated and are feasible for high-speed links. The commercially available VCSEL arrays typically have a pitch of 250 μm because that matches with the standard pitch used in fiber-ribbon cables and connectors. However, even much smaller pitches are possible as the die area required by VCSEL device is typically only in the order of 10–20 μm in diameter. Moreover, novel VCSEL-like devices, where an electroabsorption modulator is monolithically integrated above the VCSEL cavity, have shown much higher modulation bandwidths, even up to 60 GHz, with low drive currents. VCSELs are typically emitting a volcano crater-shaped beam, whose divergence is rotationally symmetric, but the intensity distribution has significant variation due to the multimodal operation. The differences in beam characteristics between the transverse and lateral direction, as well as between laser diodes in the same array, were small (Karppinen 2008).

1.1.3.3 Photodetectors

The detector to be used in optical interconnection applications should provide high bandwidth, high sensitivity (at the operation wavelength), a large photosensitive surface area (to ease optical coupling), and the possibility for high-density multichannel integration. There are three types of semiconductor photodetectors that may fulfill these requirements: P–I–N photodiodes (PINs), metal–semiconductor–metal photodiodes (MSM), and avalanche photodiodes (APD). PINs are the most commonly used in short-distance fiber-optic links. MSM is a low-capacitance device enabling high-speed operation with a larger active area than PIN. Nevertheless, the responsivity of MSM is typically lower than that of PIN due to the top electrode structure that absorbs part of the incident power. Although APD provides the highest responsivity thanks to the internal gain, it is not considered attractive for board-level interconnects due to the requirement of high-bias voltage and rather complicated, expensive, fabrication processes. The optimization of the photodiode for high speed and high sensitivity with easy optical coupling is a trade-off: in order to ease alignment and optical coupling between waveguide and detector, as large an active area as possible would be preferred. But, on the other hand, when the area of the detector increases, so its capacitance increases, which results in decreased receiver bandwidth and increased preamplifier noise. This is also means that when the channel rate is increased, the optical coupling to the detector becomes more challenging. As a reference, PINs operating up to 10 Gb/s at 850 nm typically have an active area diameter of 60.75 μm, and, the 3-dB bandwidth of the InGaAs PINs increased from 13 to 30 GHz as the diameter was reduced from 60 to 30 μm respectively. In practice, all commercially available high-speed photodiodes are optimized for fiber-optic applications today. This is a kind of shortcoming in demonstrating board-level optical interconnects as it is not always possible to find a photodiode with the optimal area. Fortunately, the high bit-rate fiber-optic links based on VCSELs use graded-index multimode fibers that have an effective core diameter of 50 or 62.5 μm, which is also in the order of the preferred waveguide core diameter for board-level interconnects. In addition, the availability of photodiode arrays is limited to the pitch of 250 μm that is commonly utilized in parallel links based on fiber-ribbon cables and MT ferrules. GaAs is a suitable material for photodiodes at wavelengths up to 850 nm, and Si is sufficient at wavelengths up to 1 μm. Both Si and GaAs photodiodes are used in fiber-optic links at around 650 and 850 nm wavelengths, but GaAs is more popular in high-speed applications. In longer wavelengths up to 1.7 μm, an InGaAs material system on InP substrates is commonly used, also in fiber-optic communications using 1.3 and 1.55-μm transmission windows. Another interesting detector material is Ge, because of its potential to monolithically integrate PINs on Si-based integrated circuits. However, discrete Ge photodiodes are not used in data transmission applications (Karppinen 2008).

1.1.3.4 Electronics

The data rate of the high-speed board-level optical interconnect is typically limited by the bandwidth of the optoelectronic circuit and devices rather than the bandwidth of the transmission channel. Therefore, similar to the fiber-optic links, it is favorable to use binary intensity modulation techniques, especially on–off keying with a nonreturn-to-zero pulse, and to have the bit-rate equal (or very close) to the symbol rate. The data signals following a standard logic level, such as the current mode logic (CML) at high bit-rates, are not suitable to drive the LD directly. The laser requires a certain bias level and, to achieve a high extinction ratio, the voltage swing must be sufficiently large. Therefore, a laser driver circuit, typically an IC, is used to do the impedance matching and transform the logic signals into the appropriate modulation of the laser current. At the receiver side, a preamplifier is needed to amplify the small photocurrent and, typically, a post/limiting amplifier is also used to adjust the preamplified signals to match the logic level. Also, a bias is required for the photodiode. The emergence of monolithic multichannel driver and receiver ICs has enabled high integration of parallel transceivers. Due to the layout restrictions of high speed traces it is very difficult to implement a high channel count with discrete electronics and ICs. It is also important to note that in order to maximize the high-speed performance of the receiver, a preamplifier, which is usually a transimpedance amplifier (TIA), should be integrated very close to the photodiode, thus minimizing parasitic impedances. Today, 10 Gb/s/channel driver and receiver ICs with, at least, up to four channels are used in commercially available fiber-ribbon links. A trend is to implement driver and receiver ICs using standard CMOS (complementary metal-oxide-semiconductor) processes, as they promise lower costs and low power consumption. For instance, a 4 $\times$ 10 Gb/s transceiver based on quad CMOS ICs and VCSEL and PIN arrays has been shown to consume only 100 mW in total. Moreover, even a low power 25 Gb/s TIA has been demonstrated using the CMOS process. To further increase integration, the monolithic integration of a photodiode on a TIA/receiver IC is aimed at. In the highest speed applications this is possible with heterostructure bipolar-transistor (HBT) processes. CMOS technology is also suitable, e.g., a 10 Gb/s integrated photodiode and TIA with low power consumption, though the responsivity is rather low. To minimize the need of additional electronics due to optical interconnects, the optoelectronic devices can be directly connected to the logic circuits. That is, the VCSELs would be driven bias-free directly with logic gates and the detectors would directly drive the logic gates. This means a trade-off in performance, but may be feasible if the link budget does not need to be maximized. Another technology to potentially improve the link bandwidth is the electronic dispersion compensation (EDC) that has been proposed to increase the data rate of multimode fiber links. This means the inter-symbol interference (ISI), which limits the channel bit-rate, is compensated by adaptive equalization circuitry in the receiver after the TIA (Karppinen 2008).

1.1.3.5 Optical Waveguides for Short-Range Optical Interconnects

Optical waveguide is another key component of optical interconnection on PCB. The accuracy of the PCB manufacturing and surface-mount technology (SMT) assembly processes suggest that convenient cross-sectional dimensions of the waveguides are in the order of 50–100 μm. That is, they are highly multimodal waveguides. The waveguide in board-level interconnects has to exhibit many challenging properties. The interconnection length can be centimeters or even dozens of centimeters. Therefore, very low attenuation is a must, requiring both a low-loss optical material and very low surface roughness of the waveguide core in order to minimize scattering losses. One of the major challenges in achieving low attenuation is the high roughness and irregular surface of the common PCBs. For instance, FR4 is an epoxy-based substrate that is re-enforced with glass fibers, which form a weave that increases the rigidity of the board. Other important optical parameters of the waveguide are numerical aperture (NA), refractive index and shape, all which should match with the properties of the source and detector, as well as the coupling optics. In addition, the embedded waveguide should provide good thermal and mechanical compatibility with the PCB material, including good adhesion. Moreover, the waveguide has to maintain good optical properties with high stability against environmental stresses during the electronics assembly processes and in the operating environment. For instance, the waveguide has to survive the reflow soldering process, typically at temperatures around 250 °C. Several different kinds of potential waveguide manufacturing technologies have been presented, such as photolithography, hot embossing, molding, printing, ion exchange, laser direct writing, and laser ablation. Of these, UV lithography with direct patterning of the optical polymer layer is probably the most common technology. Perhaps the simplest version is the lithographic waveguide fabrication using direct waveguide core photo-patterning and rib cladding process. The cladding and core layers are commonly deposited by spin coating if the substrate is small. However, screen-printing or spray-coating methods are also suitable for industrial PCB processes with large panel sizes. To achieve compatibility with PCB materials, polymer materials are favored for the board-integrated waveguides (Karppinen 2008).

For waveguides on silicon chips, there still appear to be some basic technological and fundamental challenges. There are promising ways to make waveguides using processes that are compatible with silicon processing, though there are still some issues with waveguide loss. It is encouraging that the waveguides themselves appear viable in a silicon process, and these may be interesting at least for optical clock distribution, for example, where on-chip sources are not required (Miller 2000).

1.1.3.6 Micro-optical Coupling Elements

One of the key challenges of the OE-PCB technology is to provide efficient and robust optical coupling between the waveguide and the laser diode or the detector. Micro-optical elements are often utilized to enhance the optical coupling efficiency and possibly also to loosen the mechanical alignment tolerances. For instance, the expanded beam method offers a viable solution to loosen the alignment tolerances in some locations along the light path. In practice, this means that optical elements, typically lenses, are applied to expand the light beam into a collimated beam between the elements, which are challenging to accurately align. The expanded beam method has been demonstrated, for instance, both with single lenses to loosen the alignment tolerance of a daughter-board-to-backplane connector and with microlens arrays to loosen the alignment tolerance of a ball-grid-array (BGA)-mounted component. Although additional packaging complexity is generated from the lenses, the expanded beam connector allows lateral, vertical, and longitudinal misalignments, for instance, up to several hundred micrometers for a minimal additional loss of less than 1 dB. A drawback of the expanded beam method is that it sets a lower limit on the density of channels in parallel interconnects. In many proposed and demonstrated packaging implementations the VCSEL and detectors are facing towards the waveguide, thus the optical beams have to be turned through 90°. The beam can be turned with a discrete mirror component. Different kinds of mirrors have been presented so far, such as a flip-chip bondable micro-mirror device and a surface-mountable micro-optical element. The challenge in the use of discrete micro-mirrors is that one must first form a cavity on the waveguide layer in order to expose the waveguide facet, and that facet should also be of good optical quality, i.e., not cause too much scattering. The microlens arrays are especially advantageous for parallel optical interconnects. Today, microlenses are customarily volume fabricated as 1D or 2D arrays on the wafer level for many applications. Various manufacturing technologies are possible, including even roll-to-roll-type continuous mass fabrication. Microlens arrays have been used in many hybrid-integrated multichannel optical interconnects. Microlenses enable enhancing optical coupling, extending the separation between components and loosening the required alignment accuracies. Alternatively to microlenses, the coupling between the OE device and the waveguide with mirror facet can be enhanced with an integrated vertical light guide, possibly having a pillar shape. Since the use of free-space lens optics is avoided, coupling with optical pillars, may enable very dense interconnects to a flip-chip-mounted OE device. Moreover, when fabricated from a polymer material that is flexible enough to enable bending of the pillar without deformation, the optical pillar may also compensate for small displacements in a flip-chip-mounted device. Another element that can be used to couple light into and out of the waveguide is a diffractive grating coupler. Nevertheless, it has appeared very challenging to implement high-efficiency couplers that would be feasible for board interconnect applications based on multi-mode waveguides, surface-normal incidence, and incompletely collimated beams without polarization control. An issue regarding the integration of micro-optics into electronics is the endurance of the materials during the assembly processes. Today, the micro-optical components are typically installed in the products after the electronics assembly phase. This is because the micro-optical components are often made of polymer materials, such as PMMA, that exhibit glass transition temperatures below 200 °C and thus do not survive the reflow soldering temperatures. On the other hand, such an assembly procedure is not acceptable (or at least not favorable) in the case of embedded board-level optical interconnects. Consequently, all optical components and adhesives should survive a normal reflow soldering process, which typically means peak temperatures in the order of 250 °C for around 1 min. This means that only a small part of the polymer materials commonly used in micro-optics are applicable (Karppinen 2008).

1.1.3.7 Integration and Packaging

The implementation of board-level optical interconnects using today's photonics and electronics technologies requires hybrid integration of several different material and device technologies. On the other hand, there are many functions that have to be implemented through the integration and packaging, such as optical coupling, electrical interfaces, thermal management, shielding, and protection.

Therefore, integration technologies also play a key role in defining the costs of the overall solution and, thus, its feasibility (Karppinen 2008).

Integrating the optoelectronic devices with the silicon-integrated circuits needs to use III–V optoelectronic devices at least for optical outputs in dense interconnects. A viable approach for silicon CMOS is to use hybrid integration technologies such as solder bonding. Such techniques require no modification of the basic CMOS process, using instead only a few simple additional metallization that can be done on the finished wafer. This allows the separation of the growth and processing of the optoelectronic devices from that of the silicon circuits, and allows a great deal of flexibility in the types of optoelectronic devices used (Miller 2000).

The high-alignment-accuracy hybrid integration of a microlens array with a VCSEL array achieved approximately 2 μm lateral alignment precision by the use of lithographically patterned alignment marks and precision-machined spacers. Alternatively, refractive or diffractive microlenses can be fabricated on VCSELs or photodiodes by heterogeneous integration; either on top of the device, e.g., by UV casting, or via wafer bonding, or on the rear of the chip if it has a transparent substrate such as 980-nm VCSELs have, e.g., by etching the substrate. Integrated lenses can facilitate optical coupling and ease the packaging alignment tolerances. There are several technology gaps still hindering the applicability of optical interconnects at the board level. These include mass-producible micro-mirror technology and SMT-compatible hybrid integration of optoelectronics with coupling optics (Karppinen 2008).

1.2 Fundamental Theory and Design Methodology

Optical waveguides are typically made with dielectric material structures that transmit electromagnetic waves in the direction parallel to their axis at visible or infrared frequencies. They are fundamental building components of many optical systems, including fiber-optic communications links, fiber lasers and amplifiers for high-power applications, as well as all-optical photonic integrated circuits. Optical interconnection based on low-loss waveguides is perhaps the most attractive solution proposed to overcome the electrical interconnection bottleneck especially at the board and module level. In the past decades, much effort has been devoted to advancing the theory and practice of optical waveguides. This has been mainly driven by the widespread deployment of optical fiber communication systems where system performance has been dramatically improved over the last 3 decades. In addition to the optical fibers themselves, a number of optical components, such as diode lasers, couplers, and external modulators, are formed on optical waveguide structures and their development has demanded a thorough understanding of waveguide theory. On a smaller scale, and particularly since the early 1980s, there have been growing research programs on optical sensor and short-range optical interconnection technologies which have also provided stimulus for the development of new types of waveguide devices and the associated theoretical models.

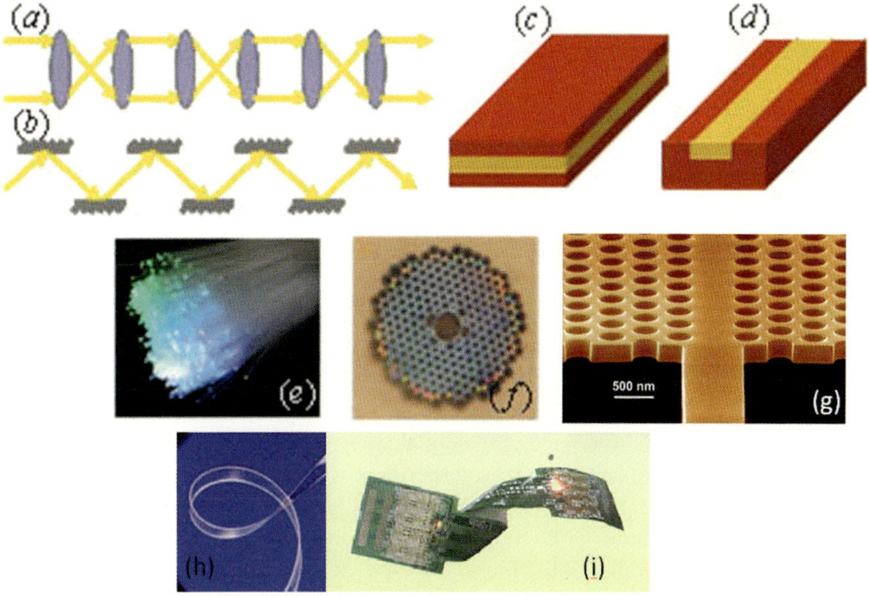

Fig. 1.5 Classification of optical waveguides based on their geometry (Litchinitser and Iakhnine 2011): (**a**) lens waveguide, (**b**) mirror waveguide, (**c**) slab waveguide, (**d**) strip waveguide, (**e**) standard optical fibers, (**f**) microstructured fiber, (**g**) photonic crystal waveguide, (**h**) flexible waveguide film and (**i**) its application in the flexible circuit board

1.2.1 Classification of Optical Waveguides

Electromagnetic waveguides can trap electromagnetic wave locally and guide it in a specific direction. In the microwave or lower frequency band, electromagnetic fields are generally distributed within a finite area surrounded by conductors, as in metal waveguides. However, it is difficult to construct a waveguide mainly from metal in the visible light wave or nearer frequency band, since the metal will behave as a substance having a complex permittivity with a large absolute value. For this reason, optical waveguides are usually constructed by combining appropriate dielectrics to confine and direct optical signals in a region of higher effective index than its surrounding media. By confining wave propagation, the waveguides provide communication between the electro-optical transmitter and receiver, creating optical interconnects. Up till now, many kinds of optical waveguides have been developed based on different application requirements.

Optical waveguides can be classified according to (Litchinitser and Iakhnine 2011):

1. Waveguide geometry as shown in Fig. 1.5: lens- and mirror-waveguides (Fig. 1.5a, b), planar, slab and strip waveguides (Fig. 1.5c, d), standard and microstructured fibers (Fig. 1.5e, f), photonic crystal waveguides (Fig. 1.5g), and flexible waveguide film (Fig. 1.5h, i).

2. Guiding mechanism: total internal reflection, antiguiding, photonic bandgap, and antiresonant guiding, as well as some more exotic mechanisms such as total external reflection in metamaterials-based waveguides.
3. Mode structure: single-mode or multimode.
4. Refractive index distribution: step or gradient index.
5. Waveguide material: glass, polymer, semiconductor, metal, artificially created materials.

Waveguides devices in photonic integrated circuits contain different shapes and sections with various functions, providing connections between various optical devices and components for the generation, focusing, splitting, combining, isolation, polarization, coupling, switching, modulation and detection of light, all on a single substrate. They can be classified into two categories (Saleh and Teich 1991): (a) passive devices, which exhibit static characteristics for optical waves; and (b) active or functional devices for optical wave control. Passive devices include a variety of 3D waveguides such as corner bent waveguides, which change optical path direction; bent and S-shaped waveguides; tapered waveguides, for smooth expansion of the guide width without mode conversion; branching waveguides, used for both dividing and combining optical power; cross waveguides and directional waveguide couplers, for distributed mode coupling. In active or functional devices, guided waves are controlled via electro-optic, magneto-optic, thermo-optic and nonlinear optic effects. They are controlled by external input signals to achieve functional waveguide devices where the guided mode undergoes amplitude modulation, phase modulation, polarization-state rotation, frequency shift or optical path direction change.

The most widely used optical waveguides have been basically designed as planar slab waveguide (2D waveguide), and channel waveguide (3D waveguide). Planar slab waveguide integrated optics is concerned with the manipulation of sheet beams. The light can propagate in any direction parallel to the surface of a high-index guiding layer, which provides optical confinement in a single direction. As shown in Fig. 1.6a, thin films basically deposited on a transparent dielectric substrate are used as optical waveguides if the film index is higher than the substrate index. The indices of the cladding layer, guiding layer and substrate are n_c, n_f, and n_s, respectively. The light cannot be guided unless $n_f > n_s > n_c$, and the thickness, T, of the guiding layer is above the critical thickness or cutoff thickness. The slab is infinite in extent in the y–z plane, but finite in the x direction. If the cladding and the substrate materials have the same index of refraction, the waveguide is called symmetric; otherwise the waveguide is called asymmetric. In the channel waveguide, the beams propagate along high-index guiding channel. These may be formed as ridges on the surface, or as buried channels. The beams are confined in two dimensions (x and y) and so can only follow predefined pathways round the chip (Gang 2005). Figure 1.6b shows cross sections through the most common geometries, such as buried channel or embedded strip guides with straight, S bend, Mach-Zchnder, directional coupler, and intersection configurations; overlaid strip waveguide; rib or ridge waveguide and strip loaded waveguide. These geometries are formed depends on the different fabrication methods.

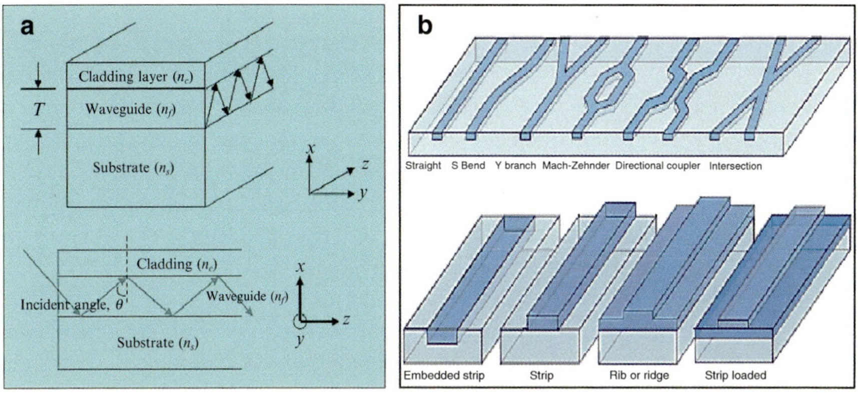

Fig. 1.6 Schematic of planar waveguide and channel waveguides: (**a**) thin film deposition structure; (**b**) various waveguide geometries (the darker the shading, the higher the refractive index)

From design and modeling viewpoints, optical guiding structures can be divided into four categories (Litchinitser and Iakhnine 2011):

1. The first class of structures includes waveguides that are invariant along their lengths. The solution of the wave equation for single frequency propagation involves calculation of the eigenfunctions, or modes of the waveguide, at a fixed frequency and the eigenvalues that correspond to the axial propagation constant of the wave in the waveguide. In several cases, modal solutions can be found analytically. However, most practical cases rely on numerical solution of the wave equation.
2. The second class of structures includes waveguides that are nonuniform in the direction of wave propagation. Examples of such structures include waveguide tapers, gratings, and photonic crystal waveguides, Y-splitters, S-bends, and helical waveguides. The accurate description of light propagation in these structures relies on efficient numerical methods.
3. The third class of problems includes studies of optical pulse propagation in nonlinear waveguides. While linear pulse propagation can be relatively easily solved in the frequency domain, its nonlinear counterpart requires more advanced numerical approaches.
4. Finally, a novel class of optical waveguides includes structures composed of metamaterials. In this case, boundary conditions should be modified to take into account the magnetic properties of the metamaterials that are not present in conventional optical waveguides. In addition, the frequency dependence of the material parameters, such as the dielectric permittivity, the magnetic permeability, and the refractive index has to be included.

Accordingly, several approaches to modeling light transmission in various types of optical waveguides have been developed. These include beam propagation method (BPM), finite-element method (FEM), finite-difference time-domain

(FDTD) method, transfer matrix method (TMM), multipole method, and so forth. Strategically, there are two basic approaches to the numerical solution of a particular research problem. The first approach includes development of a homemade numerical code. The second approach relies on adaptation of commercial or freely available Computer Aided Design (CAD) tools. Each of these approaches has advantages and disadvantages. The advantages of developing homemade codes include full access and control over the source code and therefore, flexibility with regard to code modification and improvement. On the other hand, reliable ready-to-use CAD tools are applicable to a variety of different physical problems and could be optimized for speed, memory usage, and so forth. Therefore, these CAD tools are being widely utilized in industrial product development as well as in academic research. However, all these approaches are developed based on fundamental waveguide theory.

1.2.2 Fundamental Waveguide Theory

Maxwell's equations and ray-optic approach have been applied to waveguide structures with the principal boundary conditions of the electromagnetic field components, and form the basic waveguide theory, which provides the foundation for understanding the detailed operation of a wide variety of optical components and systems based on optical waveguide technology.

There are two fundamental theories available in optics. In the classical theory, the behavior of light is described in terms of electromagnetic fields and waves with Maxwell's equations. In the newer quantum theory, light considered to be composed of photons, which are the elementary units or quanta of radiation. Quantum theory will be discussed in later chapters. Here will focus on the classical theory described by Maxwell's equations, which is particularly appropriate for the analysis of passive devices.

Maxwell's equations are a set of relations linking the values of a number of quantities that describe electric and magnetic fields. Optical fields are time-dependent fields and their behavior can be fully described by the set of Maxwell's equations. In a source free region Maxwell's equations are expressed by (Gang 2005):

$$\nabla \times \vec{E} = -\frac{\partial \vec{B}}{\partial t} \tag{1.1}$$

$$\nabla \times \vec{H} = \vec{J} - \frac{\partial \vec{D}}{\partial t} \tag{1.2}$$

$$\nabla \cdot \vec{B} = 0 \tag{1.3}$$

$$\nabla \cdot \vec{D} = \rho \tag{1.4}$$

The vector quantities are the electric field vector $\vec{E}$ (V/m), the magnetic induction vector $\vec{B}$ (Wb/m^2), the electric displacement vector $\vec{D}$ (C/m^2) and the magnetic field vector $\vec{H}$ (A/m). Equations (1.1) and (1.2) are vector equations that relate time and space derivatives of the field quantities, and (1.3) and (1.4) are scalar equations that give information about the outflow of the electric and magnetic field. In linear and isotropic media (homogeneous and lossless dielectric medium), $\vec{D}$ and $\vec{B}$ are defined as (Okamoto 2006)

$$\vec{D} = \varepsilon\vec{E} \tag{1.5}$$

$$\vec{B} = \mu\vec{H} \tag{1.6}$$

where $\varepsilon = \varepsilon_0\varepsilon_r$ and $\mu = \mu_0\mu_r$. The ε and μ define the electromagnetic properties of the medium, and are the dielectric permittivity and magnetic permeability of the medium, respectively. Also, ε_0 is the permittivity of vacuum (8.854×10^{-12} F/m^2), ε_r is the relative permittivity of the material; μ_0 is the magnetic permeability of vacuum ($4\pi \times 10^{-7}$ H/m^2) and μ_r is the magnetic permeability of the material. For nonmagnetic materials, $\mu_r = 1$. When analyzing the optical properties of a material, its refractive index, n, is used and defined as $n = \sqrt{\mu_r\,\varepsilon_r}$.

In uniform waveguides it is common to classify various wave solutions into the following types (these are not the only way to categorize the different wave solutions, but have been standardized by long usage):

(a) TEM waves: waves with no electric or magnetic field in the direction of propagation ($H_z = E_z = 0$). Plane waves and transmission-line waves are common examples.
(b) TM waves: waves with an electric field but no magnetic field in the direction of propagation ($H_z = 0$, $E_z \neq 0$). These are sometimes referred to as E waves.
(c) TE waves: waves with a magnetic field but no electric field in the direction of propagation ($H_z \neq 0$, $E_z = 0$). These are sometimes referred to as H waves.
(d) Hybrid waves: Sometimes the boundary conditions require all field components. These waves can be considered as a coupling of TE and TM modes by the boundary.

Assume that the dielectric slab waveguide is formed from materials which are linear, isotropic, and source free, as shown in Fig. 1.6a, and choose the x axis representing the transverse direction and z axis as the propagation direction, one set of solutions had $\vec{E}$ in the y direction (a direction in the plane of the film and perpendicular to the direction of propagation) and these are the TE modes. The other set with $\vec{H}$ in the y direction are the TM modes. For TE modes, assuming that there is no field variation in the y direction and the field with a periodic time dependence of angular frequency, ω and a lossless medium of dielectric constant, ε and permeability, μ_0. The vector wave equation is given by (Gang 2005)

$$\nabla^2 \vec{E} = \mu_0 \varepsilon \frac{\partial^2 \vec{E}}{\partial t^2} \tag{1.7}$$

In the Cartesian coordinates, it can be written as a set of independent scalar equations.

For simplicity, the electric field is considered to have one single component, y direction (Gang 2005):

$$\frac{\partial^2 \vec{E}_y}{\partial x^2} + \frac{\partial^2 \vec{E}_y}{\partial y^2} + \frac{\partial^2 \vec{E}_y}{\partial z^2} = \mu_0 \varepsilon_0 \varepsilon_r \frac{\partial^2 \vec{E}_y}{\partial t^2} \tag{1.8}$$

In the TE mode, the $\vec{E}$ field is polarized along the y axis, no longitudinal component along the z axis; the electric field is transverse to the plane of incidence established by the normal to the interface, and by the k vector. By assuming that the waveguide is excited by a source with frequency ω_0 and a vacuum wave vector of magnitude k_0, where $|k_0| = \omega_0/c$ and c is the speed of light or $k_0 = 2\pi/\lambda$ and λ is the wavelength of the light source. Due to the translational invariance of the structure in the z direction, the amplitude is not varying along z axis, but the phase varies. β is a propagation constant along the z direction. The wave equation reduces to a time-independent wave equation for the electric field components in each region of the waveguide can be put in the scalar form (Gang 2005):

$$\frac{\partial^2 \vec{E}_y}{\partial x^2} + \left(k_0^2 n_i^2 - \beta^2\right)\vec{E}_y = 0 \tag{1.9}$$

where $n_i = n_f$, n_s or n_c, depending on the location. Equation (1.9) is the transverse wave equation that needed to be solved for determining the eigenvalue of β and TE field distribution throughout the medium.

One of the significant properties in determining the quality of waveguide is the propagation constant. The propagation constant, or the longitudinal wave vector, β is used to identify individual modes. It is defined as the eigenvalue of the mode. Figure 1.7a shows the transverse electric field distribution in a slab waveguide for various values of β (Gang 2005).

There are three special points on the β axis, $k_0 n_c$, $k_0 n_s$, and $k_0 n_f$ (since $n_c < n_s < n_f$). For $\beta < k_0 n_c$, solutions to the wave equation in all regions of space are oscillatory. From the ray picture, the wave travels nearly perpendicular to the z axis of the waveguide, refracted at the interface, but not trapped. For $k_0 n_s < \beta < k_0 n_f$, the ray picture shows a ray total internal reflecting (TIR) at the film-cladding interface, but refracting at the lower substrate–film interface. In the wave picture, the field becomes evanescent in the cladding region. The field is still oscillatory in the film and substrate regions and this condition is called substrate mode. As β increases beyond $k_0 n_s$, the evanescent conditions are satisfied in both the cladding and substrate region, while oscillatory solutions are found in the film itself. The guided modes of the film occurred. The ray picture depicts a ray trapped

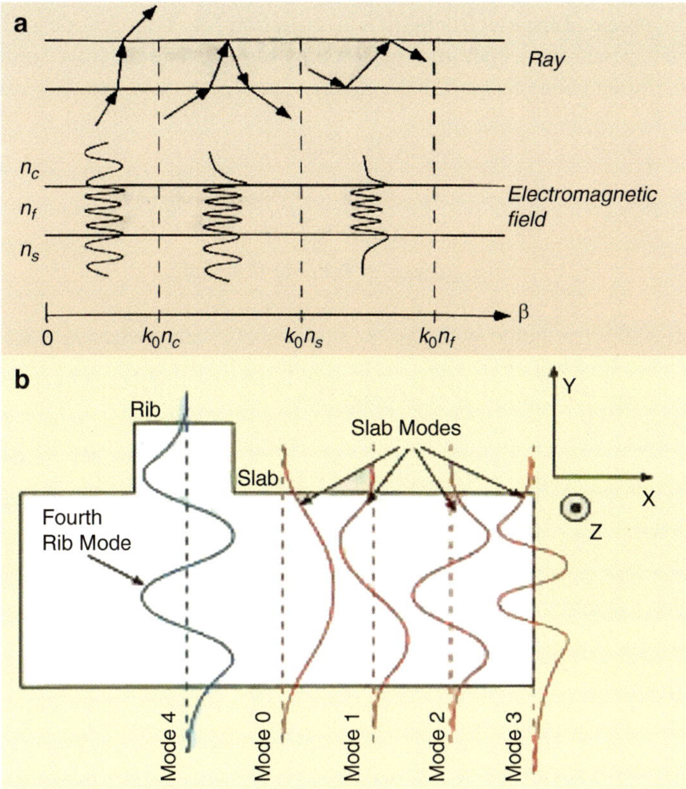

Fig. 1.7 Illustration of the transverse electric field distribution in a slab waveguide for various values of β (**a**); and the first four TE modal field patterns of the waveguide (**b**)

between the two interfaces. Therefore, a guided wave must satisfy the condition that (Gang 2005)

$$k_0n_s < \beta < k_0n_f \tag{1.10}$$

This is a universal condition for any dielectric waveguide, regardless of geometry.

The various types of wave solutions have many common features, regardless of the shape of the waveguiding structure. Examining the propagation β, corresponding to propagating waves, or exponentially decaying, or evanescent fields, the dividing line is at a frequency known as the cutoff frequency. The waveguide parameters are usually determined on the basis of cutoff of the guided modes. When the incident angle, θ becomes the critical angle θ_s, the light is no longer confined in the guiding layer, and it begins to leak into the substrate at the interface. This situation is called cutoff of the guided mode. The number of the guided modes that can be supported by a three layer slab waveguide depends on the thickness, T, of the waveguiding layer and on ω, n_c, n_s, and n_f. Figure 1.7b shows the general shape of the electric field, $\vec{E}$ distribution for several modes. The higher-order modes penetrate deeper into the cladding regions than the lower order modes. The lowest order mode has a β that is nearly parallel to the z axis (Gang 2005).

$$\beta < k_0 n_f \tag{1.11}$$

The highest order mode will have a β at nearly the critical angle.

$$\beta_{\text{highest order}} \approx k_0 n_f \cos\theta \approx k_0 n_s \tag{1.12}$$

In an asymmetric guide with $k_0 n_s > k_0 n_c$, cutoff occurs when $\beta = k_0 n_s$ and the field spreads all the way to infinity in the region with refractive index n_s. The cutoff thickness t_{co} for TE_m mode to be (Gang 2005):

$$t_{co} = \frac{\lambda}{2\pi\sqrt{\left(n_f^2 - n_s^2\right)}}\left[m\pi + \arctan\left(\frac{n_s^2 - n_c^2}{n_f^2 - n_s^2}\right)\right] \tag{1.13}$$

where $m = 0$ (fundamental mode), 1, 2, 3, ...

By calculating the cutoff thickness for certain modes, the film thickness can be determined to ensure that the waveguide is able to support the fundamental mode and to control the thickness when designing the single-mode waveguides.

1.2.3 Optical Waveguide Design Methodology

The optical waveguide systems can usually be designed and optimized by the use of wave propagation approach and ray tracing design tools. They enabled modeling the optical path losses as well as carrying out opto-mechanical tolerance analyses and optimizations (Karppinen 2008).

1.2.3.1 Wave Propagation Approach

Wave propagation techniques are essential components of waveguide and waveguide-based photonic device design. While mode-solving methods are based on the assumption that the guiding structures are uniform in the propagation direction, many practical waveguide configurations, such as tapers, bends and junctions, are actually nonuniform in the longitudinal direction. The analysis of these types of waveguides relies on propagation-type techniques. Even in uniform structures, propagation methods provide important information about beating between modes in multimode structures, mode filtering upon propagation, and mode coupling effects. In addition, propagation methods facilitate numerical studies, accurately reflecting waveguiding properties of realistic devices and providing insight into the field's evolution inside the structure. In these simulations arbitrary launch conditions are usually allowed and the evolution of such an arbitrary beam/pulse can be monitored at any point inside the structure. The major wave propagation approaches include both time-domain methods, including the FDTD method, the split-step method (SSM), and the FEM, and frequency-domain methods, such as the BPM, the TMM, and the eigenmode expansion method (EME). Of these techniques, the FDTD (implemented in a large number of free and commercially available packages not listed here) is the most general and rigorous time-domain method. It is particularly useful for a large number of optical guiding structures, such as photonic crystal waveguides, surface plasmon waveguides, high-index contrast waveguide devices, ring and disk resonators, negative index material structures, dispersive, and nonlinear materials. Modal properties can be extracted after a single propagation. However, the FDTD may become prohibitively computationally intensive and therefore impractical for long waveguides and devices (Litchinitser and Iakhnine 2011).

The SSM is a simple and straightforward technique for the modeling of linear and nonlinear pulse propagation of optical fibers. It allows the inclusion of all major effects taking place in realistic optical fibers, including group-velocity dispersion and higher-order dispersion, self-phase modulation, cross-phase modulation, four-wave mixing effects, Raman amplification and many other linear and nonlinear phenomena. The SSM method is a numerical technique used to solve nonlinear partial differential equations (PDE) like the nonlinear Schrödinger equation. The method relies on computing the solution in small steps and on taking into account the linear and nonlinear steps separately. The linear step (dispersion) can be made in either frequency or time domain, while the nonlinear step is made in the time domain. The method is widely used for studying nonlinear pulse propagation in optical fibers.

The FEM methods have been implemented in both time and frequency domains, and can be used as wave propagators and mode solvers simultaneously. However, some of the FEM software packages work only as mode solvers and should not be confused with those allowing propagation effects studies. The FEM is a method used for finding the approximate solution of PDE that handle complex geometries (and boundaries), such as waveguides with arbitrary cross-sections, with relative ease. The field region is divided into elements of various shapes, such as triangles and rectangles, allowing the use of an irregular grid. The solution approach is based either on eliminating the differential equation completely (steady state problems), or rendering the PDE into an equivalent ordinary differential equation, which is then solved using standard techniques, such as finite differences. In a context of optical waveguides, the FEM can be used for mode solving and propagation problems. Two approaches to solve waveguide problem include the variational method and the weighted residual method. Both methods lead to the same eigenvalue equation that needs to be solved.

The BPM is a widely used and well-developed method that solves wave propagation problems for a given frequency. Several advanced techniques, including semi-vectorial and vectorial implementations, wide-angle and bidirectional BPM, have been developed, making the BPM a versatile technique for a wide class of optical waveguides with arbitrary cross-sections and complex refractive index profiles. The beam propagation method was developed for the analysis of light propagation in slowly varying nonuniform guiding structures, such as tapers, bends, and couplers. Various implementations of BPM differ by the method used for calculations of the derivatives with respect to the coordinates in the propagation directions and include the finite-difference (FD)-BPM, fast Fourier transform (FFT)-BPM, and the finite-element (FE)-BPM. Historically, the FFT-BPM was widely used until the FD-BPM was developed. The standard form of implementation of the FD-BPM is the Crank–Nicholson method. The standard BPM solves the Helmholtz equation in paraxial approximation. The paraxial approximation enables BPM to analyze much longer structures (on the millimeter and centimeter scale) than would be realistic for analysis with FDTD. While the BPM is usually used for the modeling of low-index-contrast waveguides, some relatively high-index contrast structures can be studied as well. Scalar, semi-vectorial, and vectorial formulations have been developed, and several advanced features have been proposed and implemented in BPM-based software packages, including wide-angle formulation and bidirectional propagation. The BPM versions that incorporate nonlinearities, anisotropy, gain and loss have been described and implemented. In addition, two BPM-based mode solvers (based on imaginary-distance BPM and correlation technique) have been proposed. Typical BPM applications include standard, specialty, and photonic crystal fibers, WDM devices, switches, modulators, multimode interference devices, and passive splitters.

The TMM and the EME methods are relatively simple methods that are inherently bidirectional. They are especially useful for the modeling of waveguides comprising grating structures, waveguide tapers and couplers.

The FDTD technique is based on a discrete representation of time-dependent Maxwell's equations on a grid that is exact in the limit that the grid spacing goes to zero. Since the FDTD is a time-domain technique, the response of the system over a wide range of frequencies can be obtained with a single simulation. Various optical materials can be used with FDTD algorithms. Modal properties can be extracted in a single simulation by Fourier-transformation of the time-varying response of the system to some input. Then, the peaks in the resulting spectrum correspond to the eigenfrequencies. The primary limitation of FDTD is that it is computationally intensive, especially in the three-dimensional case, but the method is readily applicable to the two-dimensional structures. In practice, FDTD needs to have at least 10–20 cells per minimum wavelength in every direction, depending on the application and accuracy required. Also, even higher sampling rates may be required to reduce the cumulative numerical dispersion error, which is proportional to the length of wave propagation. This renders FDTD impractical for large devices. Various techniques have been proposed to make the FTDT method more efficient, including pseudospectral time-domain techniques, nonuniform and adaptive mesh refinement, hybrid FDTD-finite-element techniques, and hardware acceleration concepts. Different software providers take various approaches to optimize the performance of their software in solving problems that require clustering. The approaches include parallel processing features in the software and dedicated hardware to increase single system processing speeds. Typical applications for optical software based on an FDTD algorithm include the modeling of high-index-contrast waveguides, photonic-bandgap structures for photonic crystals, ring resonators, nano-plasmonic devices, gratings and other diffractive structures, dispersive, nonlinear and gain materials, and bio-photonics.

The multipole expansion method is a numerical method that can be used for full-vector modal calculations of microstructured fibers and other photonic crystal structures. It yields both the real and the imaginary parts of the mode propagation constant (thus providing information about losses) and achieves high accuracy and rapid convergence with modest computational resources. In freely available software the multipole method is implemented for circular inclusions; however, it can be extended to noncircular inclusions. It can deal with two types of MOF: a solid core MOF, surrounded by air holes, and air core MOF. Systems with large numbers of inclusions can be modeled, and for structures with discrete rotational symmetries the computational overhead is further reduced by exploitation of the symmetry properties of the modes. The method is limited to nonintersecting circular inclusions, and convergence problems arise as the spacing between the inclusions decreases. In the multipole method each dielectric boundary in the system is treated as a source of radiating fields. The essential feature of a multipole method is the application of a certain field identity that relates the regular field in the vicinity of any scatterer to fields radiated by other scatterers and external sources. The multipole method can analytically preserve symmetry and exploit the natural basis of functions for the scatterers (e.g., cylindrical harmonic functions for cylindrical scatterers). As a result, it can yield important physical insight into the scattering processes not otherwise possible with purely numerical methods. An important feature of the method is that frequency can be used as an input parameter, whereas the propagation constant follows from the calculation. This is a significant advantage when one is dealing with dispersive media: because is fixed, the appropriate refractive indices are known from the outset. Applications of the multipole methods include the modeling of photonic crystals, microstructured optical fibers, the study of the radiation properties of sources embedded in such structures, and the design of complex/composite devices, e.g., splitters, couplers, interferometers, among others devices (Litchinitser and Iakhnine 2011).

The SMT is a fully vectorial method that can be used to determine the modes of a cylindrical structure of a piecewise-homogeneous cross section. In the SMT, the fields in each homogeneous region of the cross section are approximated by the fields due to a linear combination of elementary sources placed outside of it, on curves conformal with the region boundary. The sources radiate in a homogeneous medium with the same material parameters as those of the region they enclose in the PCF cross section. Their fields, therefore, have well-known analytic expressions. The amplitudes of the elementary sources are determined so as to satisfy the continuity conditions across the media interfaces at a set of testing points. The elementary sources used are electric and magnetic current filaments, carrying longitudinally varying currents that vary with the z-coordinate. For each pair of under consideration, the vector that yields the smallest least squares error can be found. Once a low error solution is found for some pair, the current amplitudes are inserted into the linear combinations that approximate the fields, yielding the mode field patterns and any other parameter of interest. The SMT has much in common with the multipole method, in which the basic functions are employed. Although the multipole method representation is considered very compact, it is directly applicable only to cross sections composed of circular homogeneous inclusions in a homogeneous medium, while in the SMT, the homogeneous regions in the cross section can be arbitrarily shaped.

The plane wave expansion (PWE) method is a frequency-domain approach based on the expansion of the fields as definite-frequency states in some truncation of a complete basis (e.g., plane waves with a finite cutoff) and then solution of the resulting linear eigenproblem. The method is applicable to optical waveguides with arbitrary cross sections and resonators, photonic bandgap materials, and photonic crystal structures, or for calculations of optical dispersion relations and eigenstates for conventional optical fibers. One disadvantage of the PWE method is that the wavevector serves as the free parameter, whereas the frequency eigenvalues follow from the calculation. For certain applications, especially for finding waveguide modes, it is often more convenient to specify the frequency and solve for the required propagation constant, as is done in the multipole method. In particular, such an approach would make it much easier to incorporate the effects of material dispersion, which is invariably specified as a function of frequency. Also, for some problems, such as defect modes, the plane wave expansion method may be extremely time consuming.

The TMM is a simple technique that can be used for modeling for obtaining propagation characteristics, including losses for various modes of an arbitrarily graded planar waveguide structure, which may have media of complex refractive indices. The method is applicable for obtaining leakage losses and absorption losses, as well as for calculating beat length in directional couplers. The waveguide is divided into a number of layers. At a fixed frequency, one computes the transfer matrix, relating field amplitudes at one end of a unit cell, with those at the other end (via finite-difference, analytical, or other methods). This yields the transmission spectrum directly, and mode wavevectors via the eigenvalues of the matrix. TMMs may be especially attractive when the structure is decomposable into a few more-easily solvable components, and also for other cases, such as frequency-dependent dielectrics. It was also implemented for analyzing modal properties of optical fibers with a layered cladding structure. A modified TMM method capable of direct leaky mode analysis by expressing field in the outermost layer in terms of Hankel functions was proposed. This method is extremely efficient in treating multi-clad optical fibers, which would otherwise be very complex, if not computationally forbidding, to all numerical methods.

The EME is based on a rigorous solution of Maxwell's equations, representing the electromagnetic fields everywhere in terms of a basic set of local modes.

In principle an exact solution can be obtained using an infinite number of modes in our expansion. Of course, in practice the number of modes is limited and there will be numerical errors in the implementation, as in any numerical technique. To obtain higher accuracy one can simply add more modes, enabling the method to accurately compute problems that cannot be computed with other techniques. The algorithm is inherently bidirectional and utilizes the scattering matrix (S-matrix) technique to join different sections of the waveguide or to model nonuniform structures. The scattering matrix technique relates the incoming waves, i.e., the forward wave at the beginning of the section and the backward wave at the end of the section, to the outgoing waves, i.e., the backward wave at the beginning and the forward wave at the end. All reflections are taken into account in the method. If one part of a device is altered, only the S-matrix of that part needs to be recomputed. The method can simulate light propagating at any angle, even 90° to the propagation axis, simply by adding more modes. Applications include diffractive elements, directional couplers, tapers, MMIs, bend modeling, periodic structures and others. Structures with a very large cross section are less suitable for the method since computational time typically scales in a cubic fashion, with, for example, cross-section width.

The other mode solvers include the software packages containing multiple-mode-solving techniques including analytical methods, such as the effective index method and Marcatili's method, and numerical approaches, such as the finite-difference method and meshless techniques, so that depending on a problem, one of the available techniques can be used. It also includes some two-mode solvers based on the beam propagation method (imaginary distance and correlation method) and the wave-matching method (Litchinitser and Iakhnine 2011).

1.2.3.2 Ray-Trace Modeling

Optical waveguides can be modeled by ray-trace approach using a realistic 3D system model. Although the ray tracing approach is not as accurate as the wave propagation approach, it enables accurate analysis of complicated optical systems with such properties as multimode sources (e.g., VCSELs), scattering, and aperture clipping. Propagation through a multimode waveguide can be modeled as well, although the simulations tend to become time consuming due to the large number of reflections. Furthermore, the ray-trace approach enables analysis of the crosstalk in multichannel interconnections.

The ray-trace propagation through the waveguide can be included in the simulations in order to obtain more realistic characteristics of the waveguide output beam. In addition to waveguide losses, the ray tracing method allows analysis of the time jitter due to multipath dispersion. In other words, the multipath-induced impulse response of the system can be calculated from the optical path lengths of the rays propagated through the system. Furthermore, the impulse response allows estimating, for instance, the inter-symbol interference and, thus, the maximum data transfer rate of interconnect with the chosen modulation method (Karppinen 2008).

1.2.3.3 Tolerance Analysis for Optimization and Yield Estimation

In order to design an optical waveguide system that is insensitive to both component misalignments and manufacturing tolerances, two types of analyses are needed (Karppinen 2008): (1) In sensitivity analysis, each tolerance variable (*xyz*-displacements and tilts) is simulated separately and the most critical tolerances are found. Sensitivity analysis helps to optimize the optics. However, the root-sum square and worst-case analysis that can be made from the sensitivity analyses of individual misalignment effects alone is not a sufficient method because it does not take into account the simultaneous interaction of all variables. (2) In ray-trace modeling, the effect of all variables can be simultaneously analyzed by creating a large number of randomly chosen perturbed systems (with misalignments). Thus statistical information on the system performance is obtained, allowing prediction of the manufacturing yield of the system. In addition, the correlation between the loss and the tolerance parameters reveals the parameters that most affect the total system performance.

1.3 Waveguide Materials Selection and Fabrication Techniques

Optical waveguide devices have penetrated a number of markets by providing performance-competitive integrated optic and fiber-optic device solutions. This growth, in turn, requires new developments in modeling, further advances in material sciences, and innovations in integration platforms. In addition, the processing and fabrication of these new devices must be optimized in conjunction with the development of accurate and precise characterization and new testing techniques.

Many kinds of materials and the corresponding processing techniques have been developed for optical waveguides. Typical optical materials and their fabrication techniques are summarized in Table 1.1 (Kowach and Ramirez 2002). Several inorganic materials possess multiple functions, such as silica (SiO_2), silicon (Si), silicon oxynitride (SiO_xN_y), sol–gels, polymers, lithium niobate ($LiNbO_3$), indium phosphide (InP), and gallium arsenide (GaAs). These material systems have been pursued as integrated optic platforms. Silica on silicon has been a leading materials technology for application in PLCs because of its low loss and high-temperature stability. The silica system has the advantage that its index of refraction is perfectly matched to silica fiber. However, it is relatively slow (millisecond switch times) because the functionality of silica on silicon is based on the thermo-optic properties of silica. Moreover, it requires complex thermal controls to isolate the region being

Table 1.1 Typical materials and fabrication techniques for optical waveguiding systems (Kowach and Ramirez 2002)

Material	Fabrication technique	Property/ performance	Component	Function	System/network
Ag	Sputtering, evaporation, electrochemistry	Electrical and thermal conductor	Optoelectronic packaging	Thermal management and interconnection	Networks, packaging
Al	Sputtering, evaporation	Conductor, reflective (visible)	Interconnects, optical MEMS	Electrical conductor, reflective surface	Wireless, optical, packaging
AlN	Sputtering, CVD	Piezoelectric	MEMS	Sensor/transducer materials, actuation method	Wireless, optical, packaging
Au	Evaporation, sputtering	Reflective, conductor	Optical cross connect (OXC), MEMS	Mirror, reflective surface, interconnection, bonding pad	Optical, packaging, wireless, networks
Au–Sn	Evaporation of multilayers, solder	Low-melting-point alloy	Packaging	Packaging sealing and interconnection	Optical, packaging, wireless, networks
$Ba_2Ti_9O_{20}$	Ceramic sintering	High dielectric constant, low TCK	Resonator	High frequency band pass filter	Wireless, packaging
Chalcogenide	Vacuum evaporation, reactive deposition	Photoconductive	Photodetector	Optical interconnection	Optical
Cr	Sputtering, evaporation	Intermediate adhesion layer	MEMS, packaging	Adhesion layer	Optical, packaging, wireless, networks
Cu	Drawing, evaporation, electrodeposition, sputtering	Conductor	Interconnects, packaging, MEMS	Interconnection	Optical, packaging, wireless, networks
Diamond-like carbon	Sputtering, ion-beam, CVD	Hard, wear resistant, corrosion resistant	MEMS	Durable film	Optical, packaging

(continued)

Table 1.1 (continued)

Material	Fabrication technique	Property/ performance	Component	Function	System/network
Epoxy	Organic synthesis	UV curable	Modulator	Bonds optical fiber to planar waveguides	Optical, packaging
Glasses	Sintering, spin-on, CVD, ion exchange, ion implantation	Hermetic, reflective, optically transparent	Packaging, optical interconnects	Sealant, waveguide	Optical, packaging
In	Evaporation, plating, metallurgy	Low-melting-point metal	Interconnection, thermal management, packaging	Solder joints, thermal interface materials	Optical, packaging, wireless, networks
InGaAs	MBE	Small direct bandgap semiconductor, electro-optic	Photodetector	Detect light	Optical
InGaASP	MOCVD	Small direct bandgap semiconductor, electro-optic	Lasers and photodetectors	Electro-optic conversion	Optical
InP	MOCVD	Small direct bandgap semiconductor, electro-optic	Optoelectronic switch	Route optical signals	Optical
$LiNbO_3$	Czochralski growth of single crystals	Electro-optic	Modulator, attenuator	Optical interconnects	Optical
$LiTaO_3$	Czochralski growth of single crystals	Electro-optic, piezoelectric	Modulator, attenuator, surface acoustic wave	Optical interconnects, high frequency filter	Optical, wireless
Methyl methacrylate	Polymerize then drawn	Optically transparent	Plastic optical fiber core	Light waveguide	Optical
Nb_2O_5	PECVD, reactive deposition	Optically transparent	Optical waveguide	Optical interconnection	Optical
Ni	Sputtering, evaporation, plating	Magnetic	MEMS, packaging	Magnetic transducing, diffusion barrier	Optical, packaging, wireless, networks
NiTi	Sputtering	Shape memory alloy	MEMS actuation	Thermal actuation	Optical

Optical polymer	Spin-on coating, polymerization	Waveguiding, optically transparent	Waveguides	Optical interconnection	Optical
Paralenes	Gas-phase polymerization	Low-temperature conformal conductive protective film	Packaging	Protection of electrical circuitry and MEMS devices	Optical
$Pb(Zr,Ti)O_3$	Sol–gel, sputtering	Electro-optic	Modulator, switch	Imprint data stream, switch optical signals between waveguides	Optical
Pd	Plating, sputtering, evaporation	Inert noble metal	High temperature/power	Interconnection	Wireless, optical, packaging
Permalloy	Electroplated	Magnetic	Magnetic cores	Magnetic transducing	MEMS, IC, packaging
Photoresist	Spin-on	Photosensitive, structural material	Component fabrication	Masking and etching template, LIGA molds	Optical MEMS
Polyethylene	Film forming	Electrical insulator	Sheath	Provide liquid water barrier and electrical insulation	Cable
Poly-Si	CVD	Resistive, mechanical strength	MEMS structures, integrated circuits	Mechanical element in MEMS	MEMS, optical
Polystyrene	Drawn from preform, film forming	Guides light	Plastic optical fiber/waveguide cladding	Contains light in fiber/waveguide	Optical
PZT (lead zirconate titanate)	Crystal growth, sputtered	Piezoelectric	MEMS	Actuation	Optical
Si	Crystal growth	Semiconducting, mechanical element, waveguiding	Integrated circuits, MEMS, optical interconnects	Conductive path, mechanical structure, substrate, optical waveguides	Optical, wireless, MEMS
SiC	CVD, sintering	Semiconducting, hard, thermally conductive, stiff, wear resistant	MEMS, packaging	Semiconductor, sensor material, thermal management	MEMS, packaging
Silicone rubbers	Thermally cured process	Conformal protective seal	Packaging	Encapsulant	Optical, wireless

(continued)

Table 1.1 (continued)

Material	Fabrication technique	Property/ performance	Component	Function	System/network
SiN	CVD	Passivation, mechanical element	MEMS structures	Talorable passivation, protective layer	Optical, wireless
Si_3N_4	CVD	Optically transparent	Optical waveguide	Optical interconnection	Optical
SiO_2	Sol–gel, PECVD, MCVD, CVD, thermally grown, spin-on	Optically transparent, preferentially etched stable oxide, thermal and electrical insulation	Optical fiber, arrayed waveguide grating, MEMS structure, and integrated circuit material	Confines optical signal, optical fiber/waveguide cores, electrical and thermal insulation, MEMS sacrificial layer	Optical waveguides, MEMS, IC
SiO_2:Ge	MCVD	Higher refractive index than pure SiO_2, photorefractive	Core of optical fiber/waveguide, fiber Bragg gratings	Filters single wavelength or modifies pulse characteristics	Optical
Silicon oxynitride (SiO_xN_y)	CVD	Higher refractive index	Core of optical waveguide	Optical interconnection	Optical
Ta_2O_5	PECVD, reactive deposition	Optically transparent	Optical waveguide	Optical interconnection	Optical
Teflon	Polymerized then extruded	Protective sheath	Plastic fibers	Cladding	Wireless, optical
Ti	Sputtering, evaporation	Sticks to most surfaces	Metallization	Intermediate adhesion layer	Optical, wireless, MEMS
TiN	CVD, PVD	Passivation protection	MEMS	Diffusion barrier material	Optical, wireless, MEMS
W	Sputtering, evaporation	High-temperature metal	Packaging	High-temperature interconnect, diffusion barrier	Optical, wireless, MEMS
Yttrium iron garnet (YIG)	Epitaxial growth	Optically transparent, ferrimagnetic	Optical waveguide, laser, microwave filter	Optical interconnection, laser, nonlinear optics	Optical
ZnO	Reactive sputtering, CVD, epitaxial growth	Piezoelectric, waveguiding	MEMS, optical waveguide	Transducer/waveguiding material	Optical

heated, and a lot of power is needed to drive silica on silicon optical circuit. Widespread use of such circuitry could tax the power availability in a central office and would certainly lead to thermal management problems. On the other hand, because of its high absorption in the 830 nm window, silicon is limited to applications in the 1,310 and 1,550 nm transmission windows. When silica is grown on a silicon wafer, typically at high temperature by chemical vapor deposition (CVD), a high level of stress is obtained in the wafer when it cools down to room temperature due to the coefficient of thermal expansion (CTE) mismatch between silica and silicon, which stress results in polarization-dependent behavior, stress induced scattering, and high coupling loss when pig tailing to fiber blocks due to warpage in the substrate. Lithium niobate ($LiNbO_3$) has been developed for applications in high-speed electro-optic modulators and optical space switches. Lithium niobate modulators capable of 2.5 and 10 Gb/s data rates are deployed in the backbones of optical transmission networks worldwide. But lithium niobate is difficult to process and therefore costly. Furthermore, its crystalline structure prohibits monolithic integration of laser sources and detectors. The mismatch between the velocity of the light traveling in the lithium niobate waveguide and the microwave used to modulate the device limits the upper modulation frequency to about 40 GHz. Moreover, as a crystalline material, lithium niobate shows high birefringence that leads to high polarization-dependent losses. Another drawback is that lithium niobate has a considerable refractive index mismatch with silica fiber, which leads to high coupling losses of about 1.5 dB per fiber attachment point. Indium phosphide (InP) is already fairly widely used for integrated active components. However, manufacturing techniques for making integrated circuits out of this material are still quite primitive, only relatively small circuits can be made in commercial quantities, and it is still very expensive. Indium phosphide can also be quite lossy, exhibit optical nonlinearities, and temperature sensitive. Comparably, polymeric materials offer advantages over a large number of such organic and inorganic materials. They have attracted a lot of attention with regard to applications in the all-optical network, basically, because they have the potential of added optical functionality and they may be producible at low cost (Gang 2005).

The important fabrication techniques for optical waveguides include deposition technique, ion exchange, thermal diffusion, ion implantation technique, and epitaxial growth technique. The selection of an appropriate technique depends on whether the device is passive or functional. If it is functional, it further depends on which effects are to be used, such as electro-optic control, acousto-optic control, and thermo-optic control. Based on these considerations, proper materials are firstly selected, and then an appropriate technique to meet the desired specification of waveguide characteristics is chosen. For selection of materials and fabrication techniques, the following factors are usually considered (Li 2004):

(a) Waveguide thickness, refractive index increment, and its distribution.
(b) Small propagation loss (less than 1 dB/cm), that is, optically transparent, little surface corrugation, and little guided-wave scattering.
(c) The degree of purity and optical-axes setting if the material is a crystal.

(d) Stability and adhesiveness to the substrate.
(e) Reproducibility of fabrication.

A great many of fabrication techniques have been developed for optical waveguides. Methods for fabricating planar optical waveguides are solution-phase fabrication, deposition technique, and ion-exchange process. For solution-phase fabrication, doctor blading, spin coating, dip coating, and horizontal-flow coating are used. In all of these methods, a solid-phase organic (or inorganic) material is dissolved in an appropriate solvent and then deposited on an optical quality substrate. Methods for deposit fabrication are sputtering and vapor deposition. The sputtering process uses an RF-heated target as the source of thin-film atoms, which diffuse to the substrate. The thickness of the thin film grown onto the substrate is controlled largely by the duration of the process; however, other factors that affect deposition include ambient pressure and temperature. The solution and deposition processes described above produce step-index guiding films where the substrate and waveguide boundary have an abrupt change in refractive index. Conversely, ion-exchange methods produce a graded-index waveguide that has a gradual transition from the substrate to the waveguide refractive index. This is accomplished by exchanging mobile or weakly bound ions in a solid with other ions from an external source, such as exchanging Na^+ in soda lime glass with K^+ by immersing the substrate in molten KNO_3 at 365 °C. The net index variation depends on the ionic polarizability, molar volume, and the stress state created by the substitution. In the case of K^+–Na^+ exchange, as the polarizability of K^+ ions is much larger than that of Na^+ ion in the glass, the surface refractive index change is at least $\Delta n \geq 0.01$ (Yimit et al. 2005). Some typical fabrication techniques would be described as below.

1. Deposition from liquid solutions
   Some typical deposition methods from liquid solutions are shown in Fig. 1.8. These techniques are used to deposit thin films of polymer materials on a substrate, such as photoresist in acetone solvent, epoxy resin in commercial solvent, PMMA in chloroform or toluene solvent, and polyurethane in xylene. The sol–gel technique offers potential involving the production of optical films of different refractive indices, and it particularly involves two-component systems SiO_2:TiO_2, for which the refractive index, depending on the composition and course of technological processes, may reach the values ranging from 1.2 to 2.3. Therefore, such films, depending on their optical and geometrical parameters, can be applied as waveguide films, protective coating, and when they change their properties due to the influence of external factors, they can be used as sensitive films for amplitude sensors in a one-component system, where silica SiO_2 is used as the matrix binding a respective indicator. For planar sensors, waveguide films of high reflective indices are produced in a two-component system SiO_2:TiO_2 (Karasinski 2004). In spin coating, polymer materials are dissolved with solvent. Firstly, one drop is put on the substrate, then thin film is formed by rapidly rotating (spinning) the substrate. In dip coating, thin film is formed by dipping the substrate in solvent and then lifting it up. After deposition, the film is thermally processed in ovens at appropriate

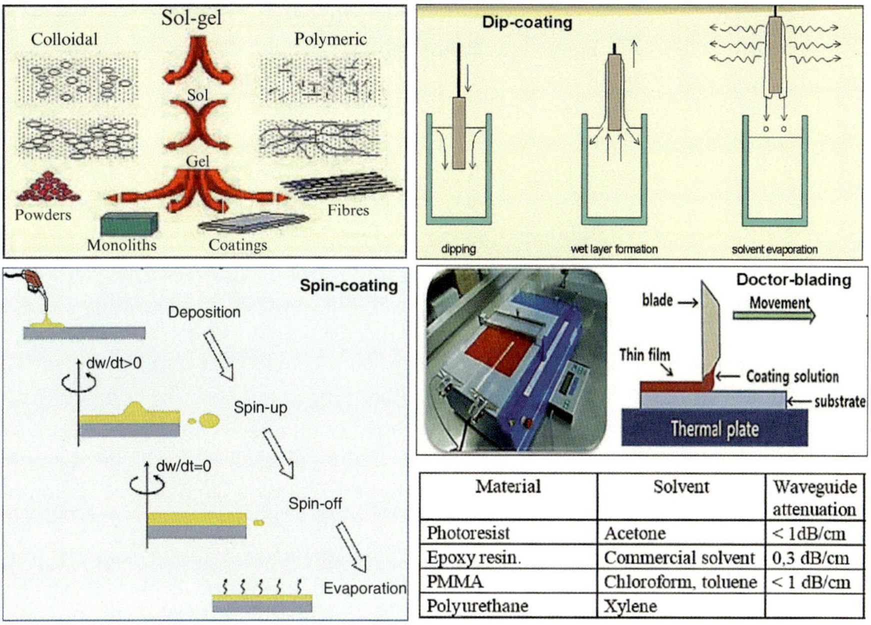

Material	Solvent	Waveguide attenuation
Photoresist	Acetone	< 1dB/cm
Epoxy resin.	Commercial solvent	0,3 dB/cm
PMMA	Chloroform, toluene	< 1 dB/cm
Polyurethane	Xylene	

Fig. 1.8 Typical deposition methods from liquid solutions

temperatures to increase film adhesiveness. This coating technique is the simplest and least expensive, but purity and uniformity of the films is rather poor. Doctor blading method imposes no restrictions on the waveguide cross-sections, which is important for multimode waveguides, and that has the unique feature of simultaneously fabricating optical waveguides together with integrated micro-mirrors and passive alignment structures for OE-module coupling. Furthermore, casting in combination with the doctor blade technique is well compatible with large area printed circuit board production technologies (Patela 2003).

2. Thermal vapor deposition
   Thin films are formed by depositing a material vaporized in a vacuum chamber (at a pressure of less than 0.00001 Torr) on a substrate. The choice of a heating method depends on the melting temperatures of the materials. In general, the resistor heating method is used for low-melting temperature materials, and electron-beam heating is used for high-melting temperature materials. Substrates are generally preheated at 200 °C by another heater for stronger adhesiveness (Li 2004).
3. Sputtering
   Like liquid deposition, and thermal vapor deposition, sputter deposition is also a physical vapor deposition (PVD) method of depositing thin films by sputtering, that is ejecting, material from a target material and then depositing onto the substrate of an optical waveguide. Sputtered atoms ejected from the

target have a wide energy distribution, typically up to tens of eV. The sputtered ions (typically only a small fraction—order 1 %—of the ejected particles are ionized) can ballistically fly from the target in straight lines and impact energetically on the substrates or vacuum chamber. Alternatively, at higher gas pressures, the ions collide with the gas atoms that act as a moderator and move diffusively, reaching the substrates or vacuum chamber wall and condensing after undergoing a random walk. The entire range from high-energy ballistic impact to low-energy thermalized motion is accessible by changing the background gas pressure. The sputtering gas is often an inert gas such as argon. For efficient momentum transfer, the atomic weight of the sputtering gas should be close to the atomic weight of the target, so for sputtering light elements neon is preferable, while for heavy elements krypton or xenon are used. Reactive gases can also be used to sputter compounds. The compound can be formed on the target surface, in-flight or on the substrate depending on the process parameters. The availability of many parameters that control sputter deposition make it a complex process, but also allow a large degree of control over the growth and microstructure of the film. Sputtering sources are usually magnetrons that utilize strong electric and magnetic fields to trap electrons close to the surface of the magnetron, which is known as the target. The electrons follow helical paths around the magnetic field lines undergoing more ionizing collisions with gaseous neutrals near the target surface than would otherwise occur. When the sputter gas is argon, the extra argon ions created as a result of these collisions leads to a higher deposition rate. It also means that the plasma can be sustained at a lower pressure. The sputtered atoms are neutrally charged and so are unaffected by the magnetic trap. Charge buildup on insulating targets can be avoided with the use of RF sputtering where the sign of the anode–cathode bias is varied at a high rate. RF sputtering works well to produce highly insulating oxide films but only with the added expense of RF power supplies and impedance matching networks. Stray magnetic fields leaking from ferromagnetic targets also disturb the sputtering process. Specially designed sputter guns with unusually strong permanent magnets must often be used in compensation.

Sputtering process can be used for high-melting temperature materials, which cannot be evaporated by thermal heating. In general, sputtered films are purer, stronger, and more uniform than thermally vaporized films.

4. Chemical vapor deposition (CVD)
The CVD method causes the chemical reaction of gases and deposits them on a substrate. The chemical reaction is carried out under various conditions, such as low temperature, high temperature, atmospheric pressure, and low pressure. The high-temperature and high-atmospheric pressure CVD is now the standard method used to make preform for optical fibers, and it is applied to make waveguides with glass substrates. The deposition rate is slow. On the other hand, the low-temperature and low-pressure RF-excitation plasma CVD method (PECVD) is now used practically for Si_3N_4-films and α-Si film formation, and will be used increasingly for waveguide fabrication. Fig. 1.9 shows

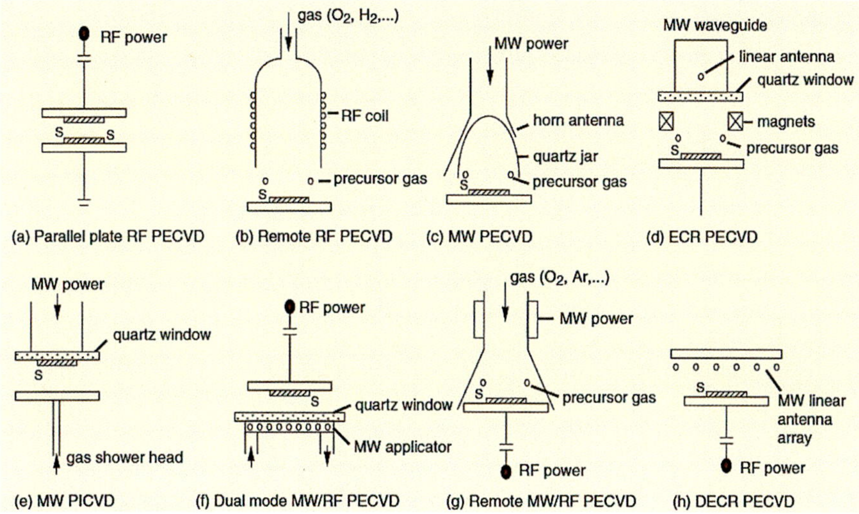

Fig. 1.9 Various PECVD configurations (Lamontagne 2007)

various PECVD configurations (Lamontagne 2007). PECVD layers usually require thermal annealing (~1,000 °C) to reduce the concentrations of Si–H and N–H bonds, and their associated absorption lines near 1,480 and 1,510 nm close to the C band (1,525–1,565 nm). The doped P and/or B cladding layers usually require multiple depositions and anneal steps to optimize glass flow. The advantages of high-contrast optical waveguides brought increased attention to PECVD SiN_x and SiO_xN_y layers. They provide the possibility of higher integration, but issues such as increased absorption losses and film stress must still be resolved (Lamontagne 2007).

5. Polymerization
   Polymerization is utilized for thin-film formation of organic polymers. Monomer is, firstly, vapor-deposited on a substrate in vacuum, and then polymerized by heating, electron beam, UV light, or plasma during deposition. Plasma polymerization uses plasma CVD with precise chemical reaction process control.
6. Thermal dopant diffusion
   A material to be diffused is deposited on a substrate, and kept in a high-temperature oven for a certain period to make the material diffused into the substrate, and become a thin surface layer with a slightly higher refractive index. The density profile of the diffusion materials along the depth direction is smooth, not stepwise (Li 2004).

   (a) In-diffusion method
      Ti:$LiNbO_3$ waveguides are usually fabricated by this method with a attenuation of 1 dB/cm. Diffusion from metallic layer (Ti) obtained by ion sputtering. Process temperature from 900 to 1,150 °C in an atmosphere of argon, nitrogen, oxygen, or air; diffusion time from 0.5 till 30 h.

(b) Out-diffusion
Out-diffusion consists in loss of lithium oxide from $LiNbO_3$ crystal: as Li_2O "goes out" of the crystal, extraordinary refraction index n_e goes up. To keep out-diffusion of LiO from the substrate surface low, diffusion is carried out in the atmosphere containing water vapor.
(c) Electric field-assisted diffusion
Diffusion is done faster at a relatively low temperature by applying an electric field during in-diffusion. This method is appropriate when the substrate is to remain unheated, and is applied, for example, to make waveguides by diffusing Cu into $LiTaO_3$ (Li 2004).

7. Ion Exchange
A substrate is put in a solvent/melt, and ions in the substrate are then exchanged by ions in the solvent/melt to form a high refractive index layer under the surface. For example, glass waveguides are made by exchanging Na^+ in glass for Ag^+, K^+, and Tl^+, and $LiNbO_3$ waveguides are made by exchanging Li^+ for H^+ in benzoic acid. In some cases, electric field is used for accelerating the ion-exchange process.
8. Ion Implantation
By implanting 20–300 keV accelerated ions into a substrate, lattice disordering takes place, resulting in a high refractive index layer. As shown in Fig. 1.10, the apparatus consists of an ion generator, an accelerator, an ion separator, and a raster-scan deflector. The whole system is expensive and quite large in size. This method has merit in that it is able to control the ion dose distribution with good accuracy considering the ion implants and the substrate material. For example, by implanting B^+ into a fused-silica substrate, and by annealing after implantation, a waveguide is fabricated in which propagation loss is as low as 0.2 dB/cm (Li 2004).
9. Epitaxial Growth
Suppose that there is some material, whose crystal structure is similar to that of the substrate crystal, and that both lattice constants are close. When the material is in the liquid or gaseous state, and it is in contact with the substrate crystal surface, then crystallized thin films are grown on the substrate. The former method is called liquid phase epitaxy (LPE), and the latter vapor phase epitaxy (VPE).

(a) LPE is the oldest technique. It was used for the preparation of the first generation of lasers. It is inexpensive but gives rather poor uniformity and control of thin layers.
(b) VPE in which the various elements including the dopants are brought to the substrates in a gaseous phase. A new version of this technique is called metal–organic chemical vapor deposition (MOCVD); it is one of the two leading techniques for growing III–V materials. One of the chief advantages of MOCVD is its ability to grow AlGaAs, which is difficult using other VPEs; however, its utilization of very toxic arsine makes it more

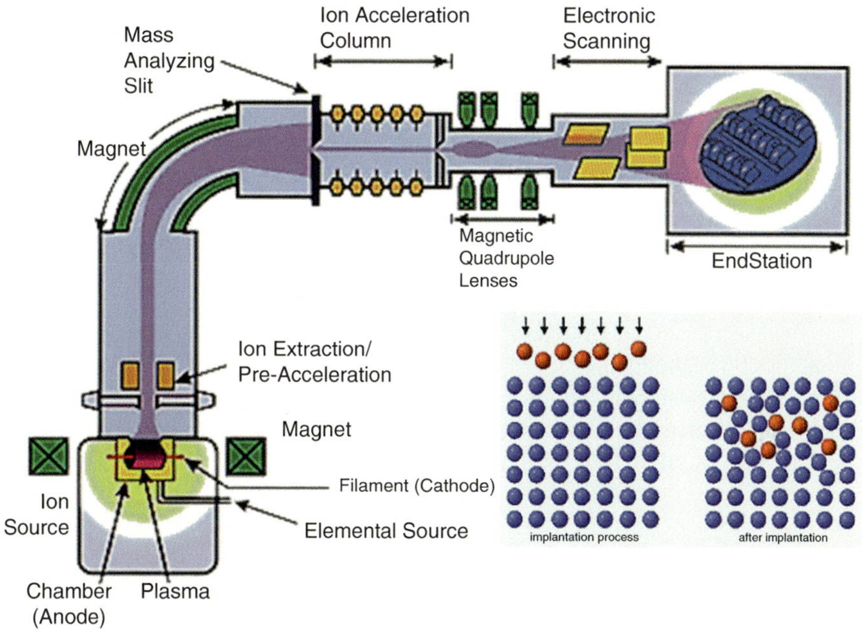

Fig. 1.10 Schematic illustration of ion implantation (Spire 2011)

difficult to implement. In a large flow of hydrogen, the reactant gases such as trimethyl gallium and arsine are brought in contact with the hot GaAs substrate.

(c) Molecular beam epitaxy (MBE) is a sophisticated evaporation or physical deposition technique. The various atomic elements are evaporated from ovens called effusion cells. Highly controlled growth can take place, assuming the critical shutter operation and the vacuum quality (base pressure: 0.000000001; -10 is the exponent to 0.00000000001; -11 is the exponent Torr) have been optimized. Such growth is rather slow (atomic layer/s) and is often monitored by a sophisticated characterization technique such as reflection high-energy electron diffraction (RHEED) to evaluate the crystal structure. Chemical beam epitaxy (CBE) is a variant of MBE that involves a chemical reaction on the substrate with a metal–organic source.

In general, the grown layer has a high quality in crystallization and uniformity and a well-controlled thickness, as well as a stepwise distribution of the refractive index. This technique is widely applied to semiconductor materials, but not to many dielectric materials. Growing $LiNbO_3$ film on $LiTaO_3$ and YiG film on $Gd_3Ga_5O_{12}$ by the LPE technique, or ZnO film on Al_2O_3 by the VPE technique have been used to fabricate waveguides.

Among the various other techniques, MBE, which is precisely controllable, is distinguishable, and the laser annealing technique and laser CVD are progressing. The waveguide fabrication with femtosecond laser pulses has been used to fabricate 3D waveguides in glass substrates. Hollow waveguides represent a novel type of planar waveguide that combines free space and integrated optics. They are of interest for new applications of planar waveguides in MOEMS, micro-fluidic, and bio-photonics applications. Hollow waveguides allow guiding light in a specific gas or liquid media present in a hollow core usually coated with reflective metal or dielectrics. The two main hollow waveguide fabrication techniques are (1) deep etching to form the core and then bonding a cover plate and (2) using a previously patterned sacrificial material to form the hollow core. It is possible to tune the optical properties by changing the dimensions of the core. Dicing hollow waveguides can be difficult because dicing debris can enter the core region. Standard lithography processes use photoresist and mask in a binary digital mode in which the photoresist is either fully exposed or not exposed. Gray-scale lithography allows the fabrication of curved surfaces and three-dimensional (3D) shapes by allowing a continuous gradient between exposed and not exposed areas; this is an analog version of lithography. Gray-scale requires specific photomasks and photoresists, or E-beam lithography. It is particularly useful for fabricating microlenses and 3D tapers for coupling waveguides to optical fibers. Pulsed plasma etching systems seem to be the next generation of dry etching systems. They permit the control of ARDE effects and mask charging effects that cause sidewall bowing or notches. High-density PECVD systems incorporate sample biasing sources (already implemented on HDP-RIE systems). These new systems allow the control of the ion energy fluxes arriving at the sample during the PECVD process and provide much better control of the structural and chemical properties of the deposited layers. Optical layers can be obtained with much lower losses at reduced deposition temperatures. Moreover, the planarization properties of the HD-PECVD layers are markedly improved (Lamontagne 2007).

1.4 Environmental Compliance of Optical Waveguide Materials

Environmental compliance must be evaluated during integrated optical waveguide design as environmental issue has become a more and more important concern. The requirements are regulated by the Restriction of the Use of Certain Hazardous Substances in Electrical and Electronic Equipment (RoHS) and Waste Electrical and Electronic Equipment (WEEE) Directives promulgated by the European Union (EU). The Directives make it critical for selections of optical waveguide materials. These environmental compliance regulations, provide forceful guideline for designing electrical products including integrated optical circuits and optical interconnect components with environmentally compliant.

Table 1.2 Restricted substances

Material/substance category	Maximum concentration in weight
Asbestos	Not detectable
Certain azo colorants and azo dyes	Not detectable
Cadmium/cadmium compounds	75 ppm
Hexavalent chromium/hexavalent chromium compounds	1,000 ppm
Lead/lead compounds	1,000 ppm 300 ppm (PVC cables only)
Mercury/mercury compounds	1,000 ppm
Ozone depleting substances (CFCs, HCFCs, HBFCs, carbon tetrachloride, etc.)	Not detectable
Polybrominated biphenyls (PBBs)	1,000 ppm
Polybrominated diphenylethers (PBDEs)	1,000 ppm
Polychlorinated biphenyls (PCBs)	Not detectable
Polychlorinated naphthalenes (more than 3 chlorine atoms)	Not detectable
Radioactive substances	Not detectable
Certain short chain chlorinated paraffins	Not detectable
Tributyl tin (TBT) and triphenyl tin (TPT)	Not detectable
Tributyl tin oxide (TBTO)	Not detectable

1.4.1 RoHS

RoHS Directive 2002/95/EC, together with WEEE Directive 2002/96/EC became European Law in February 2003, setting collection, recycling and recovery targets for all types of electrical goods. Producers must comply with all provisions of the WEEE Directive after August 13, 2005, while the RoHS Directive requires that producers prohibit any of the banned substances on the market after July 1, 2006. Producers failing to comply with the RoHS and WEEE Directives' requirements face legal penalties and potential restriction from selling products in the EU.

Table 1.2 shows currently restricted materials in the RoHS directive (Tong 2011). These substances are restricted to the ppm threshold level in all applications. All homogeneous material in purchased articles (i.e., materials, components, subassemblies, or products) must be free from the substances or cannot contain higher concentrations than the defined ppm threshold levels as listed in the table. Table 1.3 gives the banned substances in electronic products, while the substances must be reported when concentration exceeds the indicated ppm threshold level in Table 1.4 (Tong 2011). Exemptions to the maximum allowed concentrations of restricted materials are identified for cases where technology does not yet allow for substitutions, or where alternatives may have a worse impact on human health and the environment. Some exemptions include mercury in several kinds of fluorescent lamps, lead in steel, copper and aluminum alloys, lead in some types of solder, and military applications. RoHS Article 3(a) states that RoHS covers electrical and electronic equipment "which is dependent on electric currents or electromagnetic fields in order to work properly and equipment for the generation, transfer and measurement of such currents and fields falling under the categories set out in

Table 1.3 RoHS banned substances

Material/substance category	Maximum concentration in weight
Cadmium, mercury, lead, chromium VI	Total amount 100 ppm
Chlorofluorocarbons (CFCs)	Not detectable
Hydro chlorofluorocarbons (HCFCs)	Not detectable

Table 1.4 Substances that must be reported when concentration exceeds the threshold level

Material/substance category	Maximum concentration in weight (ppm)
Antimony/antimony compounds	1,000
Arsenic/arsenic compounds	1,000
Beryllium/beryllium compounds	1,000
Bismuth/bismuth compounds	1,000
Brominated flame retardants (other than PBBs or PBDEs)	1,000
Nickel (external applications only)	1,000
Certain phthalates	1,000
Selenium/selenium compounds	1,000
Polyvinyl chloride (PVC) (disclosure is limited to "is present"/ "is not present" in amounts that exceed threshold)	1,000

Annex IA to Directive 2002/96/EC (WEEE) and designed for use with a voltage rating not exceeding 1000 volts for alternating current and 1500 volts for direct current." With that said, a microwave oven would not be covered by RoHS because it cannot perform its intended function with the power switched off. On the other hand, a talking doll can still be used as a doll even when the batteries are removed, so it isn't covered by RoHS. RoHS does allow for non-compliant components after the July 1, 2006, but only as spare parts for equipment on the market before July 1, 2006.

1.4.2 WEEE

The WEEE Directive imposes the responsibility for the disposal of WEEE on the manufacturers of such equipment. Those companies should establish an infrastructure for collecting WEEE, in such a way that "Users of electrical and electronic equipment from private households should have the possibility of returning WEEE at least free of charge." Also, the companies are compelled to use the collected waste in an ecological-friendly manner, either by ecological disposal or by reuse/refurbishment of the collected WEEE. The WEEE identifies producers as any company that sells electronic and electrical equipment directly or indirectly under its own brand name. The intent of the WEEE Directive is to require producers to design products and manufacturing processes which prevent the creation of WEEE and barring that, reuse, recycle, dispose of, or incinerate WEEE. The WEEE

Directive calls for set percentages of IT and Telecommunications Equipment (Category 3 includes PCs, wireless devices, and similar devices) to be recovered and reused, or recycled (minimum 65%), incinerated (maximum 10%), or safely disposed of (maximum 25%).

Therefore, the goal of the integrated optical waveguide design and material selection should be compliant to RoHS and WEEE Directives in a minimized cost manner. Meanwhile, designing for environmental compliance must assure that time-to-market is minimized to take the product's competitive advantages.

1.5 Summary

Optical interconnects are foreseen as a potential solution to improve the performance of data transmission on chip, PCB, and system levels. Optical interconnects carry data signals as modulation of optical intensity, through an optical waveguide, thus replacing traditional electrical interconnects. Optical devices, therefore, can overcome the bottleneck imposed by the limited bandwidth of electronic circuits in areas such as computing, data storage, or telecommunication networks.

The basic element of any optical circuit is the optical waveguide, which permits to connect optically different devices. To build integrated optical circuits that substitute micro-electronic circuits, integrated optical waveguides with light confinement in a size of the order of the wavelength are mandatory. Optical waveguides are typically made with characteristic material structures that transmit electromagnetic waves in the direction parallel to their axis at visible or infrared frequencies. In the past decades, much effort has been devoted to advancing the theory and practice of optical waveguides. This has been mainly driven by the widespread deployment of optical fiber communication systems where system performance has been dramatically improved over the last 3 decades. In addition to the optical fibers themselves, a number of optical components, such as diode lasers, couplers, and external modulators, are formed on optical waveguide structures and their development has demanded a thorough understanding of waveguide theory. On a smaller scale, and particularly since the early 1980s, there have been growing research programs on optical sensor and short-range optical interconnection technologies which have also provided stimulus for the development of new types of waveguide devices and the associated theoretical models.

Optical waveguides can be classified according to their geometry (planar, strip, or fiber waveguides), mode structure (single-mode, multimode), refractive index distribution (step or gradient index) and material (glass, polymer, and semiconductor). Optical integrated waveguides are designed as energy flow only along the waveguiding structure but not perpendicular to it, so radiation losses can be avoided. Usually, optical integrated waveguides rely on the principle of total internal reflection, using materials with low absorption loss. The waveguide cross-section should be as small as possible to permit high-density integration, functionally linking devices, or systems or implementation of complex functionalities, such as splitters/combiners, couplers, AWGs, and modulators. A wide range of materials can be used, with their corresponding advantages and drawbacks. Current commercial devices are based on silicon/silica waveguides, III–V compounds, and lithium niobate waveguides. Silicon waveguides offers the possibility of mass-manufacturing and a high level of integration, which would result in cheaper chips. Photonic crystals can provide advantages to fulfill the requirements for high-density photonic integration. Waveguide optic devices have penetrated a number of markets by providing performance-competitive integrated optic and fiber-optic device solutions. This growth, in turn, requires new developments in modeling, further advances in material sciences, and innovations in integration platforms. In addition, the processing and fabrication of these new devices have been and would be optimized in conjunction with the development of accurate and precise characterization and new testing techniques.

Besides, environmental compliance must be evaluated during optical waveguide design because it has become a key advantage of competitive integrated electrical/optical products. The requirements are regulated by the RoHS and WEEE Directives. The Directives make selections of optical waveguide materials critical. These environmental compliance regulations, provide forceful guideline for designing integrated optical waveguides with environmentally compliant.

References

Gang SY (2005) Polymer based optical waveguides. Master's degree thesis, Universiti Teknologi Malaysia, Johor Bahru

Immonen M, Karppinen M, Kivilahti JK (2007) Investigation of environmental reliability of optical polymer waveguides embedded on printed circuit boards. Microelectron Reliab 47:363–371

Karasinski P (2004) Sol-gel derived optical waveguide films for planar sensors with phase modulation. Opt Appl 34(4):467–475

Karppinen M (2008) High bit-rate optical interconnects on printed wiring board: micro-optics and hybrid integration. Ph.D. dissertation, University of Oulu, VTT, Oulu

Kowach GR, Ramirez AG (2002) Advanced materials in telecommunications. In: Kutz M (ed) Handbook of materials selection. Wiley, New York

Kühner T, Schneider M (2007) Surface mounted coupling elements for PCB embedded optical interconnects. In: Proceedings of the 16th European microelectronics and packaging conference and exhibition (EMPC 2007), Oulu, 17–20 June 2007, pp 253–257

Lamontagne B (2007) Enabling fabrication technologies for planar waveguides devices. In: Calvo ML, Lakshminarayanan V (eds) Optical waveguides. Taylor & Francis Group, New York, pp 271–307

Li G (2004) Waveguides produced by ultra-short laser pulses inside lithium niobate. Master thesis, National University of Singapore. http://scholarbank.nus.edu.sg/handle/10635/14143. Accessed 26 March 2008

Litchinitser N, Iakhnine V (2011) Optical waveguides: numerical modeling. http://optical-waveguides-modeling.net. Accessed 12 Sept 2011

Ma H et al (2002) Polymer-based optical waveguides: materials, processing, and devices. Adv Mater 14(19):1339–1365

Miller DAB (2000) Rationale and challenges for optical interconnects to electronic chips. Proc IEEE 88(6):728–749

Okamoto K (2006) Fundamentals of optical waveguides, 2nd edn. Elsevier, Amsterdam

Patela S (2003) Fabrication methods of planar waveguides and related structures. http://www-old.wemif.pwr.wroc.pl/spatela/pdfy/0380.pdf. Accessed 3 March 2011

Saleh BEA, Teich MC (1991) Fundamentals of photonics. Wiley series in pure and applied optics, vol 45. Wiley, New York

Spire (2011) Ion implantation. Spire Corporation, Bedford. http://www.spirecorp.com/spire-biomedical/surface-modification-technology/ion-implantation.php. Accessed 9 Dec 2011

Tong XC (2011) Advanced materials for thermal management of electronic packaging. Springer, New York

Yimit A et al (2005) Thin film composite optical waveguides for sensor applications: a review. Talanta 65:1102–1109

Chapter 2
Characterization Methodologies of Optical Waveguides

Abstract The characterization of optical waveguides is a very important and essential step in any waveguide design and fabrication process. It is necessary to evaluate and confirm that the fabricated waveguide exhibits characteristics as designed. During materials selection and waveguide design, accurate measurements of key characteristics should be done with suitable methods. The major characteristics may include refractive index, layer thickness, optical coupling, optical loss, and nonlinear properties. Experimental evaluation and validation are necessary since these characteristics are rather difficult to determine theoretically. Such measurements provide important fundamental data to evaluate whether the waveguide is appropriate for integrated optical interconnection system, and use to specify the reason for the characteristics degradation. Therefore, the evaluation of the waveguide characteristics serves as a feedback to the design and the fabrication process, which is crucial for the modification and optimization of the waveguide performance. In this chapter, a series of optical waveguide characterization techniques will be elaborated.

2.1 Geometrical Inspection

Optical waveguides are structures that confine and direct optical signals in a region of higher effective index than its surrounding media. Beside the index difference providing the guiding region, the structure and the thickness of the fabricated waveguide plays an important role to confine the light. The minimum surface roughness and the side wall roughness give the minimum scattering loss. The light is no longer confined in the guiding layer and it begins to leak into the substrate if the thickness of the waveguide below the cutoff thickness. The thickness of the guiding layer also determines the guiding modes in the waveguide (Gang 2005). Optical specifications have been utilized throughout the design and manufacturing of a waveguide component or interconnect system to characterize how well it meets certain dimension performance requirements.

X.C. Tong, *Advanced Materials for Integrated Optical Waveguides*, Springer Series in Advanced Microelectronics 46, DOI 10.1007/978-3-319-01550-7_2,

Since the waveguides in optical integrated circuits have typical dimensions of only a few microns, observation of the waveguide across a given dimension cannot be accomplished without some equipment which featuring at least $\times 1{,}000$ magnification. The waveguides are usually inspected under high power microscope to inspect the plan view (2-D) of the waveguide. With the 2-D view, the etching quality in the fabrication process, jagged at the edges, and particles dropped in the coating film can be observed clearly. Jagged and particles and some other surface quality issues can cause significant scattering loss in the waveguides (Gang 2005).

The surface quality of an optical surface describes its cosmetic appearance and includes such defects as scratches and pits, or digs. In most cases, these surface defects are purely cosmetic and do not significantly affect system performance, though, they can cause a small loss in system throughput and a small increase in scattered light. Certain surfaces, however, are more sensitive to these effects such as (Edmund 2013): (1) surfaces at image planes because these defects are in focus or on refractive surface and (2) surfaces that see high power levels because these defects can cause increased absorption of energy and damage the optic. The scratch designation is determined by comparing the scratches on a surface to a set of standard scratches under controlled lighting conditions. Therefore the scratch designation does not describe the actual scratch itself, but rather compares it to a standardized scratch according to related optical specifications. The dig designation, however, does directly relate to the dig, or small pit in the surface. The dig designation is calculated at the diameter of the dig in microns divided by 10. Scratch-dig specifications of 80-50 are typically considered standard quality, 60-40 precision quality, and 20-10 high precision quality.

Surface flatness is a type of surface accuracy specification that measures the deviation of a flat surface such as that of a mirror, window, prism, or plano-lens. This deviation can be measured using an optical flat, which is a high quality, highly precise flat reference surface used to compare the flatness of a test piece. When the flat surface of the test optic is placed against the optical flat, fringes appear whose shape dictates the surface flatness of the optic under inspection. If the fringes are evenly spaced, straight, and parallel, then the optical surface under test is at least as flat as the reference optical flat. If the fringes are curved, the number of fringes between two imaginary lines, one tangent to the center of a fringe and one through the ends of that same fringe, indicate the flatness error. The deviations in flatness are often measured in values of waves (λ), which are multiples of the wavelength of the testing source. One fringe corresponds to ½ of a wave. Flatness with 1 of a wavelength λ is considered typical grade, ¼ of λ flatness is considered to be precision grade, and 1/20 of λ is considered high precision grade (Edmund 2013).

Power, a type of surface accuracy specification, applies to curved optical surfaces, or surfaces with power. It is tested in a fashion similar to flatness, in that a curved surface is compared against a reference surface with a highly calibrated radius of curvature. Using the same principle of interference caused by the air gaps between the two surfaces, the interferences pattern of fringes is used to describe the deviation of the test surface from the reference surface. A deviation from the reference piece will create a series of rings, known as Newton's Rings. The more rings are present, the larger the deviation. The number of dark or light rings, not the sum of both light and dark, corresponds to twice the number of waves of error (Edmund 2013).

Irregularity, a type of surface accuracy specification, describes how the shape of a surface deviates from the shape of a reference surface. It is obtained from the same measurement as power. Regularity refers to the sphericity of the circular fringes that are formed from the comparison of the test surface to the reference surface. When the power of a surface is more than 5 fringes, it is difficult to detect small irregularities of less than 1 fringe. Therefore it is common practice to specify surfaces with a ratio of power to irregularity of approximately 5:1 (Edmund 2013).

The thickness and the topography of the waveguide can be measured using the surface profiler. Normally, surface profiling can be done using scanning electron microscope (SEM) or DEKTEK stylus surface profiler. SEM is a destructive method to analysis surface morphology. Stylus of the DEKTEK surface profiler could scratch the waveguide during its scanning on the sample. A noncontact optical surface profiler had been used for the surface profiling and topography (3-D), with white light interferometry. Film thickness, waveguide width, and the surface roughness of the coating can be measured using the surface profiler.

Surface roughness or surface finish measures small scale irregularities on a surface. They are usually an unfortunate by-product of the polishing process. Rough surfaces tend to wear faster than smooth surfaces and may not be suitable for some applications, especially those with lasers or intense heat, due to possible nucleation sites that can appear in small cracks or imperfections. Manufacturing tolerances of surface finish range from 50 Å RMS for typical quality, 20 Å RMS for precision quality, and 5 Å RMS for high quality (Edmund 2013).

2.2 Refractive Index Measurements

The refractive index is one of the most important properties of an optical waveguide. The refractive index profile has been used to calculate waveguide properties such as insertion losses, and cutoff wavelengths for propagating modes and the intensity distribution of the modes themselves. The refractive index can be modified to minimize losses, to yield a specific cutoff wavelength and to optimize the mode profile. Therefore, with a more accurate refractive index measurement, a more precise calculation of the waveguide properties can be obtained.

Several techniques have been used for the measurement of the refractive index distribution of waveguides, including reflectometry and ellipsometry, surface plasmon resonance (SPR), incoherent light transmission, the refracted near-field (RNF) technique, prism coupling, as well as the propagation-mode near-field method. The former three techniques have limited resolution and require the index enhanced region to be at or near the substrate surface. Ellipsometry and prism coupling are complimentary techniques for measuring the refractive indices of thin films. The RNF technique and the propagation-mode near-field method is another pair of complimentary techniques for measuring the refractive index profiles of optical waveguides (Chien 2004).

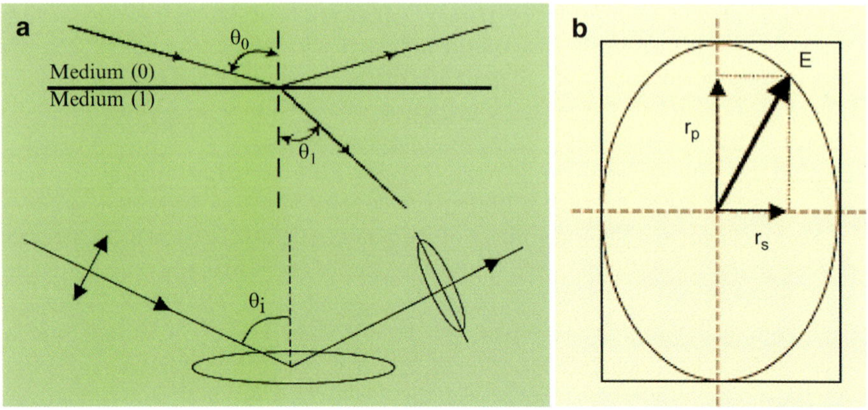

Fig. 2.1 Reflection and transmission of an incident light wave at a surface boundary (Elton 2007): (**a**) linearly polarized light reflected from a surface becomes elliptically polarized; (**b**) in plane polarized light, r_p and r_s oscillate in phase. In elliptically polarized light, there is a phase lag (Δ) between them, such that the electric field rotates and changes amplitude in time describing an ellipse

2.2.1 Reflectometry and Ellipsometry

Refractive index (n) is a complex number comprising a real refractive index and an imaginary part: the absorption (or extinction) coefficient. It may generally written as (Elton 2007)

$$n = n' - n'' \tag{2.1}$$

where n' is the real refractive index and n'' the extinction (absorption) coefficient. The real part of the refractive index describes how the speed of light changes as it enters the material. The extinction coefficient describes how light is absorbed (or scattered). In a transparent material, absorption is zero and $n = n'$. The reflectivity of an absorbing material depends on both parts of the complex refractive index, and, in fact a highly absorbing material will in general be much more reflective than a corresponding transparent material. This is why metals, for instance, make such good mirrors—the refractive index of aluminum is 1.21–6.92 (Elton 2007).

Reflectometry generally yields an approximation to the real part of the refractive index. The well-established technique of ellipsometry, by contrast, can determine both the real refractive index and the extinction coefficient. As shown in Fig. 2.1, linearly polarized light can be thought of as having two perpendicular electric field components, denoted r_p and r_s, oscillating in phase. When linearly polarized light is reflected from a surface, in general the amplitudes r_p and r_s change and so does the phase between them. The light becomes elliptically polarized. If the substrate is non-absorbing, the phase change is zero and linear polarization is preserved. Ellipsometry is well-established experimental method for analyzing the phase and amplitude of reflected polarized light in order to extract information about the surface. Various types of ellipsometers have been built up, but a common form is the rotating polarizer. In this design, linearly polarized light is incident on the sample surface, as illustrated in Fig. 2.1a, but the polarizer is rotated, so that the plane of polarization rotates about the axis of the incident light. On the detector side, the (generally) elliptically polarized light is passed through a fixed analyzer. The intensity of light on the detector oscillates with rotation of the incident polarizer according to the amplitude and phase change at the specimen. The fundamental equation of ellipsometry can be expressed as (Elton 2007)

$$\rho_e = \tan \Psi e^{\Delta} \tag{2.2}$$

$$\Psi = |R_p|/|R_s| \tag{2.3}$$

where Δ is the phase change between r_p and r_s upon reflection, Ψ is the angle whose tangent is the ratio of the intensity of the R_p and R_s components. In the case of reflection from a simple substrate, Ψ and Δ can be inverted directly to give the real refractive index and extinction coefficient. In any other situation, e.g., in the presence of one or more thin films, surface roughness, interface roughness, it is necessary to start from a model of the surface and to fit the measured values of Ψ and Δ to the model to obtain a best fit. For a non-absorbing substrate, $\Delta = 0$, and ρ_e is real. But in general, if absorption is present, ρ_e is a complex number.

With these parameters, the complex refractive index of the sample (thin film of optical waveguide) can be calculated as (Jung et al. 2004)

$$\tilde{n}_1 = \frac{\left[\sqrt{1 - 4\sin^2(\theta_0)\tan(\Psi)e^{j\Delta} + 2\tan(\Psi)e^{j\Delta}\tan^2(\Psi)e^{j\Delta}}\right]\tilde{n}_0 \sin(\theta_0)}{\cos(\theta_0)[1 + \tan(\Psi)e^{j\Delta}]} \tag{2.4}$$

where n_0 is the complex reflective index of the ambient. θ_0 is the angle of incidence. The data from the ellipsometer are values of Ψ and Δ as a function of wavelength. Using (2.4), these data can be used to calculate the complex index of refraction as a function of wavelength.

In the reflectometry measurement, the sample surface is illuminated with s- and p-polarized light. The reflected intensities R_p and R_s of the p and s polarized components are measured, and are used to calculate a refractive index using the form of the Fresnel equations for a transparent substrate (Elton 2007):

$$n_R = \sin\theta_i \sqrt{1 + \left(\frac{1-\rho}{1+\rho}\right)^2 \tan^2\theta_i} \tag{2.5}$$

where $\rho = \sqrt{R_p/R_s}$. The reflectometry analysis assumes that the material is non-absorbing. If the material truly is non-absorbing, then reflectometry will return the correct real refractive index. If the material is absorbing, then reflectometry will return an approximation to the real refractive index. This approach is a trade-off between absolute accuracy for refractive index on one hand and a desire for speed of operation and the ability to measure several other parameters (such as roughness) at the same time. For the types of material that reflectometry is aimed at, the simplified measurement of refractive index generally works very well. However, if the surface is rough, the reflected light will be scattered over a range of angles making the analysis extremely difficult.

Comparably, ellipsometry and reflectometry are closely related techniques with rather different aims. Ellipsometers are normally used for the analysis of very smooth substrates, thin films, adsorption processes, or layer growth for optical waveguides. With a suitable model of the surface, ellipsometry is a powerful technique for obtaining information on refractive index, extinction, and film thickness. Ellipsometry can be applied successfully to relatively rough surfaces, but measurement times can be fairly long. The reflectometer is generally designed specifically for determining an approximate real refractive index, which is typically accurate to about 0.01 which is probably of the same order as systematic instrumental errors due to alignment and calibration. The reflectometer can also be used to measure macroroughness, microroughness, and various gloss values. The approximations in the determination of refractive index mean the reflectometer cannot provide meaningful refractive index or microroughness results for metals or thin transparent films, although macroroughness and gloss are still valid. Both techniques have merits and limitations: although closely related, they do not ultimately do the same. If a choice is needed between the two methods, it must depend on the end applications.

Compared with prism coupling, ellipsometry and prism coupling are complementary techniques that can both be used to measure the index of bulk materials and to simultaneously measure film thickness and refractive index. In general, ellipsometry is ideal for measuring thin films such as gate oxides, pyrolytic silicon nitride, or any film whose thickness is thinner than a few thousand angstroms. For the measurement of thicker samples, the prism coupler is superior. In addition, ellipsometry requires accurate advance knowledge of the real and imaginary parts of the substrate refractive index. If these parameters are not known, or if they vary due to roughness or other surface conditions, appreciable errors can result. The prism coupling technique is only weakly sensitive to substrate optical parameters. In the case of a film over a lower index substrate, thickness and index as measured by prism coupling are weakly dependent on the real part of the substrate index only. The prism coupler can also be used to measure the bulk refractive index of the substrate directly (Chien 2004).

2.2.2 Surface Plasmon Resonance

SPR reflectivity measurement can be used to characterize thickness and index of refraction of ultra-thin films covering on surface of noble metal (Au, Ag). Plasmon is a strongly attenuated electromagnetic wave propagating along the boundary between

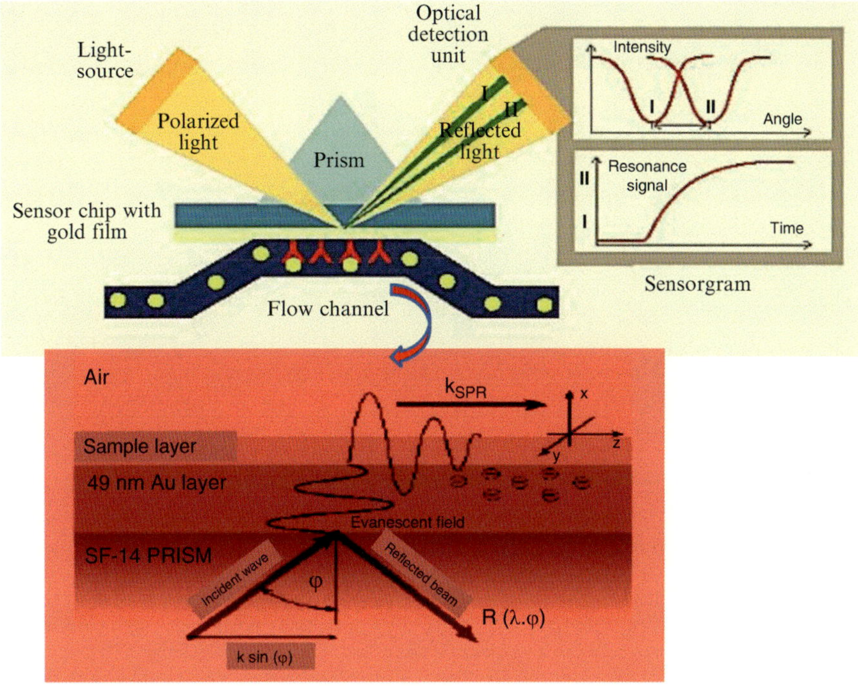

Fig. 2.2 The schematic of SPR measurement setup

the thin metal layer and the dielectric one above it, as shown in Fig. 2.2. The energy of the incident wave coupled with the surface plasmon is irrevocably dissipated in the system, so it cannot appear in the wave reflected from the metal/dielectric interface. For the plasmon resonance in the prism configuration, SPR excitation by the incident light wave at the base of the prism occurs when the wave vector of the plasmon k_{SPR} and the projection of the incident wave vector k in the direction of propagation of the plasmon are equal. The refractive index of the dielectric film can be calculated with the following equation (Tyszkieiwcz et al. 2005):

$$k \sin \varphi = k_{SPR} \tag{2.6a}$$

$$k_0 n_{prism} \sin \varphi_i = k_0 \sqrt{\frac{\varepsilon_{Au} n_s^2}{\varepsilon_{Au} + n_s^2}} \tag{2.6b}$$

where k_0 is wave vector of light in a vacuum, k_{SPR} is wave vector of the surface plasmon, n_s is a refraction index of the dielectric film above the metal layer, ε_{Au} is a real part of the dielectric complex constant of the active plasmon layer (e.g., Au), n_{prism} is a refraction index of the prism, and φ_i is an incident angle of light against the normal to the prism base.

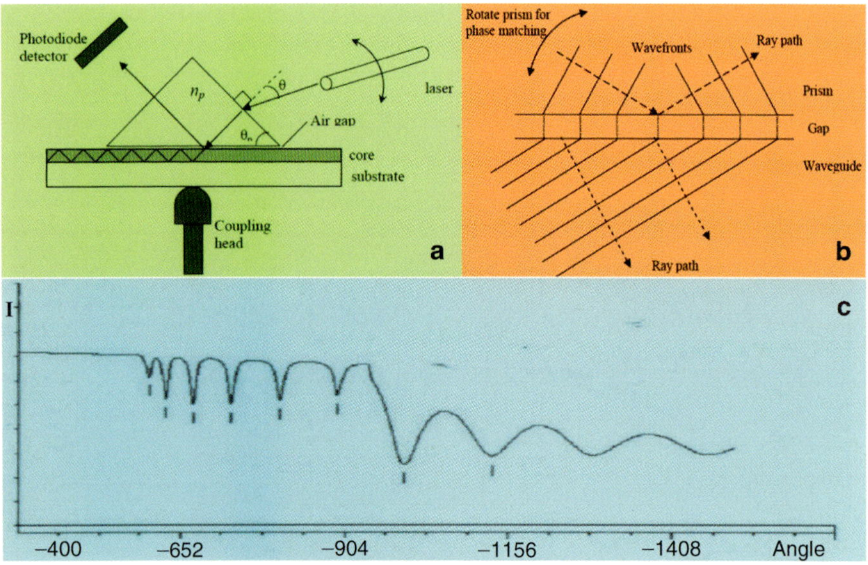

Fig. 2.3 Schematic illustration of prism coupling technique for the measurement of the refractive index and the thickness of thin films, which utilizes a high-index prism to excite a guided wave through phase matching between the incident wave and guided mode (Gang 2005): (**a**) prism coupling assembly; (**b**) phase-matching condition at prism–waveguide interface; (**c**) rotation spectra against angle of incidence θ of the prism

2.2.3 Prism Coupling

The prism coupling technique has been used for the measurement of the refractive index n and the thickness t of thin films, which utilizes a high-index prism to excite a guided wave through phase matching between the incident wave and guided mode. A prism which having the refractive index, n_{p}, higher than the refractive index of the measured material was clamped onto the film and the substrate was supported against a coupling head. The clamp created a small pressure to ensure the prism interface close to the right angle corner of the prism. This produces a small air wedge which is required to achieve efficient launching of light into modes in the film. A He–Ne laser (visible light) was used for 633 nm measurements. The laser beam was launched onto the diagonal prism face. The whole assembly was then rotated about a vertical axis until a guided mode was launched into the film (Fig. 2.3a). The effective index of the light wave along the bottom surface of the prism is adjusted to match that of a waveguide mode (Fig. 2.3b). The mode intensity was optimized by movement of the laser beam in the vertical direction. The prism was rotated until the beam was reflected back along the normal and the angular position for this was noted. This was the reference position against which the mode angles were measured. Rotating the sample and detector yields the rotation spectrum of the references against the angle of incidence, θ, at the prism as shown in Fig. 2.3c. Each sharp dip in the spectrum indicates a mode in the slab waveguide. The software of the equipment solved the relevant (2.7) and (2.8) to give the value of the refractive index and film thickness. The measured angles for the first and second modes are used to calculate the effective index for each TE mode in the film using the equation (Gang 2005)

$$n_{\mathrm{eff}(m)} = n_{\mathrm{p}} \sin\left(\theta_{\mathrm{p}} + \sin^{-1}\left(\frac{\sin\theta}{n_{\mathrm{p}}}\right)\right) \tag{2.7}$$

where n_{p} is the prism index, θ_{p} is the prism angle, and θ is the measured incident angle for mode $m = 0, 1, \ldots$. In order to calculate the bulk index for the film from the effective indices it is necessary to solve the resonance equation for mode propagation in the film (Gang 2005)

$$k_0 t(\varepsilon_0 - \varepsilon_{\mathrm{m}})^{1/2} = \tan^{-1}\left[\frac{(\varepsilon_{\mathrm{m}} - \varepsilon_{\mathrm{s}})}{(\varepsilon_0 - \varepsilon_{\mathrm{m}})}\right]^{1/2} + \tan^{-1}\left[\frac{(\varepsilon_{\mathrm{m}} - \varepsilon_{\mathrm{a}})}{(\varepsilon_0 - \varepsilon_{\mathrm{m}})}\right]^{1/2} + m\pi \tag{2.8}$$

where k_0 is the free-space wave vector and the permittivities of the materials are ε_0 for the film, ε_{a} for air, ε_{s} for the substrate, and ε_{m} for the modal effective index. t is the film thickness. The equation is solved by an iterative method.

However, the following factors limit the utilization of the prism coupling technique (Chien 2004):

(a) The film must be thick enough to permit the propagation of at least two modes. If only one mode is observed, the prism coupler may still be used to determine one of the parameters n and t if the other is known from an independent measurement.
(b) The method is a contact method and it is necessary to press the film against the base of the prism. Extra care must be taken when measuring polymers. In general, the technique is nondestructive.
(c) Alignment of small samples with the coupling spot requires a certain degree of skill and experience. Typically, the laser spot is collimated to approximately 1.0 mm^2 and this can be a challenge where the proton beam written samples are only about $2.0 \times 2.0\ \mathrm{mm}^2$ in size.

2.2.4 *Propagation-Mode Near-Field Technique*

Propagation-mode near-field method, also known as the scalar wave inversion technique, is a nondestructive technique which estimates the refractive index distribution in a waveguide from its mode field intensity by an inversion of the scalar wave equation. One of the advantages of the technique is that it requires no prior knowledge of the shape of the refractive index and does not require the waveguide to possess any form of cross sectional symmetry. However, the main limitations of the technique are (a) the waveguide has to be weakly waveguiding; and (b) only the fundamental mode of the propagation is excited. For a weakly guiding waveguide, where the peak refractive index change is small, the optical field strength of the fundamental mode of excitation, $\Psi(x, y)$, and refractive index $n(x, y)$ relationship may be described by the scalar wave equation (Chien 2004):

$$n^2(x,y) = \left(\frac{\beta}{\kappa_0}\right)^2 - \frac{\nabla_T^2 \Psi(x,y)}{\kappa_0^2 \Psi(x,y)} \tag{2.9}$$

where $\nabla_T = \frac{\partial}{\partial x} + \frac{\partial}{\partial y}$ and k_0 is the free-space wave number; $n(x, y)$ is the refractive index and β is the propagation constant. Substituting $I(x, y) = \Psi^2(x, y)$ for the intensity and introducing $n(x, y) = n_B + \Delta n(x, y)$ (where $\Delta n(x, y)$ is small), (2.9) becomes:

$$\Delta n(x,y) = \frac{\beta^2}{2n_B k_0^2} - \frac{n_B}{2} - \frac{\nabla_T^2 \sqrt{I(x,y)}}{2n_B k_0^2 \sqrt{I(x,y)}} \tag{2.10}$$

where n_B is the substrate refractive index and, β/k_0 is the effective index, N_{eff}. These two terms on the right hand side are unknown constants. Hence, the fundamental mode field intensity $I(x, y)$ can be used to uniquely determine the refractive index distribution of the waveguide to within an arbitrary constant. This constant can be found by matching the refractive index away from the waveguide core to that of the substrate. In the weak guidance approximation, Δn is small and $n_B \approx \frac{\beta}{k_0} = N_{eff}$. Hence (2.10) can be approximated to

$$\Delta n(x,y) = -\frac{\nabla_T^2 \sqrt{I(x,y)}}{2n_B k_0^2 \sqrt{I(x,y)}} \tag{2.11}$$

The substitution of $\Psi(x,y) = \sqrt{I(x,y)}$ is based on the assumption that there is no phase variation across the measured wave field. Such an approximation is valid since the mode field intensity is measured at the waveguide end face where the phase variation is negligible. Due to the second-order derivatives involved in (2.11), good mode profile data that is precisely focused on the CCD array is essential to obtaining accurate refractive index profiles. Multiple images of the near-field mode profiles of the waveguides must be taken and averaged in time (Chien 2004).

2.2.5 *Refracted Near-Field Technique*

A typical setup for the RNF technique is given in Fig. 2.4 (Goering and Rothhardt 1986). The setup is applicable to both planar or channel waveguides. Collimated light is focused into the end face of the waveguide using a microscope objective. The substrate is placed with the waveguide downwards onto the glass block, with a

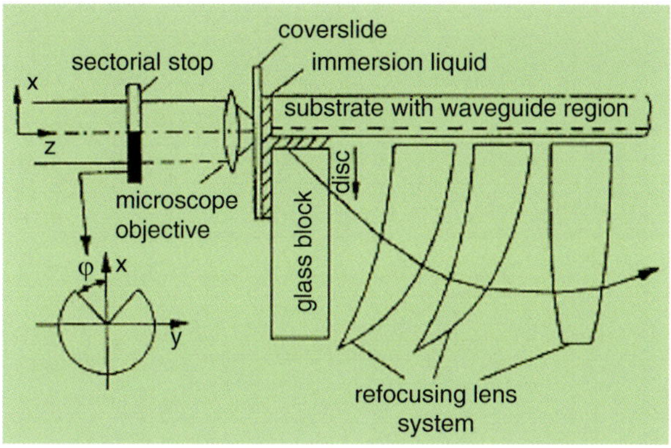

Fig. 2.4 Schematic experimental setup for the refracted near-field method (Goering and Rothhardt 1986)

film of immersion liquid film between them. Light escaping from the back surface of the glass block and passing the aperture stop (disk) is refocused by a system of hemilenses onto a large area photodiode. The detected signal is inversely proportional to the refractive index changes encountered at the waveguide end face during a scan across the sample. From the known refractive index values of the immersion liquid and the glass block, the waveguide's refractive index distribution can be determined using a linear interpolation. The experimental setup for the RNF technique is considerably more complicated compared to that for the propagation-mode near-field method. Apart from these hardware issues, there are several issues pertaining to the linearity between the measured RNF signal and the refractive index of the focal point; and the issues relating to the accuracy of RNF measurements due to the uncertainty of the calibration of the refractive indices of the immersion oil and the reference blocks. On the other hand, the setup for the propagation-mode near-field method can be readily acquired from off-the-shelves components, without the need for any custom build components; while yielding the same information on the refractive index distribution. One advantage of the RNF technique over the propagation-mode near-field method is its ability to measure the refractive index distributions of multi-mode waveguides (Chien 2004).

2.2.6 M-Line Spectroscopy (MLS)

MLS is one of the most accurate methods to determine optogeometric parameters including refractive index and film thickness of waveguiding films. It allows an evaluation of a refractive index with accuracy up to 10^{-4}. The accuracy of film thickness measurement is also as high as 10^{-2}–10^{-3} (Wu 2005). MLS also uses the same principle concept as prism coupling, and is particularly well adapted to the accurate determination of the film refractive index and thickness. In comparison with other methods, MLS has important advantages such as the requirement of only angle measurements, the opportunity to determine the anisotropy of the layer and the fact that this is a nondestructive method. The disadvantages of MLS, however,

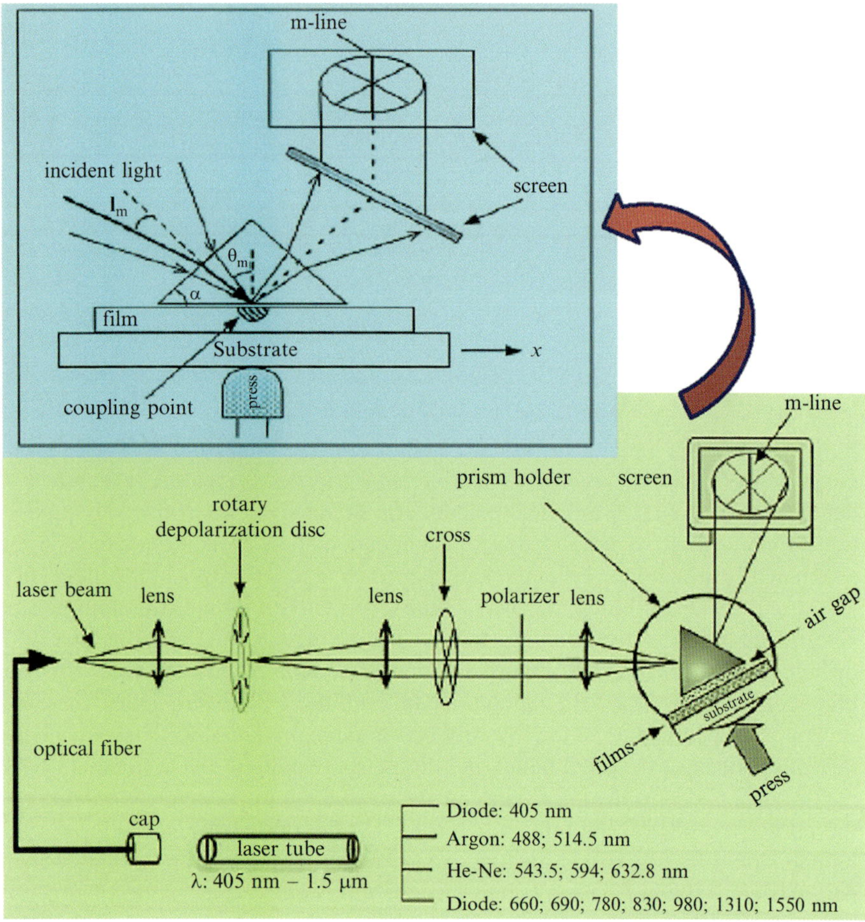

Fig. 2.5 Schematic illustration of the prism coupling technique used in MLS (Wu 2005)

include (a) the film must be thick enough to support at least two guided modes, (b) the substrate used has to be transparent at the required wavelength, (c) the layer must be hard enough to be pressed against the base of a prism and the roughness of the film must be as low as possible to increase the accuracy. Also the observation of the m-dark lines requires a certain degree of skill and experience and only an absorption up to 80 dB/cm in the direction of propagation is tolerable (although at these levels it will reduce the accuracy of the measurements).

Usually, MLS uses only one wavelength (the most popular is 632.8 nm, He–Ne laser) and the calculation is carried out on the premise of step-index model. A way to confirm the step-index assumption is to conduct MLS measurements at various wavelengths in order to verify whether the sample thickness is wavelength dependent or not. For such application, a multi-wavelength MLS (MWMLS) has been developed. The schematics of MLS and MWMLS are shown in Fig. 2.5. The refractive index (n_f) and thickness (T) of the thin film can be calculated using the step-index model (Wu 2005):

When using TE polarization

$$\frac{T}{\lambda}=\frac{1}{2\pi\sqrt{\left(n_{\mathrm{f}}^{2}-N_{m}^{2}\right)}}\left[m\pi+\tan^{-1}\left(\frac{\sqrt{\left(n_{0}^{2}-N_{m}^{2}\right)}}{\sqrt{\left(N_{m}^{2}-n_{\mathrm{f}}^{2}\right)}}\right)+\tan^{-1}\left(\frac{\sqrt{\left(n_{\mathrm{s}}^{2}-N_{m}^{2}\right)}}{\sqrt{\left(N_{m}^{2}-n_{\mathrm{f}}^{2}\right)}}\right)\right] \tag{2.12}$$

When using TM polarization

$$\frac{T}{\lambda}=\frac{1}{2\pi\sqrt{\left(n_{\mathrm{f}}^{2}-N_{m}^{2}\right)}}\left[m\pi+\tan^{-1}\left(\frac{n_{\mathrm{f}}^{2}}{n_{\mathrm{c}}^{2}}\frac{\sqrt{\left(n_{0}^{2}-N_{m}^{2}\right)}}{\sqrt{\left(N_{m}^{2}-n_{\mathrm{f}}^{2}\right)}}\right)+\tan^{-1}\left(\frac{n_{\mathrm{f}}^{2}}{n_{\mathrm{s}}^{2}}\frac{\sqrt{\left(n_{\mathrm{s}}^{2}-N_{m}^{2}\right)}}{\sqrt{\left(N_{m}^{2}-n_{\mathrm{f}}^{2}\right)}}\right)\right] \tag{2.13}$$

where $N_m = n_{\mathrm{p}} \sin\left(\alpha + \arcsin\left(\frac{\sin i_m}{n_{\mathrm{p}}}\right)\right)$.

All the parameters are known or can be measured separately. If the film is thick enough to support more than two modes on the same polarization, the method becomes a self-consistent one because the two unknown n (refractive index) and T (thickness) can be determined from more than two independent measurements. The values of n_{f} and T is just a computational problem in order to adjust these values until theoretical values of N_m are close to the experimental values of N_m. If the agreement between experimental and calculated values of N_m is in the range of 0.001 or less, the step-index model is satisfied. If the step-index model is not satisfied, the refractive index profile versus the sample thickness can be reconstructed using inverse Wentzel–Kramer–Brillouin (WKB) method which provides reliable and quick result in the case of slight refractive index variation (Wu 2005). However, at least four observable modes are required or such a calculation and a judicious choice of the exciting wavelength are necessary.

MWMLS allows studying a bulk material such as the substrate and its dispersion curve. For this latter application, just a face of the material must be polished, or bulk sol–gel must be elaborated in a very flat and smooth polymer box.

2.3 Coupling Techniques

For an integrated optical waveguide system, before exciting guided-mode propagation, introducing an incident light into the waveguide or transmitting the light out of the waveguide to the receiver is an important step because the laser beam belongs to a radiation mode which cannot propagate directly in the waveguides. Therefore, it is necessary to find a technique to effectively confine and couple the laser beam with the waveguide. Several techniques have been used for carrying out such coupling requirements, including prism coupling, end coupling, tapered and/or

Table 2.1 Comparison of some light wave coupling techniques (Christensen et al. 1992)

Techniques	Advantages	Disadvantages
Prism coupling	High efficiency; mode selective	Complex; difficult to align
End coupling	High efficiency for thick waveguides	Difficult to align for thin waveguides
Launch coupler	Tolerant of alignment	Long, thin taper required for thin waveguides
Grating coupler	Tolerant of alignment; mode selective	Lower efficiency

lunch coupling, and grating coupling. Prism coupling and end coupling are the most extensively used methods. Prism coupling is highly efficient and is mode selective, but requires placement of a prism on top of the waveguide with a carefully adjusted gap between the prism and the waveguide. This makes the prism coupling technique unsuitable for fast, alignment tolerant coupling to a disposable waveguide system. End coupling is simple and efficient, but the input beam from the source needs to be directed into the end of waveguide, requiring alignment accuracy at least as small as the thickness of the guide. For thin waveguides this may be a difficult requirement, especially if the waveguides are replaceable. The non-tapered launch coupler works well for thick waveguides (thicknesses greater than about 1 mm) and is simple in design, but its coupling efficiency drops for thinner guides. The tapered launch coupler gives better efficiency for the thinner waveguides, but fabricating the tapered end is more difficult and makes the coupler end more fragile. The grating coupler is relatively easy to fabricate once the proper release agent is determined and the molding steps work out. The coupling efficiency of the grating coupler is not as high as the launch coupler, probably due to imperfections in the replication process. The grating coupler is mode selective however, and would be a better choice than the launch coupler for a thin film, few-mode guide. The advantage and disadvantage of some coupling methods are summarized in Table 2.1 (Christensen et al. 1992).

2.3.1 Prism Coupling Method

The prism coupling method uses a high-index prism to excite an incident wave to a guided mode. As shown in Fig. 2.6, a prism with a refractive index n_p is put close to the waveguide with a thin gap layer of air (index $= n_c = 1$) between them. When the light beam is launched into the bottom of the prism at an angle θ, the propagation constant of prism β along the waveguide plan (the z-direction) is given as (Wu 2005):

$$\beta_p = k_0 n_p \sin\theta \tag{2.14}$$

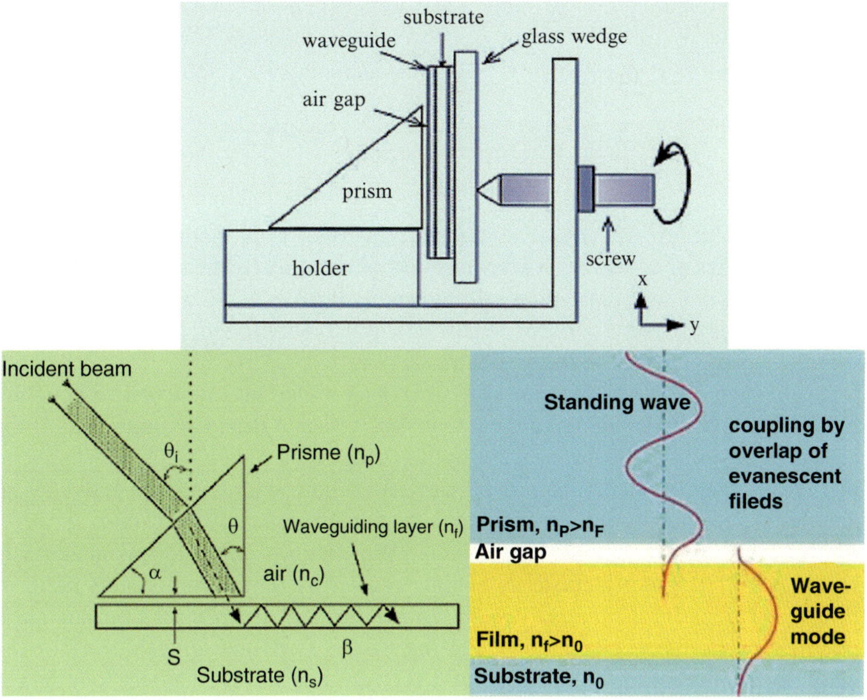

Fig. 2.6 Schematic illustration of prism coupling method uses a high-index prism to excite an incident wave to a guided mode

According to Snell's law, the incident light angle θ_i is related to the angle in the prism by

$$n_c \sin(\theta_i - \alpha) = n_p \sin(\theta - \alpha) \tag{2.15}$$

where α denotes the angle of the prism and n_p is the index of the prism. If there is no waveguide under the prism, the beam of the incident angle θ satisfying $n_p \sin\theta > n_c$ is entirely reflected at the bottom of the prism and penetrates as an evanescent wave into the air gap. When β_p adjusted by θ is consistent with the propagation constant of a guided mode β_f, the guided wave is excited through the distributed coupling resulting from phase matching between the evanescent wave and the guided mode. This condition is called matching phase. This energy transfer is effectuated through the evanescent wave created in the intermediate air gap between prism and waveguide, which is called the optical tunneling effect. Under the condition of matching phase, the effective refraction index of mth mode, N_m can be expressed as (Wu 2005)

$$N_m = n_\mathrm{w} \sin\theta_m = n_\mathrm{p} \sin\theta \tag{2.16}$$

Combined (2.15) and (2.16), N_m can be rewritten as (Wu 2005):

$$N_m = n_\mathrm{p} \sin\left(\alpha + \arcsin\left(\frac{\sin\theta_\mathrm{i}}{n_\mathrm{p}}\right)\right) \tag{2.17}$$

Thus a value for the effective refractive index of each mode can be obtained since the values of n_p and α are known and θ_i can be directly measured. In addition, the propagation constants can be inferred since $\beta_m = k_0 \cdot N_m$. Once the values of β for two modes, β_m and β_{m+1}, can be calculated, the refractive n_w, and the thickness d of the waveguide can be inferred with (2.13).

In practice, the air gap thickness (S) plays an important role in prism coupling method. In order to create the optical tunneling effect, S has to be optimized. If the thickness is too large the evanescent wave is attenuated before being able to excite the guided mode of waveguide. Conversely, a too narrow air thickness provokes an angular displacement of the guided mode so the evanescent wave cannot correctly transmit into the waveguide. These facts mean that a careful adjustment of S is required. In order to effectuate an energy transfer by the prism coupling method, three conditions must be met (Wu 2005): (a) the laser beam must have the same polarization as the excited mode; (b) the thickness of the air gap S has to be carefully controlled in the range around 100 nm; and (c) the incident light has to be in accordance with the matching phase.

The prism coupling method is particularly adapted for the planar waveguide. The implied surface of the prism has to be well polished and of good quality. The major advantages of this method include: non-destructive, and allows changing the position of the coupling point around the sample; easy manipulation and high efficiency, about 80 % for input and output couplings; ability to selectively excite any guided modes by adjusting the angle of the incident light.

2.3.2 End-Coupling Method

The end-coupling technique is the simplest method to excite a guided wave. A light wave with a profile similar to guided mode is directly launched into an end-face normal to guided-wave propagation as shown in Fig. 2.7 (Wu 2005). The light is focused into one end of the waveguide using a microscope objective lens, and propagates along the whole length to emerge as a highly divergent beam from the opposite end. The general procedure is to image the output onto a screen using a second objective lens. For planar waveguides a very bright horizontal line should be observed on the screen under correct launch conditions.

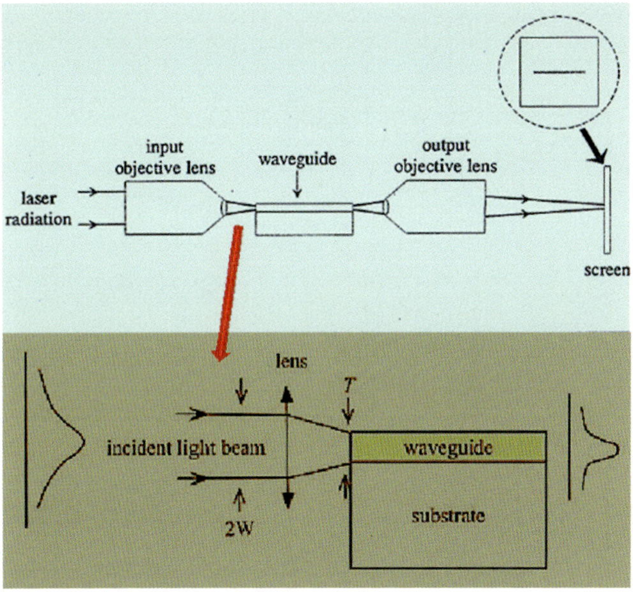

Fig. 2.7 Schematic illustration of end-coupling technique

In order to obtain a high efficiency, the incident-wave profile must be as close as possible to the profile of the guided wave. The coupling end face must exhibit extremely high optical quality and defectless which can be prepared by polishing, or cleaving treatments. End-face flatness and the alignment accuracy are strictly required. The main disadvantage of this method is that it excites all the guided modes simultaneously and it is not possible to select a particular excited mode when analyzing a multi-mode waveguide.

2.3.3 Lunch and Tapered-Coupling Method

The launch coupling basically uses a fixed lunch waveguide permanently to align with respect to the source beam to which a removable function waveguide is placed in contact. The input beam may be coupled from the source to this launch waveguide by any means: end coupling, grating coupling, or even prism coupling. The function waveguide is then placed on top of the launch coupler and light is coupled into the function guide by simple propagation. Coupling gel (or more permanent coupling polymer) can be used to increase the index match between the two guides. This scheme allows for great latitude in the placement of the

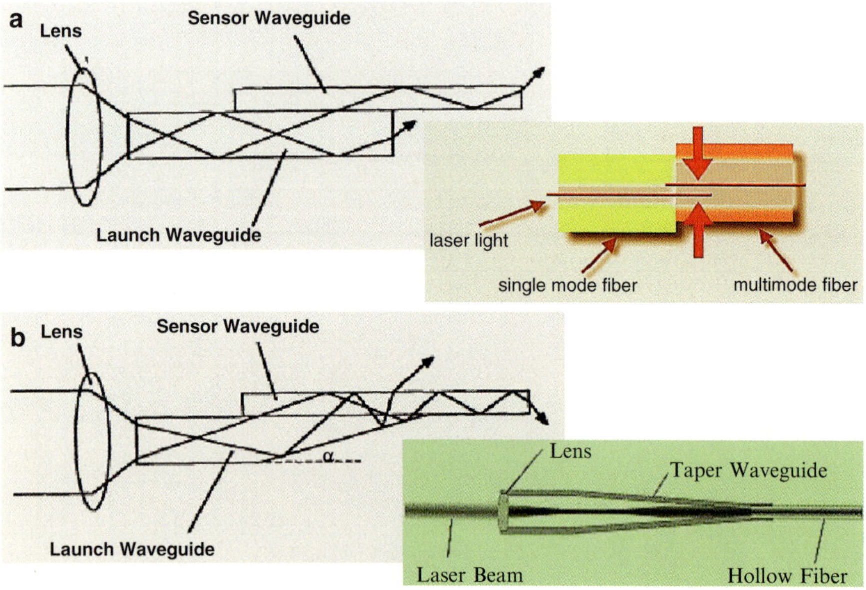

Fig. 2.8 Schematic illustration of launch coupler with another waveguide such as sensor waveguide placed on top: (**a**) no-tapered and (**b**) tapered

function waveguide with respect to the coupler. Vertical alignment is achieved automatically by the mating surfaces, and longitudinal positioning of the function waveguide is not critical (assuming an overlap distance of at least a few waveguide thicknesses). Figure 2.8 shows two versions of the launch coupler with a function waveguide in place. The first version, shown in Fig. 2.8a, uses a non-tapered launch guide. Ray theory predicts that the percentage of light initially in the launch guide that is coupled into the sensor is dependent on the ratio of the thicknesses of the two guides in the following manner (Christensen et al. 1992):

$$\%\ \text{Coupling} = t_s/\left(t_s + t_j\right) \tag{2.18}$$

where t_s is the thickness of the function waveguide and t_j is the thickness of the launch waveguide. Thus, when the two guides have the same thickness, the coupling efficiency is 50 %. When the function waveguide is much thinner than the launch guide, however, the coupling efficiency drops considerably.

For thin function waveguides, the tapered version of the coupler as shown in Fig. 2.8b, gives higher efficiency. The design is based upon the concept of transforming narrow-angle/broad-area illumination in the launch waveguide (as initially coupled in from a laser source) into broad-angle/narrow-area light in the sensor waveguide. For small taper angles, the coupling efficiency can be quite large.

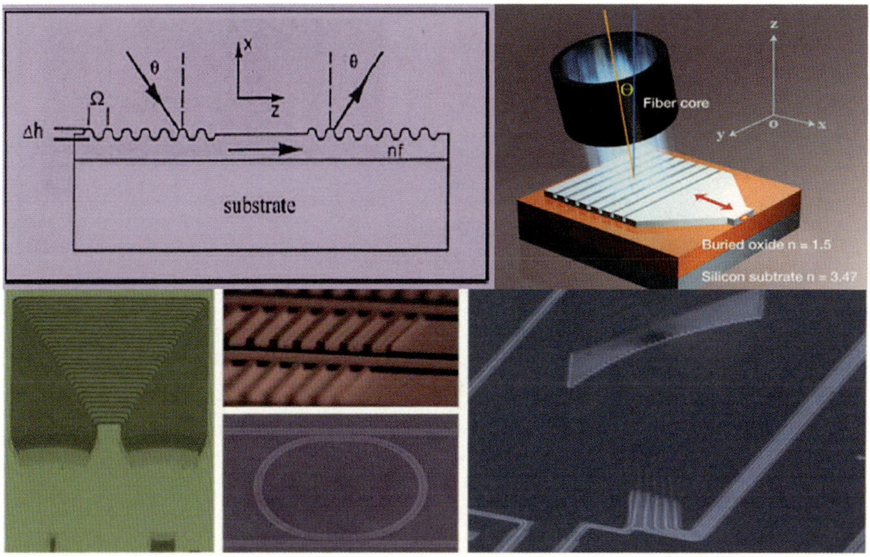

Fig. 2.9 Schematic illustration of grating coupling method and examples of some grating couplers

The tapered-coupling usually consists of tapering the end of the waveguide by a bevel length from 10 to 100 times the operating wavelength. This method is favored for high-index waveguides, such as GaAs, which require an expensive high-index prism when using the prism coupling method. When the guided mode propagates inside the waveguide that exceeds the critical angle, the total internal reflection is alternated at the tapered part and gradually reaches the cutoff condition in the substrate. An optical fiber can be coupled at the output beam position and an input coupling is also possible by supplying a laser beam which is conjugate to the output beam. This method is able to selectively excite guided modes, and has no limitation for the refractive index of waveguide. However, this method is destructive, and it is not easy to taper a waveguide precisely so it is not used frequently for experimental analyses.

2.3.4 *Grating Coupling Method*

This technique relies on a periodic grating structure on the surface of the waveguide as shown in Fig. 2.9. When a laser beam with propagation vector β_0 is incident in the grating region, the light is diffused by the gratings. According to such diffraction, the wave components with the propagation vector $\beta_g = \beta_0 + \gamma K$ is produced, called space harmonics where γ is the order of diffraction and K is the grating vector correlated to the fundamental period Ω by $K = 2\pi/\Omega$. When β_g equals to the propagation vector of guided mode, the incident light can be coupled into the waveguide. This condition is called matching phase condition which can be expressed as (Wu 2005):

$$\beta_g = \beta_0 + 2\pi\frac{\gamma}{\Omega} = k_0 N_m \tag{2.19}$$

$$N_m = n_c \sin\theta + \frac{\gamma\lambda_0}{\Omega} \tag{2.20}$$

The grating coupling method is permanent and allows selection of guided mode but the implement requires a certain degree of delicate skill. The coupling efficiency depends on the grating structure, such as grating period (Ω), grating depth (Δh). The grating coupler has showed moderate tolerance to alignment. However, the efficiency is somewhat sensitive to the lateral beam position on the grating for a focused input beam. The beam position had to be within approximately 1–2 mm of the end of the grating (toward the waveguiding region) to achieve the highest efficiency. If the beam was too far away from the grating end, light coupled into the guide would be coupled back out by the grating before reaching the guide, reducing the amount coupled.

2.4 Optical Loss

Light wave experiences attenuation or losses as it propagate through an optical waveguide. An expression for the amount of optical power remaining in the waveguide after propagating a distance x is given in the following expression (Chien 2004):

$$I(x) = I_0 10^{\left(-\frac{\alpha x}{10}\right)} \tag{2.21}$$

where I_0 is the initial power, $I(x)$ is the transmitted power through the waveguide at a distance x (cm), and α is defined as the attenuation coefficient of the waveguide, measured in decibels per centimeter (dB/cm). The loss, L in decibels (dB) is defined as (Chien 2004)

$$L(\text{dB}) = -10\log\left(\frac{I(x)}{I_0}\right) \tag{2.22}$$

In order to make a waveguide into any functional optical devices, its propagation losses, or transmission losses must be kept as low as possible. Light propagation in optical waveguides can be affected by a certain number of distortions due to the random distributions of imperfections within the waveguides. Many factors are considered to disturb the light propagation and increase the propagation losses: (a) Radiation losses due to the guided mode converse to the radiation mode.

(b) Mode conversion losses due to conversion from the excited mode to other guided modes. Optical energy can be lost from the waveguides modes by the conversion from the excited mode to other guided modes. When a waveguide is simply used for light power transmission, mode conversion does not cause substantial loss. In a device where a guided mode performs a function, mode conversion is an important factor affecting performance. (c) Absorption losses due to light absorption in the waveguide materials. Optical energy is converted into phonon energy in the form of heat. The most important losses are inter-band absorption, impurity absorption, and carrier absorption, which could either be an inherent effect due to the impurities in the material or could be an induced absorption resulting from the fabrication process. (d) Diffusion losses or scattering losses due to the imperfection of the waveguide structure. Scattering at the waveguide interfaces could also be an inherent effect due to the defects or surface roughness existing in the original material or formed after the fabrication process. Scattering losses are due to the imperfections of the waveguide structure and the quality of the material used. Voids, contaminant atoms, crystalline defects within the volume of the waveguide cause volume scattering while surface scattering loss can be significant for rough (bumps and valleys) surfaces of the waveguide. (e) Tunneling losses that only occurs in barrier optical waveguides produced by ion modification. A guide confined by a narrow optical barrier would be very susceptible to such losses.

2.4.1 Propagation Losses by Radiation

The light radiation may take place when the propagation constant of the guided mode (β) is close to that of the substrate (β_s). When $\beta \gg \beta_s$, the distribution of the electromagnetic field can be concentrated in the film center as shown in Fig. 2.10a. Otherwise, the energy transfer by the guided mode is partially dispersed into the substrate or the upper cladding. Actually, this radiation loss occurs when the refractive index of the film is close to that of the substrate or when the film thickness is close to the cutoff thickness. It is also probably induced by the imperfection existing inside the waveguide which could change the wave propagation to radiation mode and dissipate the propagation power out of the waveguide. This distortion due to the imperfections is classified as a factor of propagation losses by diffusion (Wu 2005).

2.4.2 Propagation Losses by Absorption and Mode Conversion

The propagation losses by absorption are mainly due to the light absorption by the materials of waveguides. The optical absorption can result from the anionic deficit (e.g., oxygen), defects (stoichiometry), impurities which absorb the light and reduce the light intensity as illustrated in Fig. 2.10b. This mechanism of propagation losses

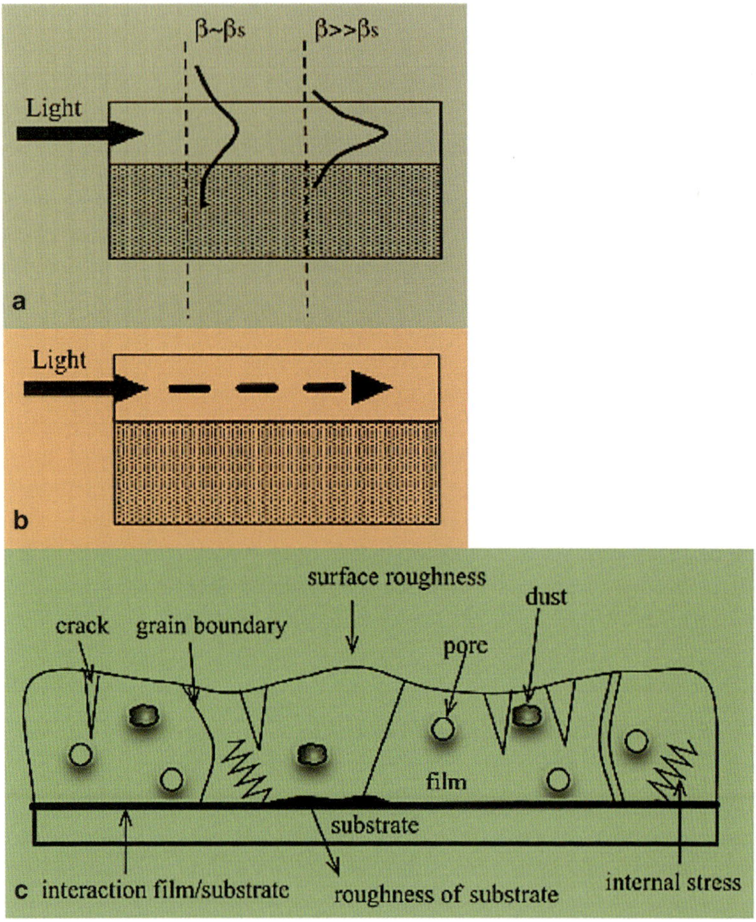

Fig. 2.10 Schematic illustration of propagation losses (Wu 2005): (**a**) the electromagnetic field distribution of waveguide; (**b**) propagation loss by absorption; and (**c**) the different structural and textural defects resulting in propagation losses by diffusion

is commonly observed in semiconductor waveguides. In addition, the impurities contained in waveguide may also induce propagation losses by absorption such as organic residues and doped ions. For example, the Tb^{4+} ions absorb in the range between 400 and 600 nm which increases largely this type of loss when the working wavelength drops into this range. The propagation losses by mode conversion result from the transformation between the excited mode and another guided mode, which involves a propagation loss by a transfer of energy. This effect is predominately caused by the existence of film imperfections such as the inhomogeneity of refractive index or a large surface roughness. This mode conversion does not create a substantial loss when a waveguide is simply used for light power transmission. However, it is still an important factor in a device where a guided mode performs a function, such as curved waveguides (Wu 2005).

2.4.3 Propagation Losses by Diffusion

The light diffusion involves mainly the imperfections at the interfaces (waveguide/substrate and waveguide/cladding layer) or inside the waveguides. These imperfections limit the light propagation because they partially diffuse of light by radiation mode. The potential imperfections appear in the waveguides in the form of porosity, cracks, grain boundaries for polycrystalline films, dust contamination, and so on. Such imperfections could be possibly produced for any deposition operation. For instance, very strict care during the sol–gel coating procedure is helpful to reduce such film imperfections. However, some of the imperfections such as pores and cracks are unavoidable, especially for sol–gel-derived waveguides. The sol–gel elaboration involves organic compounds which must be eliminated by heat treatment. With the removal of organic residues, the sol–gel waveguide is often porous and requires a high temperature of annealing treatment to densify the films, however, the cracks sometimes occur due to the difference of thermal expansion coefficient of the sol–gel layer and substrate which induces the internal stress within the film. Moreover, this light diffusion tends to increase with $1/\lambda$. According to Rayleigh's law, when the imperfection size is smaller than λ, the light diffusion is roughly proportional to $(1/\lambda)^4$. Therefore, this diffusion can be reduced at the longer wavelength range yielding weaker propagation loss. Actually, the inhomogeneities within the waveguide and the irregularity of the interface are the major causes of propagation loss by diffusion. They are mostly related to the elaboration process of waveguides. The principal types of film defects that may induce losses by diffusion are depicted in Fig. 2.10c. They may come from textural defects (cracks, pores, dust, and surface roughness), compositional defects (inhomogeneous nucleation, discontinuous index, or thickness), or structural defects (grain boundaries, internal stress, interaction between film and substrate, etc.). The size and the optical properties of these imperfections dominate their degree of participation for the diffusion loss with the working wavelength. Practically, it has been proved that the mean crystallite size of sol–gel waveguides has to be as low as possible. A nanocrystal larger than 10–15 nm drastically increases the propagation losses. It should be noted that the propagation losses always exist in different degrees whatever the elaboration methods. Through an understanding of the structural and textural imperfections which could produce the propagation losses it is possible to optimize the elaboration process to avoid or at least reduce such factors and obtain a good quality waveguide. This is an important and difficult challenge which must be solved for practical waveguide applications (Wu 2005).

2.4.4 Measurement of Propagation Losses

As the optical attenuation in the waveguide is an important parameter describing the performance of optical integrated devices, a fast and reliable loss measurement technique is a great asset when investigating loss mechanism. A few methods of loss measurement have been carried out.

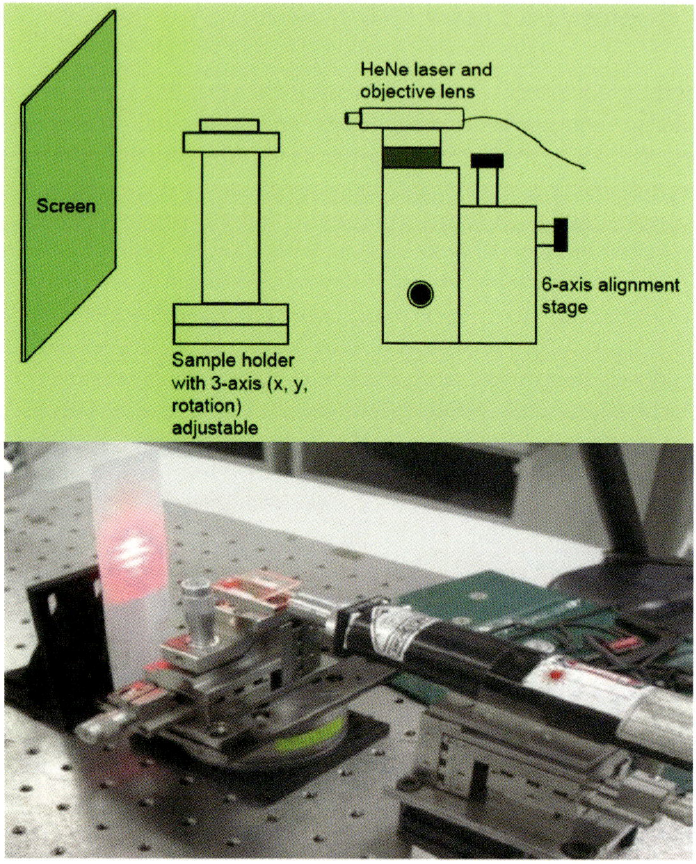

Fig. 2.11 Illustration of cutback method (Wu 2005)

2.4.4.1 Cutback Method

Cutback method is a simple method for direct measurement of transmission losses by comparing the transmittances of waveguides having different lengths. However, it is rather difficult to prepare many waveguides having equal quality and different lengths. Thus, the measurement is done by cutting a waveguide to change the length. A series of transmitted power measurements is performed starting with a relatively long waveguide sample, then after shortening the sample by cleaving and polishing the facet of the waveguide edges. For reasonable accuracy, the measurement must be carried out at each sample length under conditions of optimum coupling efficiency between the incident power and the power propagating in the waveguide. The input power must be constant. A guided wave is excited by end-fire coupling and butt-coupling. A light wave with profile similar to the guided mode profile is fed onto a waveguide end face normal to guided wave propagation. Coupling is localized at the waveguide end. For example, end-fire coupling has been carried out for the wavelength 633 nm. As shown in Fig. 2.11, the channel waveguide sample was fixed on a sample holder using double-sided tape. The stage could be adjusted on three axes, x, y, and rotation for the alignment purpose. The light source is a helium–neon (He–Ne) laser which emitting light at wavelength around 633 nm with light output 4 mW. The light emitted from the laser was focused using a $\times 10$ objective lens with 16.5 mm focal length mounted on the laser tube. The laser was set on a six axes micro-positioning alignment stage to facilitate the critical alignment required. By adjusting the laser tube from the sample to a distance which is the focal length of the lens, the laser beam was focused on the input facet of the sample. The output emission was projected on white screen. From the pattern of the emission, the coupling of light can be inspected and optimized. The output power can be measured by an optical power meter after the light was coupled into the waveguide. Since the light at wavelength 1,550 nm is invisible, it is difficult to be observed in the end-fire coupling method to optimize the coupling. The loss measurement at wavelength 1,550 nm can be done using butt-coupling, which basically consists on direct heading of the fiber optics to the waveguide. After cutting the waveguide to shorten the length, the transmittance was measured again. Transmission losses are given by (Wu 2005)

$$\alpha = \left| \frac{10 \log\left(\frac{P_1}{P_2}\right)}{\left(\frac{L_1}{L_2}\right)} \right| \quad (\mathrm{dB/cm}) \tag{2.23}$$

where L_1, L_2 (cm) and P_1, P_2 are the lengths and transmittances for before and after cutting, respectively. High quality waveguide edges must be prepared so that the equal input–output coupling efficiencies may be reproduced for each measurement. The drawback of this technique is a destructive method, application is limited to cases where a high quality waveguide edge can be prepared by cleaving, and information about the mode-order dependence losses cannot be obtained.

2.4.4.2 Sliding-Prism Method

The drawback of cutback method can be eliminated by using prism coupler. The attenuation of slab waveguide can be measured using the sliding-prism method. The measurement set up is shown in Fig. 2.12 (Wu 2005). The slab waveguide was clamped firmly below a prism. By tightening the screw, the gap between the waveguide and the prism could be minimized. In order to achieve prism coupling to either the substrate modes or the guided modes, the prism to the waveguide contact must be extremely good since the evanescent field penetration depth is only about 1 μm. The prism surface must be dirt free. In addition, optimum coupling is only achieved at the point of closest contact, where the screw clamp applies the greatest pressure to the sample, and at a point as near to the right angled edge of the prism as possible to limit the degree of return coupling to the prism. The prism coupling assembly was then set on a rotational stage which was mounted on an x–y

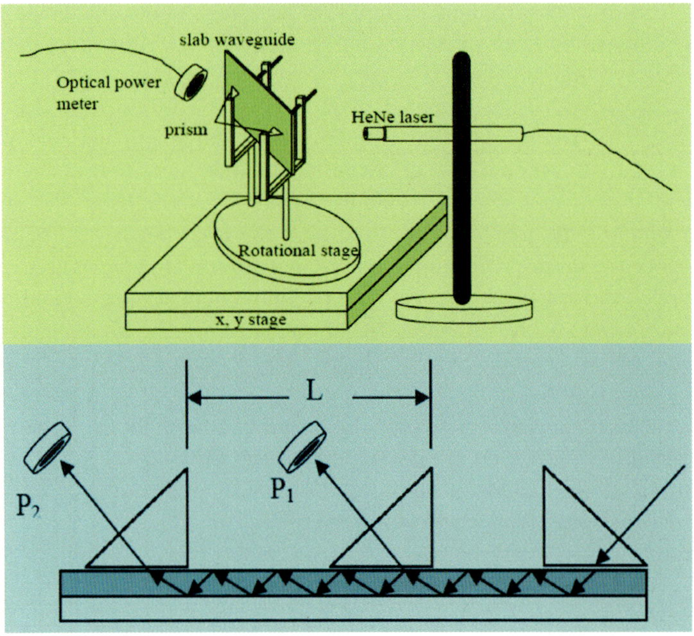

Fig. 2.12 Schematic illustration of sliding-prism method (Wu 2005)

translational stage. A He–Ne laser with an objective lens mounted on it was set at the same level with the prism. The rotational stage was adjusted to couple the light into the waveguide at certain incidence angle of the laser beam. The light was coupled into the waveguide once a bright laser line (guided mode) appeared in the waveguide. The rotational stage was locked to keep the coupling of light at that incidence angle. In order to couple out the light, a second prism was then clamped to the waveguide to interact with the guided mode at a distance of about 1–2 cm from the input prism. The light was coupled out once the optical contact was sufficiently good and the mode lines (m lines) observed on a screen. The output power was then measured using an optical power meter. The second prism was slide backward for a desired distance and the output power was measured and the detected power the two locations were compared. An important requirement is that nearly 100 % output efficiency must be maintained during the measurement in order to maintain constant coupling efficiency (Wu 2005).

2.4.4.3 Fabry–Perot Resonances

A waveguide can be seen as a Fabry–Perot cavity with two polished facets acting as two mirrors. Therefore, it is possible to measure the propagation loss using these Fabry–Perot resonances. If R is denoted as the facet reflectivity, the ratio of the output and the input powers can be given by (Yin 2009)

$$\frac{P_{out}}{P_{in}} = \frac{(1-R)^2 e^{-\alpha L}}{(1 - Re^{-\alpha L})^2 + 4Re^{-\alpha L} \sin^2(\phi/2)} \tag{2.23a}$$

where R for TM and TE modes is given by Fresnel's equations. Assuming that light is coupled into the waveguide at normal incidence, then R can be simply expressed as $R = (n - 1)^2/(n + 1)^2$. $\phi = 2nkL \cos\theta + \phi_r$ is the phase difference associated with optical path difference of the resonator, and $\theta \approx 0$ is the incident angle. Given $\zeta = \frac{I_{max}}{I_{min}} = \frac{(1 + Re^{-\alpha L})^2}{(1 - Re^{-\alpha L})^2}$ as the ratio of the maximum intensity to the minimum intensity, the propagation loss α can be expressed as (Yin 2009)

$$\alpha = -\frac{1}{L} \ln\left(\frac{1}{R} \frac{\sqrt{\zeta} - 1}{\sqrt{\zeta} + 1}\right) \tag{2.23b}$$

The measurement using Fabry–Perot resonance can be more stable or repeatable and more accurate than using the cutback method.

2.4.4.4 Scattered Light Measurement

The propagation loss of a waveguide can also be known by measuring the scattered light from the surface of the waveguide. The assumption of this method is that the amount of light scattered from the surface is proportional to the light propagating inside the waveguide. Then the rate that the scattered light decays should mimic the decaying rate of the propagated light inside the waveguide. Optical fibers can be used to collect the scattered light and can be scanned along the surface of the waveguide (Yin 2009).

2.5 Optoelectronic Characterization

2.5.1 Optical Power Meters

Optical power meter is the simplest and most basic piece of equipment for optoelectronic characterization, as shown in Fig. 2.13. Different models have different connector types and are specialized to either multi-mode or single-mode fiber or optical waveguide. There is always a wavelength switch to adjust the power readings for the particular wavelength being received. Also there is usually a range switch which determines the range of signal power expected—although this last function can be automatically determined by the meter itself (Dutton 1998).

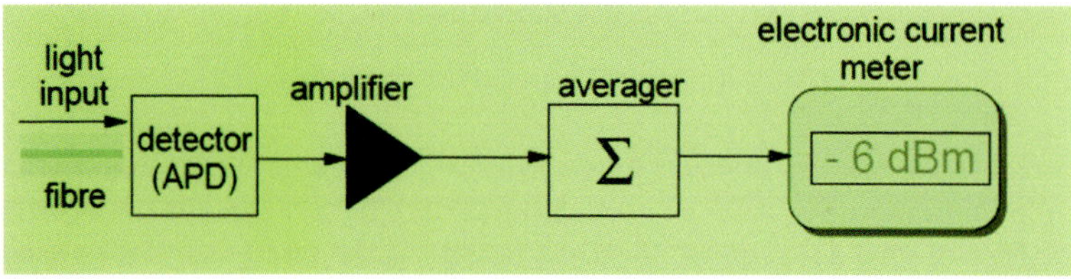

Fig. 2.13 Optical power meter—logical structure (Dutton 1998)

2.5.2 Optical Time-Domain Reflectometers (OTDRs)

The OTDR can be used to look at a fiber link from inside the fiber. In reality it is just a radar system for looking at fiber. High intensity pulses are sent into the fiber from a specialized laser and when the pulse returns its strength is displayed on an oscilloscope screen in the form of a trace. A schematic of such a display is shown in Fig. 2.14. In the trace reflections can be seen coming from all along the fiber itself. This is the result of Rayleigh scattering. Every time there is a discontinuity or imperfection in the fiber/waveguide the effect can be seen in the trace. Such events can be the presence of a connector or a splice or some more serious imperfection such as a crimp in the cable due to poor installation. OTDR can be used to quickly determine the length of the fiber, attenuation in decibel of the whole fiber link and the attenuation of separate sections of fiber (if any), attenuation characteristics of the basic fiber itself, locations of connectors, joints and faults in the cable. Many OTDRs come with additional functions such as optical power meter or laser source so that a good OTDR often has all of the function needed in the field. In addition many OTDRs offer computer output so that collect OTDR data can be collected in the form of digital readings and analyze it later on a computer (Dutton 1998).

2.5.3 Spectrum Analyzers

A spectrum analyzer scans across a range of wavelengths and provides a display showing the signal power at each wavelength. Figure 2.15 shows a simplified example of the kind of display produced by a typical spectrum analyzer. From this display the following parameters can be calculated: The power levels of each channel, spectral width of each channel, any interference between channels such as crosstalk possibilities, by connecting it in different places through the system many potential problems can be tracked such as laser drift, etc. (Dutton 1998).

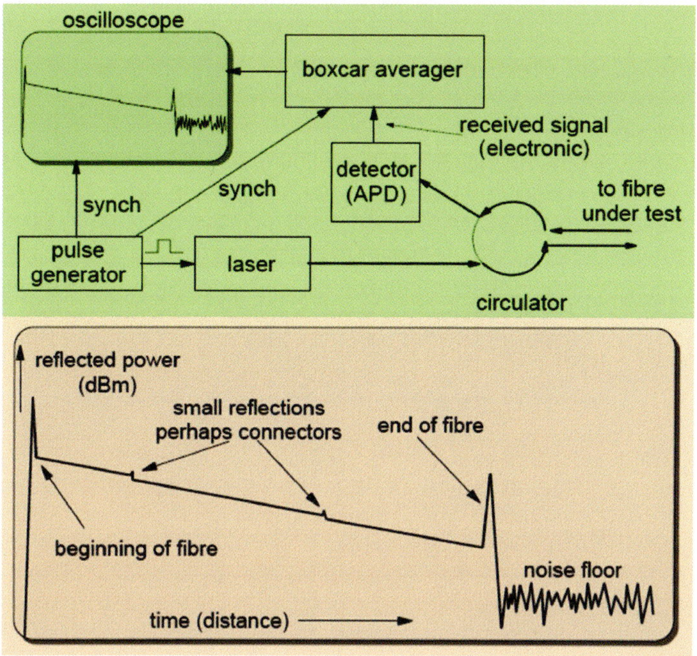

Fig. 2.14 Schematic illustration of optical time-domain reflectometer (Dutton 1998)

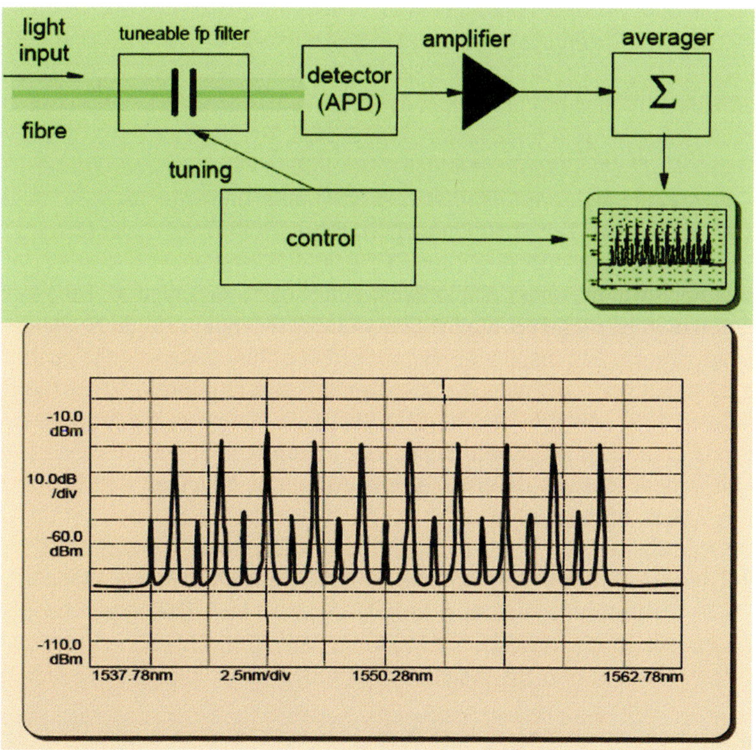

Fig. 2.15 Illustration of spectrum analyzer (Dutton 1998)

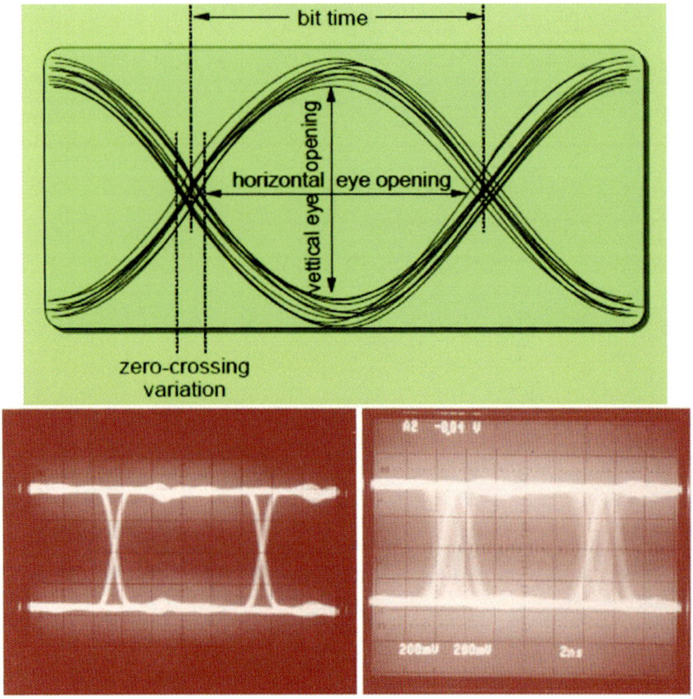

Fig. 2.16 Illustration of eye diagrams (Dutton 1998)

2.5.4 *Eye Diagrams*

The eye diagram has become the recognized way of looking at an electronic signal and determining its "goodness" as a carrier of information. It consists of many (from hundreds to millions) of instances of the signal displayed over the top of one another. In extremely fast equipment only one or two points might be got on a trace at a single sweep. But displaying them together allows assessing the quality of the received signal very well indeed. As shown in Fig. 2.16 (Dutton 1998), the diagram is produced by feeding the result of the analogue section of the receiver circuit to the *y*-axis control of an oscilloscope. The sweep is set to display one full cycle (two-bit times) and is usually triggered from the receiver's PLL circuit (the receiver's derived clock).

The eye diagram has the following important aspects (Dutton 1998): (a) The vertical eye opening indicates the amount of difference in signal level that is present to indicate the difference between one-bits and zero-bits. The bigger the difference the easier it is to discriminate between one and zero. Of course this is affected significantly by noise in the system. (b) The horizontal eye opening indicates the amount of jitter present in the signal. The wider the eye opening is on this axis the less problem likely has with jitter. (c) The thickness of the band of signals at the zero-crossing point is also a good measure of jitter in the signal. (d) The best indication of signal "goodness" is just the size of the eye opening itself. The larger it is the easier it will be to detect the signal and the lower will be the error rate. When the eye is nearly closed it will be very difficult or impossible to derive meaningful data from the signal.

2.6 Electro-optic Effects

An electro-optic effect is a change in the optical properties of a material in response to an electric field that varies slowly compared with the frequency of light. The term encompasses a number of distinct phenomena, which can be subdivided into: (a) Change of the absorption: (i) Electroabsorption—general change of the absorption constants. (ii) Franz–Keldysh effect—change in the absorption shown in some bulk semiconductors. (iii) Quantum-confined Stark effect—change in the absorption in some semiconductor quantum wells. (iv) Electrochromic effect—creation of an absorption band at some wavelengths, which gives rise to a change in color. (b) Change of the refractive index: (i) Pockels effect (or linear electro-optic effect)—change in the refractive index linearly proportional to the electric field. Only certain crystalline solids show the Pockels effect, as it requires lack of inversion symmetry. (ii) Kerr effect (or quadratic electro-optic effect, QEO effect)—change in the refractive index proportional to the square of the electric field. All materials display the Kerr effect, with varying magnitudes, but it is generally much weaker than the Pockels effect. (iii) Electro-gyration—change in the optical activity. Changes in absorption can have a strong effect on refractive index for wavelengths near the absorption edge, due to the Kramers–Kronig relation.

Using a less strict definition of the electro-optic effect allowing also electric fields oscillating at optical frequencies, one could also include nonlinear absorption (absorption depends on the light intensity) to category (a) and the optical Kerr effect (refractive index depends on the light intensity) to category (b). Combined with the photoeffect and photoconductivity, the electro-optic effect gives rise to the photorefractive effect.

An electric field E applied to a transparent material generally modifies the refractive index n_0 for a particular mode of propagation in accordance with the equation (Kaye & Laby 2012)

$$\frac{1}{n^2} = \frac{1}{n_0^2} + rE + RE^2 + \cdots \qquad (2.24)$$

In amorphous materials or centro-symmetric crystals the coefficients of odd powers of E are zero, and the first nonzero term RE^2 represents the quadratic Kerr effect, usually characterized for a particular material by the Kerr constant:

$$B = \frac{\Delta n}{\lambda_0 E^2} = \frac{n_0^3}{2\lambda_0} R \tag{2.25}$$

where λ_0 is the free-space wavelength.

For the case of a birefringent crystal, the field E modifies the index ellipsoid $\sum_{i=l}^{3}(x_i^2/n_i^2) = 1$, in the principal axis coordinate system, to (Kaye & Laby 2012):

$$\sum\nolimits_{i=1}^{3} \sum\nolimits_{j=1}^{3} \sum\nolimits_{k=1}^{3} \sum\nolimits_{l=1}^{3} \left(\frac{1}{n_{ij}^2} + r_{ijk} E_k + R_{ijkl} E_k E_l + \cdots \right) x_i x_j = 1 \tag{2.26}$$

where $n_{ij} = n_i$, for $i = j$, and $1/n_{ij} = 0$ for $i \neq j$. From a knowledge of the change in orientation and dimensions of the index ellipsoid, as given by this equation, the field-induced birefringence for a ray of given direction and polarization may be calculated.

In practice the electrooptic effect is either predominantly linear or quadratic in E, and is therefore characterized by either r_{ijk} or R_{ijkl}, depending on the material. Since r_{ijk} is symmetrical in i, j, and R_{ijkl}, symmetrical in i, j and in k, l, pairs of values i, j and k, l are denoted by indices m and n respectively, which run from 1 to 6 according to the scheme (Kaye & Laby 2012):

$$1 \leftrightarrow 1,1 \quad 2 \leftrightarrow 2,2 \quad 3 \leftrightarrow 3,3 \quad 4 \leftrightarrow 2,3 \quad 5 \leftrightarrow 1,3 \quad 6 \leftrightarrow 1,2$$

When measured at low frequencies ($<10^4$ Hz), the coefficients may include contributions from the elasto-optic effects of piezoelectric and/or electrostrictive strains. The linear electro-optic effect observed at frequencies sufficiently high for these contributions to be negligible is called the Pockels effect. The strain effects may contribute up to 50 % of the low-frequency effect.

The quadratic effect is exhibited most markedly by materials in which the permittivity is high and varies rapidly with temperature. Here, since the R_{mn} vary accordingly, it is more convenient to replace $R_{mn}\, E_k E_l$ in the equation of the index ellipsoid by $g_{mn} P_k P_l$ (where $P = D - \varepsilon_0 E$), and to express the properties of the material in terms of g_{mn}. Typical parameters for selected electrooptic materials at low frequencies are given in Table 2.2 (Kaye & Laby 2012).

The indicatrix and electrooptic coefficients are referred to the usual crystallographic coordinate system. $Ox_3 \equiv Oz$ is the fourfold axis for cubic and tetragonal symmetries, or the threefold axis for trigonal symmetry; $Ox_1 \equiv Ox$; $Ox_2 \equiv Oy$, except for the trigonal case in which Ox_1 is perpendicular to the mirror plane. For uniaxial crystals $n_1 = n_1 = n_0$, $n_3 = n_e$. The half-wave voltage V_π is used conventionally to characterize the sensitivity of an electrooptic material. It is the voltage required to obtain one half-wavelength of optical path difference between the two

Table 2.2 Properties of selected electrooptic materials at a wavelength of 633 nm (Kaye & Laby 2012)

Material	Point-group symmetry	Electrooptic coeff. and value: low freq. $\frac{r}{10^{-12}\,\text{m/V}^{-1}}$ or $\frac{g}{\text{m}^4/\text{C}^{-2}}$	High freq. value	Refractive index	Relative permittivity (LF)	Relative permittivity (HF)
$C_6H_5O_2N$ (nitrobenzene)	∞	$B = 4.4 \times 10^{-12}$ m/V^{-2}		1.55	35.7	
$pb_{0.814}La_{0.124}$–$(Ti_{0.6}Zr_{0.4})O_3$ (PLZT)	∞	$n_e^3 r_{33} - n_0^3 r_{33} = 2320$ [1]		$n_0 = 2.55$		
β-Zns	$\bar{4}3m$	$r_{41} = 2.1$	−1.6	2.35	16	12.5
ZnSe	$\bar{4}3m$	$r_{41} = 2.0$	2.0	2.60	9.1	9.1
ZnTe	$\bar{4}3m$	$r_{41} = 4.04$	4.3	2.99	10.1	10.1
$Bi_{12}SiO_{20}$	23	$r_{41} = 5.0$		2.54		
KH_2PO_4 (KDP)	$\bar{4}2m$	$r_{41} = 8$		$n_0 = 1.5074$	$\varepsilon_1, \varepsilon_2 = 42$	44
		$r_{63} = 11$		$n_e = 1.4669$	$\varepsilon_3 = 21$	21
KD_2PO_4 (KD*P)	$\bar{4}2m$	$r_{41} = 8.8$ [1]		$n_0 = 1.502$	$\varepsilon_1, \varepsilon_2 = 58$	
		$r_{63} = 24.1$		$n_e = 1.462$	$\varepsilon_3 = 50$	48
CsH_2AsO_4 (CDA)	$\bar{4}2m$	$r_{41} = 14.8$ [2]		$n_0 = 1.572$		
		$r_{63} = 18.2$ [2]		$n_e = 1.550$		
$BaTiO_3$	$m3m$	$g_{11} - g_{12} = 0.13$		2.437		4,500
$SrTiO_3$	$m3m$	$g_{11} - g_{12} = 0.14$		2.38		300
$KTa_{0.35}Nb_{0.65}O_3$ (KTN)	$m3m$	$g_{11} = 0.136$		2.29	$\approx 10^4$	
		$g_{12} = -0.038$				
		$g_{44} = 0.147$				
$Ba_{0.25}Sr_{0.75}Nb_2O_6$	$4mm$	$r_{13} = 67$		$n_0 = 2.312$	$\approx 2 \times 10^4$	3,400
		$r_{33} = 1{,}340$		$n_e = 2.299$		
		$r_{51} = 42$				
		r_c	1,090			

(continued)

Table 2.2 (continued)

Material	Point-group symmetry	Electrooptic coeff. and value: low freq. $\frac{r}{10^{-12}\,\mathrm{m/V^{-1}}}$ or $\frac{g}{\mathrm{m^4/C^{-2}}}$	High freq. value	Refractive index	Relative permittivity (LF)	Relative permittivity (HF)
$LiNbO_3$	$3m$	$r_{13} = 9.6$	8.6	$n_O = 2.286$	$\varepsilon_1, \varepsilon_2 = 78$	
		$r_{22} = 6.8$	3.4	$n_e = 2.200$		43
		$r_{33} = 30.9$	30.8		$\varepsilon_3 = 32$	28
		$r_{51} = 32.6$	28			
		$r_c = 21.1$				
$LiTaO_3$	$3m$	$r_{13} = 8.4$	7.5	$n_O = 2.176$	$\varepsilon_1, \varepsilon_2 = 51$	
		$r_{22} = -0.2$	1	$n_e = 2.180$		41
		$r_{33} = 30.5$	33		$\varepsilon_3 = 45$	43
		$r_c = 22$	–			
		r_{51}	20			
Ag_3AsS_3	$3m$	r_{22}	1.1	$n_O = 3.018$		
		r_c	3.4	$n_e = 2.739$		
$KNbO_3$	$2mm$	$r_{13} = 28$		$n_1 = 2.280$		
		$r_{23} = 1.3$		$n_2 = 2.329$		
		$r_{33} = 64$		$n_3 = 2.169$		
		$r_{42} = 380$				
		$r_{51} = 105$	270			

Table 2.3 Birefringence for commonly used configurations (Kaye & Laby 2012)

Material symmetry	Light direction	Field direction	Induced birefringence Δn
∞	Any	Transverse	$B\lambda_0 E^2$
$\overline{4}2$	Oz	Oz	$n_0^3 r_{63} E_3$
$\overline{4}3m$	Ox	Oz	$n_0^3 r_{41} E_3$
$m3m$	Ox	Oz	$n_0^3 (g_{11} - g_{12}) P_3^2/2$
$3m$	Ox	Oz	$n_3^3 \left\{ r_{33} - \left(\frac{n_1}{n_3}\right) r_{13} \right\} E_3/2 = \frac{1}{2} n_3^3 r_{33} E_3$
$4mm$	45° to Oz	45° to Oz longitudinal	$\left(n_3^3/4\sqrt{2}\right)(r_{13} - r_{33})E$

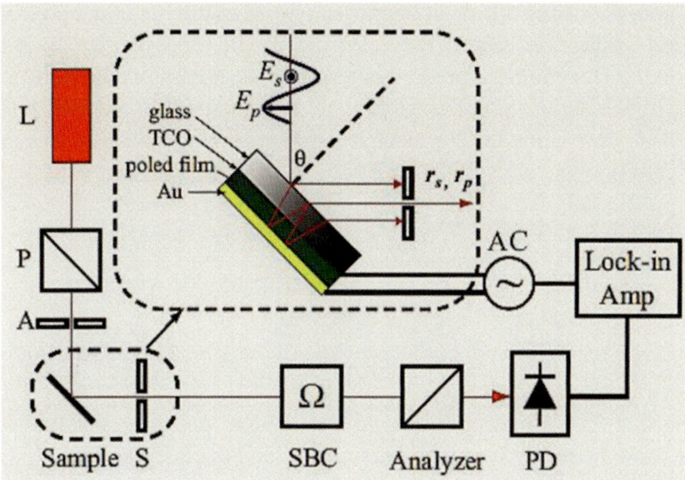

Fig. 2.17 Schematic illumination of the Teng–Man measurement setup: *L* laser, *P* polarizer, *A* aperture, *S* slit, *SBC* Soleil-Babinet Compensator, *PD* photodetector (Park 2008)

vibration components of a wavefront, using a cube of the material of side 1 cm with specified directions of light and applied field. Table 2.3 gives the induced birefringence for a number of commonly used configurations (Kaye & Laby 2012).

The electro-optic effect can be determined by measuring the change of the dielectric constant or index of refraction when a field is applied across the nonlinear electrooptical material of interest. There are various techniques for characterization of electro-optic effect, including Mach–Zehnder (MZ) interferometry, Fabry–Perot (FP) interferometry, attenuated total reflection (ATR), waveguide method, and two slit interference method. One of the most popular measurement techniques is called Teng–Man reflection method. Figure 2.17 shows the schematic illumination of the Teng–Man measurement setup (Park 2008). 45° polarized light is incident at an angle on the multilayered sample structure containing nonlinear electrooptical thin film. For the polymer material, it is usually spin-coated on a glass substrate with transparent conducting oxide (TCO), commonly indium tin oxide (ITO), and poled electrically to generate the second-order nonlinearity. The first reflection off the air–glass interface and subsequent beams resulting from reflection of the first pass on its way out at the glass/air interface back into the polymer film and out again are blocked. The remaining light reflected off the sample experiences an additional controllable phase retardation introduced by a Soleil-Babinet Compensator (SBC). The intensity of the light is detected after passing through the analyzer using a lock-in amplifier. At each angle of incidence, two different data sets, the optical bias curve I_{dc} (Ω) and the modulated intensity I_m (V, Ω) are collected. The optical bias curve is the intensity profile obtained by varying the retardation generated by the SBC with no applied voltage to the sample, while the modulation data sheet is obtained by applying an AC voltage $V\sin(\omega t)$ to the sample and using a lock-in amplifier synced to the fundamental frequency of the applied voltage to record the resulting modulation of I_{dc} (Ω) for a given retardation. I_{dc} (Ω) can be expressed in terms of the complex reflection coefficients r_s and r_p of the s- and p-polarized waves, the intensity of the incident laser I_0, and the phase retardation Ω. The electric field at the photodetector can be expressed by (Park 2008)

$$\begin{pmatrix} E_x \\ E_y \end{pmatrix} = \frac{E_0}{2}\begin{pmatrix} 1 & 1 \\ 1 & 1 \end{pmatrix}\begin{pmatrix} e^{i\Omega} & 0 \\ 0 & 1 \end{pmatrix}\begin{pmatrix} r_s & 0 \\ 0 & -r_p \end{pmatrix}\begin{pmatrix} \frac{1}{\sqrt{2}} \\ \frac{1}{\sqrt{2}} \end{pmatrix} \tag{2.27}$$

where the first 2×2 matrix represents the crossed polarizer, the second one the SBC, the third the reflectance from the sample, and the last 2×1 matrix the vector for 45° polarized light after the 45° polarizer. The first 2×2 matrix representation for the crossed polarizer doesn't look crossed compared to the input polarization. This is because the y-axis (horizontal) of the coordinate system is flipped after the reflection. For the same reason, the sign of the r_p is the minus in the third matrix for the reflectance from the sample. The measured optical intensity, I, is proportional to the square of the electric field, $|\overline{E}|^2 = E_x^2 + E_y^2$. Varying the optical bias Ω generates the optical bias curve, which is given by (Park 2008)

$$I_{dc} = \frac{I_0}{4}\left|r_s e^{i\Omega} - r_p\right|^2 = A + B\sin^2\left(\frac{\Psi_{sp} + \Omega}{2}\right) \tag{2.28}$$

where $A = \frac{I_0}{4}\left(|r_s|^2 - |r_s|^2\right)$, $B = I_0|r_s||r_p|$, $r_s = |r_s|e^{i\Psi_s}$, $r_p = |r_p|e^{i\Psi_p}$, and $\Psi_{sp} = \Psi_s - \Psi_p$.

To first order in V, the modulated intensity $I_m(V, \Omega)$ is obtained by differentiating (2.28) to get (Park 2008)

$$I_m = \delta A + \delta B \sin^2\left(\frac{\Psi_{sp} + \Omega}{2}\right) + \frac{B}{2} \sin\left(\Psi_{sp} + \Omega\right)\delta\Psi_{sp} \tag{2.29}$$

where $\delta\Psi_{sp} = \frac{I_m\left(\frac{\pi}{2}\right) - I_m\left(\frac{3\pi}{2}\right)}{2I_c}$, and $\frac{\delta B}{B} = \frac{I_m(\pi) - I_m(0)}{2I_c}$. Typically, in the Teng–Man method, I_m is measured only at points 1 and 2 as a function of voltage and the average of the difference in the slopes in the plot of I_m versus V to extract $\delta\Psi_{sp}$. Usually, I_m is positive at point 1 and negative ant point 2 or vice versa. But this is not always the case so care must be taken to note the phase of the reading on the lock-in amplifier, synced to the fundamental frequency of the applied voltage $V\sin(\omega t)$. A similar technique can be applied to obtain $\delta B/B$ from data at points 3 and 4 (Park 2008).

2.7 Thermo-optic Effects

Thermo-optic effect in the materials indicates the change of refractive index as a function of temperature change (dn/dT). It can be used in fabrication of digital optical switch, Mach–Zehnder interferometer type optical switch and optical cross-connect and waveguides. Those optical devices play a key function in communication network because of their control over the optical path by changing the temperature. Although many electro-optic devices have been well known for their fast reaction rate, their polarization change has always been a problem. That is the reason why thermo-optic devices have become of great interest in the application areas because of their advantage of polarization insensitive characteristics. The factors that decide thermo-optic effect (dn/dT) are density and electronic polarizability change with temperature (Kang et al. 2002):

$$\frac{dn}{dT} = \frac{(n^2 - 1)(n^2 + 2)}{6n}(\varphi - \alpha) \tag{2.30}$$

where φ is the temperature coefficient of the electronic polarizability, and α is the thermal expansion coefficient. When the electronic polarizability term is dominant, the refractive index becomes positive and increases with temperature. On the other hand, dn/dT becomes negative and decreases with increasing temperature when thermal expansion term is dominant. In general, dn/dT of inorganic glasses is decided by the competition between φ and α. However, in the case of optical polymers, dn/dT exclusively depends on thermal expansion term because their α is always much higher than their φ. This is the reason why most optical polymers have a negative dn/dT.

Optical waveguides for fabrication of thermo-optic switches have used the materials of silica and optical polymers. The main advantages of silica-based thermo-optic switch are their easier fiber connection, lower optical loss, and better thermal stability. However, it typically requires high switching power of about 0.4–0.5 W and exhibits long response time in the order of *ms* because silica has a low dn/dT of about 10^{-5}/°C. The polymer-based thermo-optic switch overcomes the

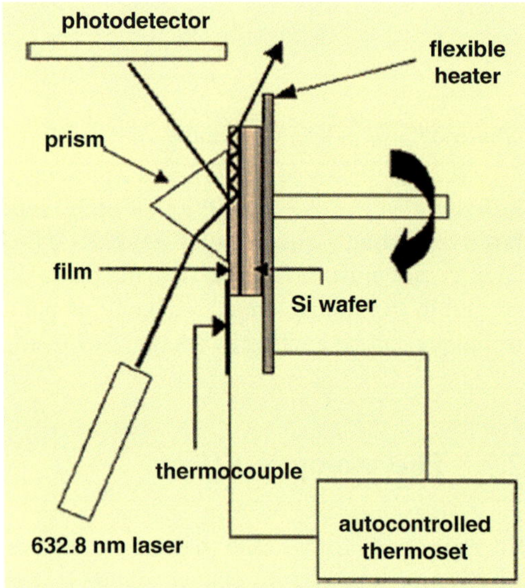

Fig. 2.18 Apparatus for measuring thermo-optic coefficients of the film (Kang et al. 2002)

high switching power of silica-based one due to its higher dn/dT of about -10^{-4}/°C, but then, its thermal stability is another major drawback of practical application. Therefore, sol–gel-derived inorganic–organic hybrid materials composed of silica and organic or polymer has been developed for application of thermo-optic switch in integrated optics. The dn/dT of sol–gel films which are composed of silica and methacrylate group are negative and the order of 10^{-4}/°C. Their high values of dn/dT make them possible in thermo-optic applications. Figure 2.18 shows a schematic diagram of the prism coupler equipped with auto-controlled hot stage to measure the refractive index of films as a function of temperature (Kang et al. 2002). The flexible heater, which is thin enough not to affect the optical coupling between the prism and the film, is used to elevate sample temperature. The film-type thermocouple is attached to sample surface and temperature variation in the prism is neglected over the range of measurement (30–100 °C). Refractive index variation due to film thickness change with temperature can be neglected because it is possible to calculate thickness and refractive index independently using a prism coupler. A He–Ne laser of 632.8 nm wavelength is used as the light source. This method using the prism coupler equipped with an auto-controlled hot stage can provide reliable measurements of dn/dT of the optical waveguide films.

2.8 Acousto-optic Effects

Acousto-optic effects are based on the change of the refractive index of optical waveguide due to the presence of sound waves in that waveguide. These variations in the refractive index, due to the pressure fluctuations, may be detected optically by refraction, diffraction, and interference effects, reflection may also be used. The acousto-optic effect is extensively used in the measurement and study of ultrasonic waves. However, the growing principal area of interest is in acousto-optical devices for the deflection, modulation, signal processing, and frequency shifting of light beams. This is due to the increasing availability and performance of lasers, which have made the acousto-optic effect easier to observe and measure. Technical progresses in both crystal growth and high frequency piezoelectric transducers have brought valuable benefits to acousto-optic components' improvements. It can be used in nondestructive testing, structural health monitoring, and biomedical applications, where optically generated and optical measurements of ultrasound give a noncontact method of imaging.

The acousto-optic interaction is a more general effect of photoelasticity consisting in the change of the material's permittivity ε under the action of a mechanical strain α. Phenomenologically, this effect is described as variations of the optical indicatrix coefficients B_i caused by the strain α_j (MSU 2013):

$$\Delta B_i = p_{ij}\alpha_j \tag{2.31}$$

where p_{ij} is the photoelastic tensor components $i, j = 1, 2, \ldots, 6$.

In case of the acousto-optic effect, the strains α_j are produced by an acoustic wave excited in a transparent waveguide. Thus, the acoustic wave is accompanied by a wave of refractive index variations. For a plane acoustic wave propagating along the axis z, the refractive index variation can be expressed as (MSU 2013)

$$n(z,t) = n + \Delta n \cos\left(2\pi ft - K_z\right) \tag{2.32}$$

where n is the undisturbed refractive index of refraction, f and K_z are the frequency and propagation constant of the acoustic wave, Δn is the amplitude variation in the refractive index generated by the acoustic wave, and is given as

$$\Delta n = -\frac{1}{2}n^3 pa \tag{2.33}$$

The generated refractive index, $n(z, t)$, gives a diffraction grating moving with the velocity given by the speed of the sound wave in the waveguide. Light which then passes through the transparent waveguide material, is diffracted due to this generated refraction index, forming a prominent diffraction pattern. This diffraction pattern corresponds with a conventional diffraction grating at angles θ_m from the original direction, and is given by (Scruby and Drain 1990)

$$\Lambda = \sin\left(\theta_m\right) = m\lambda \tag{2.34}$$

where λ is the wavelength of the optical wave, Λ is the wave length of the acoustic wave and m is the integer order maximum.

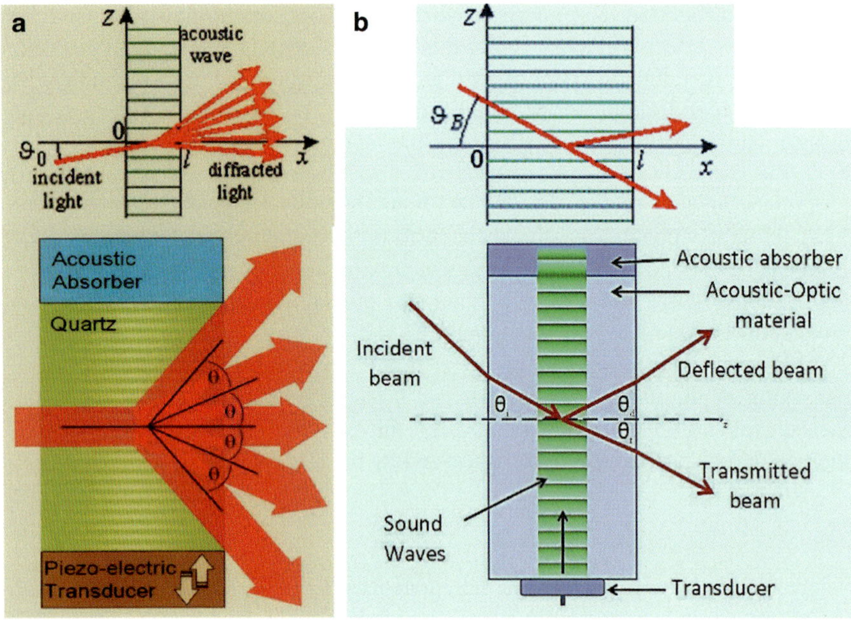

Fig. 2.19 Illustration of acousto-optic effects: (**a**) Raman–Nath diffraction; and (**b**) Bragg diffraction

Light diffracted by an acoustic wave of a single frequency produces two distinct diffraction types. These are Raman–Nath diffraction and Bragg diffraction (Fig. 2.19a, b). Raman–Nath diffraction is observed with relatively low acoustic frequencies, typically less than 10 MHz, and with a small acousto-optic interaction length, ℓ, which is typically less than 1 cm. This type of diffraction occurs at an arbitrary angle of incidence, θ_0. In contrast, Bragg diffraction occurs at higher acoustic frequencies, usually exceeding 100 MHz. The observed diffraction pattern generally consists of two diffraction maxima; these are the zeroth and the first orders. However, even these two maxima only appear at definite incidence angles close to the Bragg angle, θ_{B}. The first order maximum or the Bragg maximum is formed due to a selective reflection of the light from the wave fronts of ultrasonic wave. The Bragg angle is given by the expression (MSU 2013)

$$\sin\theta_{\mathrm{B}} = -\frac{\lambda f}{2n_{\mathrm{i}}\nu}\left[1 + \frac{\nu^2}{\lambda^2 f^2}\left(n_{\mathrm{i}}^2 - n_{\mathrm{d}}^2\right)\right] \tag{2.35}$$

where λ is the wavelength of the incident light wave (in a vacuum), f is the acoustic frequency, ν is the velocity of the acoustic wave, n_{i} is the refractive index for the incident optical wave, and n_{d} is the refractive index for the diffracted optical waves.

In general, there is no point at which Bragg diffraction takes over from Raman–Nath diffraction. It is simply a fact that as the acoustic frequency increases, the number of observed maxima is gradually reduced due to the angular selectivity of the acousto-optic interaction. Traditionally, the type of diffraction, Bragg or Raman–Nath, is determined by the conditions $Q \gg 1$ and $Q \ll 1$ respectively, where Q is given by (MSU 2013)

$$Q = \frac{2\pi\lambda l f^2}{n\nu^2} \tag{2.36}$$

which is known as the Klein–Cook parameter. Since, in general, only the first order diffraction maximum is used in acousto-optic devices, Bragg diffraction is preferable due to the lower optical losses. However, the acousto-optic requirements for Bragg diffraction limit the frequency range of acousto-optic interaction. As a consequence, the speed of operation of acousto-optic devices is also limited.

If the acousto-optic material is optically isotropic, then (2.35) is simplified to (MSU 2013)

$$\sin\theta_{\mathrm{B}} = -\frac{\lambda f}{2n\nu} \tag{2.37}$$

The utilization of laser has led the development of acousto-optics and its applications, mainly for deflection, modulation, and signal processing. Technical progresses in both crystal growth and high frequency piezoelectric transducers have brought valuable benefits to improvements of acousto-optic components. Some materials displaying acousto-optic effect include fused silica, lithium niobate, arsenic trisulfide, tellurium dioxide and tellurite glasses, lead silicate, $Ge_{55}As_{12}S_{33}$, mercury(I) chloride, lead(II) bromide, and other materials. Although acoustic interactions can be observed in liquids, practical devices use crystals or glasses as the interaction medium, with RF frequencies in the MHz to GHz range. Table 2.4 shows standard characteristics of the main materials used in acousto-optics (AA Sa 2013).

2.9 Nonlinear Optic Effects

Nonlinear effects can have a significant impact on light propagation in waveguides. While the nonlinear response of waveguides like optical fibers is relatively weak, tight confinement of light in a micron-size core together with long propagation distances (100 km and more) result in significant nonlinear interactions. Nonlinear effects can be both advantageous and detrimental for waveguide devices and transmission systems. On the one hand, nonlinear effects enable new functionalities and devices, such as switching, supercontinuum generation, wavelength conversion, fiber lasers, amplifiers, logic devices, and soliton transmission systems. At the same time, they may be detrimental for optical communication systems.

Table 2.4 Standard characteristics of the main materials used in acousto-optics (AA Sa 2013)

Material	Type	Optimum optical range for acousto-optic applications (nm)	Incident optical polarization[a]	Refractive index	Max CW laser power density (W/mm^2)	Acoustic velocity (m/s)
Ge	Crystal	2,500–11,000	Linear ‖	4 @ $\lambda = 10{,}600$ nm	5	5,500
Doped glass	Glass	500–650	Unpolarized	2.09 @ $\lambda = 633$ nm	1	3,400
$Ge_{33}As_{12}Se_{55}$	Glass	1,100–1,700	Unpolarized	2.59 @ $\lambda = 1{,}064$ nm	1	2,520
As_2S_3	Glass	700–900	Unpolarized	2.46 @ $\lambda = 1{,}150$ nm	1	2,600
$PbMoO_4$	Crystal	480–1,100	Unpolarized	2.26/2.38 @ $\lambda = 633$ nm	0.5	3,630
TeO_2	Crystal	450–1,100	Linear ⊥	2.26 @ $\lambda = 633$ nm	5	4,200
TeO_2	Crystal	350–4,500	Linear-circular	2.26 @ $\lambda = 633$ nm	5	620
SiO_2 (fused silica)	Glass	200–2,200	Linear ⊥	1.46 @ $\lambda = 633$ nm	>100	5,960
SiO_2 (fused silica)	Glass	200–2,200	Unpolarized	1.46 @ $\lambda = 633$ nm	<100	3,760

[a]‖ and ⊥ means parallel and perpendicular to the acoustic wave direction for optimum acousto-optic coupling

At high intensities, light–matter interactions become nonlinear as a result of harmonic motion of bound electrons in the applied field. As a result, the induced polarization can be written as (Litchinitser and Iakhnine 2011)

$$P(t) = \chi^{(1)}E(t) + \chi^{(2)}E^2(t) + \chi^{(3)}E^3(t) + \cdots \tag{2.38}$$

where $\chi^{(1)}$ is the linear susceptibility, and $\chi^{(2)}$ and $\chi^{(3)}$ are the second- and third-order nonlinear optical susceptibilities. For simplicity, in (2.38) it is assumed that both P and E are scalar, and the medium is dispersionless and lossless. In the most general case, $\chi^{(i)}$ are tensors of rank $j + 1$ and are frequency dependent. The second-order susceptibility, which is responsible for such effects as second-harmonic generation and sum frequency generation, vanishes in optical fibers, owing to the inversion symmetry of silica glass. On the other hand, the third-order susceptibility plays the key role.

Major nonlinear effects in optical fibers include self-phase modulation (SPM), cross-phase modulation (XPM), four-wave mixing (FWM), stimulated Raman scattering (SRS), and stimulated Brillouin scattering (SBS). Next, we briefly discuss the basics of each of these effects.

2.9.1 Self-phase Modulation

SPM is the change in the phase of an intense optical pulse due to the nonlinear change of the refractive index of the material caused by this intense pulse. Such a self-induced phase shift can be written as (Litchinitser and Iakhnine 2011)

$$\phi_{SPM} = \gamma PL \tag{2.39}$$

where $\gamma = \frac{2\pi n_2}{\lambda A_{eff}}$ is the nonlinearity coefficient, n_2 is the nonlinear coefficient related to $\chi^{(3)}$, λ is the wavelength, A_{eff} is the effective core area, P is the peak power, L is the fiber length (assuming lossless case). The SPM results in a spectral broadening of a pulse, which may be disadvantageous for multichannel communication systems, which transmit signals at multiple wavelengths simultaneously. However, the SPM is extremely useful in many applications, including supercontinuum generation, optical regeneration, and soliton systems.

2.9.2 Cross-Phase Modulation

XPM is the nonlinear phase shift of an optical field at a given frequency (or a given polarization) induced by a co-propagating field at another frequency (or another polarization). The effect of the XPM is twice as strong as that of the SPM as can be seen from the following expression (Litchinitser and Iakhnine 2011):

$$\phi_{\mathrm{SPM+XPM}} = \gamma L\left(|E_1|^2 + 2|E_2|^2\right) \tag{2.40}$$

where the first term is responsible for the SPM, and the second one is due to the XPM. One of the major effects of the XPM is asymmetric spectral broadening. The XPM leads to many fundamental nonlinear phenomena such as soliton trapping and finds applications in Kerr shutters and intensity discriminators.

2.9.3 Four-Wave Mixing

FWM is a third-order parametric process that involves nonlinear interaction of four waves. In this process, two photons at frequencies ω_1 and ω_2 are annihilated and create two photons at frequencies ω_3 and ω_4, such that (Litchinitser and Iakhnine 2011)

$$\omega_3 + \omega_4 = \omega_1 + \omega_2 \tag{2.41}$$

The FWM process described by (2.41) relies on a phase-matching condition given by (Litchinitser and Iakhnine 2011)

$$\Delta k = k_3 + k_4 - k_1 - k_2 = 0 \tag{2.42}$$

Condition of (2.42) is easily satisfied in optical fibers for the partially degenerate case when $\omega_1 = \omega_2$. The FWM effect is used in many important fiber applications, including wavelength conversion, parametric amplification, and optical regeneration, among others.

2.9.4 Stimulated Raman Scattering

SRS is a process in which the incident photon is scattered by a molecule to a lower-frequency photon and is accompanied by a transition of the molecule between two vibrational states. The SRS can strongly affect the performance of multichannel transmission systems by transferring the energy between channels. At the same time, the SRS forms a basis of broadband Raman amplifiers and tunable Raman lasers that are of foremost importance for energy restoration in optical transmission systems and have numerous other applications as well (Litchinitser and Iakhnine 2011).

2.9.5 *Stimulated Brillouin Scattering*

SBS is a nonlinear process that originates from the interaction of light with acoustic phonons and results in generation of a backward-propagating Stokes wave downshifted in frequency from the pump wave. The SBS has a relatively low threshold (lower that for the SRS) and can be detrimental for fiber-optic transmission systems. However, the SBS finds applications in Brillouin lasers and amplifiers (Litchinitser and Iakhnine 2011).

2.10 Reliability Evaluation

The reliability of an integrated optical waveguide system indicates the probability that the system will be operational within acceptable limits for a given period of time. Advanced optical waveguides are very complicated systems. The dimensions of these structures are in the micrometer scale and getting smaller. Because of the fine features and large number of parts involved in each device, the probability of system failure is high unless high reliability of each device is ensured. In addition, some optical waveguides are being used in very harsh environments. For instance, under-the-hood applications in automobiles, missiles stored in desert sand, airplanes flying in high attitude where temperatures can drop below negative 60 °C, space stations exposed to strong radiation as well as extreme temperatures, etc. These hash environments have imposed stringent requirements on the reliability of integrated waveguide systems. The waveguide system must be designed to sustain high and low temperature extremes, to survive humid and corrosive surroundings, and to be protected from ultraviolet radiation. The mean time to failure for modern waveguide system may range from several days to several decades at room temperature. Reliability tests cannot be performed for such long durations, but can simulate the real situation based on well-designed, well-understood, and thoroughly implemented accelerated testing. Reliability testing and simulating can be used to determine which failure modes apply to a given part, how probable it is that these failure modes will occur while the part is in service, and how they might be prevented during the design and manufacture of this part. Therefore, identifying and understanding the mechanisms that cause component failure is the key to make a reliable waveguide system.

2.10.1 *Failure Modes and Mechanisms*

Reliability is defined as the probability that a component functions as designed, whereas failure indicates the probability that a component does not perform the function any more. The photonic component failure model form can be expressed as (Naval 2008)

$$\lambda_{\mathrm{P}} = \pi_{\mathrm{Q}}(\lambda_{\mathrm{OB}}\pi_{\mathrm{DCO}}\pi_{\mathrm{TO}}\pi_{\mathrm{V}} + \lambda_{\mathrm{EB}}\pi_{\mathrm{DCN}}\pi_{\mathrm{TE}}\pi_{\mathrm{RH}} + \lambda_{\mathrm{TCB}}\pi_{\mathrm{CR}}\pi_{\mathrm{DT}} + \lambda_{\mathrm{ind}}) \tag{2.43}$$

where λ_{P} is predicted failure rate; π_{Q} is multiplier for quality; λ_{OB} is base failure rate from operational stresses; $\pi_{\mathrm{DCO}} = \dfrac{\mathrm{DC}}{\mathrm{DC}_{10p}}$ is failure rate multiplier for duty cycle. $\pi_{\mathrm{TO}} = \mathrm{e}^{\left(\frac{-\mathrm{Ea_{op}}}{.00008617}\left(\frac{1}{T_{\mathrm{AO}}+T_{\mathrm{R}}+273}-1/298\right)\right)}$ is factor for operating temperature. $\pi_{\mathrm{V}} = \left(\frac{V_{\mathrm{a}}+1}{V_{\mathrm{c}}}\right)^{n_{\mathrm{vib}}}$ is vibration factor; λ_{EB} is base failure rate from environmental stresses. $\pi_{\mathrm{DCN}} = \frac{1-\mathrm{DC}}{1-\mathrm{DC}_{10p}}$; $\pi_{\mathrm{TE}} = \mathrm{e}^{\left(\frac{-\mathrm{Ea_{nonop}}}{.00008617}\left(\frac{1}{T_{\mathrm{AE}}+273}-1/298\right)\right)}$ is nonoperating temperature factor. $\pi_{\mathrm{RH}} = \left(\frac{\mathrm{RH_a}+1}{\mathrm{RH_c}}\right)^{n_{\mathrm{RH}}}$ is humidity factor; $\pi_{\mathrm{CR}} = \frac{\mathrm{CR}}{\mathrm{CR_1}}$ is cycling rate factor; $\pi_{\mathrm{DT}} = \left(\frac{T_{\mathrm{AO}}+T_{\mathrm{R}}-T_{\mathrm{AE}}}{14}\right)^{n_{\mathrm{PC}}}$ is delta temperature factor; and λ_{ind} is failure rate from induced stresses.

The model parameters are defined as follows:

λ_{P} is predicted failure rate, failure per million calendar hours
π_{Q} is failure rate multiplier for quality
λ_{OB} is base failure rate, operating
DC is duty cycle (fraction of calendar time in operation)
$\mathrm{DC_{10p}} = 0.25$
π_{TO} is failure rate multiplier, temperature-operating
$\mathrm{Ea_{op}}$ is active energy-operating
T_{AO} is ambient operating temperature
T_{R} is the temperature rise above T_{AO}
π_{V} is failure rate multiplier, vibration level
V_{a} is Max vibration level applied (Grms)
$V_{\mathrm{c}} = 1$
n_{vib} is vibration exponent
λ_{EB} is base failure rate, environment
π_{DCN} is failure multiplier, duty cycle-nonoperating
π_{TE} is failure rate multiplier, temperature-environment
$\mathrm{Ea_{nonop}}$ is activation energy, nonoperating
T_{AE} is ambient environmental temperature
π_{RH} is failure rate multiplier, relative humidity
$\mathrm{RH_a}$ is relative humidity (%)
$\mathrm{RH_c} = 50\ \%$
n_{RH} is relative humidity exponent
λ_{TCB} is base failure rate, temperature cycling
π_{CR} is failure rate multiplier, cycling rate
CR = cycling rate (cycles per year)

$CR_1 = 1{,}000$
π_{DT} is failure rate multiplier, delta temperature
n_{PC} is temperature cycling exponent

By utilizing this model form, factors that account for the application and component-specific variables that affect reliability can be applied to appropriate additive failure rate term.

2.10.2 *Reliability Qualifications*

To ensure product reliability, extensive reliability tests need to be performed during product development and before a new product can be shipped. Commonly accepted accelerated tests include thermal cycling and thermal shock, steady state thermal soaking, mechanical vibration, voltage extremes and power cycling, high humidity and high pressure, and combination of the above.

2.10.2.1 Thermal Cycling and Thermal Shock

The single, dual, and triple chamber can be used for temperature cycling. In single chamber cycling, the load is placed in a stationary chamber, and is heated or cooled by introducing hot or cold air into the chamber. In dual chamber cycling, the load is placed on a moving platform that shuttles between stationary chambers maintained at fixed temperatures. In triple chamber temperature cycling, there are three chambers and the load is moved between them. This test is conducted to determine the ability of electronic packaging and thermal cooling system, including components and solder interconnects, to withstand mechanical stresses induced by alternating high and low temperature extremes. Permanent changes in electrical and/or physical characteristics can result from these mechanical stresses (JESD22-A104C). Typical testing conditions can be found from JESD22-A104C.

The thermal shock test is conducted to determine the resistance of a part to sudden exposure to extreme changes in temperature and to the effect of alternate exposures to these extremes. The conditions and recommended fluids can be found from JESD22-A106B.

Thermal cycling is an excellent test method for first order reliability assessment. However, it is not good for products with closely matched thermal coefficients of expansion because second-order effects, such as out-of-plane warpage, are not duplicated in this test method.

Comparability, power cycling or powered functional cycling is similar to thermal cycling, but more closely matches the actual service condition. It consists of a power-on transient in which the components go from room temperature to some steady state temperature in a matter of minutes. This steady state condition then lasts for several hours or longer before a power-off (cooling) transient occurs. The power-off condition then lasts for several hours. The transient conditions are difficult to determine either experimentally or analytically. For this test, the same support structure should be modeled as closely as possible. The same environment conditions, such as cooling air flow, should be duplicated. With such care this approach will yield highly accurate reliability information. However, this method is more difficult, expensive, and time consuming.

2.10.2.2 Steady State Temperature–Humidity Bias Life Test

The steady state temperature–humidity bias life test is performed for the purpose of evaluating the reliability of non-hermetic packaged electronic devices in humid environments. It employs conditions of temperature, humidity, and bias which accelerate the penetration of moisture through the external protective material (encapsulant or seal) or along the interface between the external protective material and the metallic conductors which pass through it. The test requires a temperature–humidity test chamber capable of maintaining a specified temperature and relative humidity continuously, while providing electrical connections to the devices under test in a specified biasing configuration. Test conditions consist of a temperature, relative humidity, and duration used in conjunction with an electrical bias configuration specific to the device. The temperature of 85 ± 2 °C (dry bulb) and relative humidity of 85 ± 5 % should be applied to the entire useable test area. The wet bulb temperature is 81.0 °C, and vapor pressure is 49.1 kPa, and the duration of 1,000 (−24, +168) h applied continuously except during any interim readouts. The detail testing conditions can be found from EIA/JESD22-A101-B.

2.10.2.3 Mechanical Vibration

The vibration test is usually conducted in accordance with MIL-STD-167B. The vibration frequency is swept from 4 to 22 Hz. MIL-STD-167B requires an exploratory vibration test (10-min resonance survey sweep) a variable frequency test (5-min dwell at each frequency) and a 2-h endurance test at the resonant frequency.

Vibrating tables or shake tables are usually used to test electronic packaging and thermal cooling systems to determine or demonstrate their ability to withstand vibration. These machines are capable of producing three different types of vibration profile: sine sweep, random vibration, and synthesized shock. The part under test will typically be instrumented with one or more accelerometers to measure how the component responds to the vibration input. A sine sweep vibration profile, for instance, typically starts vibrating at low frequency and increases in frequency at a set rate (measured in hertz). The amplitude (measured in gs) of the sine wave may increase or decrease as well. A sine sweep is intended to look for resonant frequencies in the part. A random vibration profile will excite different frequencies along a spectrum at different times. Significant calculation goes into making sure that all frequencies get excited to within an acceptable tolerance band. A random

Table 2.5 Telecordia GR1209 required environmental and mechanical tests

Test	Condition
Temperature–humidity aging	85 °C/85 % RH for 14 days
Temperature–humidity cycling	−40 °C to 75 °C, 10–80 % RH, 42 cycles, 14 days
Water immersion	43 °C, pH 5.5, 7 days
Vibration	10–55 Hz, 1.52 mm, 2 h
Flex test	1 lb load, 100 cycles
Twist test	1 lb load, 10 cycles
Side pull	0.5–1 lb load, 90° angle, active
Cable retention	1.2–2.2 lb load for 1 month
Impact test	6 ft. drop, 8 cycles, 3 axes

Table 2.6 Telecordia GR1221 required environmental and mechanical tests

Test	Condition
High temperature storage (damp heat)	75 °C/90 % RH for 2,500 h
High temperature storage (dry heat)	85 °C, <40 % RH for 2,500 h
Low temperature storage	−40 °C, uncontrolled RH for 2,500 h
Temperature cycling	−40 to 75 °C, dwell $\geq$ 15 min, ramp @ 1 °C/min, 500 cycles
Temperature–humidity cycling	25 °C/uncontrolled RH to 75 °C/90 %, 4–16 h dwell, ramp @ 2 °C/min, 5 cycles
Thermal shock	T = 100 °C, dwell $\geq$ 30 min, transfer 2 min, 20 cycles
Salt spray	T = 35 °C, 5 % NaCl dissolved in 95 % H_2O, 168 h
Water immersion	43 °C, pH 5.5, 340 h

vibration test can be anything as short at 30 s up to several hours. It is intended to synthesize the effect of, for example, a car driving over rough terrain or a rocket taking off. A synthesized shock pulse is a short duration high level vibration calculated as a sum of many half-sine waves covering a range of frequencies. It is intended to simulate the effects of an impact or explosion. A shock pulse test typically lasts less than a second.

2.10.2.4 GR1209 and GR1221

Requirements for telecommunication network devices have been developed with the goal of assuring operation for 20–25 years. The major governing documents for passive optical components are two monographs from Telecordia, the GR1209 and GR1221 (Ma et al 2002). The Telecordia GR1209 focus primarily on optical performance tests and short-term reliability data (several weeks) that would apply to any manufactured lot of devices (Table 2.5, GR1209-1994). The Telecordia GR1221 is a document that focuses on comprehensive reliability assurance, in particular vendor and device qualification, lot-to-lot quality and reliability control, storage and handling (Table 2.6, GR1221-1994).

References

AA Sa (2013) Acousto-optic materials. http://www.acoustooptic.com/acousto-optic-material.html. Accessed 10 March 2013

Chien ST (2004) Waveguide fabrication using proton beam writing. PhD dissertation, National University of Singapore, Singapore

Christensen D et al (1992) Comparison of robust coupling techniques for planar waveguide immunosensors. Proc SPIE 1796:20–25

Dutton HJR (1998) Understand optical communications. IBM, Research Triangle Park

Edmund (2013) Understanding optical specifications. http://www.edmundoptics.com/technical-resources-center/optics/understanding-optical-specifications/?&pagenum=3. Accessed 26 Feb 2013

Elton NJ (2007) The complex refractive index and reflectometry versus ellipsometry. Reflectometry Technical Paper No. 3. http://www.Surfoptic.com. Accessed 1 Feb 2009

Gang SY (2005) Polymer based optical waveguides. Master's degree thesis, Universiti Teknologi Malaysia, Johor Bahru

Generic reliability assurance requirement for fiber branching components, Bellcore GR-1209-CORE, Bellcore, NJ, 1994

Generic reliability assurance requirements for fiber optic branching components, Bellcore GR-1221-CORE, Bellcore, NJ, 1994

Goering R, Rothhardt M (1986) Application of the refracted-near-field technique to multimode planar and channel waveguides in glass. J Opt Commun 7:82–85

Jung J et al (2004) Ellipsometry. Aalborg University, Aalborg

Kang ES, Lee TH, Bae BS (2002) Measurement of the thermo-optic coefficients in sol-gel derived inorganic-organic hybrid material films. Appl Phys Lett 81(8):1438–1440

Kaye & Laby (2012) Tables of physical & chemical constants. 2.5.11 Electro-optic materials. Kaye & Laby Online. Version 1.0. http://www.kayelaby.npl.co.uk/general_physics/2_5/2_5_11.html. Accessed 6 Dec 2012

Litchinitser N, Iakhnine V (2011) Optical waveguides: numerical modeling. http://optical-waveguides-modeling.net. Accessed 16 March 2012

Ma H et al (2002) Polymer-based optical waveguides: materials, processing, and devices. Adv Mater 14(19):1339–1365

MSU (2013) Acousto-optic effect. http://www.mt-berlin.com/frames_ao/descriptions/ao_effect.htm. Accessed 6 Feb 2013

Naval (2008) Photonic component and subsystem reliability process. Naval Air Systems Command Control Number: 08-1125

Park DH (2008) Characterization of linear electro-optic effect of poled organic thin films. PhD dissertation, University of Maryland, College Park

Scruby CB, Drain LE (1990) Laser ultrasonics: techniques and applications. Taylor & Francis, London

Tyszkieiwcz C et al (2005) Determination of the refractive index of the SE1211 resin using an SPR spectroscopy. Mol Quantum Acoust 26:267–271

Wu YC (2005) Planar waveguides of Y2O3, Y2O3:Tb3+ and YAG prepared by sol-gel: analysis, structure and optical. PhD dissertation, Claude Bernard University, Lyon

Yin L (2009) Study of nonlinear optical effects in silicon waveguides. PhD dissertation, University of Rochester, Rochester

Chapter 3
Optoelectronic Devices Integrated with Optical Waveguides

Abstract Optoelectronics is defined as the combined use of optical and electronic devices especially involving the interchange of visual and electronic signals. Integration of multiple optical functionalities onto planar substrates is being intensively pursued for a broad range of applications. Progress and success in the optoelectronic devices, such as fundamental transverse-mode laser diodes, low loss optical waveguides, and high quantum efficiency photodetectors, comprise a new and powerful class of monolithic integrated circuits for optical communications. Potential advantages of integrating optoelectronic devices on a single chip, as opposed to using hybrid optical and electronic systems, include increased reliability and reduced size of the circuitry and the reduction in unit cost due to mass production. Moreover, replacing the electrical connections between components with optical connectors increases the available bandwidth of the system and improves immunity to noise from electromagnetic interference. This chapter will give a brief overview about the optoelectronic devices and their integration.

3.1 Optoelectronic Theory and Demonstration

Optoelectronic devices are electrical-to-optical or optical-to-electrical transducers, or instruments that use such devices in their operation to source, detect, and control light. Optoelectronic theory is based on the quantum mechanical effects of light on semiconducting materials, sometimes in the presence of electric fields. P–N junctions are an integral part of several optoelectronic devices, such as photodiodes, solar cells, light emitting diodes (LEDs) and semiconductor lasers.

A large number of optoelectronic devices consist of a P-type and N-type region, just like a regular P–N diode. The key difference is that there is an additional interaction between the electrons and holes in the semiconductor and light. This interaction is not restricted to optoelectronic devices. Regular diodes are also known to be light sensitive and in some cases also emit light. The key difference is that optoelectronic devices including photodiodes, solar cells, LEDs, and laser diodes are

X.C. Tong, *Advanced Materials for Integrated Optical Waveguides*, Springer Series in Advanced Microelectronics 46, DOI 10.1007/978-3-319-01550-7_3,

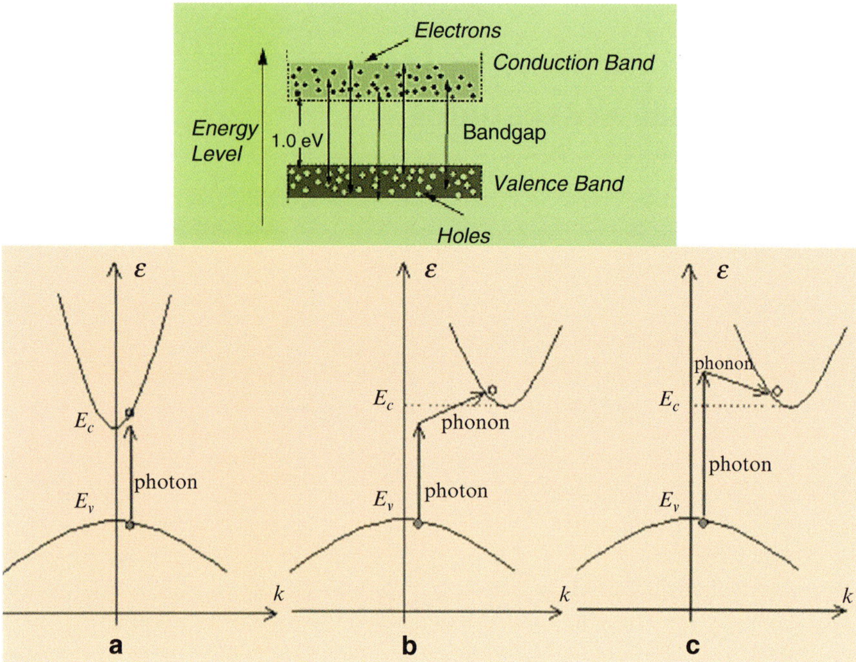

Fig. 3.1 *E–k* diagram illustrating (**a**) photon absorption in a direct bandgap semiconductor (**b**) photon absorption in an indirect bandgap semiconductor assisted by phonon absorption and (**c**) photon absorption in an indirect bandgap semiconductor assisted by phonon emission

specifically designed to optimize the light absorption and emission, resulting in a high conversion efficiency. Light absorption and emission in a semiconductor is known to be heavily dependent on the detailed band structure of the semiconductor. Direct bandgap semiconductors, i.e., semiconductors for which the minimum of the conduction band occurs at the same wavevector, *k*, as the maximum of the valence band, have a stronger absorption of light as characterized by a larger absorption coefficient. They are also the favored semiconductors when fabricating light emitting devices. Indirect bandgap semiconductors, i.e., semiconductors for which the minimum of the conduction band does not occur at the same wavevector as the maximum of the valence band, are known to have a smaller absorption coefficient and are rarely used in light emitting devices. This striking difference is further illustrated with Fig. 3.1 and can be explained based on the energy and momentum conservation required in the electron–photon interaction (Zeghbroeck 2011). The direct bandgap semiconductor, which has a vertically aligned conduction and valence band, is shown in Fig. 3.1a. Absorption of a photon is obtained if an empty state in the conduction band is available for which the energy and momentum equals that of an electron in the valence band plus that of the incident photon. Photons have little momentum relative of their energy since they travel at the speed of light.

The electron therefore makes an almost vertical transition on the *E–k* diagram. For an indirect bandgap semiconductor, the conduction band is not vertically aligned to the valence band as shown in Fig. 3.1b. Therefore a simply interaction of an incident photon with an electron in the valence band will not provide the correct energy and momentum corresponding to that of an empty state in the conduction band. As a result absorption of light requires the help of another particle, namely a photon. Since a phonon, i.e., a particle associated with lattice vibrations, has a relatively low velocity close to the speed of sound in the material, it has a small energy and large momentum compared to that of a photon. Conservation of both energy and momentum can therefore be obtained in the absorption process if a phonon is created or an existing phonon participates. The phonon assisted absorption processes are illustrated with Fig. 3.1b, c. Figure 3.1b illustrates the absorption of a photon aided by the simultaneous absorption of a phonon, while Fig. 3.1c depicts the absorption of a photon, which results in the emission of a phonon. The minimum photon energy that can be absorbed is slightly below the bandgap energy in the case of phonon absorption and has to be slightly above the bandgap energy in the case of phonon emission. Since the absorption process in an indirect bandgap semiconductor involves a phonon in addition to the electron and photon, the probability of having an interaction take place involving all three particles will be lower than a simple electron–photon interaction in a direct bandgap semiconductor. As a result it is found that absorption is much stronger in a direct bandgap material. Similarly, in the case of light emission, a direct bandgap material is also more likely to emit a photon than an indirect bandgap material. While indirect bandgap materials are occasionally used for some LEDs, they result in a low conversion efficiency. Direct bandgap materials are used exclusively for semiconductor laser diodes (Zeghbroeck 2011).

Photodiodes and crystalline solar cells are essentially the same as the P–N diodes. However, the diode is exposed to light, which yields a photocurrent in addition to the diode current so that the total diode current is given by (Zeghbroeck 2011):

$$I = I_s\left(e^{\frac{V_a}{V_t}} - 1\right) - I_{ph} \tag{3.1}$$

where the additional photocurrent, I_{ph}, is due to photogeneration of electrons and holes. These electrons and holes are pulled into the region where they are majority carriers by the electric field in the depletion region. The photo-generated carriers cause a photocurrent, which opposes the diode current under forward bias voltage. Therefore, the diode can be used as a photodetector—using a reverse or even zero bias voltage—as the measured photocurrent is proportional to the incident light intensity. The diode can also be used as a solar cell—using a forward bias to generate electrical power. The primary characteristics of a photodiode are the responsivity, the dark current, and the bandwidth. The responsivity is the photocurrent divided by the incident optical power. The maximum photocurrent in a photodiode can be expressed as (Zeghbroeck 2011):

$$I_{\mathrm{ph,\,max}} = \frac{q}{h\nu} P_{\mathrm{in}} \tag{3.2}$$

where P_{in} is the incident optical power. This maximum photocurrent occurs when each incoming photon creates one electron–hole pair, which contributes to the photocurrent. The maximum photocurrent in the presence of a reflection, R at the surface of the photodiode and an absorption over a thickness d, in a material with an absorption coefficient, α, is given by (Zeghbroeck 2011):

$$I_{\mathrm{ph}} = (1 - R)\left(1 - \mathrm{e}^{-\alpha d}\right) \frac{qP_{\mathrm{in}}}{h\nu} \tag{3.3}$$

This photocurrent is obtained by integrating the generation rate over the absorption region with thickness, d. The photocurrent is further reduced if photo-generated electron–hole pairs recombine within the photodiode instead of being swept into the regions where they are majority carriers. The dark current is the current through the diode in the absence of light. This current is due to the ideal diode current, the generation/recombination of carriers in the depletion region and any surface leakage, which occurs in the diode. The dark current obviously limits the minimum power detected by the photodiode, since a photocurrent much smaller than the dark current would be hard to measure. However, the true limitation is the shot noise generated by the current through the diode. The shot noise as quantified by the average of the square of the noise current is given by (Zeghbroeck 2011):

$$\left\langle i^2 \right\rangle = 2qI\Delta f \tag{3.4}$$

where I is the diode current and Δf is the bandwidth of the detector. The bandwidth of the diode is affected by the transit time of the photo-generated carriers through the diode and by the capacitance of the diode. The carrier transit time yields the intrinsic bandwidth of the diode while the capacitance together with the impedance of the amplifier or the transmission line connected to the diode yields a the parasitic RC delay.

Solar cells are typically illuminated with sunlight and are intended to convert the solar energy into electrical energy. The solar energy is in the form of electromagnetic radiation. The sun spectrum is consistent with that of a black body at a temperature of 5,800 K. The radiation spectrum has a peak at 0.8 eV. A significant part of the spectrum is in the visible range of the spectrum (400–700 nm). The power density is approximately 100 mW/cm^2. Only part of the solar spectrum actually makes it to the earth's surface. Scattering and absorption in the earth's atmosphere, and the incident angle affect the incident power density. Therefore, the available power density depends on the time of the day, the season, and the latitude of a specific location. Of the solar light, which does reach a solar cell, only photons with energy larger than the energy bandgap of the semiconductor generate electron–hole pairs. In addition, the voltage across the solar cell at the point where it delivers its maximum power is less than the bandgap energy in electron volt. The overall power-conversion efficiency of single-crystalline solar cells ranges from 10 to 30 % yielding 10–30 mW/cm^2 (Zeghbroeck 2011).

LEDs are P–N diodes in which the recombination of electrons and holes yields a photon. This radiative recombination process occurs primarily in direct bandgap semiconductors where the lowest conduction band minimum and the highest valence band maximum occur at $k = 0$, where k is the wavenumber. Examples of direct bandgap semiconductors are GaAs, InP, and GaN while most group IV semiconductors including Si, Ge, and SiC are indirect bandgap semiconductors. The radiative recombination process is in competition with non-radiative recombination processes such as trap-assisted recombination. Radiative recombination dominates at high minority-carrier densities. Using a quantum well (QW), a thin region with a lower bandgap, positioned at the metallurgical junction, one can obtain high carrier densities at low current densities. These quantum well LEDs have high internal quantum efficiency as almost every electron injected in the quantum well recombines with a hole and yields a photon. The external quantum efficiency of planar LEDs is much lower than unity due to total internal reflection. As the photons are generated in the semiconductor, which has a high refractive index, only photons traveling normal to the semiconductor–air interface can exit the semiconductor. For GaAs with a refractive index of 3.5, the angle for total internal reflection equals 17° so that only a few percent of the generated photons can escape the semiconductor. This effect can be avoided by having a spherical semiconductor shape, which ensures that most photons travel normal to the interface. The external quantum efficiency can thereby be increased to values larger than 50 % (Zeghbroeck 2011).

Laser diodes are very similar to LEDs since they also consist of a P–N diode with an active region where electrons and holes recombine resulting in light emission. However, a laser diode also contains an optical cavity where stimulated emission takes place. The laser cavity consists of a waveguide terminated on each end by a mirror. Photons, which are emitted into the waveguide, can travel back and forth in this waveguide provided they are reflected at the mirrors. The light in the waveguide is amplified by stimulated emission. Stimulated emission is a process where a photon triggers the radiative recombination of an electron and hole thereby creating an additional photon with the same energy and phase as the incident photon. The stimulated emission process yields an increase in photons as they travel along the waveguide. Combined with the waveguide losses, stimulated emission yields a net gain per unit length, g. The number of photons can therefore be maintained if the roundtrip amplification in a cavity of length, L, including the partial reflection at the mirrors with reflectivity R_1 and R_2 equals unity. This yields the following lasing condition (Zeghbroeck 2011):

$$\text{Roundtrip amplification} = \mathrm{e}^{2gL} R_1 R_2 \tag{3.5}$$

If the roundtrip amplification is less than one, then the number of photons steadily decreases. If the roundtrip amplification is larger than one, the number of photons increases as the photons travel back and forth in the cavity. The gain required for lasing therefore equals (Zeghbroeck 2011):

$$g = \frac{1}{2L} \ln \frac{1}{R_1 R_2} \tag{3.6}$$

Initially, the gain is negative if no current is applied to the laser diode as absorption dominates in the waveguide. As the laser current is increased, the absorption first decreases and the gain increases. The current for which the gain satisfies the lasing condition is the threshold current of the laser, I_{th}. Below the threshold current very little light is emitted by the laser structure. For an applied current larger than the threshold current, the output power, P_{out}, increases linearly with the applied current, as each additional incoming electron–hole pair is converted into an additional photon. The output power therefore equals (Zeghbroeck 2011):

$$P_{out} = \eta \frac{h\nu}{q} (I - I_{th}) \tag{3.7}$$

where $h\nu$ is the energy per photon. The factor, h, indicates that only a fraction of the generated photons contribute to the output power of the laser as photons are partially lost through the other mirror and throughout the waveguide.

In optics on the other hand, wave propagation is the fundamental physical basis and no lumped elements exist up to now. As a consequence, the design of optical and photonic components usually includes optical waveguides and element dimensions are always large with respect to the wavelengths in optics. For instance, in microwave photonic components an interaction between electrons, holes, electrical fields, and photons take place which can be regarded as microwave-optical interactions. Consequently, different technologies meet and may—in a synergetic mixture—be merged in order to develop novel concepts with great advantages.

3.2 Light Emission Devices

Light emission can take place either spontaneously or it can be stimulated by the presence of another photon of the right energy level. Spontaneous emission is really the normal case. When an electron is elevated to a high energy state this state is usually unstable and the electron will spontaneously return to a more stable state very quickly (within a few picoseconds) emitting a photon as it does so. When light is emitted spontaneously its direction and phase will be random but the wavelength will be determined by the amount of energy that the emitting electron must give up. Stimulated emission is what happens in the operation of a laser. In some situations when an electron enters a high energy (excited) state it is able to stay there for a relatively long time (a few microseconds) before it changes state spontaneously. When an electron is in this metastable high energy state it can be stimulated by the presence of a photon of light to emit its energy in the form of another photon. In this case the incident photon must have the right energy (wavelength) within quite small limits. When stimulated emission takes place the emitted photon has exactly the same wavelength, phase, and direction as that of the photon which stimulated it. Depending on the particular material in which the electron is found, it could be bound within a molecule or relatively free to move through the material. For spontaneous or stimulated emission to occur, energy must be supplied to boost the electron from its low energy state to a higher energy state. The energy can come from many sources, such as heat, electrical discharge, electric current, chemical reaction, biological reactions (bioluminescence), absorption and emission of light, and nuclear radiation (Dutton 1998).

3.2.1 Light Emitting Diodes (LEDs)

The wavelength of light emitted by the LED is inversely proportional to the bandgap energy. The higher the energy the shorter the wavelength. The formula relating electron energy to wavelength is given below (Dutton 1998):

$$\lambda = \frac{hc}{\varepsilon_{\mathrm{ph}}} = \frac{1.24}{\varepsilon_{\mathrm{ph}}} \tag{3.8}$$

where λ is wavelength in microns, h is Planck's constant $= 6.63 \times 10^{-34} = 4.14 \times 10^{-15}$ eV/s, c is speed of light (3×10^{8} m/s), $\varepsilon_{\mathrm{ph}}$ is photon energy in electron volt. This means that the materials of which the LED is made determine the wavelength of light emitted. Table 3.1 shows energies and wavelengths for commonly used materials in semiconductor LEDs and lasers (Dutton 1998). For some materials, ranges of energies and wavelengths are given, which is because the materials can be mixed in different proportions (within some limits) and get different bandgap energies. In addition, the level of dopant used is very important in determining the amount of power that can be produced and also has some effect on the wavelength. Typically the junction area has quite a high level of doping. Every time an electron recombines with a hole one photon is emitted. This means that the amount of optical energy (power) produced is equal to the number of electrons that recombine multiplied by the energy of the bandgap. The output power is directly proportional to the drive current (multiplied by some efficiency factor <1). In addition, LEDs can produce a range of wavelengths. Typically, the wavelength range is about 80 nm or so.

Figure 3.2 shows two configurations of practical LEDs. The surface-emitting LED (SLED) operates at 850 nm wavelength and the edge emitter could typically operate in the 1,310 nm region, determined by the materials used. In both types of LED a combination of insulating materials and junctions is used to (a) guide the current flow to a small active region, and (b) guide the light produced out of the device and into an easy position for coupling to a waveguide. LEDs do not produce a single light wavelength but rather a band of wavelengths. The range (or band) of wavelengths is typically about 0.05 of the wavelength (50–100 nm). The band can be reduced (and dispersion reduced) by using selective filters to

Table 3.1 Bandgap energy and possible wavelength ranges in various materials (Dutton 1998)

Material	Formula	Wavelength range λ (μm)	Bandgap energy W_g (eV)
Indium phosphide	InP	0.92	1.35
Indium arsenide	InAs	3.6	0.34
Gallium phosphide	GaP	0.55	2.24
Gallium arsenide	GaAs	0.87	1.42
Aluminum arsenide	AIAs	0.59	2.09
Gallium indium phosphide	GaInP	0.64–0.68	1.82–1.94
Aluminum gallium arsenide	AIGaAs	0.8–0.9	1.4–1.55
Indium gallium arsenide	InGaAs	1.0–1.3	0.95–1.24
Indium gallium arsenide phosphide	InGaAsP	0.9–1.7	0.73–1.35

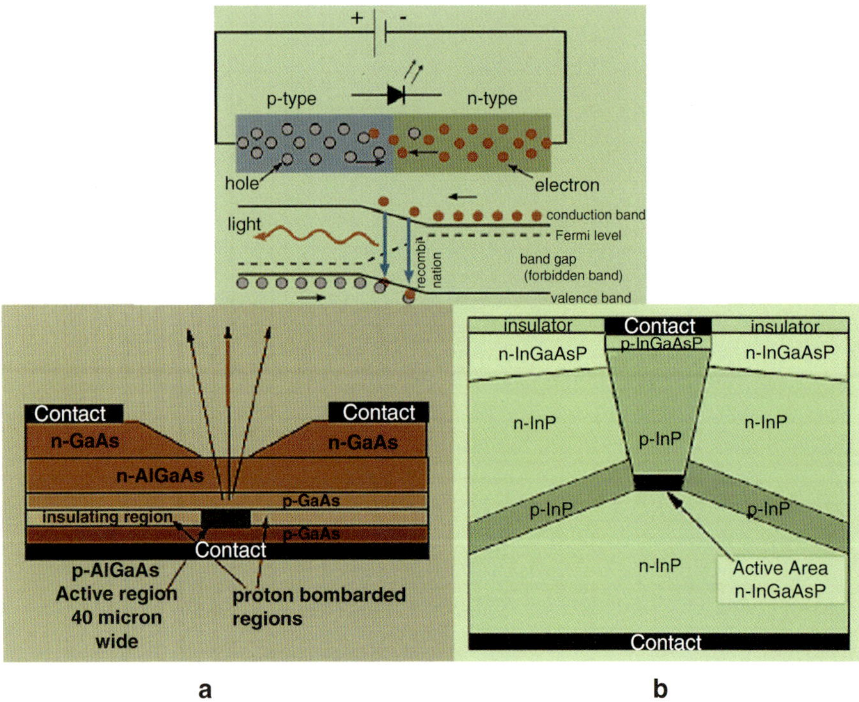

Fig. 3.2 Schematic structures of (**a**) a surface-emitting LED; and (**b**) an edge-emitting LED (Dutton 1998)

produce a narrow band of wavelengths. However, this reduces the power of the signal too. The light produced is neither directional nor coherent. A lens is needed to focus the light onto the end of a waveguide. LEDs are not suitable for use with single-mode fiber for this reason. It is too hard to get the light into the narrow core (Dutton 1998).

LEDs cannot produce pulses short enough to be used at gigabit speeds. However, systems using LEDs operate quite well at speeds of up to around 300 Mbps. LEDs can also be analog modulated quite simply by maintaining a forward bias just larger than the bandgap energy since the device response is linear with current flow. This is one advantage over lasers. While lasers can be analog modulated and are indeed used this way in some commercial situations, this is not an easy thing to do (Dutton 1998).

3.2.2 *Lasers*

A laser is a device that emits light (electromagnetic radiation) through a process of optical amplification based on the stimulated emission of photons. LASER is an acronym for "Light Amplification by the Stimulated Emission of Radiation." Ideal laser light is single wavelength only. This is related to the molecular characteristics of the material being used in the laser. It is formed in parallel beams and is in a single phase. That is, it is coherent. Lasers can be modulated (controlled) very precisely. Lasers can produce relatively high power. Indeed some types of laser can produce kilowatts of power. In communication applications, semiconductor lasers of power up to about 20 mW are available. This is many times greater power than LEDs can generate. Because laser light is produced in parallel beams, a high percentage (50–80 %) can be transferred into the waveguide. However, lasers have been quite expensive by comparison with LEDs. One of the things that causes lasers to have a high cost is that for lasers used in long distance applications temperature control and output power control is needed. Temperature control maintains a stable lasing threshold and power control ensures that the detector can see a stable signal. Both of these require added cost. A thermoelectric cooler/heater is normally used to keep the temperature stable. This also requires a thermistor to measure the device temperature. To control power a P–N diode detector is usually packaged with the laser to measure the power being produced and (via a feedback loop) control the laser's bias current.

A typical laser consists of a gain medium, a laser pumping mechanism to supply energy to it, high reflector, output coupler, and the generated laser beam. The gain medium is excited by an external source of energy into an excited state. In most lasers this medium consists of population of atoms which have been excited into such a state by means of an outside light source, or an electrical field which supplies energy for atoms to absorb and be transformed into their excited states. The gain medium of a laser is normally a material of controlled purity, size, concentration, and shape, which amplifies the beam by the process of stimulated emission. This material can be of any state: gas, liquid, solid, or plasma. The gain medium absorbs pump energy, which raises some electrons into higher energy quantum states. With different gain medium materials, a multitude of different types of lasers have been developed. Figure 3.3 shows some commercially available lasers and their wavelengths (Weber 1999).

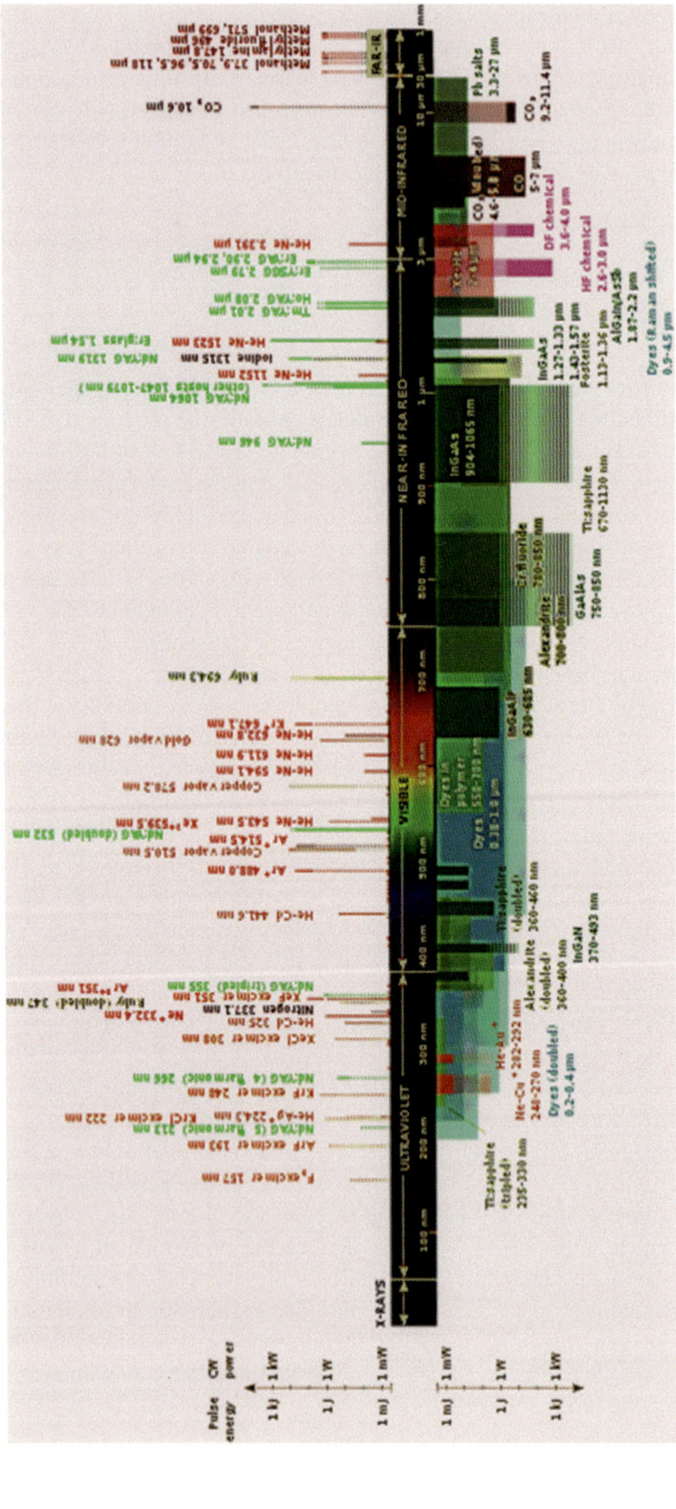

Fig. 3.3 Wavelengths of commercially available lasers. Laser types with distinct laser lines are shown above the wavelength bar, while below are shown lasers that can emit in a wavelength range. The color codifies the type of laser material (see the figure description for more details) (Weber 1999)

1. Semiconductor Laser Diodes
   Semiconductor lasers are diodes which are electrically pumped. Recombination of electrons and holes created by the applied current introduces optical gain. Reflection from the ends of the crystal form an optical resonator, although the resonator can be external to the semiconductor in some designs. Commercial laser diodes emit at wavelengths from 375 to 3,500 nm. Low to medium power laser diodes are used in laser printers and CD/DVD players. Laser diodes are also frequently used to optically pump other lasers with high efficiency. The highest power industrial laser diodes, with power up to 10 kW (70 dBm), are used in industry for cutting and welding. External-cavity semiconductor lasers have a semiconductor active medium in a larger cavity. These devices can generate high-power outputs with good beam quality, wavelength-tunable narrow-linewidth radiation, or ultrashort laser pulses.
   Vertical cavity surface-emitting lasers (VCSELs) are semiconductor lasers whose emission direction is perpendicular to the surface of the wafer. VCSEL devices typically have a more circular output beam than conventional laser diodes, and potentially could be much cheaper to manufacture. Quantum cascade lasers are semiconductor lasers that have an active transition between energy sub-bands of an electron in a structure containing several quantum wells. Silicon laser has been developed for integrated circuits, and so electronic and silicon photonic components (such as optical waveguides/interconnects) could be fabricated on the same chip. The most common communication laser is called the Fabry–Perot laser. In many situations this type of laser gives good service. However, because it produces a relatively wide spectral width, it is not considered suitable for applications requiring extended distances, coherent reception, or wavelength multiplexing. The Fabry–Perot laser can be modified by placing something in the cavity that will disperse unwanted frequencies before they reach the lasing threshold. There are a number of alternatives, but a common way is to place a diffraction grating within the cavity. When this is done, the laser can produce a very narrow spectral linewidth. Lasers using this principle are called Distributed Feedback (DFB) or Distributed Bragg Reflector (DBR) lasers. Other lasers use a cavity external to the device itself—these are called external-cavity lasers. This allows a long cavity, and if you put a diffraction grating on one of the end mirrors a very narrow linewidth indeed can be obtained.

The Fabry–Perot laser is conceptually just an LED with a pair of end mirrors. The mirrors are needed to create the right conditions for lasing to occur. A laser is considered to be "Fabry–Perot" when it has a relatively short cavity (in relation to the wavelength of the light produced). Wavelengths produced are related to the distance between the mirrors by the following formula (Dutton 1998):

$$C_l = \frac{\lambda x}{2n} \tag{3.9}$$

where λ is wavelength, C_l is length of the cavity, x is an arbitrary integer—1, 2, 3, 4, ... n is refractive index of active medium. Since the speed of propagation in

Fig. 3.4 Directing the light in a Fabry–Perot laser (Dutton 1998): (**a**) unguided; (**b**) grain guided; and (**c**) index guided

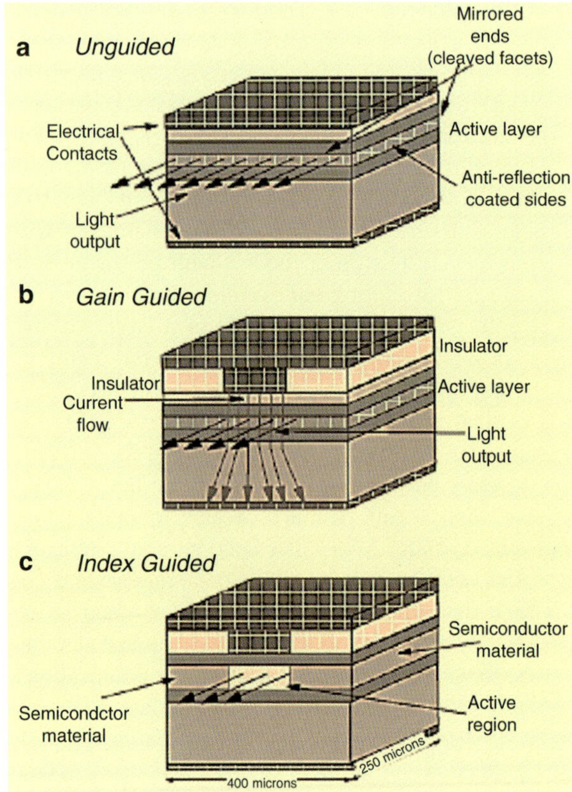

the cavity is a lot lower than c (the speed of light) the wavelength is a lot shorter than it would be in free space. The adjustment factor is the refractive index. As shown in Fig. 3.4a, in its basic form an Fabry–Perot laser is just an edge-emitting LED with mirrors on the ends of cavity. In order to get lasing on a particular path, the gain along the path needs to be controlled to exceed the loss with the gain guided operation by controlling the entry of power (current in the form of electrons and holes) to the active region. The typical technique used to guide power into the active region is to limit the area of electrical contact on the surface of the device. As shown in Fig. 3.4b, power is applied along a stripe on the top of the device for instance, current will flow predominantly along the path of least resistance (the shortest path) as shown by the gray arrows. When this happens power is delivered into the active layer in a long stripe. There will easily be sufficient gain along this path (longitudinal mode) for lasing to occur but transverse modes and longitudinal modes outside the region of power delivery may not have sufficient gain to sustain lasing. Thus a narrow beam of light issuing is obtained from the center of active region. Gain guided Fabry–Perot lasers produce a spectral width of between 5 and 8 nm consisting of between 8 and 20 or so lines. Linewidth is typically around 0.005 nm. Gain guidance can be further improved

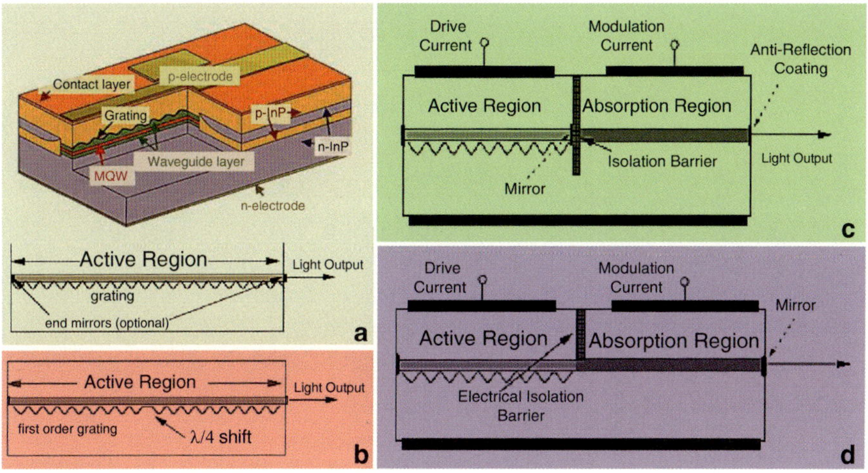

Fig. 3.5 Directing the light in a Fabry–Perot laser (Dutton 1998): (**a**) schematic structure of DFB laser; (**b**) DFB laser with phase shifted grating; (**c**) DFB laser with integrated electro-absorption modulator; (**d**) DFB laser with integrated electro-absorption modulator within the cavity

with index-guided operation if stripes of semiconductor material with a high bandgap energy are put beside the active region. Since the active region is bounded on all sides by material of a lower refractive index (generally about 0.1 lower) mirrored surfaces are formed and this serves to guide the light much better than gain guidance along. Index-guided Fabry–Perot lasers produce a spectral width of between 1 and 3 nm with usually between 1 and 5 lines (Fig. 3.4c). Linewidth is generally around 0.001 nm, which is much better than the gain guided case.

2. Distributed Feedback (DFB) Lasers
Regular Fabry–Perot lasers give rise to mode partition noise, and have too many wavelengths, and too great a spectral width for long distance communication application. DFB lasers have no this problem. As shown in Fig. 3.5a, in principle a DFB laser doesn't need end mirrors. The grating can be made strong enough to produce sufficient feedback (reflection) for lasing to take place. However, in a perfect DFB laser there are actually two lines produced (one at each side of the Bragg wavelength). Only one line is needed. A way of achieving this and improving the efficiency of the device is to place a high reflectance end mirror at one end of the cavity and either an AR coating or just a cleaved facet at the output end. In this case the grating doesn't need to be very strong—just sufficient to ensure that a single mode dominates. The added reflections (from the end mirrors) act to make the device asymmetric and suppress one of the two spectral lines. Unfortunately they also act to increase the linewidth. DFB lasers are very effective and widely used but they have a problem with chirp. There are two main sources of chirp (Dutton 1998): (1) When the current is switched on the

charge carrier (electron and hole) flux in the cavity changes very rapidly. This causes a change in the refractive index. A change in refractive index changes the resonant wavelength of the grating and the wavelength of the laser output changes (typically the wavelength gets longer) in well less than a single bit time. (2) During lasing the cavity heats up. This also happens very quickly (in a lot less than a bit time). This heating has two principal effects: (a) It causes the RI of the cavity to change. (b) It changes the electron energy gap in the material.

In an FP laser (as distinct from a DBR or DFB laser) this change in the energy gap dominates other effects and is the predominant cause of chirp. In the DFB laser the energy gap change is irrelevant. This is because the energy gap covers a range of energies and the DFB resonant wavelength is determined by the grating spacing and the cavity RI. So long as the range of energies in the gap extends to cover the resonant wavelength then the device will lase (Dutton 1998).

Sometimes DFB lasers are constructed with a quarter-wave phase shift in the middle section of the grating as shown in Fig. 3.5b. This phase shift introduces a sharp transmission "fringe" into the grating reflection band. The fringe acts to narrow the linewidth of the laser significantly (Dutton 1998).

DFB lasers have a number of significant advantages over FP types: (a) They can exhibit very narrow linewidths (of the order of 50 kHz). (b) They have quite low chirp. (c) They typically have a very low Relative Intensity Noise (RIN). However, DBR lasers are extremely sensitive to reflections. Any reflection entering the cavity will disturb the laser's stable resonance. This causes a widening of the linewidth. To minimize the effects of this problem DFB lasers are often packaged with an isolator integrated within the assembly. However, these don't always suppress all reflections and additional steps must be taken in system design to minimize the problem. They are sensitive to temperature variations in two ways: (a) The stable (average) temperature of the device has a very strong influence on wavelength. Wavelength variation on a scale of many seconds or longer doesn't have much detrimental effect on a single channel long distance communication system but it is a critical issue in WDM systems. The device requires temperature control for stable operation. This is usually provided by including a thermoelectric cooler in the laser package. (b) During transmission (in even one bit time) the cavity heats up. Moreover, varying conditions produce significant fluctuations in laser output power. They have a relatively high cost. As seen above, to get stable operation it is almost always needed for proper temperature control, power control, and optical isolation. All this adds to the cost (Dutton 1998).

One of the problems with all solid-state lasers is the maximum possible speed of modulation. Capacitance effects and effects caused by the (relatively slow) movement of charge carriers within the semiconductor limit the speed of modulation. External modulators are used at high speeds but this adds cost and complexity. Also signal strength is lost in the additional couplings required. Figure 3.5c illustrates a DFB laser with an integrated absorption modulator. Basically, the DFB laser and the modulator are separate devices built together on the same chip. A mirror is placed between the absorption layer and the laser cavity and the AR coating on the end facet of the device. The laser is run in the ON state all the time and modulation is accomplished by varying the absorption in the modulator. The absorption region is reverse biased and there is only a tiny leakage current. This results in much faster modulation than can be achieved by turning the drive current to the active region on and off. In addition it almost completely removes the problem of chirp because the laser cavity is operating at the same level of output power all the time. In Fig. 3.5d the absorber is placed within the laser cavity itself. When the material is in an absorbent state it prevents lasing. When it is transparent (nonabsorbent) the laser can produce light. This laser produces very short high-power pulses. When the laser is in the OFF state (the absorber is absorbing the light) the active medium gets completely pumped up and starts fluorescing. When the laser is then turned ON lasing commences very quickly and produces a short high-power pulse. This method of operation is called "*Q*-switching" because the *Q* of the laser cavity is changed to initiate the pulse (Dutton 1998).

3. Mode-Locking and Self-pulsating Lasers
   Under some conditions some types of laser will produce a very fast sequence of regularly spaced separate pulses. In many cases this pulse stream is a natural result of the construction of the laser. In other cases the behavior can be induced by modulation of the drive current. When a laser produces such a pulse stream it is said to be mode-locked or self-pulsating. There are two general types of mode-locking: Active mode-locking and passive mode-locking. In a passive mode-locked device, pulses are produced at intervals corresponding to the roundtrip time of the fundamental mode of the laser's cavity. In semiconductor lasers the cavity is generally too short for this type of pulsation to be of much use as a generator of separate pulses. In external-cavity lasers and in fiber lasers however the cavity is quite long enough for mode-locked pulses with good separation to be generated. These can be used to produce the very fast streams of short pulses. In a passive device a substance is placed into the optical feedback path which acts as a saturable absorber, which absorbs light at a fixed maximum rate. At low light levels it absorbs everything and suppresses lasing. At high light levels it saturates and allows high-intensity pulses to pass. Thus after the drive current is switched on, the saturable absorber absorbs light and keeps the cavity *Q* low. Lasing is suppressed and a large inversion is built up in the active region. At some stage a single photon will happen to be in the right direction and will build up a pulse large enough to pass the absorber. This pulse then oscillates back and forth in the cavity producing output pulses at a fixed interval determined by the cavity length. In an actively mode-locked device there is something placed into the optical path within the cavity that acts as a shutter. Thus the action is very similar to *Q*-switching with a regular modulated triggering mechanism. The *Q*-switched laser will do this very well when its absorption modulator is driven with a regular square wave train. The usual way this is done is to use only shallow variations in the drive current at the same frequency as the resonant frequency of the laser cavity (Dutton 1998).

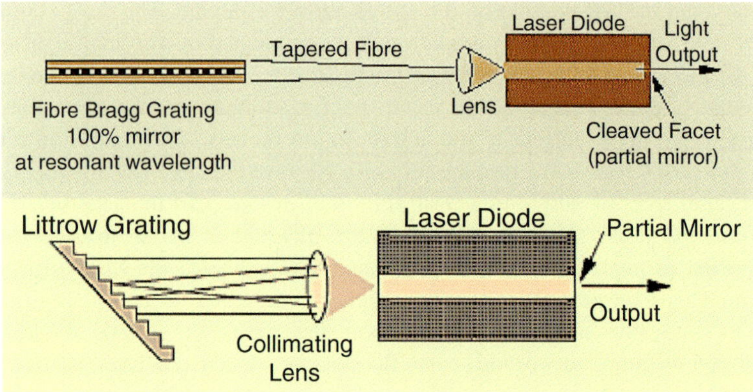

Fig. 3.6 Schematic structure of external-cavity DBR laser

4. Distributed Bragg Reflector (DBR) Lasers
   DBR lasers are very similar to the DFB lasers. The major difference is that where DFB lasers have a grating within the active region of the cavity, DBR lasers have a partitioned cavity with the grating in a region that is not active (amplifying). The reason for this structure is that the refractive index within the cavity of a laser changes during operation due to changes in temperature and electron flux. Change of refractive index in the grating region of course changes the operational wavelength of the device. If the grating is put into an inactive extension of the cavity then there is a lot less wavelength variation from these causes, because the characteristics of the material immediately adjacent to the grating are not being changed by the laser's operation. Figure 3.6 shows a schematic of this structure. As in the DFB laser, end mirrors are not always needed as the grating can provide sufficient reflection. However, end mirrors are almost always used in an asymmetric configuration. Typically this is 30 % mirror on the back facet and either just a cleaved facet or a 4 % mirror on the front (output) facet. DBR lasers typically produce a single line only with a linewidth of around 0.0001 nm. The major problem is that there can be significant absorption in the inactive region near the grating(s). This causes a loss in efficiency (Dutton 1998).
5. Quantum Wells
   DBR and DFB lasers are often built using a quantum well (QW) structure. When light is confined into a cavity smaller than its wavelength it behaves as a particle (quantum) rather than as a wave. In the case of semiconductor lasers if the size of the cavity is restricted, quantum behavior changes the operation of the laser in a dramatic and fundamental way. By their nature, most semiconductor lasers are very thin (20 μm or so) in the vertical direction but this is not thin enough to cause quantum behavior. In QW lasers cavity height is reduced to around 10 or 20 nm. The width of the cavity does not need to be this restricted but it should be narrow enough to prevent unwanted lateral modes from forming. Cavity width is generally from 5 to 20 μm. The cavity has to be many wavelengths long to get sufficient gain. In addition, it is difficult to manufacture lasers with cavities shorter

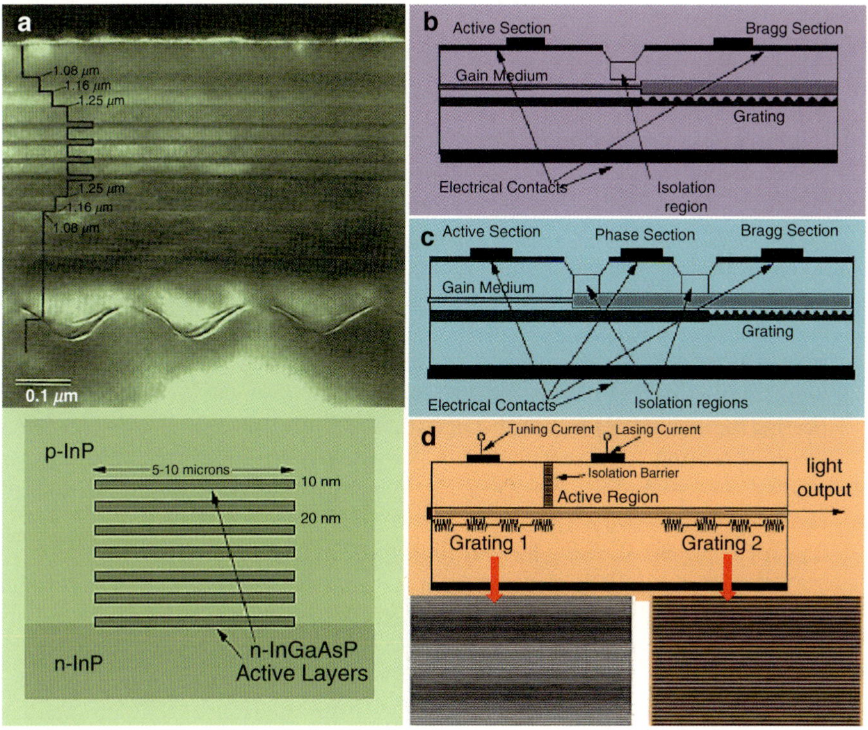

Fig. 3.7 Schematic of tunable DFB laser (Dutton 1998): (**a**) cavity structure of a multiple quantum well laser; (**b**) tunable two-section DBR laser; (**c**) tunable three-section DBR laser; and (**d**) sampled grating tunable DFB laser

than 50 μm. (As with other semiconductor lasers the cavity length is typically 200–250 μm.) This cavity geometry is called a quantum well. The most obvious change in laser characteristic that this brings is that the amount of material in the active region is substantially reduced. This reduces the amount of energy needed to achieve lasing and thus the lasing threshold. The result is a higher gain characteristic but a lower maximum output power than conventional (non-QW) devices. In addition, quantum wells have a much reduced sensitivity to temperature change (compared to non-QW DBR and DFB structures). In the very narrow cavity available there is much less space for energy and momentum effects to occur. The quantum well structure prevents lateral modes forming and ensures that lasing produces only one line. In addition the single line that is produced tends to have a narrower linewidth than for non-QW structures. As with all DFB and DBR structures, the sides of the cavity are made of a material of lower refractive index than that of the cavity itself. In some forms of QW laser the index at the side of the cavity is graded to further confine lateral modes. Lasers with a single active region are called single quantum well (SQW) lasers. Often a number of quantum wells are used one on top of another as shown in Fig. 3.7a (Dutton 1998). These are called multiple quantum well (MQW) structures. MQW structures retain many of the desirable characteristics of SQW ones (such as reduced lasing threshold) but produce higher gain and greater total power. A disadvantage is that MQW lasers produce a broader linewidth than SQW ones (but still narrower than comparable non-QW structures).

One MQW laser called Strained-Layer Quantum Wells has been developed based on that the atoms in the thin layer are pulled apart (or forced together) to conform to the crystalline structure of the surrounding material. This means that the chemical bonds (electron orbits) holding the material together become longer or shorter than their natural length. This has two very beneficial effects (Dutton 1998): (a) It allows the use of a much wider range of materials in the heterojunction. The added flexibility allows the design of more efficient lasers. (b) The "strain" in the lattice of the thinner material changes its properties in a number of highly desirable ways. Most obviously, it changes the bandgap energy levels. By engineering the amount of strain the bandgap (laser wavelength) could be tuned without affecting too much the semiconductor properties of the junction. In addition, the strain allows the device to operate at significantly lower threshold currents than are possible in unstrained MQW lasers. Devices have been constructed with very low threshold currents allowing for very high speed modulation.

There are several situations where it is desirable to have the ability to change (tune) the laser's wavelength. And different systems require different kinds of tunability (range and speed). Figure 3.7b, c show two electronically tunable DBR lasers. One is a two-section device and the other has three sections. The three-section device can be continuously tuned over the same range. A disadvantage of the three-section device is that it requires relatively complex electronics for control.

To extend the tuning range DBR lasers employing a sampled grating have been constructed, as shown in Fig. 3.7d. A tunable DBR laser relies on a change in the RI of the cavity adjacent to the grating to change the resonant wavelength. The problem is that the achievable RI change is not very great. A sampled grating consists of a number of short grating sections with periodic blanking between them. Two sampled gratings are used in such a way that the interaction between them causes a small change in RI to produce a large change in wavelength. The major drawback of this approach is that the tuning is discontinuous. One section is tuned until one of its resonances is the same as one of the possibilities of the other grating. This then becomes the lasing wavelength. Gratings with periodic blanking have the interesting property of producing reflection peaks not only at the Bragg wavelength but also at other wavelengths at either side of the Bragg wavelength. Thus a grating with multiple (precisely determined) reflection peaks can be obtained. The separation between these peaks is inversely proportional to the blanking (sampling) period.

An interesting form of DBR laser can be constructed where the resonant cavity extends outside the laser chip itself. This runs counter to intuition because as the cavity length is extended so the possible number of longitudinal modes increases exponentially. This would seem to make matters worse. The key reason for the external cavity is so that a much stronger (more selective) grating can be used. The back mirror of a regular index-guided FP laser is removed and replaced with a lens and Littrow grating assembly. Such lasers produce a single line only with a very narrow linewidth (a few tens of kilohertz). The major problem here is that these devices are significantly more expensive than other types of high quality laser. A useful feature is that the grating can be mount on a moveable platform (such as a piezoelectric crystal) and mechanically tune the device by tilting the grating. This provides a very wide tuning range but a relatively slow tuning speed.

One interesting option for providing narrow linewidth, wavelength stable light is to use a simple Fabry–Perot laser diode controlled by an external grating written within the fiber itself. This type of grating is called an "In-Fiber Bragg Grating." It reflects light (is a mirror) at one very specific wavelength. This grating should reflect between 50 and 65 % of light at the selected wavelength. The laser cavity therefore now extends into the fiber and the laser end mirror is formed by the grating. The key characteristic here is that the grating only reflects light at a very specific wavelength. Thus the cavity cannot resonate at any other wavelength because there will be no feedback at any other wavelength (Dutton 1998).

6. Vertical Cavity Surface-Emitting Lasers (VCSELs)
VCSELs are semiconductor lasers whose emission direction is perpendicular to the surface of the wafer. VCSEL devices typically have a more circular output beam than conventional laser diodes, and potentially could be much cheaper to manufacture. VCSELs have been around in various forms since the late 1970s. However in 1991 there was a major development in construction techniques reported and in 1996 the first commercial devices became available. As of 2005, only 850 nm VCSELs are widely available, with 1,300 nm VCSELs beginning to be commercialized, and 1,550 nm devices an area of research. VCSELs are light sources based on the stimulated emission of photons inside a short cavity consisting of multiple-layer quarter wavelength stacked mirrors. As is the case for all laser diodes, these devices have a threshold current below which there is no coherent light emission. Typical threshold currents for VCSELs are of the order of 1 mA. Above the threshold the device can work very efficiently and can produce a diffraction-limited light beam with a very low numerical aperture (NA). Nevertheless, many VCSELs operate in a multimode regime which leads to some broadening of the far field. The emission spectrum is narrow.

VCSELs are constructed by laying down a very large number (perhaps 500) of relatively thin layers of semiconductor material. The device emits light vertically through the stack of material layers, as shown in Fig. 3.8. As in any laser the overall structure is one of two end mirrors on each side of an active region which produces the light. The key points are (Dutton 1998): (a) The mirrors are made of alternating layers of material of different refractive indices with carefully controlled thickness. The stack forms a Bragg grating which is a wavelength-selective mirror. (b) The sides of the laser are formed by cutting the material out. (c) The laser dimensions can be such that no lateral modes are possible.

In fact the laser is so confined that it forms a quantum well in which light behaves as individual photons rather than as waves or rays. Typical dimensions

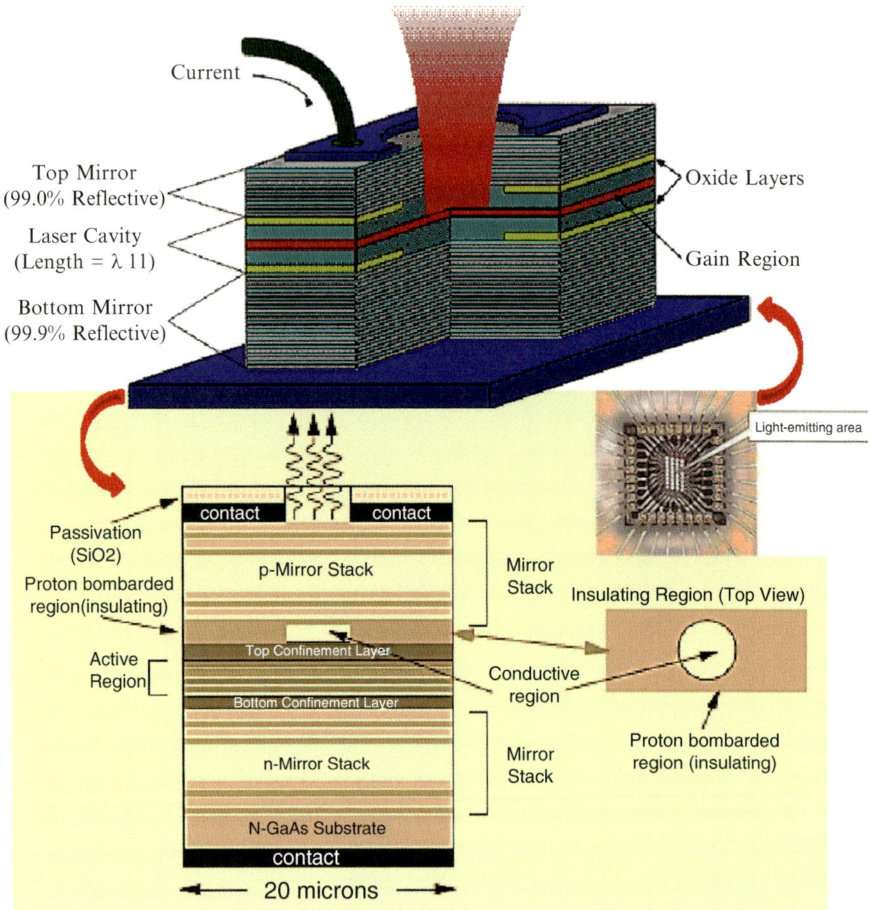

Fig. 3.8 Schematic illustration of VCSEL structure

are about 12 μm in diameter (for single-mode operation) and 20 μm (for multimode operation). The active region is very short compared to other types of semiconductor laser. This means that the mirrors have to have a relatively high reflectivity. The operational characteristics of VCSELs combine many of the desirable properties of lasers and LEDs. Typical coupled output power can be somewhere around a milliwatt. This is very good for LAN type communications as a maximum power of the eye safety limit (−4 dBm) is only needed to transmit up. Typical LEDs are a lot lower in power and require the use of expensive InGaAs pin diode receivers. VCSELs in the LAN environment can work well with simple, low cost Si diode receivers. VCSELs with a large diameter (20 μm or so) have multiple transverse modes. This makes the device very suitable for use with multimode fiber. The low coherence of output light produced by a multi-transverse mode VCSEL leads to insensitivity to mode selective loss and minimizes the problem of modal noise. A low divergence circular light beam is produced which allows for easy and efficient coupling to a fiber. Typical VCSELs have very low threshold currents (less than 5 mA). Very low power dissipation and low modulation current requirements mean that special driver circuitry is not required. VCSELs have very high modulation bandwidths. This is well in excess of what can be achieved with much more expensive LEDs. Devices are very stable and generally do not need a monitor photodiode or feedback power control as is conventionally needed for most communications lasers and high-end LEDs. Therefore, optimized VCSELs have many advantages (Thienpont and Baukens 2002): (a) a high incremental efficiency in converting electrical power to optical power (typical efficiencies are of the order of a few tens of percent), (b) a high efficiency in the delivery of this optical power to the receiver because of the high spatial coherence and low numerical aperture and (c) a high modulation bandwidth, at least if the device is driven at a current far enough above threshold. Modulation bandwidths of over 10 GHz have been achieved.

7. Fiber Lasers

   Solid-state lasers or laser amplifiers where the light is guided due to the total internal reflection in a single-mode optical fiber are instead called fiber lasers. Guiding of light allows extremely long gain regions providing good cooling conditions; fibers have high surface area to volume ratio which allows efficient cooling. In addition, the fiber's waveguiding properties tend to reduce thermal distortion of the beam. Erbium and ytterbium ions are common active species in such lasers. Quite often, the fiber laser is designed as a double-clad fiber. This type of fiber consists of a fiber core, an inner cladding, and an outer cladding. The index of the three concentric layers is chosen so that the fiber core acts as a single-mode fiber for the laser emission while the outer cladding acts as a highly multimode core for the pump laser. This lets the pump propagate a large amount of power into and through the active inner core region, while still having a high numerical aperture (NA) to have easy launching conditions. Pump light can be used more efficiently by creating a fiber disk laser, or a stack of such lasers. Fiber lasers have a fundamental limit in that the intensity of the light in the fiber cannot be so high that optical nonlinearities induced by the local electric field strength can become dominant and prevent laser operation and/or lead to the material destruction of the fiber. This effect is called photodarkening. In bulk laser materials, the cooling is not so efficient, and it is difficult to separate the effects of photodarkening from the thermal effects, but the experiments in fibers show that the photodarkening can be attributed to the formation of long-living color centers (Wikipedia 2013a).

8. Photonic Crystal Lasers

   Photonic crystal lasers are lasers based on nano-structures that provide the mode confinement and the density of optical states (DOS) structure required for the feedback to take place. They are typical micrometer-size and tunable on the bands of the photonic crystals (Wikipedia 2013a).

9. Gas Lasers

   Gas lasers are well known for the very high power that can be produced. Low powered gas lasers are also used in more mundane applications. In operation gas lasers are somewhat different from their solid-state counterparts. In a gas laser, individual atoms (or ions) interact with their electrons quite unaffected by the surrounding material. In a semiconductor structure the characteristics of the material are all important. Gas lasers are not used very often in modern communications. However they have a minor use as a transmitter for constructing outdoor optical links. Quite high speed, short distance, line-of-sight connections can be made this way. Unfortunately, heavy rain and smoke cause attenuation and scattering which significantly limit the usefulness of this kind of communication. Carbon dioxide (CO_2) lasers operating at a wavelength in the far infrared (10.6 nm) are often used for this application.

   Gas lasers using many different gases have been built and used for many purposes. The helium–neon laser (HeNe) is able to operate at a number of different wavelengths, however the vast majority are engineered to lase at 633 nm. Commercial carbon dioxide (CO_2) lasers can emit many hundreds of watts in a single spatial mode which can be concentrated into a tiny spot. This emission is in the thermal infrared at 10.6 μm; such lasers are regularly used in industry for cutting and welding. The efficiency of a CO_2 laser is unusually high: over 10 %. Argon-ion lasers can operate at a number of lasing transitions between 351 and 528.7 nm. Depending on the optical design one or more of these transitions can be lasing simultaneously; the most commonly used lines are 458, 488, and 514.5 nm. A nitrogen transverse electrical discharge in gas at atmospheric pressure (TEA) laser is an inexpensive gas laser, often home-built by hobbyists, which produces rather incoherent UV light at 337.1 nm. Metal ion lasers are gas lasers that generate deep ultraviolet wavelengths. Helium–silver (HeAg) 224 nm and neon–copper (NeCu) 248 nm are two examples. Like all low-pressure gas lasers, the gain media of these lasers have quite narrow oscillation linewidths, less than 3 GHz (0.5 pm), making them candidates for use in fluorescence suppressed Raman spectroscopy (Wikipedia 2013a).

3.3 Optical Modulators and Drivers

Optical modulator devices are used to modulate a beam of light. The beam may be carried over free space, or propagated through an optical waveguide. Depending on the parameter of a light beam which is manipulated, modulators may be categorized into amplitude modulators, phase modulators, polarization modulators etc. Often the easiest way to obtain modulation of intensity of a light beam, is to modulate the current driving the light source, e.g., a laser diode. This sort of modulation is called direct modulation, as opposed to the external modulation performed by a light modulator. For this reason light modulators are, e.g., in fiber-optic communications, called external light modulators. With laser diodes where narrow linewidth is required, direct modulation is avoided due to a high bandwidth chirping effect when applying and removing the current to the laser (Wikipedia 2012).

According to the properties of the material that are used to modulate the light beam, modulators are divided into two groups: absorptive modulators and refractive modulators. In absorptive modulators absorption coefficient of the material is changed, in refractive modulators refractive index of the material is changed. The absorption coefficient of the material in the modulator can be manipulated by the Franz–Keldysh effect, the quantum-confined Stark effect, excitonic absorption, changes of Fermi level, or changes of free carrier concentration. Usually, if several such effects appear together, the modulator is called an electro-absorptive modulator. Refractive modulators most often make use of an electro-optic effect. Some modulators utilize an acousto-optic effect or magneto-optic effect or take advantage of polarization changes in liquid crystals. The refractive modulators are named by the respective effect: i.e., electro-optic modulators and acousto-optic modulators. The effect of a refractive modulator of any of the types mentioned above is to change the phase of a light beam. The phase modulation can be converted into amplitude modulation using an interferometer or directional coupler (Wikipedia 2012).

Separate case of modulators are spatial light modulators (SLMs). The role of SLM is modification two dimensional distribution of amplitude and/or phase of an optical wave.

Modulators based on the quantum-confined Stark effect work through an applied voltage shifts the absorption edge of the device to another wavelength. Since this shift is rather small, a modulator only functions over a wavelength interval of a few nanometers. Modulators differ from LEDs and VCSELs in that it is an external light source that is modulated. This way, the heat dissipation caused by the electrical to optical power conversion is eliminated from the dense array of modulating devices. Furthermore, the voltage is applied to a reverse-biased diode so that the power dissipation of the device itself is negligible when the modulated light output stays constant. Only when switching, significant power is required to charge the capacitors. The speed of modulators can be very large and is mainly limited by resistance/capacitance effects. The main problem with modulators is their temperature sensitivity: they only operate within a very limited temperature range (Neefs 2000).

Based on the quantum-confined Stark effect, the MQW InGaAs/GaAs P–I–N diode structure can behave as a modulator. A buffer layer of InGaAs is grown on the GaAs substrate with a graded concentration of In. Relaxation of this layer provided a pseudo-substrate, with the required lattice constant, for growth of the strain-balanced MQW structure. These materials are chosen so as to give an operating wavelength corresponding to a convenient solid-state laser (1,047-nm, diode-pumped Nd:YLF laser) and lying within a spectral region in which the GaAs was transparent, ensuring access to the devices through the substrate after flip-chip assembly. Photo-lithographic etching creates arrays of mesas which formed independent diodes capable of acting as modulators. These same diodes can also act as photodetectors. For the cross-bar switch, for instance, a chip is designed containing both modulators and photodiodes and this is flip-chip bonded to CMOS using solder bumps attached directly on top of the diode mesas. The optoelectronic chip comprised 4,096 photodiodes, for input signals, plus 128 modulator diodes, linked in pairs to provide 64 differential optical outputs. The modulators are reverse biased and the drivers provide a voltage signal capable of varying the electro-absorption field. Most light source drivers consist of a small buffer driving a CMOS modulation transistor serving as a modulated current source and contain a CMOS transistor providing in an adjustable current source for biasing. In contrast to light sources which need a constant supply of current, MQW modulators behave as capacitive loads. Nevertheless, here current sources provided by CMOS transistors can also be used to charge these capacitive loads. An important aspect of the optical link, which has repercussions on the need of a data conditioning circuit in the drivers, is whether DC-coupling or AC-coupling is used (Neefs 2000).

In optical communications, three wavelength ranges are currently of interest: long wavelength infrared 1,300–1,550 nm), short wavelength infrared (850–980 nm) and visible light (mainly around 650 nm). The long wavelength IR range has traditionally been important in long-haul optical fiber communications, because it spans the absorption minimum (1,550 nm) and the dispersion minimum (1,300 nm) of glass optical fibers. For short distance optical interconnect these properties are of lesser importance. Short wavelength IR has so far been the workhorse of data link products, mainly because the technology of the corresponding AlGaAs-based components is quite mature. Moreover, these wavelengths are compatible with traditional Si photodetectors. The wavelength 980 nm is quite suited for bottom-emitting devices, because the GaAs substrate is transparent at that wavelength. At 850 nm, however, the GaAs substrate is absorbing and therefore needs to be removed somehow in bottom-emitting components. In the context of optical interconnect, the advantage of bottom-emitting devices lies in their compatibility with flip-chip bonding to CMOS. For transmission over PMMA-based plastic optical fiber, visible light is more suited, especially in the wavelength range of 650 nm, coinciding with the fiber's absorption minimum. Short wavelengths have an advantage when it comes to detecting light in a CMOS integrated detector (Neefs 2000).

3.4 Optical Detectors

The predominant types of light detector used in communications systems rely on the principle of ionization in a semiconductor material. There are a number of different kinds of device but they can all be viewed as variations on a central principle rather than as devices that involve radically different principles. There are four important parameters to be considered (Dutton 1998): (a) Detector responsivity—the ratio of output current to input optical power. Hence this is the efficiency of the device. (b) Spectral response range—the range of wavelengths over

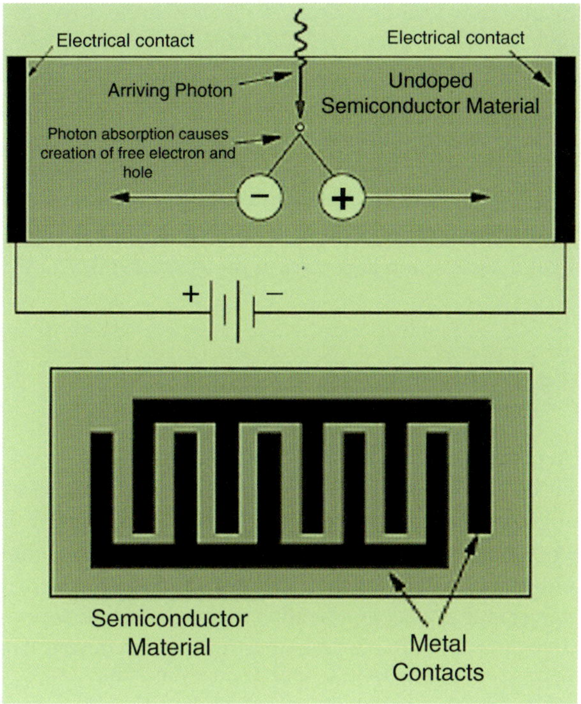

Fig. 3.9 Principle and schematic structure of photoconductor (Dutton 1998)

which the device will operate. (c) Response time—a measure of how quickly the detector can respond to variations in the input light intensity. (d) Noise characteristics—the level of noise produced in the device is critical to its operation at low levels of input light.

3.4.1 *Photoconductors*

Photoconductors are the simplest conceivable optical detector. The device consists of a piece of (undoped) semiconductor material with electrical contacts attached. A voltage is applied across the contacts. When a photon arrives in the semiconductor it is absorbed and an electron/hole pair is created. Under the influence of the electric field between the two contacts the electron and the hole each migrate towards one of the contacts. The electron to the positive contact and the hole to the negative contact. Thus the resistance of the device varies with the amount of light falling on it. Schematic illustration of photoconductive detectors are shown in Fig. 3.9 (Dutton 1998). A flat piece of semiconductor has metal contacts plated onto it. The contact areas are interdigitized so that there is only a short distance between the contacts. This is done because the charge carriers drift quite slowly through the semiconductor material and this limits the speed of the device. As electron drift is much faster than hole drift, electrons tend to arrive at the positive contact before a hole arrives at the other side. This then causes the negative electrode to emit an electron into the semiconductor. This electron in turn may arrive at the positive electrode before the hole arrives at the negative electrode and yet another electron is emitted into the material. Classic examples of photoconductive materials include the conductive polymer polyvinylcarbazole, used extensively in photocopying (xerography); lead sulfide, used in infrared detection applications, such as the U.S. Sidewinder and Russian Atoll heat-seeking missiles; and selenium, employed in early television and xerography (Law 1993).

3.4.2 Photodiodes

A photodiode is a type of photodetector capable of converting light into either current or voltage, depending upon the mode of operation. The common, traditional solar cell used to generate electric solar power is a large area photodiode. Photodiodes are similar to regular semiconductor diodes except that they may be either exposed (to detect vacuum UV or X-rays) or packaged with a window or waveguide connection to allow light to reach the sensitive part of the device. When a photon of sufficient energy strikes the diode, it excites an electron, thereby creating a free electron (and a positively charged electron hole). This mechanism is also known as the inner photoelectric effect. If the absorption occurs in the junction's depletion region, or one diffusion length away from it, these carriers are swept from the junction by the built-in field of the depletion region. Thus holes move towards the anode, and electrons towards the cathode, and a photocurrent is produced. This photocurrent is the sum of both the dark current (without light) and the light current, so the dark current must be minimized to enhance the sensitivity of the device. When used in zero bias or photovoltaic mode, the flow of photocurrent out of the device is restricted and a voltage builds up. This mode exploits the photovoltaic effect, which is the basis for solar cells—a traditional solar cell is just a large area photodiode. In photoconductive mode the diode is often reverse biased (with the cathode positive), dramatically reducing the response time at the expense of increased noise. This increases the width of the depletion layer, which decreases the junction's capacitance resulting in faster response times. The reverse bias induces only a small amount of current (known as saturation or back current) along its direction while the photocurrent remains virtually the same. For a given spectral distribution, the photocurrent is linearly proportional to the illuminance (and to the irradiance). Although this mode is faster, the photoconductive mode tends to exhibit more electronic noise. The leakage current of a good PIN diode is so low (<1 nA) that the Johnson–Nyquist noise of the load resistance in a typical circuit often dominates. Other modes of operation, such as avalanche photodiodes (APDs), have a similar structure to regular photodiodes, but they are operated with much higher reverse bias. This allows each photo-generated carrier to be multiplied by avalanche breakdown, resulting in internal gain within the photodiode, which

Table 3.2 Materials commonly used to produce photodiodes

Materials	Electromagnetic spectrum wavelength range (nm)
Silicon	190–1,100
Germanium	400–1,700
Indium gallium arsenide	800–2,600
Lead (II) sulfide	<1,000–3,500

increases the effective responsivity of the device. A phototransistor is in essence a bipolar transistor encased in a transparent case so that light can reach the base–collector junction. The electrons that are generated by photons in the base–collector junction are injected into the base, and this photodiode current is amplified by the transistor's current gain β. If the emitter is left unconnected, the phototransistor becomes a photodiode. While phototransistors have a higher responsivity for light they are not able to detect low levels of light any better than photodiodes. Phototransistors also have significantly longer response times (Wikipedia 2013b).

The material used to make a photodiode is critical to defining its properties, because only photons with sufficient energy to excite electrons across the material's bandgap will produce significant photocurrents. Materials commonly used to produce photodiodes are shown in Table 3.2 (Held 2008).

A photodiode can be a P–N junction or PIN structure. Many diodes designed for use specifically as a photodiode use a PIN junction rather than a P–N junction, to increase the speed of response. A photodiode is designed to operate in reverse bias (Cox 2001).

The principle involved in a P–N diode is simply the principle of the LED in reverse. That is, light is absorbed at a P–N junction rather than emitted. A reverse-biased p–n junction passes no current unless something happens to promote electrons from the valence band to the conduction band within the depletion zone. At ordinary room temperatures there is a small amount of this due to the action of heat and thus there is a very small current. Photons of sufficient energy can be absorbed and cause the promotion of an electron from the valence band to the conduction band. The free electron and hole created by the photon absorption now are attracted to their opposite charges at either side of the junction and current flows. The big problem here is that the depletion zone in a P–N junction is extremely thin. Most light passes through without being absorbed in the junction and instead is absorbed in the doped material at either side of the junction. Ultimately many of the electron–hole pairs created outside the junction do end up being attracted to the junction and creating current. However the process is quite slow and P–N junction devices are not fast enough for current communications applications. P–N photodiodes are used in similar applications to other photodetectors, such as photoconductors, charge-coupled devices, and photomultiplier tubes. They may be used to generate an output which is dependent upon the illumination (analog; for measurement and the like), or to change the state of circuitry (digital; either for control and switching, or digital signal processing). Photodiodes are used in consumer electronics devices such as compact disc players, smoke detectors, and the

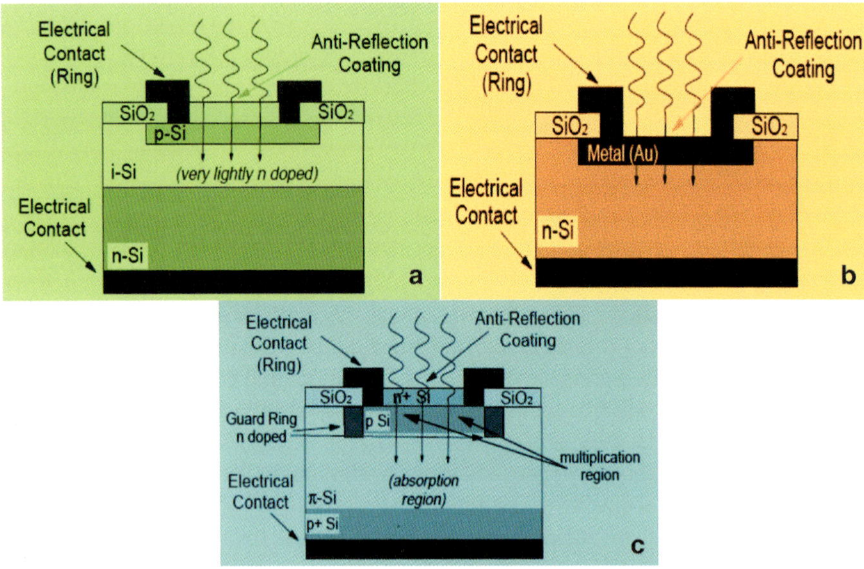

Fig. 3.10 Schematic illustration of several photodiodes (Dutton 1998): (**a**) typical silicon P–I–N diode schematic; (**b**) Schottky-barrier photodiode schematic; (**c**) avalanche photodiode (APD)

receivers for infrared remote control devices used to control equipment from televisions to air conditioners. For many applications either photodiodes or photoconductors may be used. Either type of photosensor may be used for light measurement, as in camera light meters, or to respond to light levels, as in switching on street lighting after dark. Photosensors of all types may be used to respond to incident light, or to a source of light which is part of the same circuit or system. A photodiode is often combined into a single component with an emitter of light, usually a LED, either to detect the presence of a mechanical obstruction to the beam (slotted optical switch), or to couple two digital or analog circuits while maintaining extremely high electrical isolation between them, often for safety (optocoupler). Photodiodes are often used for accurate measurement of light intensity in science and industry. They generally have a more linear response than photoconductors. They are also widely used in various medical applications, such as detectors for computed tomography (coupled with scintillators), instruments to analyze samples (immunoassay), and pulse oximeters. P–N photodiodes are not used to measure extremely low light intensities. Instead, if high sensitivity is needed, APDs, intensified charge-coupled devices or photomultiplier tubes are used for applications such as astronomy, spectroscopy, night vision equipment, and laser range finding (Dutton 1998; Wikipedia 2013b).

PIN diodes are much faster and more sensitive than P–N junction diodes, and hence are often used for optical communications and in lighting regulation. The P–I–N junction is extended by the addition of a very lightly doped layer called the intrinsic zone between the P- and N-doped zones. Thus the device is called a P–I–N diode rather than a P–N diode. This is illustrated in Fig. 3.10a (Dutton 1998).

The wide intrinsic (I) layer has only a very small amount of dopant and acts as a very wide depletion layer. There are a number of improvements here: (a) It increases the chances of an entering photon being absorbed because the volume of absorbent material is significantly increased. (b) Because it makes the junction wider it reduces the capacitance across the junction. The lower the capacitance of the junction the faster the device response.

There are two ways of current carriage across the junction: diffusion and drift. Increasing the width of the depletion layer favors current carriage by the drift process which is faster than the diffusion process. The result is that the addition of the "I" layer increases the responsivity and decreases the response time of the detector to around a few tens of picoseconds. The key to operation of a PIN diode is that the energy of the absorbed photon must be sufficient to promote an electron across the bandgap. However, the material will absorb photons of any energy higher than its bandgap energy. Typically PIN diodes will operate at any wavelength shorter than the cutoff wavelength. A material with a low bandgap energy is usually used for all PIN diodes regardless of the wavelength. Moreover, the materials used for PIN diode construction are different depending on the band of wavelengths for which it is to be used. However, this restriction is nowhere near as stringent as it is for lasers and LEDs where the characteristics of the material restrict the device to a very narrow range. The optimal way is to choose a material with a bandgap energy slightly lower than the energy of the longest wanted wavelength. These crystalline semiconductor materials are transparent at wavelengths longer than their cutoff. Typical materials used in the three communication wavelength "windows" are as follows (Dutton 1998):

(a) 500–1,000 nm Band
Silicon PIN diodes operate over a range of 500–1,120 nm as silicon has a bandgap energy of 1.11 eV. Since silicon technology is very low cost silicon is the material of choice in this band. However, silicon is an indirect bandgap material (at the interested wavelengths) and this makes it relatively inefficient. (Silicon PIN diodes are not as sensitive as PIN diodes made from other materials in this band.) This is the same characteristic that prevents the use of silicon for practical lasers.
(b) 1,300 nm (1,250–1,400 nm) Band
In this band indium gallium arsenide phosphide (InGaAsP) and germanium can be used. Germanium has a lower bandgap energy (0.67 versus 0.89 eV for InGaAsP) and hence it can theoretically be used at longer wavelengths. However, other effects in Ge limit it to wavelengths below 1,400 nm. InGaAsP is significantly more expensive than Ge but it is also significantly more efficient (devices are more sensitive).
(c) 1,550 nm Band (1,500–1,600 nm)
The material used here is usually InGaAs (indium gallium arsenide). InGaAs has a bandgap energy of 0.77 eV.

In many circumstances the junction between a metal and a semiconductor can display some of the properties of a semiconductor P–N junction, Schottky-barrier photodiodes make use of this effect. A thin metal layer replaces one half of the

P–N junction as shown in Fig. 3.10b. Schottky photodiodes have a number of characteristics (Dutton 1998): (a) There are a number of semiconductors available which promise higher efficiency operations but cannot be used in regular P–N or P–I–N configurations because they cannot be doped to both P and N characteristics. In addition with normal heterostructure devices the crystal lattice have to be matched on both sides of the junction and this severely limits the choice of materials. There would be no these problems when one side of the junction is metal. (b) The metal layer is a good conductor and so electrons are conducted away from the junction immediately. This means that recombination effects are minimized thus improving the efficiency. In addition it means faster operation.

APDs amplify the signal during the detection process. As shown in Fig. 3.10c, in its basic form an APD is just a P–I–N diode with a very high reverse bias. A reverse bias of 50 V is typical for these devices compared with regular P–I–N diodes used in the photoconductive mode which is reverse biased to only around 3 V (or less). The main structural difference between an APD and a P–I–N diode are that the *i* zone (which is lightly N-doped in a P–I–N structure) is lightly P-doped and renamed the π layer. It is typically thicker than an i zone and the device is carefully designed to ensure a uniform electric field across the whole layer. The guard ring shown in the figure serves to prevent unwanted interactions around the edges of the multiplication region. The device operates as follows (Dutton 1998): (a) Arriving photons generally pass straight through the N^+–P junction (because it is very thin) and are absorbed in the π layer. This absorption produces a free electron in the conduction band and a hole in the valence band. (b) The electric potential across the π layer is sufficient to attract the electrons towards one contact and the holes towards the other. In the figure electrons are attracted towards the N^+ layer at the top of the device because being reverse biased it carries the positive charge. The potential gradient across the π layer is not sufficient for the charge carriers to gain enough energy for multiplication to take place. (c) Around the junction between the N^+ and P layers the electric field is so intense that the charge carriers (in this case electrons only) are strongly accelerated and pick up energy. When these electrons (now moving with a high energy) collide with other atoms in the lattice they produce new electron–hole pairs. This process is called impact ionization. (d) The newly released charge carriers (both electrons and holes) are themselves accelerated (in opposite directions) and may collide again. Both electrons and holes can now contribute to the multiplication process. In silicon, electrons are the dominant carrier. In III–V alloy materials holes are often employed. The result of the above process is that a single arriving photon can result in the production of between 10 and 100 or so electron–hole pairs. As with P–I–N diodes different materials are usually employed for each of the three important wavelength bands (Dutton 1998):

(a) 800 nm to 1 μm Band
  In this band silicon is usually employed although germanium will also work reasonably well. Germanium devices however produce higher noise levels than silicon ones. Silicon has a relatively high bandgap energy and is therefore used only for wavelengths shorter than about 1 μm.

(b) 1,310 nm Band
This is important as it is the band used by most existing long distance communications systems. Germanium APDs are used extensively but III–V semiconductor alloys are increasing in usage because of the high noise levels in germanium.

(c) 1,550 nm Band
III–V APDs are used widely in the 1,550 nm band. The most common materials system in use is InGaAs/InP where the majority carrier is holes rather than electrons.

3.4.3 *Photodetectors*

3.4.3.1 Hetero-interface Photodetectors

Of available materials, APDs made with silicon have the best response, the highest gain and the lowest noise. However, silicon cannot detect at wavelengths longer than about 1 μm because of its bandgap energy, and cannot absorb light at longer wavelengths. The idea of a heterojunction APD is to replace the P-layer silicon material with some material that will absorb light in the long wavelength bands. InGaAs is one such material. The electric fields inside the device are organized such that the detection takes place in the InGaAs material but multiplication is allowed only in the silicon i-layer (this minimizes noise). Illumination can be from either side although illumination through the substrate material (silicon) is generally preferred. The big problem is how to match the lattices of the Si i-layer and the InGaAs material. This has been done using a process called wafer fusion. The resultant device is called a Silicon Hetero-Interface Photodetector (SHIP). This structure can provide a much higher gain-bandwidth product with lower noise than existing APDs (Dutton 1998).

3.4.3.2 Traveling-Wave Photodetectors

The principle of a traveling wave detector is illustrated in Fig. 3.11a. A number of P–I–N detectors are integrated into an optical waveguide such that light not absorbed in one will travel on to the next. In the TW configuration, light entering the waveguide on the left of the device encounters the P–I–N detectors one after another. Each detector would normally have a simple integrated preamplifier in this configuration. Output of the detectors is placed on an electrical waveguide on which the electrical signal travels in the same direction as the optical signal. The whole idea here is to match the speed of light propagation in the optical waveguide to the speed of electrical propagation in the electric waveguide. Thus the outputs of all the detectors add-up in phase as each detector places its output signal onto the waveguide. It is actually a significant design challenge to match

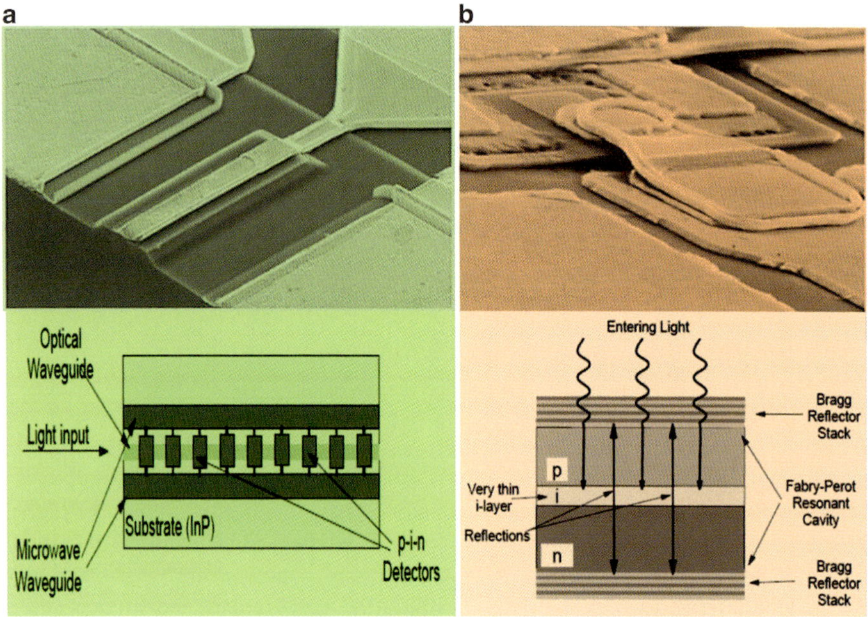

Fig. 3.11 Schematic structure of photodetectors: (**a**) traveling wave photodetector; and (**b**) resonant-cavity photodetector

the electronic propagation speed to the optical one but it can be done. Although devices such as this still cannot provide as good a quantum efficiency as a regular P–I–N diode at slower speeds. Nevertheless traveling wave devices give a very significant improvement in QE over single-stage P–I–N diodes at extremely high speeds (Dutton 1998).

3.4.3.3 Resonant-Cavity Photodetectors

An alternative to traveling wave devices for getting higher quantum efficiency at very high speeds is the Resonant-Cavity Photodetector (RECAP) illustrated in Fig. 3.11b. The principle here is just to put a P–I–N detector with a very thin i-layer inside a resonant FP cavity. The idea is that since the i-layer needs to be very thin to get fast response the light can be passed through it multiple times to give it a bigger chance of being absorbed. An FP cavity is of course a resonant cavity and is highly wavelength selective—it is a wavelength-selective filter. Thus this type of detector is highly wavelength specific and could be used in some situations as a WDM demultiplexor. Devices for the 1,550 nm wavelength band are quite difficult to build as the InP/InGaAsP system must be used and this gives very low RI contrast steps with which to build the mirrors (Dutton 1998).

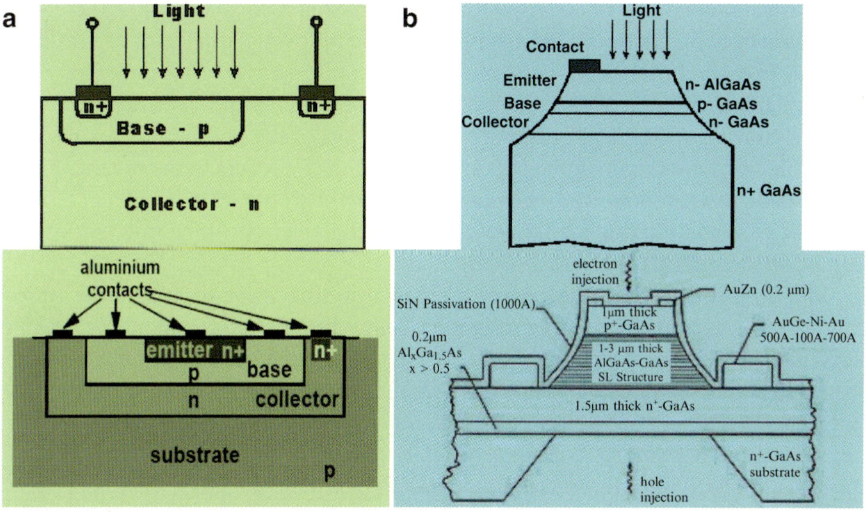

Fig. 3.12 Schematic phototransistor structure (Poole 2013): (**a**) homo-junction planar phototransistor structure; (**b**) heterojunction mesa-structure phototransistor

3.4.3.4 Phototransistors

While photodiodes fulfill many requirements, phototransistors are more suitable in some applications for providing high levels of gain, and low cost of standard devices. Although ordinary transistors exhibit the photosensitive effects if they are exposed to light, the structure of the phototransistor is specifically optimized for photo applications. The phototransistor has much larger base and collector areas than would be used for a normal transistor. These devices are generally made using diffusion or ion implantation (Poole 2013).

Early phototransistors used germanium or silicon throughout the device giving a homo-junction structure, as shown in Fig. 3.12a. The more modern phototransistors use type III–V materials such as gallium arsenide and the like. Figure 3.12b shows heterostructures that use different materials either side of the P–N junction provide a high conversion efficiency. These are generally fabricated using epitaxial growth of materials that have matching lattice structures. These phototransistors generally use a mesa structure. Sometimes a Schottky (metal semiconductor) junction can be used for the collector within a phototransistor, although this practice is less common because other structures offer better levels of performance. In order to ensure the optimum conversion and hence sensitivity, the emitter contact is often offset within the phototransistor structure. This ensures that the maximum amount of light reaches the active region within the phototransistor (Poole 2013).

Photo transistors are operated in their active regime, although the base connection is left open circuit or disconnected because it is not required. The base of the phototransistor would only be used to bias the transistor so that additional collector current is flowing and this would mask any current flowing as a result of the photo-action. For operation the bias conditions are quite simple. The collector of an N–P–N transistor is made positive with respect to the emitter or negative for a P–N–P transistor. The light enters the base region of the phototransistor where it causes hole electron pairs to be generated. This mainly occurs in the reverse-biased base–collector junction. The hole–electron pairs move under the influence of the electric field and provide the base current, causing electrons to be injected into the emitter (Poole 2013).

Phototransistor has a high level of gain resulting from the transistor action. For homo-structures, i.e., ones using the same material throughout the device, this may be of the order of about 50 up to a few hundred. However for the heterostructure devices, the levels of gain may rise to 10,000. Despite their high level of gain the heterostructure devices are not widely used because they are considerably more costly to manufacture. A further advantage of all phototransistors when compared to the APD, is that the phototransistor has a much lower level of noise (Poole 2013).

One of the main disadvantages of the phototransistor is the fact that it does not have a particularly good high-frequency response. This arises from the large capacitance associated with the base–collector junction. This junction is designed to be relatively large to enable it to pick up sufficient quantities of light. For a typical homostructure device the bandwidth may be limited to about 250 kHz. Heterojunction devices have a much higher limit and some can be operated at frequencies as high as 1 GHz (Poole 2013).

The characteristics of the phototransistor are very similar to those of a conventional bipolar transistor, but with the different levels of base current replaced by the different levels of light intensity. There is a small amount of current that flows in the phototransistor even when no light is present. This is so-called the dark current, and represents the small number of carriers that are injected into the emitter. Like the photo-generated carriers this is also subject to the amplification by the transistor action.

The major use of phototransistors is in noncommunications applications using visible (or near visible) light. Alarm systems (light beam detection) and remote controls for TV sets and automobiles are among the most common uses. Phototransistors are occasionally built as part of an integrated circuit. In this configuration they are referred to as Integrated Preamplifier Detectors (IPDs) (Dutton 1998).

3.5 Optical Receivers

A typical optical receiver is composed of an optical photodetector, a transimpedance amplifier (TIA), a limiting amplifier, and a clock-data recovery block. A schematic optical receiver model is shown in Fig. 3.13 (Rasheed et al. 2012). The received

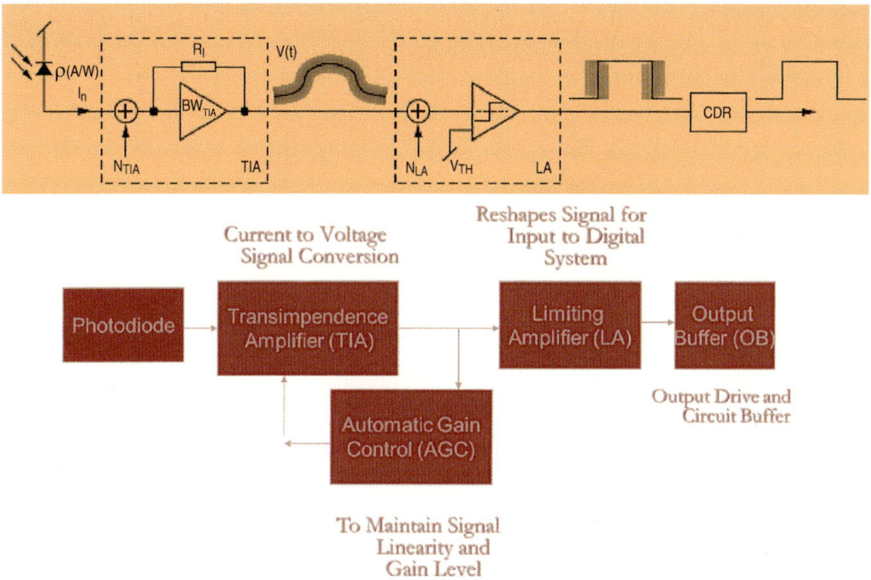

Fig. 3.13 Schematic optical receiver model (Rasheed et al. 2012)

optical signal is first converted into photocurrent and amplified by the TIA. The limiting amplifier acts as a decision circuit, where the sampled voltage $v(t)$ is compared with the decision threshold V_{TH}.

Almost all receivers used are single-ended (as opposed to differential) TIAs, with three notable exceptions: the clocked sense amplifier, the receiver of minimal change, and a differential amplifier. Special attention has to be paid to the design of the receiver arrays demonstrated in a digital environment, because this environment induces all kinds of noise, e.g., on the supply or references, which can easily interfere with the correct operation of the sensitive amplifiers. On top of this, the receivers in the array also introduce noise capable of influencing each other. Besides the properties of bandwidth, power dissipation, sensitivity and area, an important additional requirement put on an optical link is its robustness, expressed by its dynamic range and its adaptability to signal strength variations. These are directly related to source variability and to mechanical tolerances and positioning accuracy and reproducibility of the optical pathway (Neefs 2000). Different receivers have been used and will be described as below.

3.5.1 *Transimpedance Amplifiers*

A TIA is an amplifier that converts current to voltage. Its input ideally has zero impedance, and the input signal is a current. Its output may have low impedance, or in high-frequency applications, may be matched to a driven transmission line; the output signal is measured as a voltage. Because the output is a voltage and the input is a current, the gain, or ratio of output to input, is expressed in units of ohms. TIAs are commonly used in receivers for optical communications to convert the current generated by a photodetector into a voltage signal for further amplification. The current generated by a photodetector generates photo voltage, but in a nonlinear fashion. Therefore the amplifier has to prevent any large voltage by its low input impedance and generate either a 50 Ω signal (considered low impedance by many) to drive a coaxial cable or a voltage signal for further amplification (Wikipedia 2013c).

3.5.2 *Clocked Sense Amplifier and the Receiver of Minimal Change*

For parallel optical interconnects between chips and modules, transmission of DC data is desired, hence DC-coupling of detector with receiver circuitry appears to be desired.

Most traditional receivers rely on AC detector coupling, filtering out DC data and DC problems (like bad transistor matching). A third road has been explored in the receiver of minimal change, somewhere in between both options, giving the advantages of both.

Some of the major properties of this new receiver concept are highlighted (Neefs 2000):

(a) No threshold level: Different receivers in the array may operate with different light input powers. This leads to improved immunity to heterogeneity in light sources, in light channels, and in light focusing when considering the total link.
(b) No bias of any terminal is required: The electrical terminals required are Vcc, Ground, and Digital Output.
(c) Significant mismatches in transistor pairs are tolerated.
(d) Many tens of millivolts of mismatch in the thresholds of all transistor pairs are allowed with very limited penalty.
(e) DC transmission of data: Data can be transmitted without DC-balance requirement, allowing Non-Return-to Zero coding (NRZ). Signaling rate equals bandwidth.
(f) A high sensitivity (average received optical power 6.8 W) is measured at signaling rate of 250 Mbaud/channel.
(g) Small size: The whole receiver fits in a $100 \times 100\ \mu m^2$ area (100 channels/mm^2), of which one quarter is detector area (0.6 μm) technology. In a 0.25-μm technology, this receiver cell is expected to shrink to $55 \times 55\ \mu m^2$ including detector, and operating at 1 Gbaud/channel. The channel density is then 330/mm^2.
(h) Low power: The array is operated at 3.3 V and consumes 2.3 mW/channel (not counting the driver).

(i) A high-power supply rejection ratio is realized, as the detector is differential, and all subsequent amplification stages are fully symmetric and differential.

Currently, arrays of detector/receivers are available (with a near-100 % yield in standard CMOS) for demonstrating the link principle.

3.6 Optical Pathways

The approaches to the optical pathways taken in optical interconnects including inter-chip as well as intra chip are diverse. Some have implemented the free-space concept one way or another, while others have preferred guided wave approaches. For inter-chip interconnects, for instance, maintaining the 10 μm alignment of the chips relative to the pathway is difficult. Since interconnect distances are of the order of several centimeters and different materials are involved, different thermal expansion coefficients could be difficult to match. So a simple rigid free-space optical interconnect will not suffice for reliable long-term operation; some active adaptability to compensate for alignment errors has to be engineered. Without this active alignment adaptation in the free-space approach, the case for robust inter-chip interconnects is clearly in favor of the flexible guided wave approach. For intra-chip interconnects, the case remains open. The best choice of an optical pathway will not only depend on the optical performance but also on reliability, robustness, ruggedness, practicality, and obviously cost (Neefs 2000).

3.6.1 Free-Space Approaches

Free-space optics (FSO), is a reemerging technology using modulated optical beams to establish short, medium, or long reach wireless data transmission. Due to the high carrier frequency of 300 THz and the consequently high bandwidth, the most prominent advantage of free-space optical (FSO) communication links may be their potential for very high data rates of several Gbps up to 40 Gbps. Other advantages like license-free operation, easy installation, commercial availability, and insensitivity to electromagnetic interference, jamming, or wiretapping make FSO interesting for applications like last mile access, airborne and satellite communication, temporary mobile links, and permanent connections between buildings (Forin et al. 2010).

In many cases, the adoption of FSO is needed when a physical connection is not a practicable solution and where is requested to handle an high bandwidth. As a matter of fact, FSO is the only technology, in the wireless scenario, able to grant bandwidth of several Gigabits per second. The interest in this technology is also due to the low initial CAPEX (Capital Expenditure) requirement, to the intrinsic high-level data protection and security, to the good flexibility and great scalability innate in this solution. For these reasons FSO possible applications cover a wide range. Thus this technology generates interest in several markets: the first/last mile in dense urban areas, network access for isolated premises, high speed LAN-to-LAN (Local Area Networks) and even chip-to-chip connections, transitional and temporary network connection, undersea and space communication. Furthermore FSO can be used as an alternative or upgrade add-on to existing wireless technologies when the climatic conditions permit its full usage. In spite of the growing interest in space and undersea applications, in fact, the terrestrial FSO still remains of primary significance and the performance of such links is highly dependent on different weather conditions. Atmospheric effects affect the distance and the availability of the optical wireless links so not all the geographical sites are suitable for this kind of broadband solution. Links as long as 7 km are in operation, but prior to the deployment of wireless optical links the average weather condition must be evaluated to estimate the expected outage time on the link in that area. The outage will depend on the link length and on the persistence of adverse weather conditions. So in general, it can be affirmed that short enough links, hundreds of meters, will be operational also with the worst possible weather conditions. Besides commercially available opaque systems, where the optical signal is terminated to an electronic receiver and subsequently sent through the atmosphere by means of a dedicated laser, a fully transparent configuration has been developed. The bandwidth achievable in the system is comparable to the optical fiber one. For transparency, in fact, is meant launching and collecting power directly through single-mode optical fibers. Such extremely high bandwidth wireless systems are gaining more and more interest especially in the last mile scenarios. In space communications, FSO links are considered the ultimate media to establish high bit-rate data links between satellites. They use diffraction-limited laser beams with large-aperture optics (telescopes) and require accurate pointing and tracking technology. Using these space laser-communication technologies and terrestrial fiber-optic components, FSO terminal that provides seamless connectivity with terrestrial fiber-optic networks for link distances of up to 1 km. Free-space optics are additionally used for communications between spacecraft. The optical links can be implemented using infrared laser light, although low-data rate communication over short distances is possible using LEDs. Maximum range for terrestrial links is in the order of 2–3 km, but the stability and quality of the link is highly dependent on atmospheric factors such as rain, fog, dust, and heat. In outer space, the communication range of free-space optical communication is currently in the order of several thousand kilometers, but has the potential to bridge interplanetary distances of millions of kilometers, using optical telescopes as beam expanders (Forin et al. 2010).

Relying on an unobstructed line-of-sight, FSO links are strongly influenced by atmospheric conditions reducing or influencing visibility, such as fog, precipitation, haze, and scintillation. Fog, as one can expect, is the most critical effect affecting attenuation and, subsequently, availability of the FSO link. Attenuation is caused by scattering, resulting from the fact that the size of the fog particles is in the order of the wavelength of optical and near-infrared waves (as they are used for FSO). Consequently, link distances in coastal or metropolitan environments which are prone to fog are limited to a few hundred meters. Radio Frequency (RF) links on the other hand show almost negligible fog attenuation if the carrier frequency is chosen accordingly, while they usually suffer from other precipitation types like rain and wet snow. Combining these two technologies to an FSO/RF hybrid network may increase overall availability significantly, guaranteeing quality-of-service and broadband connectivity regardless of atmospheric conditions. Regarding to increasing availability, bandwidth efficiency, or minimizing system complexity, hybrid systems can be categorized in three different groups (Forin et al. 2010): redundant, load-balancing, and switch-over hybrids.

Redundant systems: These systems duplicate data and transmit it simultaneously over both the RF and the FSO link. As a consequence, the data rate of both links has to be equal, resulting in either a requirement for very high frequencies on the RF link or a relatively high FSO underutilization. Moreover, systems which duplicate and recombine data are necessary. Redundant systems provide a high reliability, but suffer from the fact that both links have to be active all the time, wasting a significant amount of energy (Forin et al. 2010).

Switch-over systems: These systems transmit data only over one link, which is chosen according to link availability. Usually, since the FSO link allows higher data rates, it is chosen as a primary link whereas the RF link acts a backup. Consequently, data rates of both links need not be identical, if one accepts a reduced bandwidth during fog events. Switch-over systems require multiplexers on both ends, algorithms choosing the active link, synchronization, and accurate, timely measurement data of the optical signal strength. However, these algorithms save energy by transmitting over one link only, and can be connected to standard network equipment without protocol overhead (Forin et al. 2010).

Load-balancing systems: These most sophisticated algorithms distribute traffic among the links according to the quality of their connectivity, thus exploiting the full available bandwidth each time. Besides a measurement of the link quality, these systems require recombination systems on either sides of the hybrid link, often resulting in either significant protocol overhead or high complexity codes which automatically distribute data among different links (Forin et al. 2010).

3.6.2 Guided Wave Approaches

The guided wave approach to the optical pathway can be POF (plastic optical fibers) ribbons, imaging fiber bundle, and the on-chip rigid wave guides.

3.6.2.1 POF Ribbons

Optoelectronic FPGA demonstrates a reliable, practical, and manufacturable solution to the inter-chip optical interconnect problem: an array of POFs. A standard PGA package is augmented with a carrier tile and a spacer to get very accurate alignment. To this package, POF ribbons can be easily attached and detached, possibly by the end user. Inter-MCM interconnects, and possibly inter-chip interconnects, require a flexible transmission medium, preferably with possibilities for connectorization. The flexibility of the interconnect medium, in combination with low bending radii (due to headroom limitations) and the parallel nature of the interconnect approach all point in the direction of a fairly new product: small-diameter plastic optical fiber. The numerical aperture of this fiber is high (typically 0.5), facilitating tight bending without excessive loss, and the diameter can be as small as 125 μm (120 μm core). The main drawback of this medium is the rather high absorption (about 12 dB/m at 980 nm, about 6 dB/m at 850 nm). For bending radii larger than 2 mm, excess losses do not exceed 0.5 dB/90 (Neefs 2000).

3.6.2.2 Imaging Fiber Bundles

An imaging fiber bundle is a waveguiding medium that consists of tens of thousands of microcore glass fibers (with a pitch of 8–20 μm). Due to the manufacturing process these fiber cores are arranged in a correlated way in a hexagonal pattern at the entrance and exit surfaces, while in between they are loose and hence very flexible. Usually they are used for transferring images from the input plane to the output plane, e.g., in medical applications or in high energy physics. The physical properties of imaging fiber bundles can be tuned to meet the demands of massively parallel inter-chip interconnects. Several inherent advantages make them unique for the envisioned application: massive parallelism, large numerical aperture, flexibility, variable dimensions, and robustness (Neefs 2000).

3.6.2.3 On-Chip Rigid Wave Guides

These include rigid GaAs/AlGaAs-based wave guides, polymer-based plane wave guides etc. Polyimide have been investigated as an alternative.

3.6.3 Reconfigurable Optical Pathways

Reconfigurable optical networks have been studied extensively to adapt routing of high bandwidths between inputs and outputs, ideally with low power, low cross talk, low latency between input and output, fast and high reconfigurability, in a reliable, scalable, and easily manufacturable manner. In reality, all these requirements are difficult to realize together, and trade-offs have to be made. The term reconfigurable optical networks describes a whole spectrum of approaches differing in emphasis on the optical aspect: from the purely optically reconfigurable network where no electronics are involved at all, over electrical reconfigurability but still with optical routing, to the almost completely electrically reconfigurable network with electrical routing, where only inputs and outputs are optical interconnections to overcome the I/O bottleneck. Micromirrors can be used to steer light beams, but reconfiguration is still electrical. Two instances of reconfigurable optical networks were studied (Neefs 2000): (a) the cross-bar system, where optical broadcast is used to its advantage, but reconfigurability and routing takes place electronically, and (b) an optical polarization-based bypass-exchange network, where routing is implemented optically and reconfiguration is electrical. The latter network consists of bypass-exchange elements, and is based on polarization because this approach allows high light efficiency. The requirement of high light efficiency is dictated by scalability requirements, which amounts to the addition of more cascaded levels of switching elements. Most realizations of reconfigurable optical networks involve free-space optics. The approach typically consists of a very compact, multilayer system, where each layer consists of a planar configuration. The central element in each layer is a ferro-electric liquid crystal cell, around which four polarization sensitive holographic optical elements are concentrated in a folded, compact approach. Cross talk between channels is measured to be 1:100. Improvements in misalignment and spurious reflections could increase this number to 1:1,000. An important advantage of the optical routing approach is the low latency between inputs and outputs which is not hampered by the reconfigurability, as would be the case in electrical routing approaches. Just like all optical routing networks known to the present day, reconfiguration is slow—limited by the liquid crystal (40 kHz). For many application areas this is acceptable, for some it isn't. The telecom area is one of the big application areas of reconfigurable optical networks, e.g., to reconfigure interconnection topologies and thus dynamically balance traffic among communication lines. The latter application area does not necessarily require very fast reconfiguration. Still, the faster the reconfiguration, the smaller the down time of the optical links involved, the better. Multiprocessing is another application area. It has been shown that even a slowly reconfigurable interconnection topology can be advantageous to performance. Traditional processor load-balancing techniques use process migration or task stealing. The technique of interconnection reconfiguration allows dynamic traffic balancing by changing the interconnection topology, thus possibly reducing congestion and reducing routing latency. The time scale at which such reconfiguration takes place may be orders of magnitude slower than the processor clock frequency (Neefs 2000).

3.6.4 Guided Wave Versus Free Space Optics

A key aspect of an optical interconnect system is the optical pathway between light sources and detectors. It has a strong impact on the performance of the interconnect system in terms of power transfer efficiency, interconnect density, and bandwidth. Moreover the type of optical pathway has a profound impact on the size and geometric topology of the system, on the assembly method and the required assembly precision and therefore on the cost. There are many types of optical pathways possible, and at present there is no clear consensus on which one to choose. In general, optical pathways can be divided into free space and guided wave solutions (Neefs 2000).

In free space solutions one uses optics similar to the optics used in cameras: the 1D or 2D array of light sources is imaged onto the 1D or 2D array of detectors by means of a lens system, possibly including mirrors. In such a system one makes extensive use of the fact that photon beams can cross each other without interfering. The interconnect density is in principle only limited by the diffraction limit, although most systems have a density much lower than that. The main problem associated with free space systems is that the detector array needs to be aligned to the image of the source array with an accuracy which is a small fraction of the pitch of the array. In case the pitch is small and the distance between source and detector plane is relatively large, this leads to a complex mechanical system design if only passive alignment methods are used or if the system needs to work over an extended temperature range. While free space optics can in principle operate with both incoherent (e.g., LEDs) and coherent light sources (e.g., VCSELs), the optical design is much simpler in the case of coherent sources because it allows to work with optics with a relatively small numerical aperture. In some free-space systems one uses special types of imaging optics such as GRIN lens arrays or arrays of microlenses. These systems are very scalable in the size of the array being imaged. A special case is the relay system in which the diverging beams from the light sources are collimated by a microlens array and then propagate as parallel and spatially separated beams across a certain distance up to a second microlens array which focuses the beams onto a detector array. This type of free space system does actually have some similarities with a guided wave system (Neefs 2000).

Guided wave systems are based on an array of optical wave guides, which guide light by total internal reflection. In these systems light beams are generally kept separate throughout the pathway (although wave guides can in principle cross each other without leading to signal interference). In general one uses one waveguide per light beam. There are many variants: the wave guides can be single-mode or multimode waveguides. Single-mode waveguides have a smaller cross section and therefore allow for a higher interconnect density but they require tighter alignment tolerance between light source and waveguide and between waveguide and detector. Furthermore single-mode waveguides are only practical when coherent light sources are being used. Multi-mode waveguides however can equally well be used with either coherent or incoherent sources. Against that comes the fact that the relatively large size of multi-mode wave guides enforces the use of relatively large detectors, which limits receiver bandwidth. There are many waveguide technologies: planar waveguide technologies in which the wave guides are patterned by microelectronic techniques at the surface of a substrate; alternatively one that can use bundles of optical fibers, either glass or plastic optical fibers. Fibers are flexible and this has the important advantage that the light source array does not need to be aligned with respect to the detector array. One only needs local alignment between sources and waveguides and between waveguides and detectors. Because of this, fiber bundle-based systems are much more similar to conventional electronic systems than systems with other types of optical pathways. A very special type of guided wave pathway is the image fiber. This is a high-density bundle of fibers—either glass or plastic—as used in endoscopic medical procedures and it can be used as an imaging system whereby each light beam is transported through a number of fibers. Finally some optical pathways are hybrid variants somewhere between free space and guided wave systems. The best known example is the parallel plate system, in which light beams zigzag within a relatively thick transparent plate. These beams undergo free space diffraction effects but are nevertheless guided by the plate. The plate can incorporate diffraction gratings to deflect light towards detector arrays (Neefs 2000).

3.7 Optoelectronic Device Hybridization and Integration

As silicon is an indirect band gap material and its light emission is difficult, light sources are usually made in other materials, mostly III–V (e.g., GaAs), which then have to be integrated with the silicon system chip. Furthermore, for some light sources, the wavelength is not compatible with good absorption in silicon so that detector diodes are also often made in the better absorbing III–V materials. In that case, detector diodes should also be integrated, one way or another, with the silicon system chip. Some integration options have been investigated, such as flip-chip bonding and monolithic hetero-epitaxial integration.

3.7.1 Bonding Techniques

Arrays of light sources (or modulators) and sometimes arrays of detector diodes can been flip-chip bonded onto silicon chips. For example in the optoelectronic FPGA, the detector and VCSEL chips were flip-chip bonded to the silicon CMOS chips by solder reflow. Within the limits of the measurement techniques, it was shown that there was essentially no difference in the performance of the detector or VCSEL chips after mounting, compared to the probed values observed before solder reflow. Furthermore, underneath the flip-chip paths, no precious silicon estate is lost, there is functional and operating circuitry present. Earlier tests with ring oscillators also showed no or minimal influence on their functionality and performance. In order to lower the cost of flip-chip bonding, instead of bonding on a chip per chip base, wafer scale bonding was investigated. Bonding was demonstrated of a fully processed silicon wafer with an entire GaAs wafer, on which epitaxial structures for optoelectronic components have been grown. After selective backside thinning of the III–V wafer and removal of the unused layers, the only process left is contact formation for the III–V devices and their metallic interconnection with the underlying Si devices. Essential to the performance and the reliability of the components is the low temperature at which the wafer bonding process takes place.

Another very important measure of the process is a high yield. The wafer bonding approach provides a technology for high-performance integrated O/Is, excellent manufacturability and full compatibility with the mainstream Si CMOS technology. In particular, it concerns process steps which can be easily added without interfering at all with the conventional CMOS fabrication process. The main advantages of the wafer approach, of bonding entire GaAs and Si wafers that makes it attractive for industrial use in the future, are (Neefs 2000): (a) Once the fabrication of the silicon electronic part, including metallization, has been finished there is no return of the wafers to the silicon foundry. So there is no restriction on the silicon technology imposed by further processing. (b) The growth of the MQW layers on a GaAs substrate is done completely independently, so there is no concern for mutual contamination with silicon, taken also into account that the bonding process will be realized at low temperature. (c) The wafer bonding process offers the possibility for high integration of Si and III–V devices since the pad sizes used for the interconnection of the devices can be significantly reduced.

Three critical points determine the applicability and manufacturability, and therefore the success, of this low-temperature wafer bonding approach (Neefs 2000):

(a) Keep the bonding temperature below 450 °C to prevent aluminum interconnect degradation of the already fabricated CMOS circuits.
(b) Avoid structural quality degradation (crystal defects, cracking) of the GaAs material, which would results from the large difference in the thermal expansion coefficients of Si and GaAs, when the wafer bonding procedure requires heating to high temperatures such as 500 °C typically used for the conventional bonding in a hydrogen atmosphere.
(c) Achieve a GaAs/Si bonded system which has the mechanical integrity to successfully withstand the additional processing steps for the fabrication of optoelectronic devices. The most difficult problem which has to be overcome in bonding of GaAs to Si is the large difference between the thermal expansion coefficients (TECs) of Si and GaAs (TEC of GaAs is almost twice that of Si). This usually leads to de-bonding at thermal annealing over 160 °C and, as a consequence of such a low annealing temperature, the surface energy is limited to 0.20–0.30 J/m^2. SOG-(silicate glass) spin coated Si wafers were room temperature bonded with semi-insulating 4″ GaAs (100) wafers (625 μm thick). The surface energy of the room temperature bonded wafers was in the range of 0.3–0.5 J/m^2. This value is very high compared with the surface energy in the case of silicon-to-silicon hydrophilic bonding (0.1–0.155 J/m^2). Measurements showed that, after annealing, the bow of the GaAs/Si bonded comes back to zero without any hysteresis and without affecting the bonded interface. The maximum temperature at which the GaAs/Si wafer pair can be annealed was determined to be about 280 °C, a higher temperature made the wafers de-bond and/or shatter. The GaAs wafers were ground down to 30–40 μm and polished down to 10 μm. Thinning one of the wafers down to some tens of microns, can increase the maximum temperature limit to values of about 450 °C. The surface energy achieved after annealing at 200 °C was about 1.75 J/m^2 which allows the

4″ GaAs/Si bonded samples to sustain harsh mechanical processing like grinding and chemical mechanical polishing (CMP). SOG was demonstrated to be suitable as adhesion layer. After a planarization process, CMOS test wafers were bonded to plain GaAs wafers. The bonding quality was investigated by ultrasonic microscopy. It shown that due to the initial wafer bow the polishing was more effective at the outer region of the wafer, where the bonding quality is good. In the less unbonded area one can guess the underneath Al structures of the CMOS wafer. The dishing effect at the polishing stage make bonding being effective only in the highest regions (Al interconnections). The laser diodes and photodetector devices are etched in the same epitaxial layer, after completing the "epitaxial GaAs"/"VLSI Si" wafer bonding and the backside GaAs substrate removal. The optoelectronic devices are based on a single-MBE-grown structure, consisting of GaAs/AlGaAs MQW-separate confinement heterostructures. An RIE-based process for defining the laterally integrated O/E devices with minimum degradation of etched mirror lasers was employed. The optimum RIE process was selected to achieve etched mirror laser diodes, with good quality and reproducibility (Neefs 2000).

3.7.2 Monolithic Integration

Monolithic integration is the other approach to integrate GaAs devices with silicon-based devices. The monolithic integration method is usually based on the use of a new conformal epitaxy technique, known to produce low defect density hetero-epitaxial III–V layers on silicon. It has been shown that high quality optoelectronic devices can be integrated monolithically on processed silicon ICs and that the technique is compatible with standard (1 μm) CMOS technology. An efficient monolithic integration technique could be a more flexible and powerful tool than the volume and yield-limited, labor-intensive hybrid approaches, for the realization of future optoelectronic ICs. However, wafer scale flip-chip bonding is a strong rival. An optimized monolithic integration procedure can achieve (Neefs 2000):

(a) Very low roughness seed layers with a quality fulfilling the conformal growth requirements were obtained with a low-temperature process (below 600 °C).
(b) The conformal growth temperature was decreased up to the range 720–750 °C with a sufficiently high growth rate (up to 8 μm/h), so that the maximum total thermal budget equivalent to the integration procedure is 5 h at 750 °C.

The monolithic integration technique could be used to integrate various optoelectronic devices on complex optically interconnected systems. The development of industrial IIIV epitaxial equipment compatible with large substrates (at least 6″ or 8″) will be very useful to fully exploit the results obtained and the demonstrated potentialities of the conformal growth technique. The conformal growth is further developed with MOCVD, which enables to use only one epitaxial technique (MOCVD) for the three epitaxial steps of the conformal integration

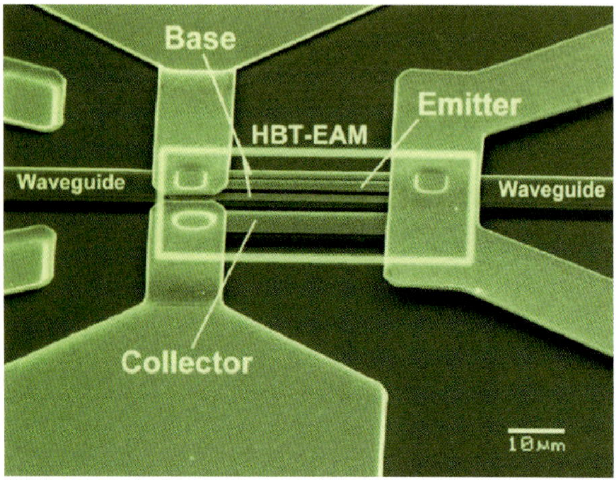

Fig. 3.14 Schematic illustration of silicon-based light emission device (Jager 2007)

technique (seed, conformal layer, and device structure growths). This will also enable to further decrease the temperature used for conformal growth (in the range of 680–700 °C). An additional decrease of about 50 °C can also be obtained in the case of integrated InP-based materials and devices. Additional improvements of the quality of the conformal layers and of the device structures regrown on it could also be made, leading to electro-optical properties for the Si-integrated III–V devices fully comparable with homo-epitaxial ones (Neefs 2000).

3.7.3 Silicon-Based Light Emission

Ideally silicon-based light emission would represent the best solution of the interconnects problem if reliable and efficient Si-based light emitting devices were available. In an optoelectronic traveling-wave device, as shown in Fig. 3.14, an optical waveguide (e.g., a strip loaded or a buried waveguide) is used for optical wave propagation and an electrical transmission line (e.g., microstrip or coplanar waveguide) for guiding the microwave, usually in the same direction (Jager 2007). In the region where the electrical fields overlap, the optoelectronic interaction occurs. From a physical point of view the interaction is a nonlinear or active process. The photodetector and the laser diode are basic examples of two-port devices where optical power is converted into electrical power and vice versa. Typical three-port devices are electrically controlled optical modulators/switches or optically controlled microwave modulators/switches. Due to the inherent nonlinearity these devices are further used for optoelectronic mixing of input electrical and/or optical signals where the output signal can be electrical or optical.

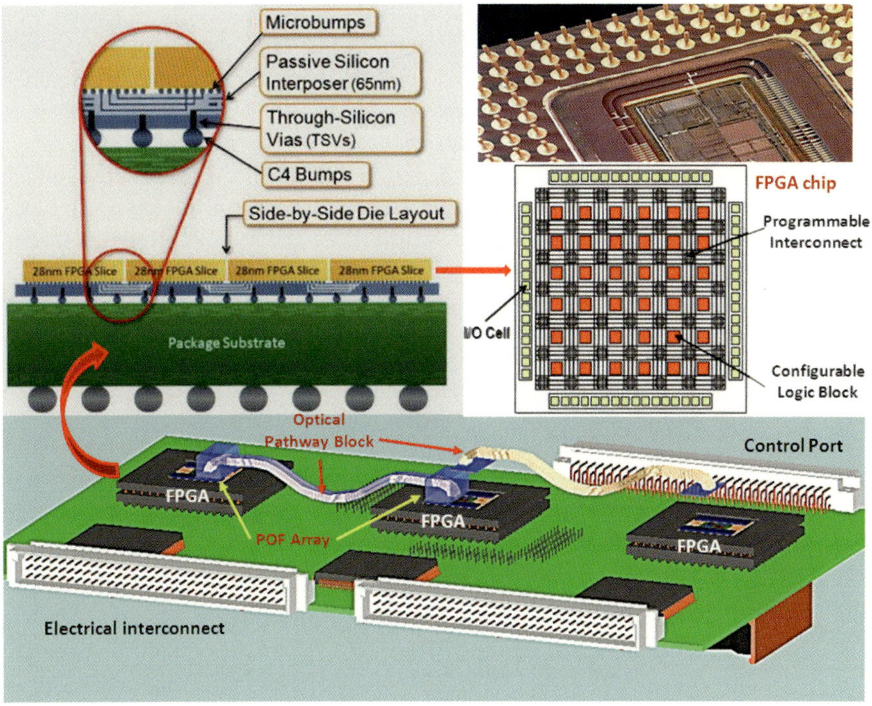

Fig. 3.15 Schematic illustration of optoelectronic FPGA modules (Neefs 2000)

As a result, such microwave-optical interaction devices show a variety of optoelectronic functions where in special devices different functions may be achieved simultaneously (Jager 2007).

3.7.4 Multifunctional Device

A multifunctional device, for instance, consists of an electro-absorption modulator integrated into the structure of a hetero-bipolar transistor (HBT). Such a device includes the common transistor characteristics as well as the modulator behavior leading to a novel approach to optoelectronic integrated devices and circuits. A further key integration technology is the optical coupling of the light between the optical waveguide and the external connection which usually is the optical fiber. This technique of fiber chip coupling includes the processing of V-grooves into the substrate or mother board and the tapering of the optical fiber. A further important technology is related to the adjustment and fixing of the fiber with respect to the optical waveguide. A practical, manufacturable and economically viable system should deal with the issue of packaging. For instance, Fig. 3.15 shows a manufacturable approach based on a standard PGA package with detachable POF arrays. The key problem in this arrangement is that alignment between the POF array and the optoelectronic device arrays is the result of a cascade of alignment steps (Neefs 2000): (a) mask level alignment errors between active area and solder bumps, (b) solder bump misalignment, (c) solder-bump positioning error on the CMOS, (d) CMOS chip positioning error, (e) carrier MT-hole alignment, (f) MT-pin alignment tolerance, (g) hole positioning error in the pathway block, (h) positioning error of the POFs in the optical pathway block. A passive alignment scheme would result in excessive misalignment levels, in turn implying a severe optical power penalty and intolerable cross talk levels. Therefore, a special alignment process has been developed, allowing to reduce lateral and height misalignment to about 20 μm. This approach is based on index alignment of a spacer through the use of a master pickup tool.

The application areas of optical interconnects are closely related to the advantages they provide: optical interconnects are good in providing huge bandwidths. So optical interconnects are interesting in applications which demand huge bandwidths and/or high interconnectivity. Precisely at these high bandwidths, electrical interconnects (of several centimeters) cope with problems: cross talk, lines needing to be properly terminated and EMI. Electrical interconnects also struggle with high interconnectivity problems which require dense interconnects, the so-called escape perimeter problem. This problem emerges at the chip boundary or for electrical area I/O at the MCM/PCB level. Hence optical interconnects are useful to interconnect chips with ultrahigh bandwidth demands. The carefully laid out and terminated electrical MCM/PCB wiring on tens of layers, connecting two chips, can be replaced by one small connectorized optical cable consisting of a bundle of wave guides. One application which requires both huge bandwidths and high interconnectivity is high-performance switching and routing, involving crossbar switches and multistage networks. High-performance switching and routing is relevant to a number of application fields. The first and probably best known field is telecommunications and the enormous demands of the Internet. Another interesting and very important application is in multiprocessing. Here too, high interconnectivity and/or high bandwidths are required, e.g., to keep caches coherent or in realizing a low-latency interconnect between processors and/or memories in large multiprocessor systems. In particular in the emerging (internet) servers, huge memories and very high bandwidths are required. At a much lower architectural level, an interesting application of massive optical inter-chip interconnects is in connecting several electrical FPGAs into one large optoelectronic FPGA. In general, all applications which require multichip solutions with high interconnect densities requiring high signaling rates are potential applications. A caveat for this kind of application however is that latency must not be critical, as the latency of current-day optical interconnects still is several nanoseconds. Even deeper down in the architectural hierarchy, at the intra-chip level, electrical interconnects have no problem with bandwidth. However, the latency of long on-chip interconnects becomes problematic. The latency of current optical interconnects is still relatively high, but future optical interconnects with lower latency will be developed. It remains an interesting but unanswered question as to whether optical interconnects will attain a latency lower than cross-chip electrical interconnects. Although some argue this will be the case, only the future can tell (Neefs 2000).

3.8 Nanomaterials for Optoelectronic Devices

Semiconductor nanowires (NWs), nanocrystals, and carbon nanotubes offer many opportunities for the assembly of nanoscale devices and arrays by the bottom-up paradigm. Moreover, these nanomaterials demonstrate new and/or enhanced functions crucial to many areas of technology. Central to realizing applications through a bottom-up paradigm is the rational control of key nanomaterial parameters, including chemical composition, structure, size, morphology, and doping. It is these parameters that determine, for example, electronic and optoelectronic properties critical to predictable device function. Significantly, semiconductor NWs represent the nanomaterial system where these key parameters have been best controlled to date. First, an underlying conceptual framework has been developed to enable the growth of nanowires of virtually any uniform composition and structure, with the wide range of reported nanowires confirming these models. Second, in many cases controlled P- and N-type doping, which is critical to almost any active device application, has been demonstrated. Third, the control over nanowire growth has enabled the electronic and optoelectronic devices impact many areas of society, from simple household appliances and multimedia systems to communications, computing, and medical instruments. Given the demand for ever more compact and powerful systems, there is growing interest in the development of nanoscale devices that could enable new functions and/or greatly enhanced performance. Semiconductor nanowires are emerging as a powerful class of materials that, through controlled growth and organization, are opening up substantial opportunities for novel nanoscale photonic and electronic devices (Li et al. 2006).

While homogeneous doped NWs represent key building blocks for a variety of electronic and optoelectronic devices, the ability to prepare controlled and diverse axial and radial heterostructures sets NWs apart from other nanomaterials, such as carbon nanotubes, and as represents a substantial advantage for the development of increasingly powerful and unique nanoscale electronic and optoelectronic devices crucial to future applications. Homogeneous doped NWs represent key building blocks for a variety of electronic and optoelectronic devices. A prototypical example of a homogeneous doped NW device with broad potential for applications is the NW field-effect transistor (FET) and, moreover, studies of FETs enable evaluation of the performance level of NWs compared with corresponding planar devices. Representative NW materials include Si, Ge, InAS, and GaN can be prepared with complementary N- and P-type doping. The high performance of these homogeneous NW devices has been further verified by low-temperature measurements. For example, proximity-induced superconductivity has been realized in InAs NWs contacted with Al-based superconductor electrodes. The results indicate Schottky-barrier-free contacts between NWs and metals, and show that the phase-coherence length for electron propagation in these NWs is up to hundreds of nanometers.

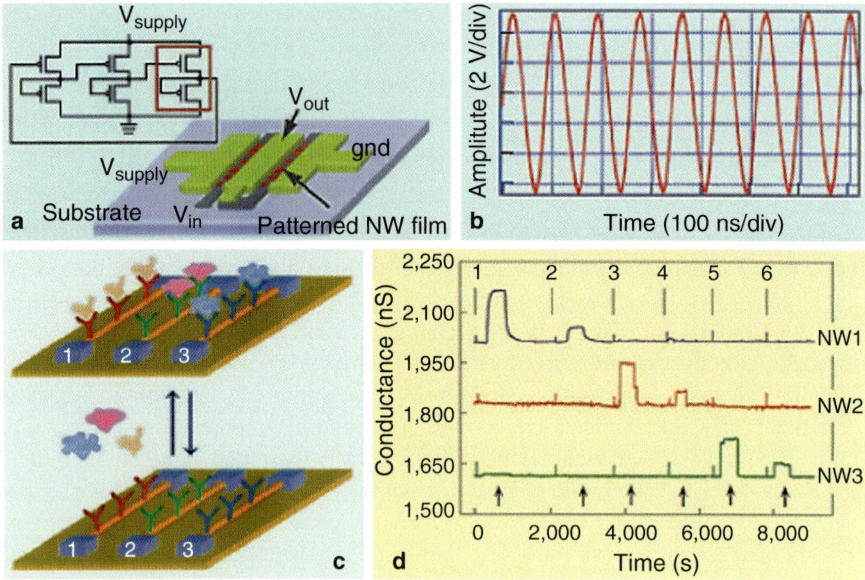

Fig. 3.16 Applications of Si NW FETs (Li et al. 2006): (**a**) a multi-NW inverter on a glass substrate and a circuit diagram for a NW ring oscillator consisting of three inverters in series; (**b**) oscillation at 11.7 MHz for a P-Si NW ring oscillator with $V_{supply} = 43$ V; (**c**) multiplexed protein detection by three Si NW devices in an array (1, 2, 3); (**d**) conductance versus time data recorded for the simultaneous detection of prostate specific antigen (PSA), carcinoembryonic antigen (CEA), and mucin-1 on a P-Si NW array in which NW1, NW2, and NW3 were functionalized with monoclonal receptors for PSA, CEA, and mucin-1, respectively

In addition, Si NWs have demonstrated as a clean system with little or no structural/dopant variation on this length scale. In contrast, lithographically defined Si NWs have much greater structural and/or dopant fluctuations and yield a length scale for electronically distinct regions that is over an order of magnitude smaller. Notably, coherent transport has been observed in molecular scale Si NWs down to the last few charges, which further demonstrates the high quality of the NW material, long carrier mean free-paths, and the potential to serve as a unique building block for both low- and room-temperature applications (Li et al. 2006).

An attractive feature of NW FETs is that there is a separation of the high-temperature growth processes, which are used to prepare high quality single-crystalline material, and the low-temperature assembly and contact deposition, which enables rapid design and fabrication of a host of single- and multi-NW device structures on virtually any substrate. Two distinct applications of this key concept include the development of high-performance, multi-NW devices and circuits on noncrystalline substrates and arrays of single-NW FETs for sensing. Moreover, NW FETs and inverters can be configured on flexible plastics with properties comparable to high-performance single-crystal planar devices. It is possible to assemble more complex ring oscillator circuits (Fig. 3.16a) by fluid-based assembly and patterning. The necessary on-chip device integration is achieved during fabrication without the need for external wiring because of the high reproducibility of Si NW FETs. Significantly, characterization of these NW ring oscillators demonstrates very stable and self-sustained output voltage oscillations with a frequency of 11.7 MHz (Fig. 3.16b), which is substantially better than organic and amorphous semiconductor devices processed at low temperatures. In addition, NW FETs have emerged as extremely powerful sensors for ultrasensitive, direct, label-free detection of biological and chemical species. Binding to the surface of an NWFET is analogous to applying a gate voltage, which leads to the depletion or accumulation of carriers and subsequent changes in the NW conductance (Fig. 3.16c). The small diameters and high performance of NW FETs yield high sensitivity, with the detection of single virus particles representative of their unique power. NW FET sensors can also be readily integrated into electrically addressable sensor arrays, which demonstrate multiplexed, real-time detection of multiple disease marker proteins at the femtomolar level (Fig. 3.16d). This work offers potential for powerful sensors that could significantly improve healthcare in the future (Li et al. 2006).

NW building blocks and device architectures more complex than the NW FETs can open up new opportunities that differentiate NW-based devices from conventional paradigms. The crossed NW architecture is a clear example since the key device properties are defined by assembly of the two nanowire components and not by lithography. Hence, the dimensions of the crossed NW device are limited only by the NW diameters, which makes the architecture readily scalable for high-density integration and, depending on the choice of NWs, the structure can yield a variety of critical device elements, including transistors and diodes. For example, crossed NW FETs can be configured from one NW as the active channel and the second crossed NW as the gate electrode separated by a thin SiO_2 dielectric shell on the Si NW surface, with the gate on the surface of one or both of the crossed NWs. This concept was first demonstrated using Si NWs as the channel and GaN NWs as the gate electrodes, including the integration of multiple crossed NWFETs on a single Si NW channel to demonstrate both NOR logic-gate structures (Fig. 3.17a) and basic computation. Selective chemical modification is used to differentiate specific cross points in a four-by-four crossed Si NW FET array (Fig. 3.17b), thus allowing selective addressing of the four individual outputs (Fig. 3.17c). Significantly, these results provide a proof-of-concept that assembled crossed NW arrays can serve as the basis for addressable integrated nanosystems in which signals are restored at the nanoscale. The crossed NW concept has also been used to create nanoscale P–N diodes by crossing P- and N-type NWs. This concept was first demonstrated for P–N crossed InP NW junctions and subsequently extended to crossed NW P–N diode junctions with P-Si/N-GaN_{60}, P-GaN/N-GaN_{31}, and other systems. Transport measurements have shown that nanoscale junctions formed in this way exhibit the expected rectifying behavior and, moreover, band-edge emission at the nanoscale cross points in forward bias (Fig. 3.18a). Significantly, the capability to assemble a wide range of different n-type direct bandgap NWs, including GaN (ultraviolet), CdS (green), and CdSe (near infrared), with P-type Si NWs as a common p-type indirect bandgap material has enabled the facile creation of multicolor LEDs on a single substrate in a manner not possible with conventional

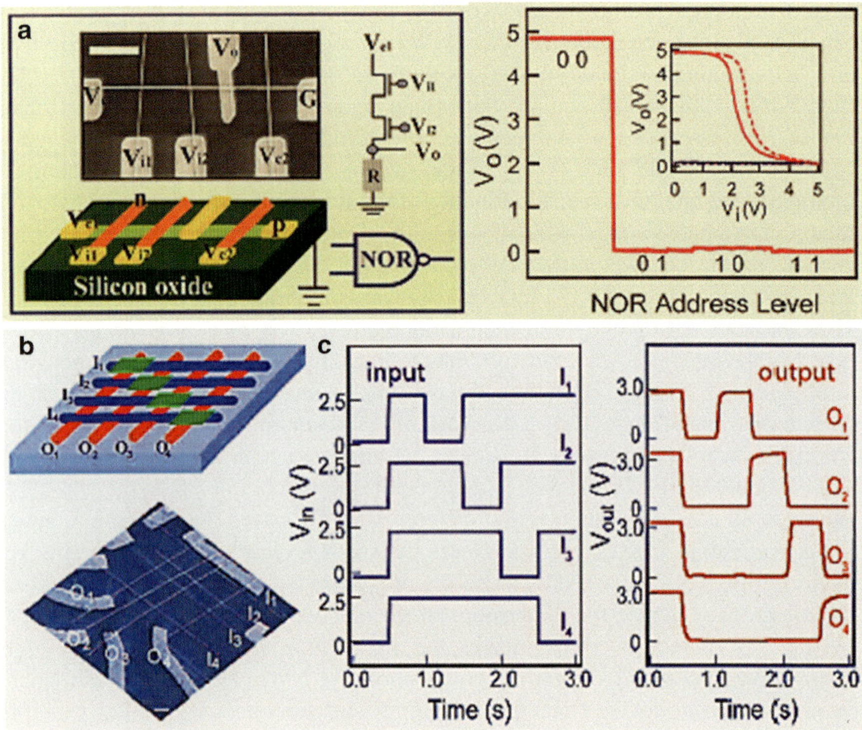

Fig. 3.17 Crossed NW electronic devices (Li et al. 2006): (**a**) a logic NOR gate constructed from a one-by-three crossed NW junction array using one Si NW and three GaN NWs, and output voltage versus the four possible logic address level inputs; (**b**) a four-by-four crossed Si NW address decoder with four horizontal NWs (I1–I4) as inputs and four vertical NWs (O1–O4) as signal outputs; (**c**) real-time monitoring of the V_g inputs (*blue*) and signal outputs (*red*) for the four-by-four decoder

planar technology. Crossed NW architecture can also generalize hybrid devices consisting of n-type CdS NWs assembled onto p-type Si electrodes defined in heavily p-doped planar substrates (Fig. 3.18b). When the injection current increases above the threshold, these hybrid NW devices show a superlinear increase in the electroluminescence (EL) intensity at the end of the nanowire, as well as simultaneous peak narrowing to a single-mode emission with instrument-resolution limited width, corresponding to the first demonstration of a nanoscale electronic injection laser (Fig. 3.18c). In addition to nanoscale light sources, crossed NW P–N junctions can also be configured as photodetectors critical for integrated photonics. For example, Fig. 3.18d shows avalanche multiplication of the photocurrent in nanoscale P–N diodes consisting of crossed Si/CdS NWs. These NW avalanche photodiodes (nanoAPDs) exhibit ultrahigh sensitivity with detection limits of less than 100 photons and subwavelength spatial resolution of 250 nm. Moreover, the elements in nanoAPD arrays can be addressed independently without electrical cross talk (Fig. 3.18e; Li et al. 2006).

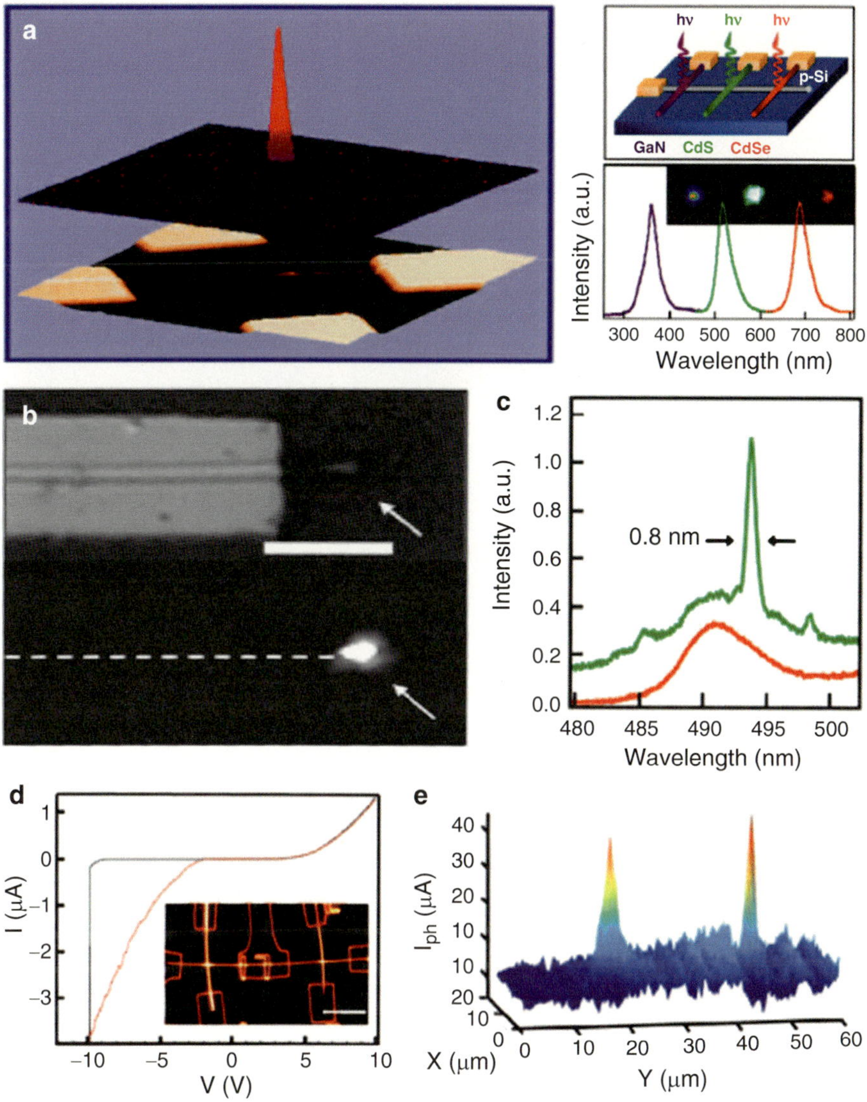

Fig. 3.18 Crossed NW photonic devices (Li et al. 2006): (**a**) a typical N-InP/P-InP crossed NW device, overlaid with corresponding spatially resolved EL image showing the light emission from the cross point, and a tricolor nano-LED array, consisting of a common P-type Si NW crossed with N-type GaN, CdS, and CdSe NWs; (**b**) a nanolaser device fabricated by assembling N-CdS NWs on a heavily doped P-Si substrate; (**c**) EL spectra obtained from the end of the nanolaser with injection currents below and above the lasing threshold; (**d**) *I*–*V* characteristic of a N-CdS/P-Si crossed NW APD in dark and under illumination; and (**e**) spatially resolved photocurrent measured from the NW APD array as in the inset of (**d**)

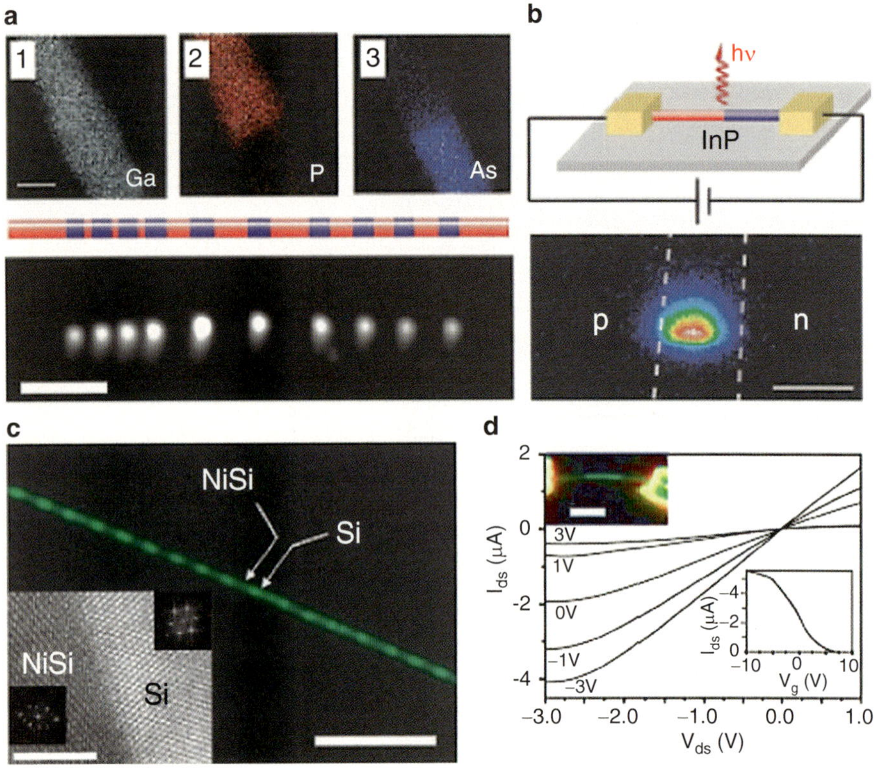

Fig. 3.19 Axial NW heterostructures (Li et al. 2006): (**a**) TEM elemental mapping of a signal GaAs/GaP nanowire heterojunction, and photoluminescence image of a 21-layer (GaP/GaAs) 10GaP nanowire superlattice; (**b**) a modulation-doped InP NW LED and image of the emission from the device; (**c**) dark-field optical image of a single NiSi/Si NW superlattice heterostructure; and (**d**) I_{ds}–V_{ds} curves of a NiSi/P-Si/NiSi heterojunction NW FET fabricated using a 30 nm diameter P-SiNW

The integration of device function at the nanoscale can also be carried out during NW synthesis by varying the composition and/or doping during axial elongation, whereby the resulting axial junctions can yield controlled nanoscale device function without the need for lithography. A representative example is a GaAs/GaP compositionally modulated axial heterostructures (Fig. 3.19a). Since GaAs is a direct bandgap semiconductor and GaP has an indirect bandgap, these NW heterostructures can be patterned synthetically and emit light as nanoscale barcodes. In addition, P–N junctions formed within individual NWs can also be prepared in a similar way. Forward-biased N-InP/P-InP single-NW devices function as nanoscale LEDs with light emission at the P–N interface as shown in Fig. 3.19b. Figure 3.19c shows the selective transformation of Si NWs into metallic NiSi NWs and NiSi/Si NW heterostructures by thermal annealing as-made SiNWs with Ni. Significantly, this method yielded the first example of atomically sharp metal–semiconductor interfaces between single metallic (NiSi) and semiconductor (Si) nanowires. In these heterostructures, Si NW FET source-drain contacts are defined by the metallic NiSi NW regions, which function as excellent ohmic contacts at room temperature (Fig. 3.19d), and thus provide an integrated solution for nanoscale contacts and interconnects. The concept of modulating axial doping has also been demonstrated for Si NWs, thereby providing another method for introducing rich function at the initial stage of building block synthesis. In addition, the synthetic control of the size and separation of modulation-doped regions can be exploited to define quantum dot (QD) structures, where the band offset caused by variations in dopant concentration produces potential barriers confining the QD. A single Coulomb oscillation period consistent with two weakly coupled QDs when the barrier n^2 is large and, as this barrier is reduced (through synthesis), the tunneling conductance between QDs is enhanced (Li et al. 2006).

Radial composition and doping modulation in NW structures represent another approach for enhancing performance and/or enabling new function through synthesis versus lithography. Figure 3.20a shows a one-dimensional hole gas system based on an undoped epitaxial Ge/Si core/shell structure. The valence band offset of ~500 meV between Ge and Si at the heterostructure interface serves as a confinement potential for the quantum well. Free holes accumulate in the Ge channel when the Fermi level lies below the valance band edge of the Ge core. Low-temperature electrical transport studies have shown distinct conductance plateaus corresponding to transport through the first four sub-bands in the Ge/Si NW (Fig. 3.20b), where the subband spacing (Fig. 3.20c), $E_{1,2} = 25$ mV and $E_{2,3} = 30$ mV. Notably, the conductance exhibits little temperature dependence, suggesting that transport is ballistic even at room temperature. The unique transport characteristics of Ge/Si core/shell NW heterostructures make them excellent building blocks for high-performance NW FETs and potential alternatives to planar metal-oxide semiconductor field-effect transistors (MOSFETs). Figure 3.20d shows Ge/Si nanowire devices with scaled transconductance (3.3 mS/μm) and on-current (2.1 mA/μm) values that are 3–4 times greater than state-of-the-art MOSFETs and the highest obtained on NW FETs. Figure 3.20e shows a clear speed advantage at a given channel length, L, for the Ge/Si NW FETs versus Si p-MOSFETs. The generality of band-structure engineering for creating NW carrier gases has been further reinforced by an electron gas in dopant-free GaN/AlN/AlGaN radial NW heterostructures. Achieving both hole and electron gases is important because they are required to enable high-performance complementary nanoelectronics, and to explore the fundamental properties of both one-dimensional electron and hole gases (Li et al. 2006).

The radial NW concept also offers substantial opportunities for NW optoelectronics since the required n- and p-type active materials can be incorporated as the core and shell, which enables carrier injection or collection over a much larger area than possible in crossed NW devices and axial NW heterostructures. A general strategy for realizing these structures through the synthesis of well-defined doped III-nitride-based core–multi-shell (CMS) NW heterostructures (Fig. 3.21a) have been developed. In these materials, an n-type GaN core and p-type GaN outer shell

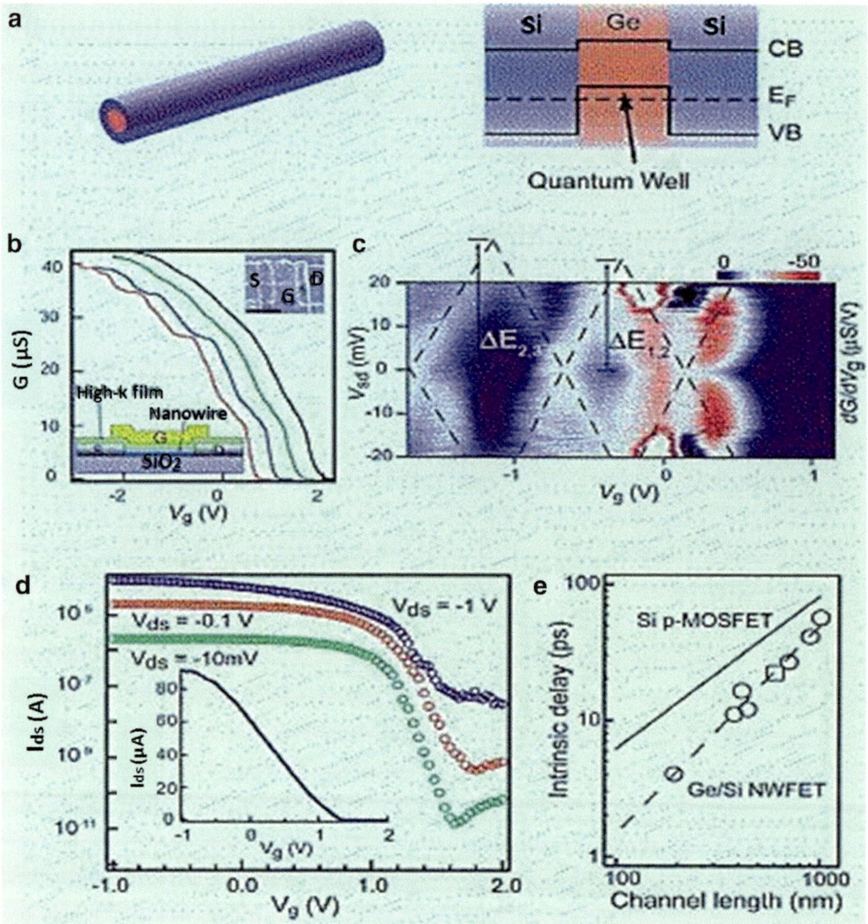

Fig. 3.20 Ge/Si core–shell NW FETs (Li et al. 2006): (**a**) an undoped Ge/Si core–shell NW and correspond diagram; (**b**) G–V_g recorded at different temperatures on a 400 nm long top-gated device; (**c**) transconductance dG/dV_g as a function of V_{ds} and V_g; (**d**) I_{da}–V_g data for a Ge/Si NW FET; and (**e**) intrinsic delay versus channel length for seven different Ge/Si nanowire devices

serve as electron and hole injection layers, an $In_xGa_{1-x}N$ shell provides a tunable band gap quantum well for efficient radiative recombination of injected carriers, and an AlGaN shell is incorporated to enhance confinement of both carriers and photons in the InGaN active layer. Current versus voltage characteristics of CMS NW devices with separate contacts to the N-type core and P-type outer shell show the expected P–N diode current rectification (Fig. 3.21b). In forward bias, the devices yield strong light emission with the LED color dependent on the In composition, defined during synthesis, in the CMS NW heterostructure (Fig. 3.21c, d). Significantly, LED spectra collected from CMS NW devices with intentionally increasing

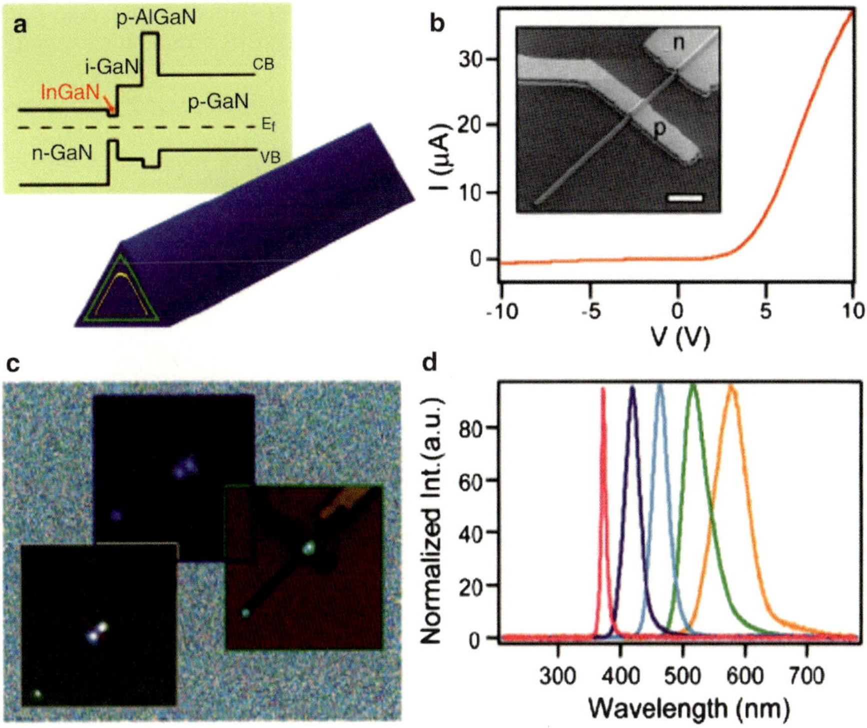

Fig. 3.21 Multicolor III-nitride CMS NW LEDs (Li et al. 2006): (**a**) schematic and corresponding band diagram for an N-GaN/InGaN/GaN/P-AlGaN/P-GaN CMS nanowire; (**b**) *I–V* data recorded from a typical CMS NW device with contacts to the N-core and P-shell. (**c**) EL images of three forward-biased CMS NW LEDs; and (**d**) normalized EL spectra recorded from five multicolor CMS NW LEDs

In composition demonstrate a systematic redshift of the emission from 367 to 577 nm, covering the short wavelength region of the visible spectrum. Notably, preliminary data recorded from these new CMS structures exhibit an external quantum efficiency that is comparable to InGaN-based SQW thin-film LEDs at similar emission wavelengths and substantially better than previous crossed NW LEDs. The efficient injection and radiative recombination of carriers, as well as synthetically tunable emission wavelength of these radial NW devices, represent a clear advance in nano-LED sources and thus a promising pathway to multicolor NW injection lasers in the future (Li et al. 2006).

Semiconductor NWs offer many opportunities for the assembly of nanoscale electronic and optoelectronic devices and arrays by the bottom-up paradigm. Efforts at realizing applications has been the rational control of key NW parameters during growth, including chemical composition, structure, size, morphology, and doping, since it is these parameters that determine predictable device function. Continued advances in controlling the structural/compositional complexity of NWs during growth, which correspondingly determines the functional complexity of the building blocks, together with advances in organizing them into larger integrated arrays, will lead to increasingly unique nanoelectronic and optoelectronic circuits and systems that will create the technologies of the future (Li et al. 2006).

References

Cox J (2001) Fundamentals of linear electronics: integrated and discrete, 2nd edn. Cengage Learning, Stamford

Dutton HJR (1998) Understand optical communications. IBM, Research Triangle Park

Forin DM et al (2010) Free space optical technologies. In: Bouras CJ (ed) Trends in telecommunications technologies. InTech, Rijeka

Held G (2008) Introduction to light emitting diode technology and applications. CRC Press, Boca Raton

Jager D (2007) Microwave photonics—from concepts to devices and applications. In: Lee CH (ed) Microwave photonics. Taylor & Francis Group, Boca Raton, pp 1–27

Law KY (1993) Organic photoconductive materials: recent trends and developments. Chem Rev 93:449–486

Li Y et al (2006) Nanowire electronic and optoelectronic devices. Mater Today 9(10):18–27

Neefs H (2000) Optoelectronic interconnects for integrated circuits: achievements 1996–2000. Gent University-IMEC, Ghent Office for Official Publications of the European Communities. http://bookshop.europa.eu/en/optoelectronic-interconnects-for-integrated-circuits-pbCD2599883/downloads/CD-25-99-883-EN-C/CD2599883ENC_001.pdf;pgid=y8dIS7GUWMdSR0EAlMEUUsWb0000strbWlUw;sid=LINlgaSMDbllXvXx9j77JsapYeB78-nvMvE=?FileName=CD2599883ENC_001.pdf&SKU=CD2599883ENC_PDF&CatalogueNumber=CD-25-99-883-EN-C. Accessed 1 June 2012.

Poole I (2013) Phototransistor. http://www.radio-electronics.com/info/data/semicond/phototransistor/photo_transistor.php. Accessed 21 Feb 2013

Rasheed I et al (2012) Evaluation of optical receiver sensitivity–bit error rate (BER)/Q factor. In: International conference on computer and communication technologies (ICCCT'2012), 26–27 May 2012, Phuket, pp 123–126

Thienpont H, Baukens V (2002) Short-distance optical interconnections with VCSELs. In: Guenther AH (ed) International trends in applied optics. SPIE - The International Society for Optical Engineering, Bellingham, Washington, pp 301–334

Weber MJ (1999) Handbook of laser wavelengths. CRC Press, Boca Raton

Wikipedia (2012) Optical modulator. http://en.wikipedia.org/wiki/Optical_modulator. Accessed 26 Dec 2012

Wikipedia (2013a) Laser. http://en.wikipedia.org/wiki/Laser. Accessed 8 Feb 2013

Wikipedia (2013b) Photodiode. http://en.wikipedia.org/wiki/Photodiode. Accessed 12 Feb 2013

Wikipedia (2013c) Current-to-voltage converter. http://en.wikipedia.org/wiki/Current-to-voltage_converter. Accessed 23 Feb 2013

Zeghbroeck BV (2011) Principles of electronic devices. http://ecee.colorado.edu/~bart/book/book/title.htm. Accessed 6 June 2012

Chapter 4
Optical Fibers

Abstract Optical fiber has been extensively used as a waveguide medium for telecommunication and computer networking because it is flexible and can be bundled as cables. It is especially advantageous for long-distance communications, because light propagates through the fiber with little attenuation compared to electrical cables. This allows long distances to be spanned with few repeaters. Additionally, the per-channel light signals propagating in the fiber have been modulated at rates as high as 100 or higher gigabits per second, although 10 or 40 Gbit/s is typical in deployed systems. Innovations in optical fiber technology are revolutionizing communication, and data processing systems. Newly developed fiber amplifiers, for instance, allow for direct transmission of high-speed signals over transcontinental distances without the need for electronic regeneration. Optical fiber technology including fiber materials, devices, and systems has become a rapidly evolving field. This chapter will give a brief review about basics, structures, materials, fabrication processes, and applications of optical fibers.

4.1 Historical Perspective

Optical fibers about the size of a human hair can carry several orders of magnitude more information than copper wires many times larger in diameter, and they are stronger, lighter, and cheaper. In addition, the signal-carrying load of already installed optical fiber cables can be readily increased as signal-processing technologies improve. The invention of the optical fibers was based on principles that were extended from waveguide theory and practice into the optical region; and creation of clad-glass fibers with a minimum of defects and impurities that attenuate (i.e., scatter or absorb) a light signal.

Although light-guiding optical fibers had been developed for medical endoscopes, instrument panel lighting, and other uses in the 1950s and 1960s, the clearest existing fibers could carry detectable light only a few meters. In the mid-1960s, the target was to reduce attenuation to 20 dB/km or less, which was nearly two orders of magnitude

X.C. Tong, *Advanced Materials for Integrated Optical Waveguides*, Springer Series in Advanced Microelectronics 46, DOI 10.1007/978-3-319-01550-7_4,

less than attenuation in the clearest optical glass available at that time. An attenuation of 20 dB (99 %) over a kilometer meant that signal regenerators or repeaters could be placed about 1 km apart, making optical fibers potentially competitive with copper transmission lines. The breakthrough occurred at Corning Glass in 1970, where researchers adapted chemical vapor deposition techniques for making bulk fused silica to making high-silica fibers that had the required physical and chemical characteristics. The first fiber was still far from suitable for mass production and commercial use, however, and major improvements were made by the original inventors and others, notably Bell Laboratories, during an intensive period of research and development before optical fibers went into mass production in the early 1980s. R&D on optical fibers accelerated rapidly after the initial demonstration that low-loss fibers were feasible, resulting in a long period of continuous improvements in materials and especially in fabrication techniques and processes that led to maturation of the technology in the mid- to late-1980s. The innovation process was strongly driven by technological competition. The initial focus on reducing sources of light loss in the fibers was broadened to include improvements in the strength and durability of the fibers and in related technologies, especially fiber coating, splicing, and cabling techniques and cable connectors. A major bottleneck limited faster transmission for at least a decade after the breakthrough at Corning in 1970 was not the fiber itself, but the component devices that hang on the ends. For example, although it was recognized from the beginning that single-mode fibers (SMFs) would have the largest bandwidth, the cores of SMFs were too small for efficient splicing and coupling into early lasers. Accordingly, the main emphasis during the 1973–1980 period was the development of multimode fibers. Their large cores were easier to splice and couple with existing lasers, and were also suitable for coupling with light-emitting diodes (LEDs), which were much more reliable and cheaper. At the same time, there were intensive and ultimately successful efforts to develop reliable semiconductor lasers that were not only small enough to work with SMFs but also that worked at the longer wavelengths at which attenuation in optical fibers was lowest. An additional concern was the ability to splice small core, SMFs with low loss and high reliability in a field situation. The coupling from laser to fiber could be accomplished in a more controlled situation than could field splicing, so research during this period also emphasized these and other kinds of problems associated with actual conditions during installation and maintenance of commercial lines. After the Corning research team demonstrated the first optical fiber with loss less than 20 dB/km, there ensued a long period of continuous improvements and adaptations to the concurrent developments in lasers and other components of fiber-optic communications systems. By 1984, attenuation had dropped to 0.20 dB/km in mass-produced fibers (0.16 dB/km in the laboratory), two orders of magnitude better than the first experimental fibers. The lower attenuation was due in part to improved fabrication techniques that reduced impurities, but it also stemmed from the development of supporting technologies (e.g., lasers, detectors, and other components) that operated at higher frequencies where intrinsic loss in silica fibers was lowest.

Simultaneously, there were numerous improvements in the manufacturing process–e.g., in yields, deposition rates, preform size, draw rates, and size tolerances–that resulted in much higher production rates, lower unit costs, and better quality (Scribd 2012).

Field trials of fiber-optic telephone systems began in 1976. Corning and Western Electric opened full-scale production plants in 1980, the same year that the first sea trial of a commercial optical fiber cable was undertaken. The field trials were very successful, and fiber-optic cables began to go into regular service. AT&T's fiber-optic cable system between Washington, DC, and New York entered service in 1983. By 1982, component development (e.g., lasers, detectors, couplers) and splicing techniques had proceeded to the point that telephone companies began to switch from multimode to higher performance SMFs. Production lengths of commercial fiber increased from 2 km in 1982 to 25 km in 1987, while at the same time the price per meter fell sixfold. A fiber-optic cable could carry the same information as a copper wire cable 4 times larger and 8 times heavier. By the end of 1988, more than ten million kilometers of fiber were installed, and more than 90 % of long distance telephone traffic in North America was carried on optical fibers. TAT-8, the first transatlantic optical fiber cable, was laid in 1988. The use of fiber optics in telecommunications and wide area networking has been common since then, but more recently fiber optics have become increasingly prevalent in industrial data communications systems as well. High data rate capabilities, noise rejection, and electrical isolation are just a few of the important characteristics that make fiber-optic technology ideal for use in industrial and commercial systems. Most often used for point-to-point connections, fiber-optic links are being used to extend the distance limitations of RS-232, RS-422/485, and Ethernet systems while ensuring high data rates and minimizing electrical interference. Conventional electrical data signals are converted into a modulated light beam, introduced into the fiber and transported via a very small diameter glass or plastic fiber to a receiver that converts the light back into electrical signals. Fiber's ability to carry the light signal, with very low losses, is based on some fundamental physics associated with the refraction and reflection of light. Even though optical fiber seems quite flexible, it is made of glass, which cannot withstand sharp bending or longitudinal stress. Therefore when fiber is placed inside complete cables special construction techniques are employed to allow the fiber to move freely within a tube. Usually fiber-optic cables contain several fibers, a strong central strength member and one or more metal sheaths for mechanical protection. Some cables also include copper pairs for auxiliary applications. Advantages of Fiber-Optic Cables can be summarized as (Scribd 2012):

- Noise or EMI Immunity
  Noise immunity is one of the most useful features of fiber optics in industrial applications. In environments where electromagnetic interference is prominent and unavoidable, optical fibers are unaffected. While cables are normally contained in protective sheaths and often run inside conduit, there is no need

to physically isolate fiber-optic cables from electrical cables. This makes cable routing simpler.

- Electrical Isolation
  The problem of ground loop noise and common mode potential differences is eliminated by the use of fiber-optic cables. Field signals, generated by devices floating at high potentials, can be coupled to other equipment at much lower potentials without the risk of damage. This is particularly desirable in industrial applications.
- Low Error Rates
  When properly designed to provide adequate signal levels at the receiving end of the link, a fiber-optic system provides very low bit error rates.
- Safe for Use in Hazardous Areas
  Fiber-optic links can be used to couple signals into areas with potentially explosive atmospheres without a risk to delivering or storing sufficient energy to ignite an explosion. This makes fiber-optic technology particularly useful when designing intrinsically safe systems.
- Wide Bandwidth
  Fiber-optic cables can carry very wide bandwidth signals, well into the GHz range. Many individual, lower bandwidth signals can be multiplexed onto the same cable. In commercial systems fiber-optic cable often carries a mixture of signal types, including voice, video, and data all on the same fiber.
- Low Signal Attenuation
  Optical fibers do exhibit some attenuation due to absorption and scattering. However, this attenuation is relatively independent of frequency, a factor that is significant in copper cables.
- Light Weight, Small Diameter
  Because many signals can be multiplexed onto one fiber, cables tend to be smaller and lighter. This makes installation easier.
- No Crosstalk
  Since fibers do not pick up electromagnetic interference, signals on adjacent cables are not coupled together.
- Inherent Signal Security
  For applications where signal security is a concern, optical fiber is an excellent solution. Fiber-optic cables do not generate electromagnetic fields that could be picked up by external sensors. It is also more difficult to "steal" signals by spicing into optical fibers than it might be with conventional copper wiring.

Fiber optics has become the transmission medium of choice for most communications. Its high speed and long-distance capability make it the most cost effective communications medium for most applications. While higher performance and lower cost components are in continual development, it has also become critical to train competent personnel to design, install, and maintain state-of-the-art fiber-optic networks.

4.2 Fiber Optical Principles

The simplest optical fiber is a dielectric cylindrical waveguide consisting of a central core with a higher refractive index that is surrounded by a cladding with lower refractive index. The cladding is usually enclosed in a protective jacket. As other waveguides, optical fibers can be single-mode and multimode. The important parameter that defines a number of modes is the V parameter, given by (Litchinitser and Iakhnine 2011)

$$V = \frac{2\pi a}{\lambda}\sqrt{\left(n_1^2 - n_2^2\right)} \approx \frac{2\pi a}{\lambda} n_1 (2\Delta)^{1/2} \tag{4.1}$$

where a is the radius of the core, λ is a free space wavelength of light, n_1 is the refractive index of the core, n_2 is the refractive index of the cladding, and Δ is the relative difference of the refractive indices. The fiber supports only one mode if $V < 2.405$. In standard telecom fibers, the difference between n_1 and n_2 is typically small, so that the relative difference, $\Delta = (n_1 - n_2)/n_2 \ll 1$. Optical fibers satisfying this condition are called weakly guiding. The typical value for a SMF is $\Delta \sim 0.003$.

Total internal reflection (TIR) is the most common guiding mechanism in optical fibers, although other mechanisms are becoming more and more widespread with the emergence of microstructured optical fibers (MOFs), such as photonic crystal fibers (PCFs). The TIR is an optical phenomenon that occurs when a ray of light strikes a medium boundary at an angle larger than the critical angle with respect to the normal to the surface. The critical angle is given by (Litchinitser and Iakhnine 2011)

$$\theta_c = \sin^{-1}\left(\frac{n_2}{n_1}\right) \tag{4.2}$$

If the refractive index is lower on the other side of the boundary, then no light can pass through, so effectively all of the light is reflected. Other important fiber/waveguide parameters are the numerical aperture defined as $\mathrm{NA} = \sqrt{\left(n_1^2 - n_2^2\right)}$ and the acceptance angle given by

$$\theta_c = \sin^{-1}(\mathrm{NA}) \tag{4.3}$$

which determine the cone of external rays that can be guided by the fiber and, therefore, is an essential parameter for coupling light in and out of the fiber. Light rays that impinge fiber at angles greater than the acceptance angle are refracted into the fiber, but are not guided for a long distance, as they are not totally reflected at the core-cladding interface, but rather are partially refracted into the cladding. Major phenomena that have an effect on light propagation in optical fibers include (a) dispersion, (b) nonlinear interactions, (c) the effect of loss, and (d) the effects of gain and noise. Modes are fields that maintain the same transverse distribution and polarization at all distances along the waveguide axis. They can be regarded as transverse resonances of the fields in the waveguide.

4.2.1 Fiber Modes

Fiber modes can be briefly described as below.

1. Transverse Electric (TE) Modes
   TE modes exist when the electric field is perpendicular to the direction of propagation (the z-direction) but there is a small z-component of the magnetic field. Here most of the magnetic field is also perpendicular to the z-direction but a small z-component exists. This implies that the wave is not traveling quite straight but is reflecting from the sides of the fiber waveguide. However, this also implies that the "ray" path is meridional (it passes through the center or axis of the waveguide). It is not circular or skewed (Dutton 1998).
2. Transverse Magnetic (TM) Modes
   In a TM mode the magnetic field is perpendicular to the direction of propagation (z) but there is a small component of the electric field in this direction and most of it is perpendicular to the z-axis. In other words, the orientation of the electric field is only a few degrees away from being perpendicular to the z-axis.
3. Transverse Electromagnetic (TEM) Modes
   Transverse electromagnetic (TEM) is a mode of propagation where the electric and magnetic field lines are all restricted to directions normal (transverse) to the direction of propagation. In other words, in the TEM mode both the electric and magnetic fields are perpendicular to the z-direction. The TEM mode is the only mode of a SMF.
4. Helical (Skew) Modes (HE and EH)
   In a fiber, most modes actually travel in a circular path of some kind. In this case components of both magnetic and electric fields are in the z-direction (the direction of propagation). These modes are designated as either HE or EH (H = magnetic) depending on which field contributes the most to the z-direction (Dutton 1998).
5. Linearly Polarized (LP) Modes
   It turns out that because the RI difference between core and cladding is quite small. In fiber propagation a single-mode designation can be used to approximate all of the others. Thus TE, TM, HE, and EH modes can all be summarized and explained using only a single set of LP modes (Dutton 1998).

   Characteristics of some fiber modes are shown in Fig. 4.1. Figure 4.1a shows energy distribution of some TEM modes. The numbering system used here applies to TE, TM, and TEM modes. It is conventional to number the TE and TM modes according to the number of nulls in their energy pattern across the waveguide. Thus mode TE00 would have a single energy spot in the center of the

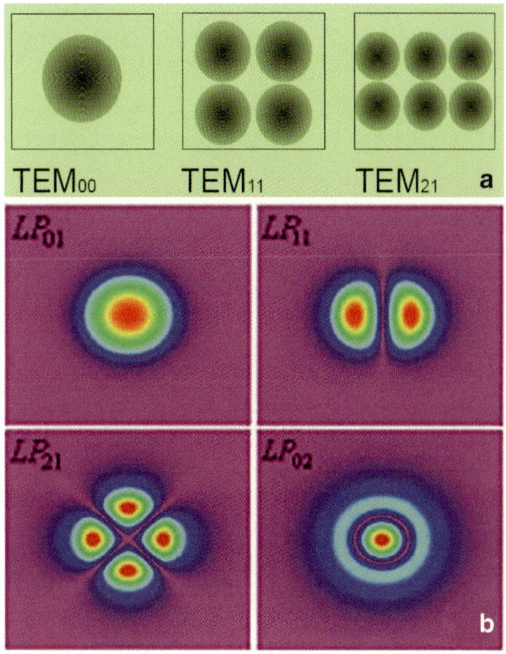

Fig. 4.1 Characteristics of some fiber modes (Litchinitser and Iakhnine 2011): (**a**) energy distribution of some TEM modes; (**b**) energy distribution of some LP modes

waveguide and no others. (This would be the same mode as TEM.) Mode TEM21 would have two nulls (three energy spots) in one direction and a single null (two energy spots) in the other. When the subscript numbers are low (0, 1, 2) the modes are often referred to as "low order" modes. When the subscripts are high numbers the modes are referred to as "high order" modes.

Energy distribution of some LP Modes in fiber is shown in Fig. 4.1b. LP mode numbering is different from TE or TM mode numbering. LP modes are designated LP*lm* where *m* is the number of maxima along a radius of the fiber (note here the number of maxima rather than the number of nulls as before). *l* here is half the number of maxima around the circumference. Roughly, *m* is related to the angle of incidence of the ray with the core-cladding interface. *l* tells us how tight the spiral (helix) is. The fundamental mode (straight down the center) is thus referred to in different ways: LP01, HE11, or TEM00. In practice, TE and TM designations are usually used when discussing lasers and planar waveguides. LP is used when discussing multimode fiber (MMF) propagation.

Fiber modes can also be classified into the bound modes and the radiation modes. Then, the electric and magnetic field vectors can be written as (Litchinitser and Iakhnine 2011)

$$E(x, y, z) = \sum_j q_j E_j(x, y, z) + \sum_j q_{-j} E_{-j}(x, y, z) + E_{\text{rad}}(x, y, z) \tag{4.4a}$$

$$H(x, y, z) = \sum_j q_j H_j(x, y, z) + \sum_j q_{-j} H_{-j}(x, y, z) + H_{\text{rad}}(x, y, z) \tag{4.4b}$$

where $j = 1, 2, \ldots, M$; E_j, E_{-j}, E_{rad} and H_j, H_{-j}, H_{rad} are the forward, backward, and radiation electric and magnetic fields, respectively, and q_j and q_{-j} are the modal amplitudes. If the waveguide has a refractive index profile that does not change along its length z, that is, $n = f(x, y)$, such waveguide is called translationally invariant. Then the electric and magnetic fields can be written in the form

$$E_j(x, y, z) = e_j(x, y)\exp(i\beta_j z) \tag{4.5a}$$

$$H_j(x, y, z) = h_j(x, y)\exp(i\beta_j z) \tag{4.5b}$$

where β_j is the propagation constant. At a given frequency ω, the value of β_j is calculated from an eigenvalue equation. The eigenvalue equation, also called dispersion relation or characteristic equation, is a transverse resonance condition, resulting from the field solutions that are bounded everywhere, approach zero sufficiently fast at infinity, and satisfy all boundary conditions at the core-cladding interface. In weakly guiding fiber, a small but nonzero index difference maintains TIR, but the medium is nearly homogeneous as far as polarization effects are concerned. Then, the modes of weakly guiding fibers are nearly TEM waves, with the longitudinal components of the electric and magnetic fields much weaker than the transverse components. For weakly guiding fibers, the eigenvalue equation is given by (Litchinitser and Iakhnine 2011)

$$X\frac{J_{i\pm1}(X)}{J_i\,(X)} = \pm Y\frac{K_{i\pm1}(Y)}{K_i(Y)} \tag{4.6}$$

where $X^2 + Y^2 = V^2$, and $X = a(n_1^2k^2 - \beta^2)^{1/2}$. The eigenvalue equation (4.6) is a transcendental equation and may be solved graphically by plotting its right- and left-hand sides versus X. The range of allowed β is bounded by (Litchinitser and Iakhnine 2011)

$$kn_2 < \beta \leq kn_1 \tag{4.7}$$

where $k = 2\pi/\lambda$. The lower limit $\beta = kn_2$ of allowed values of the bound mode propagation constant is called modal cutoff. Far from cutoff, a mode is well confined in the core, while close to cutoff the mode significantly extends into the cladding region. In contrast to planar waveguides, not all fiber modes escape from the core completely at the cutoff as shown in Fig. 4.1b.

4.2.2 *Dispersive Properties*

Dispersion refers to any phenomenon in which the velocity of propagation of an electromagnetic wave is wavelength dependent. Therefore, dispersion plays a critical role in many applications that involve short pulse propagation in optical fibers. Owing to dispersion, different frequencies within the pulse spectrum propagate at slightly different velocities along the optical fiber. In the so-called normal dispersion regime, red frequency components propagate faster than blue components. In the anomalous dispersion case, blue components propagate faster. As a result, different frequencies arrive with different delays, leading to broadening and chirping of the output pulse. Dispersion-induced broadening can be detrimental for fiber-optic transmission systems. At certain point, two neighboring pulses may overlap and their amplitudes may reduce to a level when they cannot be detected by a receiver at the end of the transmission system. On the other hand, in many cases, dispersive effects can be advantageous. For example, in conjunction with a nonlinear effect of self-phase modulation, fiber dispersion gives rise to a pulse that propagates without any temporal and spectral changes, called soliton. There are different types of dispersion taking place in optical fibers, including a group-velocity and higher order fiber dispersion (also called chromatic dispersion or intramodal dispersion), waveguide dispersion, and modal dispersion in multimode fibers (MMFs) (Litchinitser and Iakhnine 2011).

1. Chromatic dispersion
   Both lasers and LEDs produce a range of optical wavelengths (a band of light) rather than a single narrow wavelength. The fiber has different refractive index characteristics at different wavelengths and therefore each wavelength will travel at a different speed in the fiber. Thus, some wavelengths arrive before others and a signal pulse disperses (or smears out) (Dutton 1998). Chromatic dispersion arises for two reasons: (a) the refractive index of the material used to make optical fiber, is frequency dependent. Thus different frequency components travel at different speeds in silica. This component of chromatic dispersion is called material dispersion. (b) Although material dispersion is the principle component of chromatic dispersion for most fibers, there is a second component, called waveguide dispersion.
2. Material dispersion
   Material dispersion is the phenomena whereby materials cause a bundle of light to spread out as it propagates. For instance, a laser pulse, while almost monochromatic, actually contains a continuum of wavelengths in a small range. The index of refraction (IOR) of a material is dependent on the wavelength, so each frequency component actually travels at a slightly different speed. As the distance increases, the pulse becomes broader as a result. In optical communications, material dispersion limits how much data can be sent, as the pulses will overlap and information will be lost. Methods such as graded-index fibers and optical temporal solitons have been used to reduce the effects of material dispersion.
3. Waveguide dispersion
   Waveguide dispersion is a very complex effect and is caused by the shape and different refractive indexes of the core and cladding of an optical fiber. However, this can be controlled by careful design and, in fact, waveguide dispersion can be used to counteract material dispersion. Regardless of the nature of the light source

and optical fiber, some light travels in the cladding, as well as the core. Assuming a step-index fiber, the core is of one highly consistent IOR, and the cladding is of another, although sometimes the cladding is of several layers of glass, each with a sharp step in IOR. As the IOR of glass varies as the wavelength varies, with longer wavelengths propagating at higher velocities, as no light pulse has a perfectly narrow spectral width, and as multiple layers of glass of different properties make up the core and cladding, different wavelengths of light propagate at different velocities in the different layers. So, the optical pulse can disperse, or spread, over a distance, which clearly can confuse the light detector at the far end of the fiber. Waveguide dispersion is one factor contributing to chromatic dispersion, both of which are issues in long haul fiber-optic transmission systems (FOTS) employing SMF of step-index construction.

4. Modal dispersion
   Modal dispersion is a distortion mechanism occurring in MMFs and other waveguides, in which the signal is spread in time because the propagation velocity of the optical signal is not the same for all modes. When using MMF, the light is able to take many different paths or modes as it travels within the fiber. The distance traveled by light in each mode is different from the distance traveled in other modes. When a pulse is sent, parts of that pulse (rays or quanta) take many different modes (usually all available modes). Therefore, some components of the pulse will arrive before others. The light arrival time difference taking the fastest mode versus the slowest obviously gets greater as the distance gets greater. In the ray optics analogy, modal dispersion in a step-index optical fiber may be compared to multipath propagation of a radio signal. Rays of light enter the fiber with different angles to the fiber axis, up to the fiber's acceptance angle. Rays that enter with a shallower angle travel by a more direct path, and arrive sooner than rays those enter at a steeper angle (which reflects many more times off the boundaries of the core as they travel the length of the fiber). The arrival of different components of the signal at different times distorts the shape. Modal dispersion limits the bandwidth of MMFs. Modal dispersion may be considerably reduced, but never completely eliminated, by the use of a core having a graded refractive index profile. However, multimode graded-index fibers having bandwidths exceeding 3.5 GHz km at 850 nm are now commonly manufactured for use in 10 Gbit/s data links. Modal dispersion should not be confused with chromatic dispersion, a distortion that results due to the differences in propagation velocity of different wavelengths of light. Modal dispersion occurs even with an ideal, monochromatic light source. A special case of modal dispersion is polarization mode dispersion (PMD), a fiber dispersion phenomena usually associated with SMFs. PMD results when two modes that normally travel at the same speed due to fiber core geometric and stress symmetry (for example, two orthogonal polarizations in a waveguide of circular or square cross section), travel at different speeds due to random imperfections that break the symmetry (Wikipedia 2012).

On a fundamental level, the response of any dielectric medium is related to the resonance frequencies at which the medium absorbs the incoming electromagnetic wave and, as a result, the response is frequency dependent. Resulting chromatic dispersion reveals itself in the frequency dependence of the refractive index $n(\omega)$. The effect of group-velocity dispersion (GVD) can be described by expanding the mode propagation constant in a Taylor series (Litchinitser and Iakhnine 2011)

$$\beta(\omega) = n(\omega)\frac{\omega}{c} = \beta_0 + \beta_1(\omega - \omega_0) + \frac{1}{2}\beta_2(\omega - \omega_0)^2 \tag{4.8}$$

where $m = 1, 2, \ldots$

$$\beta_m = \left(\frac{\partial^m \beta}{\partial \omega^m}\right)_{\omega=\omega_0} \tag{4.9}$$

In (4.8) and (4.9), $\beta_1 = 1/v_g$ is inversely proportional to the group velocity, a velocity of the pulse envelope, and β_2 describes the GVD responsible for pulse broadening. These terms can expressed as

$$\beta_1 = \frac{1}{c}\left(n + \omega\frac{dn}{d\omega}\right) \tag{4.10a}$$

$$\beta_2 = \frac{1}{c}\left(2\frac{dn}{d\omega} + \omega\frac{d^2n}{d\omega^2}\right) \tag{4.10b}$$

Higher order terms in (4.8) and (4.9) may also be important, especially as the pulse width decreases; however, in many practical cases, the effect of the GVD dominates. In silica fibers, the GVD vanishes at around 1.3 μm. This wavelength is often referred to as a zero-dispersion wavelength. Therefore, it seems logical to design the transmission systems around this wavelength. Unfortunately, fiber losses are not minimized at 1.3 μm; however, they minimize at 1.55 μm. As a result, modern systems operate in the 1.55 μm transmission window. Waveguide dispersion is usually relatively small in comparison to material dispersion in standard telecom fibers (except near the zero-dispersion wavelength) and its effect is to shift the zero-dispersion wavelength to slightly longer wavelengths. However, a significant advantage of the waveguide dispersion is that it depends on fiber design and, therefore, can be controlled at least to some degree. In particular, it was used to shift the zero-dispersion wavelength from 1.3 to 1.55 μm in so-called dispersion-shifted fibers. In addition, emergence of MOFs opened new opportunities in waveguide dispersion design. Besides, there is another dispersive effect that results from the difference in propagation velocities of light in the orthogonal principal polarization states of the fiber, called PMD. As a result of the PMD, different polarization components within the optical pulse arrive at different times, degrading the received optical signal (Litchinitser and Iakhnine 2011).

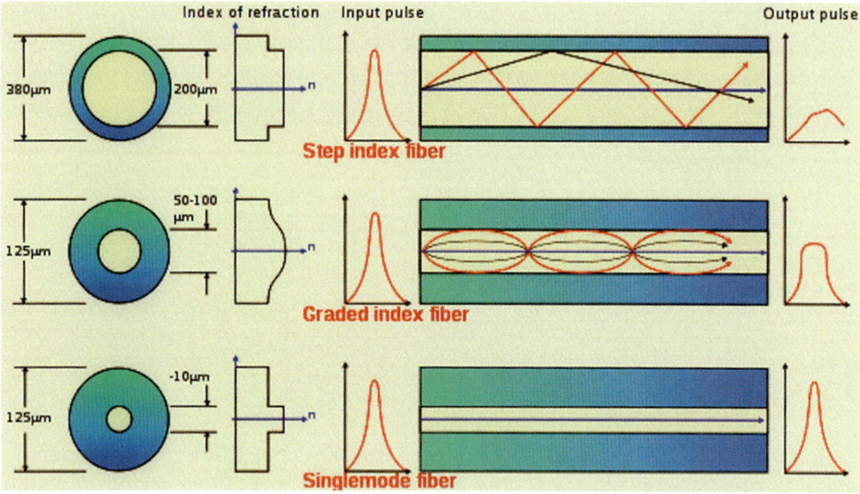

Fig. 4.2 Schematic illustration of optical fiber types that are manufactured: multimode step-index, multimode graded-index, and single mode

4.2.3 Type of Optical Fibers

Optical fibers are manufactured in three main types: multimode step-index, multimode graded-index, and single mode, as shown in Fig. 4.2. The difference between them is in the way light travels along the fiber. The top section of the figure shows the operation of MMF. There are two different parts to the fiber. For the graded-index fiber for instance, there is a core of 50–100 μm in diameter and a cladding of 125 μm in diameter. The cladding surrounds the core. The cladding glass has a different (lower) refractive index than that of the core, and the boundary forms a mirror. Light is transmitted (with very low loss) down the fiber by reflection from the mirror boundary between the core and the cladding. This phenomenon is so called TIR. Perhaps the most important characteristic is that the fiber will bend around corners to a radius of only a few centimeters without any loss of the light.

1. Multimode Step-Index Fiber
   There are only a finite number of possible paths for the light to take through the fiber. These paths are so called modes and identify the general characteristic of the light transmission system being used. Fiber that has a core diameter large enough for the light used to find multiple paths is called MMF. For a fiber with a core diameter of 62.5 μm using light of wavelength 1,300 nm, the number of modes is around 400 depending on the difference in refractive index between the core and the cladding. The problem with multimode operation is that some of the paths taken by particular modes are longer than other paths. This means that light will arrive at different times according to the path taken. Therefore the pulse tends to disperse (spread out) as it travels through the fiber. This effect is one

cause of intersymbol interference. This restricts the distance that a pulse can be usefully sent over MMF. In comparison, multimode step-index fiber has the largest diameter core (typically 50–100 μm). The larger distance between interfaces allows the light rays to travel the most distance when bouncing through the cable. Multimode fibers typically carry signals with wavelengths of 850 or 1,300 nm. Multimode step-index fiber is comparatively easy to splice and terminate due to the large diameter fiber. It is also relatively inexpensive to manufacture compared to other types. However, it tends to be too slow for most purposes and it not common in modern systems.

2. Multimode Graded-Index Fiber

   One way around the problem of modal dispersion in MMF is to do something to the glass such that the refractive index of the core changes gradually from the center to the edge. Light traveling down the center of the fiber experiences a higher refractive index than light that travels further out towards the cladding. Thus light on the physically shorter paths (modes) travels more slowly than light on physically longer paths. The aim of this is to keep the speed of propagation of light on each path the same with respect to the axis of the fiber. Thus a pulse of light composed of many modes stays together as it travels through the fiber. This allows transmission for longer distances than does regular multimode transmission. This type of fiber is so called graded-index (GI) fiber. Within a GI fiber light typically travels in around 400 modes (at a wavelength of 1,300 nm) or 800 modes (in the 800 nm band). Note that only the refractive index of the core is graded. There is still a cladding of lower refractive index than the outer part of the core. Multimode graded-index fiber is constructed in such a way that the refractive index between the core and cladding changes gradually. This causes the light rays to bend gradually, as well. The resulting pattern of reflections tends to be more uniform and dispersion is reduced. This provides improved performance for a moderate increase in cost. Graded-index fibers provide wider bandwidth than step-index fibers (Dutton 1998).

3. Single-Mode Fiber

   The core diameter is typically between 8 and 10 μm while the diameter of the cladding is 125 μm. If the fiber core is very narrow compared to the wavelength of the light in use then the light cannot travel in different modes and thus the fiber is called single-mode or monomode. There is no longer any reflection from the core-cladding boundary but rather the electromagnetic wave is tightly held to travel down the axis of the fiber. It seems obvious that the longer the wavelength of light in use, the larger the diameter of fiber that can be used and still have light travel in a single-mode (Dutton 1998). A significant proportion (up to 20 %) of the light in a SMF actually travels in the cladding. For this reason the apparent diameter of the core (the region in which most of the light travels) is somewhat wider than the core itself. The region in which light travels in a SMF is often called the mode field and the mode field diameter is quoted instead of the core diameter. The mode field varies in diameter depending on the relative refractive indices of core and cladding. Core diameter is a compromise. The core cannot be made too narrow because of losses at bends in the fiber. As the core diameter decreases compared to the wavelength (the core gets narrower or the wavelength gets longer), the minimum radius that the fiber can be bent without loss increases. If a bend is too sharp, the light just comes out of the core into the outer parts of the cladding and is lost. SMFs give the highest performance of the three types. Manufactured using a very small diameter fiber (typically 8 μm), when light is introduced into the fiber reflections are kept to a minimum by the dimensions of the core. Light travels virtually straight through the core and pulses introduced at one end are reproduced at the other end with very little dispersion. Typically, SMFs carry signals with wavelengths of 1,320 or 1,550 nm. SMF is relatively expensive, however, and is more difficult to splice and terminate since the core must be aligned very accurately. SMFs offer much lower attenuation than multimode fibers. At typical SMF will attenuate a 1,310 nm signal less than 0.5 dB/km. A typical multimode graded-index fiber will attenuate the same signal about 3 dB/km. SMF is most often used in applications with high bandwidth requirements over long distances. Some Ethernet fiber-optic equipment can increase distances from 2 km using multimode fiber to about 70 km over SMF.

SMF can be fabricated by (a) Making the core thin enough, (b) Making the refractive index difference between the core and the cladding small enough, and (c) Using a longer wavelength. SMF usually has significantly lower attenuation than multimode (about half). This has nothing to do with fiber geometry or manufacture. SMFs have a significantly smaller difference in refractive index between core and cladding. This means that less dopant is needed to modify the refractive index as dopant is a major source of attenuation. There is a SMF characteristic called the cutoff wavelength. This is typically around 1,100 nm for SMF with a core diameter of 9 μm. The cutoff wavelength is the shortest wavelength at which the fiber remains single-mode. At wavelengths shorter than the cutoff the fiber is multimode. When light is introduced to the end of a fiber there is a critical angle of acceptance. Light entering at a greater angle passes into the cladding and is lost. At a smaller angle the light travels down the fiber. If this is considered in three dimensions, a cone is formed around the end of the fiber within which all rays are contained. The sine of this angle is called the numerical aperture and is one of the important characteristics of a given fiber. Commercial SMF has a core diameter of 4–10 μm (8 μm is typical). MMF can have many core diameters but in the last few years the core diameter of 62.5 μm in the US and 50 μm outside the US has become predominant. However, the use of 62.5 μm fiber outside the US is gaining popularity—mainly due to the availability of equipment (designed for the US) that uses this type of fiber (Dutton 1998).

4.3 Fiber Materials

Optical fibers are kinds of long and flexible optical waveguides. They are essentially based either on some glass or on polymers (plastic optical fibers). Photonic crystal fiber (PCF) is a new class of optical fiber based on the properties of photonic crystals, although the base materials used for PCF are still glass or polymers. Because of its ability to confine light in hollow cores or with confinement characteristics not possible in conventional optical fiber, PCF is now finding applications in fiber-optic communications, fiber lasers, nonlinear devices, high-power transmission, highly sensitive gas sensors, and other areas. In addition, optical nano-fibers including nano-engineered fibers and optical fiber nanowires have attracted increasing interest because of their tremendous potential in a wide range of fields ranging from telecommunications to chemistry, biology, lasers, and sensors.

4.3.1 Glasses

Along with the growing interest in optical fiber and other waveguide materials for optoelectronic and telecommunication applications, it has been demanded to understand any modification or variation in their properties resulting from the transformation from bulk material to thin film form for planar waveguide writing using laser exposure or to optical fiber. On another hands, glasses have played an important role among these waveguide materials. The development of advanced glass optical materials focus on three aspects (Carlie et al. 2009): (a) identification of glass compositions suitable for fiberization or planarization and an understanding of the materials' properties; (b) fiber drawing or film deposition step followed by the fiber and film characterization; and (c) optimization of the composition for low optical loss, resistance to optical damage and strength. Most studies about glasses and their applications have been carried out on silicate and quartz glasses, which transmit the radiation in visible range of the electromagnetic spectra. The application in optics, photonics, and optoelectronics increased the demand in glasses which can transmit a radiation in infrared range up to the wavelength of ~2 μm. The limit wavelengths of the optical transmission for SiO_2 and special glasses are shown in Fig. 4.3 (Lezal 2003). The chemical and physical properties of special glasses have been used to overcome the limitation of silicate glasses. The special glasses can be divided into three groups, including (a) Fluoride glasses based on ZrF_4 or HfF_4; (b) Chalcogenide glasses—glasses based on chalcogens (S, Se, Te): e.g., As–S, As–Se, Ge–Ga–S, Ge–Ga–As–S, Ge–As–Se, Ge–Se–Te, As–Se–Te; and (c) Heavy metal oxide (HMO) glasses: e.g., GeO_2–PbO, TeO_2–PbO, $PbCl_2$, TeO_2, ZnO, Ga_2O_3–PbO, and Bi_2O_3.

Glass fiber-optic waveguides are ideal for use in the visible spectrum and allow the use of higher power lamps than can be used with the polymer fiber (up to 250 °C). The glasses can be used for the preparation of optical fibers both for passive and active applications. The glasses are stable, but some fluoride and HMO glasses have the tendency to the crystallization and are hygroscopic. For preparation of highly pure glasses with diminished OH, oxide and low physical defect concentrations, advanced purification methods have been developed based on the preparation in a reactive atmosphere. In this way, the average concentration of hydride impurities can be lowered below 5×10^{-5} mol. HMO glasses are more

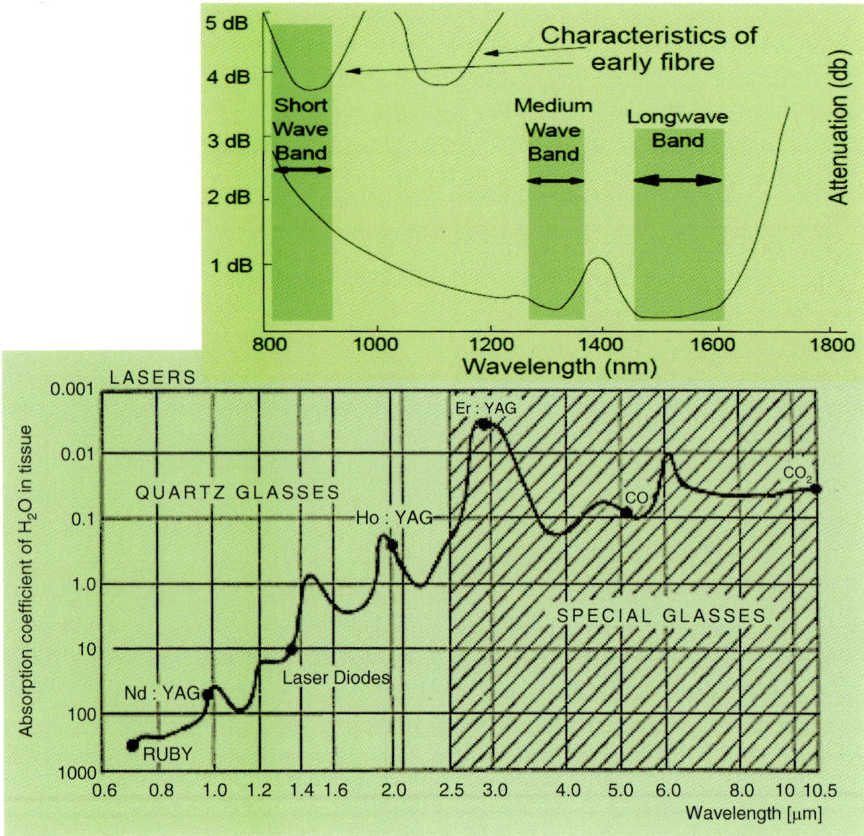

Fig. 4.3 The limit wavelengths of the optical absorption for SiO_2 and special glasses (Lezal 2003)

difficult to be prepared with a high purity. The OH group concentration is about 1–2 orders higher than in fluoride glasses. This is the reason why the treatment in reactive atmosphere is less effective. Optical losses of fluoride optical fibers are about 1 dB/km (the theoretical value is 10–2 dB/km); optical losses of chalcogenide and HMO fibers are many times higher (0.1–1 dB/m), but the glasses are good for photonics applications on short distance. Rare earth (RE) ions are introduced into the base glasses in concentrations from 500 to 6,000 wt.ppm. High RE and OH group concentration favors the growth of clusters. Sulphide glasses exhibit quantum efficiency of fluorescence of about 80 %, while for fluoride and HMO glasses the efficiency is about 15–22 % (Lezal 2003).

4.3.1.1 Silica Glass Fibers

The inventions of the low-loss silica glass optical fiber using chemical vapor deposition, and continuous wave semiconductor laser operating at room temperature in

1970 have opened new horizon in communication technology, and led to the possibility of ultra-high capacity information flow, marked the inception of the photonics era. Owing to excellent transmission properties of low-loss and high bandwidth, silica optical fiber is the status quo broadband telecommunications media such as transoceanic cables, terrestrial cables, metro-loops, and customer premises loops. Real time video and audio applications for advanced broadband systems prompted optical manufacturers to develop high volume mass production facilities of fibers such as large perform fabrication and high-speed drawing and consequently lower the cost of the fiber. As a result, fiber fabrication technology has produced various types of fibers and fiber devices that prompted all-fiber platform in telecommunications, sensing, and metrology (Oh and Paek 2012).

Among the glasses, fused silica (amorphous silicon dioxide, SiO_2) is the predominant material (particularly for optical fiber communications), because it has a number of very favorable properties (Gambling 2000):

(a) Silica has a wide wavelength range with good optical transparency. In the near-infrared spectral region, particularly around 1.5 μm wavelength, silica can have extremely low absorption and scattering losses of the order of 0.2 dB/km, when the fiber preform is fabricated in very pure form. Whereas a high transparency in the 1.4-μm region is achieved by keeping the concentration of hydroxyl groups (OH) low, a high OH concentration is better for good transparency in the ultraviolet region. (b) Silica can be drawn into fibers at reasonably high temperatures, and has a conveniently broad glass transition. (c) Fusion splicing and cleaving of silica fibers works fairly well. (d) A silica fiber has an amazingly high mechanical strength against pulling and even bending, provided that the fiber is not too thick and that the surfaces are well prepared. The mechanical strength of a fiber can be further improved with a suitable polymer jacket. Even simple cleaving (breaking) of silica fiber ends can provide nicely flat surfaces with sufficient optical quality. (e) Silica is chemically very stable. In particular, it is not hygroscopic. (f) Silica glass can be doped with various materials. One purpose of doping can be to raise the refractive index (e.g., with GeO_2 or Al_2O_3) or to lower it (e.g., with fluorine or B_2O_3). Doping is also possible with laser-active ions, such as rare-earth-doped fibers, in order to obtain active fibers, to be used, for instance, in fiber amplifiers or fiber lasers. The fiber core and sometimes the fiber cladding are doped, so that the material is effectively an aluminosilicate, germanosilicate, phosphosilicate, or borosilicate glass for example. For active fibers, pure silica is usually not a very suitable host glass, because it exhibits a low solubility for rare earth ions. This leads to quenching effects due to clustering of dopant ions, even for moderate doping concentrations. Aluminosilicates are much better in this respect. (g) Silica has a high damage threshold, i.e., a low tendency for phenomena such as laser-induced breakdown. This is important for, for instance, fiber amplifiers when applied for the amplification of short pulses. (h) Silica has a particularly low Kerr nonlinearity, which is beneficial in many cases where nonlinear effects can be detrimental.

Major research efforts began with achieving low-loss fibers to extend the optical link distance. After the demonstration of fiber with loss of 20 dB/km in 1970, the fiber process advanced very quickly. By 1973, the loss of less than 5 dB/km was reported at 850 nm, approaching the intrinsic scattering limit at that wavelength. Researchers started to explore longer wavelengths where the intrinsic scattering loss is lower. In 1976, the first fiber with loss of 0.47 dB/km at 1,200 nm was reported. Within 3 years, the fiber attenuation reached 0.2 dB/km at 1,550 nm, close to the theoretical limit of 0.15 dB/km. Along with the efforts to reduce the optical loss, there have been intensive researches to control the waveguide properties of optical fibers in order to control the modal dispersion and the chromatic dispersion. As a first step, two types of fibers have been developed in parallel; MMF and SMF. Starting from a step index fiber, MMFs quickly evolved to graded-index profile MMF to increase the bandwidth for 850 nm light-emitting diode (LED) light sources. Graded-index MMF diameter started from 50 μm, and then increased to 62.5 μm to accommodate 1,300 nm LED. As high-speed narrow linewidth light sources such as vertical cavity surface emitting lasers (VCSELs) and laser diodes (LDs) replaced LEDs, graded-index MMF diameter once again reduced to 50 μm with further more tailored graded refractive index profile near the center to secure high bandwidths. Presently 10 Gbit/s signals can be transmitted over a several 100 m long, using graded-index 50 μm core MMFs and further increase in both data rate and link distance is being explored. In the case of single-mode optical fibers the evolution routes are more complicated due to introduction of wavelength division multiplexing (WDM) based on optical amplifiers such as erbium-doped fiber amplifier (EDFA) and Raman fiber amplifier (RFA) and their combinations. In WDM-SMF links multitudes of channels are carried within a certain wavelength range where the channels are amplified to restore the attenuated intensity. However chromatic dispersion linearly accumulates and the signal pulses spread out in the time domain to result in intersymbol interferences and subsequent bit errors in the data streams. Therefore it is imperative that chromatic dispersion be properly managed or compensated in long haul WDM systems, which opened various avenues of dispersion controlled fibers such as dispersion-shifted fiber (DSF), dispersion flattened fiber (DFF), nonzero dispersion-shifted fiber (NZDSF), and dispersion compensating fiber (DCF). Dense WDM systems have also introduced new concepts in fiber design to manage the high optical power and consequent optical nonlinearity, which brought standardization of effective mode areas, and nonlinear coefficients. Nonlinear effects in optical fibers can be significantly suppressed if the modal area is large enough and large mode area (LMA)-SMFs have been developed with novel refractive index profiles (Oh and Paek 2012).

Silica fibers dominate many applications, such as optical fiber communications (except for very short distances with plastic optical fiber), most fiber lasers and amplifiers, and fiber-optic sensors. The large efforts which have been invested in the development of various kinds of silica fibers have further increased the performance advantages of such fibers over fibers based on other materials. For instance, pure silica has been used to make PCFs, containing tiny air holes. Here, the guidance (waveguide function) is achieved either by a reduced effective index of the cladding (caused by a larger fraction of air) or by a photonic bandgap (PBG) effect. For special applications, certain non-silica fibers are required (Gambling 2000): (a) Phosphate glass fibers can be advantageous when a high doping concentration of rare earth ions is required. (b) Fluoride fibers containing heavy metals can be used for the transmission of mid-infrared light, and for fiber lasers or amplifiers where low-energy laser transitions with long metastable level lifetimes are required. (c) Plastic optical fibers (polymer fibers) can be cheaper and thicker than silica fibers and are used for illumination purposes and for short-range data transmission.

4.3.1.2 Halide Glass Fibers

Halide glasses can mainly be divided into three subcategories: Single component halide glasses, heavy metal fluoride glasses (HMFGs), and heavier halide glasses. Single component halide glasses (SCHGs) have low attenuation due to monovalent ion behavior; for example, BeF_2- and $ZnCl_2$-based halide glasses have the attenuation loss of 10^{-2} to 10^{-5} dB km^{-1}. These glasses have some drawbacks; for example, beryllium in BeF_2 is highly toxic, while $ZnCl_2$ highly deliquescent. Single-anion multicomponent or HMFGs have a wide transparency range (from 0.3 μm to about 9 μm, that is, mid-infrared to near ultraviolet). Attenuation losses of these glasses are on the order of 5 to 8×10^{-3} dB km^{-1} at the wavelength of 2.4 μm. This high loss is mainly because of impurities—such as three-dimensional transition metal additions (mainly iron and nickel), which absorb in the 2–3 μm range, and rare earths, and because of a fundamental stretching of O–H absorption at about 3 μm. These glasses have a tendency to recrystallize from the surface as a result of attack by moisture and oxygen, so special jackets or hermetic coatings are required for fibers made of these glasses. HMFGs have relatively narrow glass-forming regions because viscosities are low (for example, 0.4 Pa at 490 °C). HMFGs also have thermal instability tendencies, leading to devitrification; they also have a high susceptibility to attack in aqueous environments (Lewis 1990).

For instance, silver-halide fibers crystalline materials have the advantage of a better long wavelength transmission compared to mid-infrared glasses, but suffer from a difficult fabrication process. The first crystalline fibers were made of hot extruded KRS-5 (Tellurium-Bromide-Iodide). Today, poly-crystalline silver-halide (Silver-Chloride-Bromide) is most widely used. Silver-halide fibers show good transmission up to almost 20 μm. Poly-crystalline fibers are made of crystal-like solid solutions of Thallium halides, e.g., KRS-5, or silver halides, e.g., AgClBr. These materials offer such properties as ductility, low melting point, and isotropy. Crystalline fibers can be fabricated via plastic deformation by extrusion from a preform. The rod-in-tube preform is realized by a rod of AgClBr as core and a Cl-rich AgClBr crystal as cladding. The perform can also be made by the casting method or by preform growth methods. The preform is placed in a heated chamber and the fiber is extruded to its final form through a polished die. Typical extrusion temperatures are in the range of 50–80 % of the melting temperature, which is 457 °C for AgCl, for example. The pressure within the container ranges from a few to some ten tons per square centimeter. To achieve optimum results, extremely homogeneous crystals have been used and different preform manufacturing techniques have been applied, including mechanical combination, preform growth,

Table 4.1 Common laser-active ions and host glasses and important emission wavelength (Paschotta 2012)

Ion	Common host glasses	Important emission wavelengths
Neodymium (Nd^{3+})	Silicate and phosphate glasses	1.03–1.1 μm, 0.9–0.95 μm, 1.32–1.35 μm
Ytterbium (Yb^{3+})	Silicate glass	1.0–1.1 μm
Erbium (Er^{3+})	Silicate and phosphate glasses, fluoride glasses	1.5–1.6 μm, 2.7 μm, 0.55 μm
Thulium (Tm^{3+})	Silicate and germanate glasses, fluoride glasses	1.7–2.1 μm, 1.45–1.53 μm, 0.48 μm, 0.8 μm
Praseodymium (Pr^{3+})	Silicate and fluoride glasses	1.3 μm, 0.635 μm, 0.6 μm, 0.52 μm, 0.49 μm
Holmium (Ho^{3+})	Silicate glasses, fluorozirconate glasses	2.1 μm, 2.9 μm

capillary drop, and capillary suction. Single-mode operation has been successfully demonstrated at a wavelength of 10.6 μm for a fiber with a core-clad composition of $AgCl_{75}Br_{25}/AgCl_{60}Br_{40}$, a core diameter of 20 μm, a cladding diameter of 500 μm, and a single-mode cutoff wavelength of 5.8 μm. An improved fiber has been developed with a double-step index profile. The core composition was $AgCl_{30}Br_{70}$, the composition of the first cladding $AgCl_{32}Br_{68}$ and that of the second cladding$AgCl_{5}Br_{95}$. The core diameter was 50 μm, the diameter of the first cladding 250 μm and that of the second cladding 900 μm. The outer cladding was blackened by exposure to UV radiation (Labadie and Wallner 2009).

Heavier halide glasses are based on chloride, bromide, and iodide mixtures, although the latter two types give rise to glasses of low stability and hence are unsuitable for fiber-optic applications. The chloride-based glasses have major problems with impurity ions and are water-soluble.

4.3.1.3 Rare-Earth-Doped Glass Fibers

Fiber lasers and fiber amplifiers are nearly always based on glass fibers which are doped with laser-active rare earth ions (normally only in the fiber core). These ions absorb pump light, typically at a shorter wavelength than the laser or amplifier wavelength (except in up-conversion lasers), which excites them into some metastable levels. This allows for light amplification via stimulated emission. Such fibers are often called active fibers. They are gain media with particularly high gain efficiency, resulting mainly from the strong optical confinement in the fiber's waveguide structure. The most common laser-active ions and host glasses and also typical emission wavelength ranges of rare-earth-doped fibers are shown in Table 4.1 (Paschotta 2012). The technologically most important rare-earth-doped fibers are erbium-doped fibers for erbium-doped fiber amplifiers and ytterbium-doped fibers for high-power fiber lasers and amplifiers. The chemical composition of the glass for the laser-active fiber core has many important influences on the possible performance and practical use of an active fiber (Digonnet 2001): (a) The limited transparency range may exclude the use of certain laser transitions. For example, mid-infrared lasers cannot be realized with silicate fibers, which are strongly absorbing for wavelengths above ≈2 μm. (b) The glass composition strongly influences the maximum concentration of the dopant ions that can be incorporated without excessive clustering, which would resulting in quenching effects and possibly increased propagation losses. (c) It also influences in various ways the optical transitions of the rare earth ions, in particular the emission and absorption cross sections, absorption and emission bandwidth, total transition rates and thus the metastable level lifetimes, etc. (d) The rate constants for energy transfers between different ions also depend on the chemistry. (e) Mainly the maximum phonon energy of the host glass determines the rate of multiphonon emission processes, thus the speed of nonradiative transfers between certain levels. This effect can be strong: certain levels may be long lived (multiple milliseconds) in heavy metal fluoride glasses, but very short-lived (few microseconds) in silicate glasses. (f) Some glasses (e.g., fluoride glasses) tend to be difficult and expensive to fabricate and handle. Clean fiber cleaves are not always easy to obtain, and often require modified methods. (g) Some glasses are photosensitive, allowing the fabrication of fiber Bragg gratings with ultraviolet light. The photosensitivity can strongly depend on certain dopants. (h) Glasses differ very much in their optical nonlinearities and optical damage threshold.

For rare-earth-doped fibers, the core composition is normally modified by additional dopants. For example, one rarely uses pure silicate glass, but rather some aluminosilicate, germanosilicate, or phosphosilicate glass. Some co-dopants such as aluminum improve the solubility of rare earth ions and thus allow for higher rare-earth doping concentrations without quenching of the upper-state lifetime. Others have effects on the refractive index, on the spectral shape of the optical transitions, or the rate of energy transfers (Paschotta 2012).

For many up-conversion lasers and visible fiber lasers, some kind of fluoride glass is required where the phonon energies are lowered so that the metastable level lifetimes are long enough (no quenching via multiphonon transitions). Such fluoride fibers also exhibit good transmission in the mid-infrared and are therefore used for mid-infrared laser sources. Some fibers are intentionally doped with two different kinds of rare earth ions. Most popular is the combination of erbium and ytterbium (erbium–ytterbium fibers)—normally with a significantly higher concentration of ytterbium. When such a fiber is pumped e.g., around 980 nm, most of the pump light is absorbed by ytterbium ions (called sensitizer ions), bringing these into their excited states. From there, the energy can be efficiently transferred to the erbium ions, which then provide laser gain in the 1.5-μm spectral region. Compared with purely erbium-doped fibers, Er:Yb fibers offer much higher pump absorption per unit length and can therefore be used for fiber devices with much shorter lengths, such as distributed-feedback lasers. For example, this is useful for making robust single-frequency fiber lasers of a few centimeters length, or for double-clad fiber devices with a moderate length. For the energy transfer to be efficient, the doping concentrations have to be well balanced, and the core composition must be suitable (Paschotta 2012).

Ytterbium co-doping can also be used for other gain systems, such as in praseodymium-doped up-conversion lasers. This allows for, e.g., red, orange, or blue emission with single-wavelength pumping (instead of dual-wavelength pumping for purely praseodymium-doped fibers). Co-dopants can also be used for quenching the lower-state population in gain systems with self-terminating laser transitions. For example, praseodymium co-doping allows for relatively efficient operation of 2.7-μm erbium fiber lasers (Paschotta 2012).

For high-power fiber lasers and amplifiers, double-clad fibers are used. These have a highly multimode inner cladding, into which the pump light is launched, and a fiber core which is either single-mode or supports only a few modes. Only the core (or sometimes a ring around the core) is rare earth doped. Such fibers allow for a high beam quality of the laser or amplifier output, whereas the pump beam quality can be very low. The resulting devices are often called brightness converters, since the brightness of the output can be much higher than that of the pump source. In addition to all the properties of a passive (un-doped) optical fiber, such as the guiding properties (effective mode area, numerical aperture, cutoff wavelength, bend losses), nonlinearities, etc., active fibers can be characterized with respect to several other properties (Digonnet 2001): (a) One of the most important parameters is the rare-earth doping concentration, most often specified in wt. ppm (parts per million by weight). A higher doping concentration allows for efficient pump absorption in a shorter length and thus also reduces the effect of nonlinearities in high peak power devices. However, it can also lead to concentration quenching. (b) Wavelength-dependent effective absorption and emission cross sections (and possibly ESA cross sections) together with the upper-state lifetime (and possibly lifetimes of intermediate levels) are required for calculating the wavelength tuning behavior, power efficiency, etc. (c) Parameters for quantifying the speed of energy transfer processes are important particularly for co-doped fibers.

As an alternative method, so-called Giles parameters can be specified, which depend on the doping concentration, effective mode area, and effective cross sections. For such characterization, a variety of measurement techniques are used. White-light absorption spectra can be used for finding absorption cross sections (for known doping concentrations). Emission cross sections are obtained from fluorescence spectra, with scaling e.g., via the reciprocity method (→ McCumber theory) or the metastable level lifetimes (→ Füchtbauer–Ladenburg equation). Upper-state lifetimes are often obtained from fluorescence measurements with pulsed pumping, and ESA parameters can be obtained in experiments with a modulated pump power. The resulting set of data can be used e.g., in laser and amplifier models based on rate equations. Such models allow one, e.g., to predict or check the performance of fiber laser or amplifier devices, the effect of possible modifications. Further characterization may be required for quantifying effects such as photodarkening, which can sometimes seriously degrade the efficiency of active fiber devices (Paschotta 2012).

4.3.1.4 Chalcogenide Glass Fibers

Chalcogenide glasses are based on the chalcogen elements S, Se and Te, and the addition of other elements such as Ge, As, and Sb leads to the formation of stable glasses. The addition of halides leads to the formation of chalcohalide glasses. Rare earth doping for active applications and consequently alternative glasses have been developed. Since the chalcogenide glasses transmit to longer wavelengths in the IR than silica and fluoride glasses, there are numerous potential applications in the civil, medical, and military areas. These can be essentially divided into two groups, namely passive and active applications. The passive applications utilize chalcogenide fibers as a light conduit from one location to another without changing the optical properties, other than that due to scattering, absorption, and end face reflection losses associated with the fiber. Active applications of chalcogenide glass fibers are where the initial light propagating through the fiber is modified by a process other than that due to scattering, absorption, and end face reflection losses associated with the fiber. Examples of these include fiber lasers, amplifiers, bright sources, gratings and nonlinear effects (Aggarwal and Sanghera 2002).

Chalcogenide glasses are nonoxide glasses with covalent bonds, the maximum fraction of ionic bonds being 9 %. Compared to oxide glasses, namely SiO_2 and silicates, chalcogenide glasses contain one or more chalcogen elements, sulphur, selenium, or tellurium, in a combination with other elements namely from IV-th, V-th, and VI-th group of the periodical table. In 1950, As_2S_3 was first studied for near and middle infrared region of the electromagnetic spectrum. At the beginning of 1970s the research was split in two directions: the first one was dedicated to the optical materials for middle and far infrared region of the electromagnetic spectrum (for example As_2Se_3, Ge-As-Se, Ge-Se-Te) and, the second one was dedicated to the materials exhibiting threshold and memory switching phenomenon (this direction started with amorphous layers containing tellurium). And then the attention is paid to the study of infrared optical fibers for the transmission of radiation and power delivery of Er: YAG, CO, and CO_2 lasers, active laser fiber, and nonlinear optics. Among chalcogen elements only selenium and sulphur can be prepared in glassy state. Tellurium cannot be prepared in glassy form by cooling the melt. It can be obtained only as thin amorphous layer by deposition methods (Lezal 2003).

Chalcogenide glasses can be prepared by various ways both from the liquid and from the vapor phase. Most known methods are: cooling of melt, sol–gel, chemical vapor deposition, evaporation, and sputtering. Bulk glasses are mostly prepared by cooling of melts and partly by sol–gel method. The CVD methods are mostly used for the preparation of glassy thin layers. Modified chemical vapor deposition method (MCVD) has been developed for the preparation of preforms for fiber's drawing. The evaporation or the sputtering is used for the preparation of thin layers for switching devices, xerography, lithography, memory films. The applications in optoelectronics demand high chemical and physical purity of the glasses. The chemical purity is determined by the concentration of chemical impurities, while physical purity is defined by the concentration of physical defects as: heterogeneous particles, microcrystalline phase separation, cracks, inhomogenities, etc. Chemical impurities cause the extrinsic absorption of radiation, while physical defects cause the scattering radiation. In order to approach the values of optical losses to the theoretical ones the concentration of impurities should be 10^{-5} mol% for hydrides, 10^{-6} mol% for oxides, 5×10^{-5} mol% for silicon and carbon, while the concentration of the physical defects should be about 10^2–10^3 cm^{-3}. Chalcogenide glasses belong to substances which have an incongruent melting point, exhibit a high partial vapor pressure during melting and are susceptible to oxidation and hydrolysis and, therefore, the synthesis must be carried out in sealed evacuated quartz ampoules. The synthesis conditions are widely varied—they depend on the glass composition, glass-forming region, and glass-forming ability. Thus, the glasses containing arsenic can be prepared at a temperature situated in the range 700–750 °C, the glasses with germanium are melted at 900–950 °C, while silicon melts at the temperature of 1,150 °C. Further increase of the temperature of synthesis is limited by the softening temperature of quartz ampoule, which is around 1,200–1,250 °C. The melt-cooling regime is also very dependent on the glass-forming ability of the system. The melt can be cooled slowly (self-cooling furnace) for good stable glasses (As_2S_3, As_2Se_3). Faster cooling rates (quenching) for unstable glasses can be achieved by cooling in air or by plunging the sealed ampoule into cold water or even liquid nitrogen (telluride). The temperature of annealing is very important. To achieve the value of the chemical purity suitable for power delivery of laser energy and for the preparation of active and laser fibers doped with rare earth ions (Pr, Er, Dy, Nd, etc.) advanced methods have been developed because in glasses prepared only by direct synthesis from starting elements the average concentration of hydride and oxide impurities is around 10^{-3} mol% and the physical defects around 10^4 cm^{-3}. The purification of the glasses must be done in many steps, namely, before the synthesis and during synthesis. The major sources of impurities seem to be the starting elements S, Se, Te, As, Ge, Ga, etc. and, therefore, volatile elements are repurified by sublimation in reactive atmosphere (SCl_2) or under vacuum. Ge, Ga, etc. are melted in a high vacuum at the temperature about of 950 °C. The technological procedures involve the distillation of S or Se or As. All these operations are carried out in special quartz ampoules to diminish the danger of contamination of the atmosphere during the preparation procedure. Some physical properties of chalcogenide glasses are shown in Table 4.2. The transparency range of these glasses depends on the composition and is situated in the range from 0.5 to 7 μm for glasses on the base of sulphides, from 0.8 to 12 μm for selenide and 1.2 to 16 μm for telluride glasses.

Refraction indices (n) and the values of transparencies (%) also depend on glass composition as can be seen in Table 4.2. The region of the transparency is limited by an intrinsic absorption at short and long. In the transparency range (from short to long wavelength cut off) it can be observed absorption bands caused by impurities, which are called extrinsic absorption. Extrinsic absorption determines the intensity of absorption bands of impurities built up in the glass network, for well determined wavelengths. The concentration of impurities can be estimated from the intensity of extrinsic absorption bands. The extrinsic absorption in chalcogenide glasses can be

Table 4.2 Some physical properties of chalcogenide glasses (Lezal 2003)

Glass	Phonon energy (cm^{-1})	Transmission range (μm)	Refraction index (n)
As_2S_3	350	0.5–6	2.35
As_2Se_3	360	0.8–10	2.70
$Ge_{25}Ga_{10}S_{65}$	380	0.6–7	2.55
$Ge_{25}Ga_5As_5S_{65}$	330	0.6–7	2.58
Fluoride glass	440–650	0.25–6.5	1.52
Heavy metal oxide glass	590–850	0.4–7	2.65
SiO_2	1,150	0.25–2	1.148

caused by two kinds of impurities (Lezal 2003): (a) oxide impurities—the presence of O–X bonds; (b) hydride impurities—the presence of H–X bonds.

The manufacturing techniques for optical fibers are determined by the different physical properties of fiber materials. In the mid-infrared, it can be distinguished between glassy fibers and crystalline fibers. For both types, a preform is created by mechanical combination of the core and cladding material. Glassy fibers are drawn from the preform, whereas crystalline fibers are extruded from the preform. Chalcogenide glass was the first material used to produce mid-infrared fibers. A two or more component glass is formed by combining one or more chalcogene elements such as sulphur, selenium, or tellurium with one or more elements such as arsenic, germanium, phosphorus, antimony, gallium, aluminum, or silicon. The properties of the glass change drastically with composition. In general, chalcogenide glasses are very stable, durable, and insensitive to moisture. A distinctive difference between these and other mid-infrared fiber glasses is that they do not transmit well in the whole visible region and that their refractive indices are quite high. Arsenic-Trisulfide (As_2S_3) fibers have a transmission range from 0.7 to about 12 μm. Longer wavelengths can be transmitted by adding heavier elements like Tellurium, Germanium, or Selenium. Chalcogenide glass formation can be achieved when the chalcogenide elements are melted and quenched in an evacuated silica glass ampoule with one or more elements. After heating in a rocking furnace for longer than 10 h at a temperature between 700 and 900 °C, the ampoule is quenched in air and annealed to room temperature. The preform is fabricated by using rod-in-tube (core rod is placed within the cladding tube) or casting methods (melted core material is cast into the cladding rod). The two methods in use for fiber drawing are preform drawing and crucible drawing. In the case of fibers with no glass cladding or only a polymer cladding, e.g., teflon, the conventional preform drawing method with a tube furnace and a narrow heat zone within an inert gas atmosphere is used. Fibers with chalcogenide cladding are drawn in a similar apparatus. The core rod and cladding tubes are placed into a silicon muffle furnace, zonally heated, and drawn into a fiber. Core-cladding fibers can also be produced by a double-crucible technique or by modified crucible drawing, a technique offering the advantages of both crucible drawing and preform drawing. SMFs made of chalcogenide glass have been and are currently developed for ESA and NASA in the framework of the planned missions DARWIN and TPF-I. The fibers are intended as spatial wavefront filters for the wavelength range from about 4 to 12 μm. The Naval Research Laboratory (United States) developed for NASA chalcogenide glass SMFs for use at wavelengths up to about 11 μm. The fibers had a core diameter of 23 μm, a cladding diameter of 127 μm, and core and cladding refractive indices of 2.725 and 2.714 (Labadie and Wallner 2009).

4.3.2 Plastic Optical Fibers

Plastic optical fibers (POFs), also referred to as polymer optical fibers, have many of the same advantages as conventional silica optical fibers. These include low weight, immunity to electromagnetic interference and multiplexing capabilities. In general, POFs provide a much lower cost alternative to silica optical fibers, albeit with higher transmission losses. POFs have thus been applied for data transmission over short distances, e.g., local to home Internet connections and automotive applications (Peters 2011).

POFs have the same geometry as silica optical fibers, with a core, cladding and sometimes a jacket. The very first research on optical fiber transmission in 1955 was done with plastic fibers. Since then the development of silica fiber has all but eliminated plastic from consideration. It survives in niche applications for short distance connections in medical instruments, some industrial instrumentation and in up-market consumer Hi-Fi equipment. In addition new research of POFs has opened up a number of other significant possibilities. The most important characteristic of POF is its thickness. This is because it is the thickness that makes it easy to fit connectors and from this comes low installation cost. Low system cost is the only advantage POF has over glass fibers. Step index POF has a core diameter of 980 μm and a cladding of 20 μm for a fiber diameter of 1 mm. This is a core diameter 100 times larger than SMF. The total diameter of the fiber is 8 times the diameter of typical glass fiber (Dutton 1998).

A variety of optical polymers are used in the fabrication of POFs, including polymethyl-methacrylate (PMMA), amorphous fluorinated polymer (CYTOP®, Asahi Glass), polystyrene (PS), and polycarbonate (PC). Unlike silica optical fibers, POFs are primarily available as multimode fibers, which have larger diameters and propagate multiple, interacting modes. Since the larger diameter of multimode POFs makes them easier to handle, cleave and connect, multimode POF sensors are often promoted as less expensive and easier to install than their silica counterparts. While this is a true benefit for multimode POFs, care should be taken as the same does not hold for single-mode POFs (Peters 2011).

Hard polymer clad fiber (HPCF) has been developed to use a relatively thick glass fiber as a core and a hard plastic coating as the cladding. This has significantly less attenuation than POF and is claimed to cost about the same and to be just as easy to join. It is thinner than POF and therefore suffers from much less modal dispersion. Attenuation is relatively high compared to other types of glass fiber but this is because the wavelength used is in the high-attenuation part of the glass absorption spectrum (650 nm). This wavelength was chosen to be the same as that used for POF. The light sources used are LEDs rather than lasers partly due to a concern with modal noise if lasers were used. HPCF is step index with a pure silica core and plastic cladding. The specifications for HPCF are intentionally very close to those of POF with the exception that HPCF has about 20 times lower attenuation. Also the core diameter of SI HPCF is significantly less than that of POF and therefore the fiber has significantly lower levels of dispersion than POF (Dutton 1998). HPCF shows the best performance when it is applied to short to intermediate, a few 100–1,000 feet, fiber-optic data transmission. Hard polymer (HP) cladding made of fluoroacrylate has a role of protective coating of glass core in addition to a role of fiber clad. HPCF is hard to break if it is bent tightly or treated roughly in the process of installation or termination. The hard polymer cladding provides higher tensile strength and greater resistance to moisture than the conventional claddings. These features together create a fiber well suited for a wide variety of communications, industrial applications, as well as near-IR spectroscopy (Rayscience 2013).

4.3.3 Photonic Crystal Fibers

PCFs are made of dielectric structures with periodically varying refractive index at the wavelength scale. These structures exhibit PBGs, which are frequency intervals within which light propagation is forbidden in certain directions. PCF are formed from a dielectric rod containing tubes of a material of lower refractive index or air, arranged in a special pattern and running along the fiber axis. Two types of PCFs may be distinguished (Fig. 4.4a): those with an index-guiding structure and those with a PBG-guiding structure. The pattern of air or low refractive index tubes determines the waveguiding properties. Single-mode operation is possible for both types of PCFs, although with different single-mode bandwidth and different propagation properties (Labadie and Wallner 2009).

Index-guiding PCFs are formed by a not necessarily periodic structure of air or low refractive index tubes at wavelength scale, embedded in some dielectric material. A defect which confines the light and therefore acts as the core is introduced by omitting one tube. Wave guidance can be explained by considering that the cladding region provides (due to the tubes) an effective index which is smaller than the core index. Depending on the wavelength, the effective index approaches the material index of the core, if the field is mainly guided within the material, or the index of the tubes, if the field penetrates the cladding. With the effective cladding index approach, the index-guiding PCF can be modeled as a single-step index fiber with an effective normalized frequency of $V_{\mathrm{eff}} = \pi\Lambda/\lambda(n^2{}_m - n^2{}_{\mathrm{eff}})^{1/2}$, where the core diameter is given by the center-to-center spacing Λ between two adjacent tubes. In contrast to conventional single-step index fibers, V_{eff} approaches a constant value for large Λ/λ. This means that by proper choice of the hole diameter to hole spacing ratio d/Λ, such a fiber can be single-mode for any wavelength (Birks et al. 1997).

PBG guiding PCFs are obtained by locally breaking the periodicity of a photonic crystal by introducing a well-defined defect, e.g., in the form of an extra tube. The

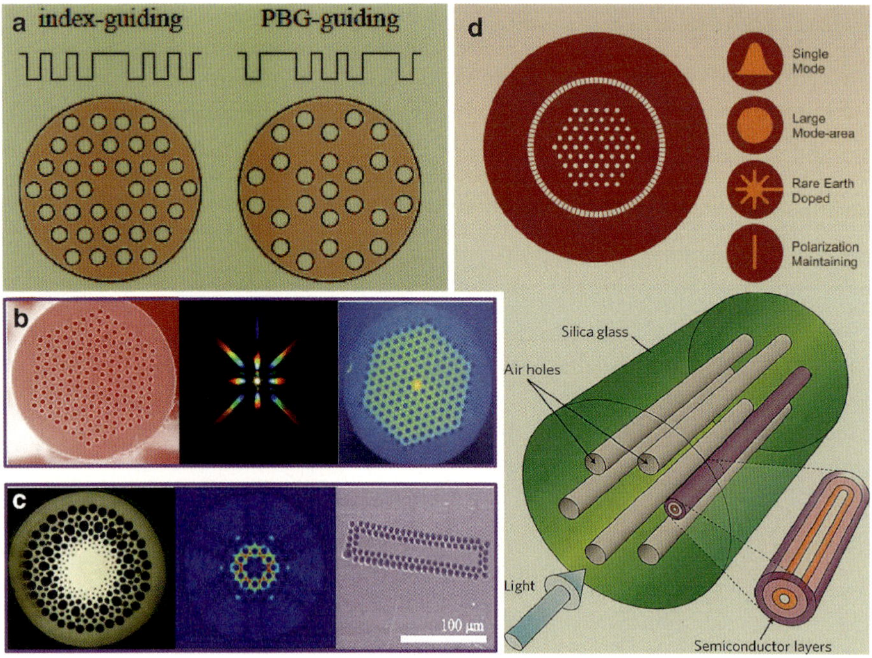

Fig. 4.4 Schematic illustration of different photonic crystal fibers (PCF): (**a**) refractive index profiles of index-guiding and photonic bandgap (PBG) guiding PCF; (**b**) single-mode PCF; (**c**) graded-index PCF; (**d**) different modes and 3D sketch

light then is confined and thus guided, provided that the surrounding photonic crystal cladding exhibits a PBG at the operation wavelength. A remarkable difference between

PBG-guiding PCFs and conventional fibers or index-guiding PCFs are that they allow wave guidance with propagation constants (i.e., PBGs) below the effective cladding index. The PBG regions, i.e., the transmission windows, are determined by the cladding structure only. They are usually very narrow. The behavior of the guided modes within the PBGs is solely determined by the characteristic of the defect. By varying the size and shape of the defect, the frequency of the guided mode can be positioned precisely within a PBG (Labadie and Wallner 2009).

In fact, PCFs typically utilize multiple air holes that run longitudinally along the fiber to guide light. After the first demonstration in 1996 of a silica single-mode PCF, it was quickly realized that a wide variety of optical properties could be achieved by the variety of microstructures possible, and many reports followed on solid-core and hollow-core PCF. The use of air holes meant that the fiber could be made from a single material. The technology could thus be transferred from silica to any other suitably transparent, drawable material to access a wider range of mechanical and/or optical properties. The first such non-silica single-mode PCF reported was made using polymer—it was a microstructured polymer optical fiber (mPOF) made of PMMA, a common material used in the manufacturing of polymer optical fiber (POF). The change to polymers had significant potential given the range of polymers and processing methods available, and the low processing temperatures which would allow organic and inorganic dopants to be used that would not otherwise survive the higher processing temperatures of silica. The addition of microstructure to polymer fibers led to application-specific fibers with potential for use in sensing, engineering, medicine, textiles, and communications. The surrounding work also contributed to understanding the physics of MOFs, with many firsts demonstrated in mPOF, and thus contributed to the larger silica-dominated field of microstructured fibers as a whole (Argyros 2009).

The general fabrication scheme employed for mPOF is a two-step process. A large diameter ~8 cm primary preform is first fabricated with the desired hole arrangement and stretched to ~6 mm diameter cane. The cane is sleeved to increase the outer diameter to ~12 mm, forming the secondary preform from which fiber is drawn. The details can vary between different fibers, and e.g., fiber can be drawn directly from the primary preform. As the first step in the process, much effort has focused on the primary performs, which have been fabricated by drilling, stacking, casting/molding, billet extrusion, and solvent deposition, with the properties of polymers being crucial to the success of some of these methods. Drilling was the first method employed to fabricate mPOF. Apart from the ease with which polymers can be machined, it allowed a variety of designs of greater or lesser complexity to be fabricated easily. Hole positions are not restricted to the hexagonal arrangement typical of stacked preforms, leading to the first nonhexagonal microstructured fiber—a ring-structured fiber with the holes placed on concentric circles. The flexibility of drilling thus made it suitable for research purposes and resulted in its wide use, however it is limited by the short performs produced. The high air fractions required for some fiber designs led to the investigation of tube stacking for mPOF, as used in silica PCF fabrication. Thin-walled tubes of 4 mm outer diameter and 3.7 mm inner diameter are typically used and preform lengths of >40 cm are possible. Large performs can also be produced through casting/molding by polymerizing the polymer in a mold with the holes defined by steel rods, which are removed once the polymerization is complete. Extrusion is a common fabrication method for conventional step and graded-index POF as it allows for continuous manufacture, rather than batch fabrication (where one preform produces one batch of fiber). To date, a batch-approach billet extrusion has been demonstrated for mPOF preforms. Microstructured polymer fibers containing a second polymer (rather than air holes) have been made by solvent deposition. The first polymer is dissolved in a nonsolvent of the second and vice-versa. Layers of each polymer can be deposited and dried without damaging the remaining preform. Once the preform is completed it can be drawn to cane and/or fiber. Drawing must be done above the glass transition temperature Tg of the material. For PMMA, with a Tg of 115 °C, the drawing temperatures are 180–230 °C. Tension of 0–2 kg for cane draws and 0–100 g for fiber is also typical (typical dimensions are 6 mm diameter for cane and 100s of microns diameter for fibers). Positive or negative pressure can be applied at any stage of the process to some or all of the microstructure as required. Extensive modeling of the draw process of microstructured fibers has identified the role of various process parameters and material properties such as feed and draw rates, viscosity, and surface tension. As a result, a large range of behavior such as hole expansion or hole collapse (relative to the outer diameter of the cane/fiber) can be achieved solely by varying the draw conditions and without the use of vacuum or pressure. The behavior of drawing mPOF in this sense can vary from that of silica PCF where the combination of material properties and draw parameters tend to favor hole collapse (Argyros 2009).

The first and most commonly used polymer for mPOF is PMMA, being the most transparent ordinary polymer (i.e., not deuterated or fluorinated), low cost, widely available and already in use for POF. PMMA is generally transparent in the visible up to 850 nm. PMMA-POF typically operates at the 0.15 dB/m loss minimum at 650 nm. In the infrared the transparency is compromised by absorption bands arising from C–H bonds. Other polymers with specific properties have also been used for mPOF, although this typically results in higher material absorption or cost. Among such examples is the cyclo-olefin copolymer (COC) Topas which displays lower water absorption than PMMA; polystyrene (PS) with a higher refractive index; polycarbonate (PC) with a higher Tg; cellulose which is biodegradable and fluorinated materials which have lower material absorption. The low processing temperature of PMMA and other polymers mean that a variety of organic and inorganic dopants can be used in mPOF fabrication. Such dopants would be damaged by the draw temperatures used for PCF of other materials. Dopants can be incorporated into a polymer at the polymerization stage, by casting from a solution of both the polymer and dopant, or by mixing with or diffusion through the solid polymer. In solution doping a solution of the dopant is inserted into the holes of an mPOF preform or cane. The small solvent molecules diffuse into the polymer and swell it, creating voids through which the dopant can enter the polymer matrix. The solvent is then removed by evaporation. The concentration of the dopant is determined by the concentration of the solution and the residence time. Diffusion of the dopant upon heating results in a uniform distribution across the core. The "solvent" is necessarily a small, volatile molecule such as methanol or acetone, in which the dopant is soluble. The residence time of the solution in the holes depends on the extent to which this solvent swells or dissolves the polymer. Methanol does not dissolve PMMA, and the PMMA can become completely saturated so long as the swelling does not close the holes of the microstructure. Acetone dissolves PMMA and residence times are limited to ~30 s. This deposits a layer of dopant on the surface of the hole, which is transported through the polymer by diffusion upon heating. Solution doping is limited to dopant molecules that will diffuse through the swollen polymer such as organic dye molecules, whereas inorganic ionic species are typically excluded. Ionic species may be stabilized by organic ligands, but these make the resulting complex too large and, although compatible with the polymer, prevent its diffusion. Other doping methods such as adding the dopant at the polymerization stage or to a solution of the polymer must be used in these cases. Particulate dopants can be added to PMMA powder and mixed using a ball mill to achieve a homogeneous distribution. The doped powder is consolidated into a rod using heat and vacuum and the doped rod can be inserted into position into a drilled or stacked preform and drawn to fiber. This method has been demonstrated using rhodamine-doped silica nanoparticles and quantum dots, cobalt nanoparticles, as well as to produce blends of polymers e.g., PMMA-PS. The inherent disadvantage of this method is scattering that may arise from the particulate dopants and any air bubbles formed during consolidation of the doped powder (Argyros 2009).

Although silica fiber technology is abundant in SMFs, POF technology has focused almost exclusively on large-core step- and graded-index MMFs, some examples are shown in Fig. 4.4b–d. Single-mode POFs have been fabricated, but they are typically single-mode only in the infrared, suffer from high loss (e.g., scattering at the core-cladding interface), and are generally not commercially available. The use of microstructure however, meant that single-mode polymer fibers could be easily fabricated to operate at any wavelength. The loss of single-mode mPOF is typically 1 dB/m across the visible, and the core sizes vary from 1 to 15 μm depending on the wavelength to be guided. Alternative single-mode designs have also been fabricated, but the hexagonal arrangement of holes is found to be most robust and is most commonly used. Large core, large diameter, and flexible MMFs operating in the visible are desirable as that could facilitate low-cost connectors and user installation. The transfer of this research area to mPOF resulted in some of the complexities of multimode PCF (Argyros 2009): (a) the variety of structures possible meant that a very large parameter space needed to be searched and optimal properties of designs had to be defined. (b) modeling the guided modes of PCF was typically limited to SMFs, whereas now thousands of (leaky) modes would need to be processed. And (c) the transfer of guided light between different modes along the length of the fiber had to be taken into consideration. The use of polymer for PCF allowed a wide variety of fiber structures to be considered, which in turn created a necessity for developing methods of fabricating them. Evolutionary algorithms were developed for describing and optimizing this wider variety of complex structures, and the ABC method was developed to model multimode fibers with thousands of modes and arbitrary index profiles. The ability to consider thousands of modes contributed to understanding the physics behind e.g., air-clad high-NA fibers and hollow-core fibers guiding by the inhibited coupling mechanism. Combining polymers with microstructure extends what is possible with polymer fibers, from simply making single-mode polymer fibers available for gratings applications or through extending the wavelengths and materials that can be used through hollow-core fibers, to more exotic applications in textiles. For sensing applications, it is possible to produce robust, extended lengths of fiber in which the core is exposed. Potentially this allows the deposition of (smooth) metal films on the fiber core, extending the surface modifications already demonstrated. A wider variety of refractive indices, thermal and mechanical properties are available through an appropriate choice of polymer, as well as dopants. Now that more preliminary issues such as fabrication have been largely addressed, PCF technology shows promise in exploring these and other applications (Argyros 2009).

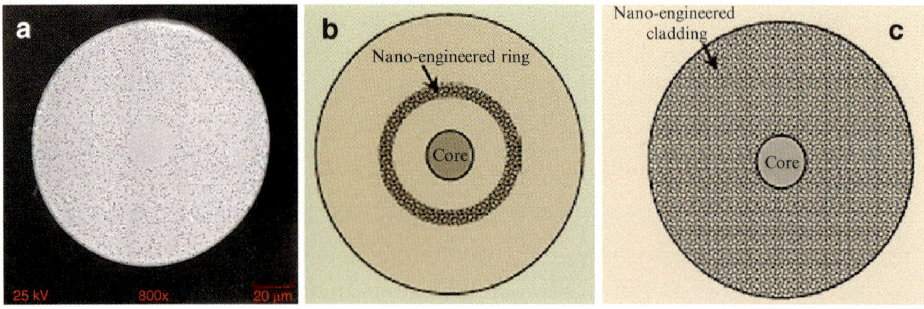

Fig. 4.5 Illustration of nano-engineered fibers (Li et al. 2010): (**a**) SEM image of a fiber with nano-engineered cladding; (**b**) nano-engineered ring; (**c**) nano-engineered cladding

4.3.4 Nano-Fibers

4.3.4.1 Nano-Engineered Fibers

Nano-engineered fibers contain nanometer-sized gas filled voids in the cladding that are incorporated in the glass during the fiber processing. The fiber usually consists of ~20 μm diameter void-free silica core, with nonperiodically distributed voids in the silica cladding, as shown in Fig. 4.5a. The cross sections of the voids are circular and have diameters ranging from several dozens to several hundreds of nanometers. The void fill fraction can be designed to be between 1 and 10 %. The voids are sealed and nonperiodically distributed along the fiber length with void lengths ranging from less than 1 m to several meters. The size characteristics of voids and void fill fraction significantly affect the optical properties of the nano-engineered region, thereby influencing the fiber performance. The nano-engineered glass fibers offer several advantages compared to conventional glass fibers. First, the wavelength dependence of glass refractive index having nanometer-sized features is very different from that of glass with conventional index lowering dopants (such as fluorine) used in fiber manufacturing. A second advantage of nano-engineered fibers is that large negative index changes can be made with nanometer-sized features. A relative index change as high as several percent can be achieved by using a nano-engineered design. Such a high index change is very difficult to realize using the conventional fluorine doping technology. This feature can be used to design fibers with high numerical apertures. Thirdly the scattering property of a glass having nanometer-sized voids also has strong wavelength dependence. Light at shorter wavelengths has higher scattering losses than at longer wavelengths. This feature facilitates the suppression of higher order modes in nano-engineered ring-assisted bend-insensitive fiber designs. The most important advantage of nano-engineered fiber is that they are much easier to make than PCFs using the stack and draw process. A nano-engineered fiber can be made using outside vapor deposition (OVD) process. To make a nano-engineered glass cladding, silica soot is deposited on the glass core cane first. Then the soot preform is sintered in a consolidation furnace at fast sinter rates in a low permeability gas atmosphere at a peak sinter temperature of around 1,500 °C to form a glass cladding containing voids. Additional silica soot can be deposited on the preform and then consolidated in a helium atmosphere to produce void-free outer region of the final optical perform. In the draw process, the voids are stretched into thin elongated voids. Because the process for making nano-engineered fibers is compatible with the OVD process, it is suitable for large scale production (Li et al. 2010).

The unique features of nano-engineered glass allow design fibers for different applications. Figure 4.5b, c show schematics of two fibers designs. The first design (Fig. 4.5b) consists of a germania-doped core and a nano-engineered ring in the cladding. This design offers very low bending loss while keeping the other optical parameters fully compliant with the ITU-TG.652 standards, which is suitable for making bend-insensitive fiber for FTTH applications. Modeling results show that the nano-engineered fiber has 1,550 nm bending loss of less than 0.1 dB/turn at 5 mm bend radius, about 10 times lower than the typical fluorine trench fiber. Fibers made with OVD process demonstrated this design is suitable for large scale production. Typical measured optical properties of the fiber are fully compliant with the G.652 standards. Typical bending loss at 5 mm radius is 0.03 dB/turn at 1,550 nm wavelength. The nano-engineered fiber has about 500 times lower bending loss than the standard SMF, and 6–10 times lower than fluorine trench fibers. The second design shown in Fig. 4.5c has a pure silica core and a nano-engineered cladding. This design does not require any conventional dopants in the core and the cladding regions. Both single-mode or MMFs can be designed by changing the core size and void fill fraction in the cladding. For single-mode designs, the wavelength-dependent effective index property of nano-engineered cladding can be used to design a fiber with very wide single-mode operating window. The nano-engineered glass material exhibits unique features that are interesting for designing fibers for different applications (Li et al. 2010).

4.3.4.2 Optical Fiber Nanowires and Microwires

Optical fiber nanowires and microwires (OFNM) offer a variety of enabling properties, including large evanescent fields, flexibility, configurability, high confinement, robustness, and compactness. These distinctive features have been exploited in a wealth of applications ranging from telecommunication devices to sensors, from optical manipulation to high Q resonators. OFNM have been manufactured from a range of amorphous materials which include phosphate, tellurite, lead silicate, bismuthate and chalcogenide glasses and a variety of polymers. OFNM attract considerable interest because they offer a number of exciting enabling optical and mechanical properties, including (Brambilla 2010):

(a) Large evanescent fields: for small radii, a considerable fraction of the transmitted power can propagate in the evanescent field outside the OFNM physical boundary. This property is extremely useful for the manufacture of microfluidic sensors and high Q resonators, which can be easily fabricated by coiling the OFNM. Large evanescent fields can also be exploited to facilitate the launch of light into high Q microresonators.
(b) Strong confinement: when an OFNM has a diameter comparable to half of the wavelength of the light transmitted in it, the propagating beam is confined to its minimum waist diameter, allowing the observation of spectacular nonlinear effects such as supercontinuum generation and nonlinear optical switching.
(c) Configurability: OFNM are fabricated by adiabatically stretching optical fibers and thus preserve the original optical fiber dimensions at their input and output ends; this allows for low-loss splicing to standard fibers and easy interconnection to other optical fibers and fiberized components. Moreover, because of their flexibility, OFNM can be bent with a small bending radius (typically a few microns) to yield extremely compact devices.
(d) Robustness: optical OFNM can have a negligible surface roughness and thus a superior mechanical strength. This allows for relatively easy handling of a nanowire with macroscopic tools and equipment typical of the macroscopic world. Different top-down fabrication techniques have been used for OFNM fabrication. In contrast to bottom-up techniques, which grow nanowires from a seed of few nanometres, top-down techniques manufacture nanowires by reducing the size of macroscopic samples, and thus can provide much longer nanowires.

Four top-down techniques have emerged for the fabrication of glass nanowires and microwires (Brambilla 2010):

(a) Self-modulated taper-drawing
This technique is a two-step process: first, an optical fiber is tapered to a diameter of several micrometers using the conventional flame-brushing technique; then, the taper is broken into two halves and one fiber pigtail is wrapped onto a small hot sapphire rod and drawn further to a submicron diameter. The sapphire tip is heated with a flame at a distance from the fiber; thus it confines the heat to a small volume and helps to stabilize the temperature distribution. Although this methodology has a complex fabrication procedure and a relatively high loss (at least one order of magnitude higher than for other techniques), the self-modulated taper-drawing technique has the great advantage of providing wires with extremely small diameter: with a careful control of the manufacturing parameters, nanowires with radii as small as 10 nm have been fabricated.

(b) The flame-brushing technique
The flame-brushing technique was initially developed for the manufacture of fiber tapers and couplers and it is based on a small flame moving under an optical fiber which is being stretched. Both the burner and the optical fiber extremities are fixed onto stages and connected to a computer. By accurately controlling the fiber stretch and the flame movement, the taper shape can be set to an extremely high degree of accuracy. This technique was used to fabricate nanowires with radii as small as 30 nm; moreover, it provides the longest nanowires with the lowest measured loss. Finally, the flame-brushing technique provides microwires and nanowires with both ends pigtailed: this is extremely important for practical applications, where connectivity can become an issue.

(c) The modified flame-brushing technique
In the modified flame-brushing technique the setup resembles the original flame-brushing technique with the only exception being that the flame is replaced by a different heat source: a microheater or a sapphire capillary tube heated by a CO_2 laser beam. A microheater is generally a resistive element whose temperature can be tuned by setting the current level. If a sapphire tube/ CO_2 laser approach is used, the temperature is controlled by changing the degree of focusing of the laser beam onto the sapphire tube. The modified flame-brushing technique adds flexibility to the flame-brushing technique by changing the processing temperature: this technique can be used to manufacture microwires/nanowires from a wide range of low softening temperature glasses. If used on silica optical fibers, this technique will fabricate nanowires with extremely low OH content (more than three orders of magnitude smaller than for those manufactured by the flame-brushing technique, which has water vapor as a combustion by-product).

(d) Direct drawing from the bulk
If the material used to manufacture nanowires is not available in fiber form, it is possible to pull nanowires straight from the bulk. In this approach, a small sapphire rod is heated and then brought into contact with a bulk glass, which then experiences a localized softening. If the sapphire rod is promptly moved away, a glass filament with micrometric/sub-micrometric diameter is formed. This technique is extremely flexible and does not require expensive equipment, but the nanowire uniformity and diameter are difficult to control. This technique has been successfully used to manufacture microwires and nanowires from tellurite and phosphate glasses and from polymers.

Because of their flexibility, easy connectivity, and extraordinary properties, OFNM have been used in a wide range of applications, which have been grouped according to the property that they exploit (Brambilla 2010): transition regions, confinement or the evanescent field. Transition regions are the conical connections between OFNM and optical fiber pigtails: in these regions, mode guidance changes from core bound to cladding bound. Applications based on this property include ones in mode filters, mode converters, interferometric sensors in taper tips, and comb-like filters for tunable lasers. Confinement relates to the possibility of channeling light into the minimum possible section, at the diffraction limit. Applications exploiting the confinement properties benefit from the extremely high intensity achieved in the minimum waist region and include nonlinear optics (supercontinuum and higher harmonic generation), particle trapping, subwavelength light sources and nanotip sensors. The evanescent field represents the fraction of power propagating outside the OFNM physical boundary. Applications exploiting the evanescent nature of the propagating mode represent by far the largest group and include, among others, those in high Q resonators (loop, knot, and microcoil), particle manipulation, sensors (chemical, refractometric, biological, gyroscopes), lasers, add–drop filters, comb-like filters, atom trapping and guiding, and interferometers. Although still in their infancy, devices based on OFNM seem to promise a tremendous impact on a wide range of technological applications in areas ranging from optical communications to optical sources, lasers, sensing, chemistry, biology, and microelectronics.

4.4 Fiber Fabrication

In manufacturing fiber there are two major challenges (Dutton 1998): (a) Purifying the materials (especially silica) to an extreme degree. Levels of heavy metal impurity of one in a billion are quite sufficient to have an adverse effect on the performance of the fiber. (b) Achieving extreme precision in fiber dimensions and tolerances. The size of the core, its position and the outer dimension of the cladding must be precise to tolerances of 1 μm or less. Refractive indices must also be precise especially in the case of graded-index fibers.

4.4.1 Purifying Silica

For a good fiber the maximum tolerable level of impurity due to the presence of the transition metals (Fe, Cu, Ni, V, Cr, Mn) is about 1 part in 10^9. This is about 1,000 times better (more pure) than is usually achieved with traditional chemical purification techniques. The tolerable level of OH (water) impurity is higher than this—it is about 1 part in 10^8 but this is also very hard to achieve. The extreme levels of purity required are achieved by making use of the same principle for manufacturing whisky. When distilling a mixture of alcohol and water (and other things) the alcohol reaches boiling point at a lower temperature than the water. The fermented mixture is heated to a temperature where the alcohol (along with many impurities that impart the flavor) evaporates from the mixture but the water does not. The vapor is collected and cooled and the resulting liquid has a much higher concentration of alcohol than the original mixture.

A very good first stage bulk purification of silica can be achieved by a process very much like making alcoholic spirits. Sand is heated (with an electric arc) to just above the boiling point of silica (2,230 °C). Silica evaporates and is condensed into a liquid at about 2,000 °C. Most of the metallic impurities are less volatile and will be left behind. Some impurities are actually more volatile and will not condense at such a high temperature. Thus the silica can be separated in a relatively pure. This process is difficult however, because of the very high temperatures involved (Dutton 1998).

Another process is to heat the sand to an extremely high temperature (around 7,000 °C). Silica is a gas at this temperature but most transition metal compounds decompose. So if sand is passed through a very high temperature electric arc and then condenses the silica relatively quickly afterward, most of the impurities decompose and are carried away in the exhaust gasses.

At the input to the fiber manufacturing process chemicals are only pure to about 1 part in a million. The fiber manufacturing process itself achieves the necessary additional purification.

In most fiber manufacture the glass material (silica with required dopants) is created at high temperature from reaction between gasses. The following reactions are typical of the ones used in the MCVD process. There are many alternative reactions—for example the chloride form is used here but the fluoride form may be used as an alternative (Dutton 1998)

$$SiCl_4 + O_2 \rightarrow SiO_2 + 2Cl_2 \tag{4.11}$$

$$GeCl_4 + O_2 \rightarrow GeO_2 + 2Cl_2 \tag{4.12}$$

$$4POCl_3 + 3O_2 \rightarrow 2P_2O_5 + 6Cl_2 \tag{4.13}$$

Gasses are mixed in a carefully controlled way in a chamber. Silicon chloride (in gaseous form) reacts with oxygen to produce silica (silicon dioxide) in solid form and chlorine. This process is called vapor deposition as solid materials (or rather viscous liquids) are formed from the gaseous state and are deposited where they are wanted to be. In general most of the impurities are carried away in the exhaust. The above reactions involve oxidation. In the OVD and vapor axial deposition (VAD) processes hydrolysis reactions are used instead (Dutton 1998):

$$SiCl_4 + O_2 + 2H_2 \rightarrow SiO_2 + 4HCl \tag{4.14}$$

$$GeCl_4 + O_2 + 2H_2 \rightarrow GeO_2 + 4HCl \tag{4.15}$$

Fiber manufacturing techniques usually make use of vapor deposition in one form or another.

4.4.2 Drawing the Fiber

The challenge in making a fiber is the construction of a precise cross section with tiny dimensions. Fiber manufacture is generally done by drawing the fiber from a large cross section to an extremely small one. The large cross section can be

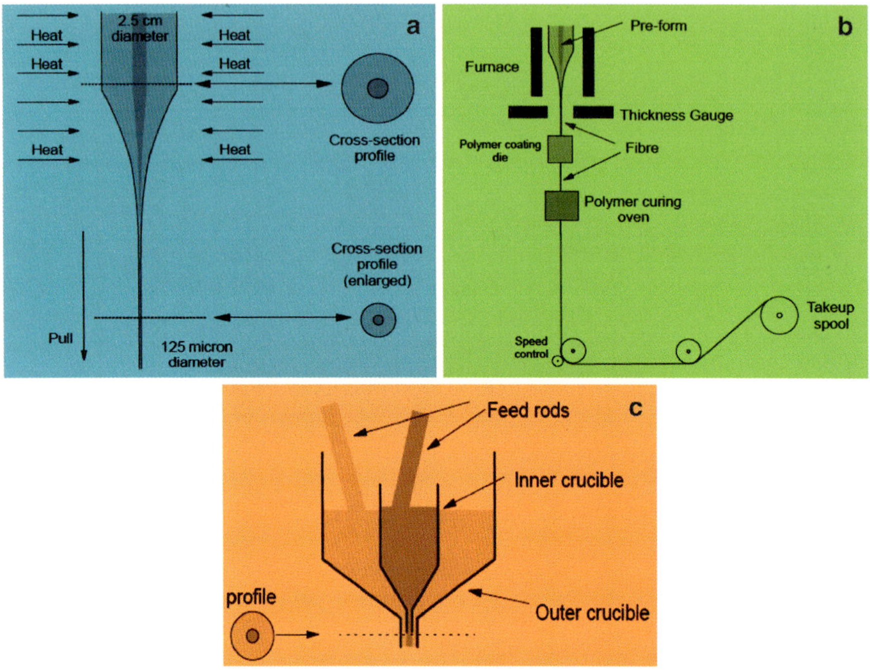

Fig. 4.6 Schematic illustration of optical fiber drawing methods (Dutton 1998): (**a**) principle of drawing a fiber; (**b**) drawing tower; (**c**) double-crucible method

(usually is) in the form of a preformed rod of material or it could be in a viscous form fed from a reservoir. The key fact here is that when drawing a fiber as shown in Fig. 4.6a, the cross section of the fiber so formed is an exact miniaturized replica of the cross section of the rod from which the fiber was drawn. That is, the proportions of the fiber profile are exactly the same as the proportions of the rod from which it was drawn. This is only true of course if the rod is uniformly hot (and therefore has a uniform viscosity) through its entire cross section.

Fiber is made using a tower illustrated schematically in Fig. 4.6b. This tower is usually about 6 m high to allow the fiber time to cool down before being spooled. The process starts with a rod of material called a preform. This rod has exactly the same composition and cross-sectional profile as the fiber to be produced but is between 1 and 2.5 cm in diameter. The end of the preform rod is heated in an electrical resistance furnace until it reaches melting point across its whole cross section. The required temperature here is typically between 1,850 and 2,000 °C. This initial heating can take up to an hour because of the need to ensure that the rod is uniformly hot at its end. Fiber is drawn from the end of the preform mostly by the force of gravity. A small pulling force is used to enable control of the rate of fiber production (and hence of the fiber's diameter). The fiber cools and solidifies very quickly (within a few centimeters of the furnace). It then passes through a station which monitors the fiber diameter and controls the pulling speed. This ensures very close control of the fiber's diameter to the desired 125 μm. The device used here usually comprises a laser and one or more detectors. The laser produces a very narrow beam which is focused on the fiber itself. A diffraction pattern is produced from interaction of the laser beam and the fiber. This diffraction pattern changes with changes in the diameter of the fiber. The pattern is sensed with an optical detector (or detectors) and monitored for changes. Thus an automatic control is achieved which pulls the fiber a bit faster when it is too thick and a bit slower when it is too thin. The fiber is now quite cool and it is passed through a vat of polymer. At the exit point from the vat there is an extrusion die which ensures a uniform coating of polymer onto the fiber. This is done as soon as possible after the fiber is drawn to minimize the absorption of water from the atmosphere. The coated fiber then passes through a curing oven where the polymer is cured and set. The fiber is now 250 μm in diameter. The fiber is then reeled onto a spool with a diameter of 20 cm or so. This is actually a critical part of the process as the fiber must be evenly spooled. If fibers cross one another on the spool microbends can result in this can do long-term damage to the fiber (Dutton 1998).

Figure 4.6c shows a double-crucible method, a pair of platinum crucibles set one inside the other hold molten silica containing the desired dopants etc. The whole assembly is housed inside a furnace to keep the temperature at the optimum level. This is often done in an inert (nitrogen) atmosphere. The inner crucible holds the core glass and the outer crucible, the cladding. The crucibles are continually replenished by the insertion of feed rods of silica. Fiber of the desired cross section is drawn from an outlet in the base of the assembly. The double-crucible method of fiber manufacture was important in the past but is not in commercial use today. It was very attractive because it is a continuous process and therefore offers potentially very low-cost operation. However, it has not been able to deliver the required precision in control of fiber dimensions and characteristics (Dutton 1998).

4.4.3 Vapor Deposition Techniques

The major methods of fiber manufacture create a solid rod (called a preform) from which the fiber is drawn in a batch process. Preforms are created by depositing silica (including dopants) from reactions between gasses at high temperatures. There are four general processes (Dutton 1998): (a) OVD; (b) VAD; (c) MCVD; and (c) Plasma-activated Chemical Vapor Deposition (PCVD). Of these deposition methods, VAD and OVD are the most widely used processes for making SMF although MCVD is also sometimes used. MCVD is used extensively for specialty fibers such as in fiber made for pig tailing and in doped fibers for amplifiers etc. MCVD is also extensively used in making GI MMF. In manufacturing SMF, the VAD process is almost exclusively used in Japan. OVD and MCVD are used in Europe and America (Dutton 1998).

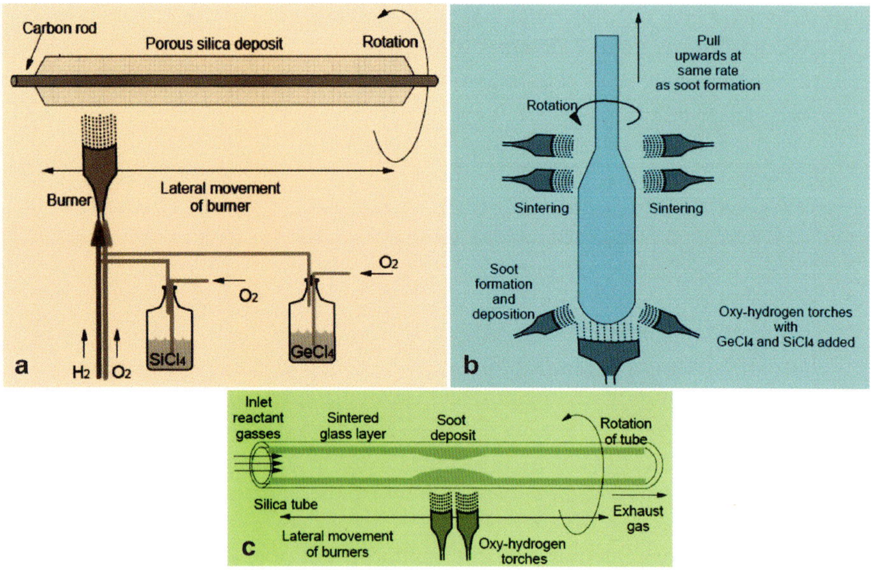

Fig. 4.7 Schematic illustration of vapor deposition techniques (Dutton 1998): (**a**) outside vapor deposition; (**b**) vapor axial deposition; and (**c**) modified chemical deposition

As shown in Fig. 4.7a, the basic process involved in OVD is called flame hydrolysis. $SiCl_4$ reacts with oxygen to produce silica SiO_2 and HCl. This reaction takes place within an oxy-hydrogen flame. At normal temperatures $SiCl_4$ is a relatively volatile liquid. Oxygen is passed through the silicon chloride to form a mixture of $SiCl_4$ vapor and oxygen. The $SiCl_4$–O_2 mixture is then fed to an oxy-hydrogen burner such that it is introduced into the flame. Minute particles of molten silica (called soot) are formed in the flame. A rod of either carbon or a metal is rotated in a lathe and the silica depositing flame is moved back and forth along it. The silica is deposited evenly along the rod. Dopants such as germanium are introduced into the flame to create the required variations in the refractive index. When sufficient material has built up on the rod the process is stopped and the rod removed. The glass body so formed is a porous conglomerate of silica particles stuck together. The preform is then heated (sintered) to coalesce the material into a solid glass tube. The tube is then heated further and collapsed to form a rod. In principle the fiber can be drawn directly from the collapsed silica preform. In practice it is a waste of time to deposit all of the cladding in this way. When the preform has built up a sufficient thickness for the core and a small part of the cladding the process is usually stopped. This preform is then sintered and inserted into a silica tube. The assembly is then heated and collapsed together creating a preform with the appropriate profile. Of course the above process must take place in either a vacuum or an inert atmosphere. Since the process produces HCl and water a mist of hydrochloric acid is also produced which must be neutralized and disposed of (Dutton 1998). Fiber produced using OVD process is purely synthetic, exhibits enhanced reliability, and allows for precise geometrical and optical consistency. The OVD process produces well-controlled fiber profiles and geometry, both of which lead to a more consistent fiber. Fiber to fiber consistency is especially important when fibers from different manufacturing periods are joined, though splicing and connectorization, to form an optical system (Corning 2013).

Vapor-phase axial deposition (VAD) is a very important process and accounts for a large proportion of world fiber production. It was originally intended to be a continuous process which would have a lot lower cost than the batch processes. The basic mechanism used in VAD is flame hydrolysis (like OVD). It was discovered that if $GeCl_4$ is mixed as a dopant into the $SiCl_4$–O_2 feed the proportion of germania (GeO_2) deposited with the silica varies with the temperature of the flame. By controlling the flame temperature, the proportion of dopant deposited can be controlled. Further if a wide flame is used with a temperature gradient across it, a graded proportion of germania deposited can be obtained. The process is shown in Fig. 4.7b. At the bottom of the figure flame hydrolysis torches produce silica soot (including dopants) which is deposited across the base of a relatively wide preform. Other dopants can be included instead of or in addition to germanium. $POCl_3$ and BBr_3 are often used. As the soot is deposited the preform is pulled upwards so that the torches themselves do not move. Further up the preform other torches (or sometimes an electric furnace) are used to sinter the preform and form a solid glass preform rod. The whole process is of course enclosed and an atmosphere of pure oxygen with $SOCl_2$ vapor is used to capture the water produced by the torches and prevent it from entering the glass. The resultant rod is used as the feed for the fiber drawing process (Dutton 1998).

In the beginning there was a process called Inside Vapor Deposition (IVD). This process was significantly improved and renamed MCVD. Presently, MCVD accounts for the majority of world fiber production. In the MCVD process silica (along with the required dopants) is formed in a gaseous reaction inside a silica tube. The reaction takes place in the gas phase. Silica soot is deposited on the inside of the tube as shown in Fig. 4.7c. The reaction between the glass-forming gasses takes place in a hot region of the tube and the silica particles so formed stick to the walls of the tube. The burners traverse the tube many times at temperatures up to 1,600 °C. This sinters the soot and produces a highly controllable refractive index profile on the inside of the tube. When sufficient material has been built up the burners are moved to one end of the tube, the tube is evacuated and the heat turned up (to around 1,800 °C). This causes the sides of the tube to collapse on themselves as the burners travel slowly along it.

The big advantages of MCVD are (Dutton 1998): (a) The gaseous reaction does not produce water so the problem of OH contamination is significantly reduced. (b) The RI profile can be controlled very accurately. (c) Compared to other processes it is relatively fast (albeit it is still a batch process). (d) When using Ge as the core dopant in case a dip is obtained in the refractive index profile along the axis of the fiber. This is caused by the Ge evaporating off at the high temperatures needed to collapse the tube. This RI profile dip is not much of a problem in SM fibers but in some contexts with GI MM fiber it can be a significant problem.

The PCVD process is very similar in principle to the MCVD process. Instead of heating the outside of the silica tube the energy source is provided by a high-power microwave field (the same principle as a microwave oven). The microwave field is provided through a magnetron cavity which surrounds the silica tube. This microwave field can be moved very quickly along the tube as it heats the gas plasma directly and doesn't have to heat up the silica tube itself. This means that the tube can be traversed thousands of times depositing extremely thin layers at each pass. The result is much better control of the RI profile than that obtained with MCVD. As it happens the tube is kept hot by another set of heaters but this is only at 1,000 °C rather than the 1,600 °C used in MCVD. Heating to create the gaseous reaction comes from the microwave field not from the tube. Silica is deposited on the inside of the tube uniformly without the need to rotate it. The reaction is nearly 100 % efficient and proceeds several times faster than MCVD. In addition the process can produce large preforms capable of producing a few 100 km of fiber (Dutton 1998).

4.4.4 Joining Fibers

The diameter of the core in an optical fiber is very small and any irregularity (such as a join) can result in significant loss of power. To get the maximum light transfer from the cut end of one fiber into another, both ends must be cut precisely square and polished flat. They must then be butted together so that there is minimal air (or water) between the butted ends and these ends must match up nearly exactly. Light traveling in a fiber is a guided wave and the fiber is a waveguide. Any imperfection or irregularity is a potential source of loss and of noise. The problem is that the dimensions of a fiber are tiny and accuracy of alignment is critical. There are three general ways of joining fibers (Dutton 1998): (a) By fusion splicing (a type of weld); (b) Use of index matching epoxy glues; and (c) With mechanical connectors of different types. The common requirement of all three methods is that the cores must be aligned. The major improvement in connector losses is due generally to better fiber manufacture as much as to better connectors and connection techniques.

4.5 Optical Fiber Cables

To make practical use of fiber for communication the fiber needs to be enclosed in a cable. Fiber cables vary widely in their characteristics due to the differing environments in which they are installed and requirements they must fulfill. Fiber cables are made to suit the application they are to perform and there are hundreds, perhaps thousands of types. The objective of the cable is to protect the installed fiber from anything that may damage it (Dutton 1998):

(a) Tensile Stress
While fiber itself is quite strong under tension, stress causes a significant increase in attenuation and a number of other undesirable effects. The fiber needs to be protected from any kind of stress.

(b) Bends
Bends in the fiber that are too small in radius cause signal loss. Microbends in the fiber caused by crimping of the cable also cause signal loss. One function of the cable is to prevent the fiber being bent to a radius where loss may occur. With long distance outdoor or undersea cables this is not a big problem. Such cables often have a minimum bend radius of a few feet.

(c) Physical Damage from Environmental Conditions
Just what is needed to protect the cable varies with the particular environment. In many indoor environments vermin (rats, etc.) chew cable (they usually find electrical cable unpleasant but fiber is less so). In outdoor ones, gophers and termites also like eating cable. Heavy earth-moving equipment also has very little respect for cable integrity. One major hazard for outdoor cables is cable-laying machines. In many countries cables are laid along defined cable "rights-of-way." When someone comes to lay a new cable along a route where there are already other cables, the existing ones tend to get cut.

(d) Damage in the Cable Installation Process
Cable doesn't just have to operate satisfactorily in its installed environment but it must withstand the stresses of being installed. In some cases these stresses can be quite severe, for example, being lifted up the core of a multistory building or dragged through a long conduit.

(e) Water
It sounds illogical, but waterproofing is often more important in the fiber optical environment than it is in the electrical world. Glass immersed in water gradually picks up hydroxyl ions. The presence of hydroxyl ions greatly increases the absorption of light. In addition, the presence of water causes micro-cracking in the glass and this causes the scattering of light. The micro-cracks also weaken the fiber significantly. Water is the worst enemy of an optical fiber system.

(f) Lightning Protection
Lightning is a problem for all outdoor cables containing conductive materials. This, of course, depends on which part of the world the cable is installed. In some places lighting can hit the ground and sever an underground telecommunications cable up to 10 m away. In addition there are other functions that need to be supported in some environments. For example in some cable situations (especially undersea) it is necessary to provide power for repeaters or amplifiers along a long-distance cable route. One example of this is in submarine cables. Electrical power cabling is often included to deliver power to the repeaters.

4.5.1 Cabling Environments

Cables are specially designed for each environment (Dutton 1998): (a) *Outdoor Buried Cable (Long Distance)*—Typical outdoor buried cables contain a large number of SMFs. They contain very extensive waterproofing, strength members, and often armoring. (b) *Outdoor Buried Cable (Campus Area)*—These are typically lighter than the long-distance variety and contain both multimode and SMFs. These usually have good waterproofing and protection but it is often not as strong as long-distance varieties. In some places cable is installed in a conduit such as a 2-inch diameter steel pipe. In this case strong armoring is not needed. (c) *Outdoor Overhead Cable*—Cable intended for overhead use needs to have very great tensile strength to prevent the fibers being stressed. Typically they have a separate support member which takes the stress outside of the cable itself. (d) *Outdoor Overhead (High-Voltage Earthwire) Cable*—One very popular and creative place to put optical fiber is inside the earth wire of a high-voltage electrical transmission system. A common system of this kind might operate at 132,000 V. The earth wire is usually the top wire on the tower (relatively safe from vandalism). These cables are called "Optical Ground Wire" (OPGW) cables. (e) *Undersea Cable*—The undersea environment is the most difficult cabling environment imaginable. Keeping high pressure salt water out of the cable poses a very significant challenge. In contrast to the large numbers of fibers in terrestrial long-distance cables there are usually only a small number (6–20) of fibers in an undersea cable. Undersea cables also often carry electric wires to provide power to regenerators (repeaters or amplifiers) in the cable. (f) *Indoor Cabling*—Typically indoor cables have a very small number of fibers (most often only two) and these are generally multimode. In the indoor office environment there is less need for waterproofing or armoring than in the outdoor environment. However, some protection is needed from vermin (such as rats) and from accidental damage both during and after installation. In addition, it is often desirable to make the cable from materials that don't give off toxic fumes when they burn. This costs more but may save lives in the case of a fire. Indoor cables are typically short distance (up to 300 m or so) and need to be relatively light and flexible for installation in the office environment. They are usually terminated with pluggable connectors. (g) *Jumper Leads and Fly Cables*—One of the harshest environments for a fiber cable is as an end-system connector in the office. These often tend to run across the surface of a floor. They get stepped on and desks and chairs occasionally run their wheels over them. These are the best ways known to science of breaking a fiber.

4.5.2 Fiber Coating

The basic form of optical fiber is called "primary coated optical fiber" or PCOF, as shown in Fig. 4.8a. This is the fiber as it emerges from the drawing tower. In this form the overall diameter is 250 μm. In many applications (such as in "loose-tube"

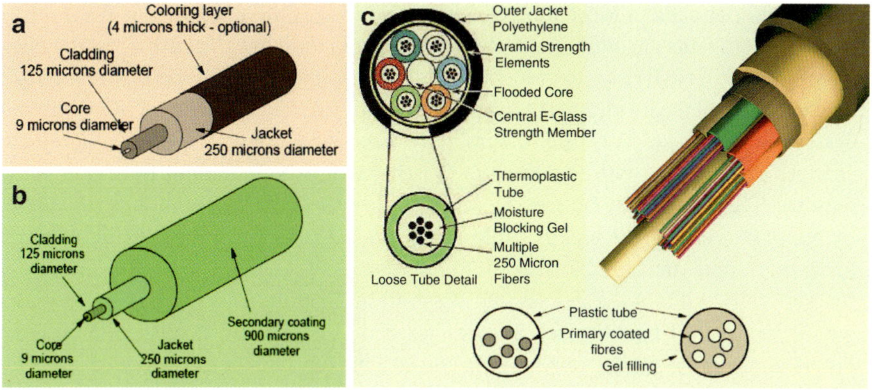

Fig. 4.8 Different optical fiber assembly structures (Dutton 1998): (**a**) primary coated optical fiber (PCOF); (**b**) secondary coated optical fiber (SCOF); (**c**) loose tube construction

cables) there is a need to identify the fiber within its cable. To allow this, either the jacket itself is colored or there is a further very thin colored coating added immediately prior to building the cable. Since there are no standards for color coding, cable makers tend to prefer to color the fiber at cabling time to allow for flexibility in satisfying customer specifications. In end user environments PCOF is never used without further encasement in a cable. For cabling within office environments, PCOF is usually further coated with a secondary coating as shown in Fig. 4.8b. This is called "Secondary Coated Optical Fiber" (SCOF). The secondary coating forms a tight bond with the primary coating (Dutton 1998).

4.5.3 Basic Cable Construction

Cables can be classified into three types depending on how the fiber is encased within them (Dutton 1998): (a) *Tight Buffered Construction*—Tight buffering is where the secondary coated fiber (SCOF) is encased firmly in surrounding material as shown in Fig. 4.8b. This is similar to electrical cabling. Construction of this type is usually used for indoor applications where the number of fibers needed in the cable is low and distances are relatively short. However, it is also used for medium distance outdoor applications such as around a campus. (b) *Loose Tube Construction*—In loose tube construction a small number of PCOF fibers are carried inside a plastic (PVC) tube of 4–6 mm in diameter, as shown in Fig. 4.8c. Typically between one and eight primary coated fibers are carried in a single tube. There is plenty of room in the tube for the fibers to move loosely within it. Fiber coils around inside the tube in a helical pattern. If the cable is stretched or bent then the fibers inside do not experience tension. (c) *Loose Tube with Gel Filler*—In most buried outdoor cable a loose tube construction is used where the tube is filled with a jelly. This prevents the ingress of water from faults in the cable. It also buffers the fibers from one another and helps to prevent losses due to microbends caused by irregularities on the surface of the insides of the tubes. The composition of the gel used is a significant design issue. Historically, petroleum jelly was used but this exhibits significant changes in viscosity with temperature. Viscosity is very important for a number of reasons: The fiber must be free to move within the tube to counter stress caused by temperature changes and/or cable lying. The viscosity needs to be high enough to provide some mechanical stability to the tube. And it should allow the cable to be run vertically without the gel settling down to the bottom. It needs to be reasonably well-behaved in the field when a cable has been severed and needs to be rejoined. For example, if the gel became very mobile on a hot day and the cable was severed, the gel could run out of the tubes for a long distance. Of course the gel needs to be stable during the process of cable manufacture where it may be subject to high temperatures.

Silicone gels are a lot better generally than petroleum ones but today specially designed synthetic gels are used which all but eliminate the early problems with changes in viscosity.

4.5.4 Indoor Cables

The most basic form of indoor optical cable is shown in Fig. 4.9a. This is simply a single strand of SCOF with a layer of strengthening aramid (or fiberglass) fibers and an outer PVC jacket added. Single-core cable of this nature is used in short lengths as jumper or fly cables but is almost never used for fixed cabling. This is because it is lower in cost to have a cable with many fibers in it than it is to have many single-core cables. The construction shown in Fig. 4.9b is a very common low-cost indoor cable construction. Two basic single-core cables are carried together in a common "figure-8" sheath. Figure 4.9c shows the cross section of a typical heavy-duty, tight buffered indoor cable. Such a cable might be installed vertically in a building riser connecting many floors. Many simple single-cored cables are encased in a common sheath. The central strength member (in this case plastic) supports the weight of the total cable. Tight buffering ensures that individual fibers are not put under tension due to their own weight. Cables of this kind commonly come with up to 12 fibers each of which is individually sheathed and colored (Dutton 1998).

4.5.5 Air Blown Fibers

In a "blown" fiber installation instead of installing fiber cables narrow plastic tubes or conduits are installed. Later, very lightly clad fiber bundles are installed into the tubes (blown in) using a system of compressed air (or compressed nitrogen). There are various blown fiber systems available but one of the popular ones allows bundles of

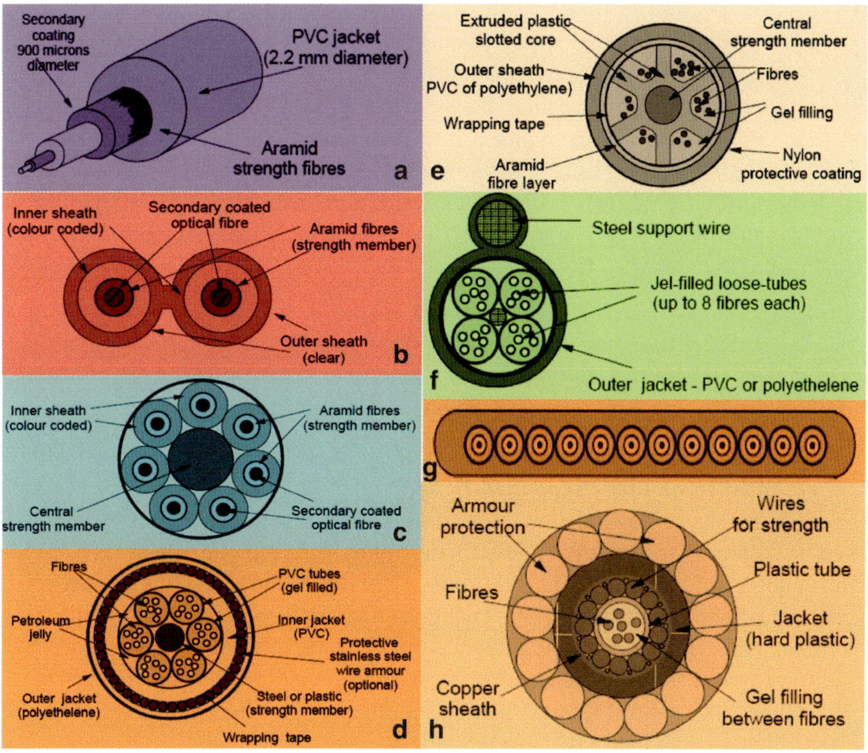

Fig. 4.9 Typical types of optical cables (Dutton 1998): (**a**) single-core cable; (**b**) dual indoor cable; (**c**) 6-core tight buffered indoor cable; (**d**) outdoor fiber cable (loose tube—gel filled); (**e**) segmented-core cable design; (**f**) outdoor aerial cable; (**g**) 12-core flat cable; and (**h**) undersea cable

between 2 and 18 fiber strands (either SM or MM) to be inserted into the already installed tubes. This can be achieved for distances of up to about 2 km (or about 6,000 feet). The fiber bundles blown into the tube conduits are really very lightweight cables specifically designed for their aerodynamics. In the installation process the fiber floats within the tube and there is very little contact between the bundle and the tube. The installation process is quite fast with drawing speed approaching 50 m per minute. There are some very important advantages in the blown fiber technique (Dutton 1998): (a) Typical tubes can be joined using very simple "push-fit" connectors within specially designed junction boxes. In some office environments it is very difficult to install a long, unbroken fiber cable. In this situation installation of the conduit tube in sections can save substantial cost. When the fiber is "blown in" each single strand is unbroken from end-to-end. This point is particularly important in the case where some parts of the fiber connection are indoor and other parts outdoor. Once the conduits are connected between the indoor and outdoor sections, the fiber is blown in as a single unbroken cable eliminating the need for making fiber joins (or having connectors) between indoor and outdoor sections. (b) Tubes with multiple cavities can be installed so that additional fibers can be installed later as the occasion demands. This saves some fiber cost but is more significant in the additional flexibility provided. (c) It is possible to remove fiber from the tubes and reinstall it into other tubes on other routes as demands change. Since installing the tube (or regular fiber cable) is the major part of the installation cost, this allows for very low-cost changes as the installation evolves. (d) As speeds increase the bandwidth limitations of GI MM fiber become more and more restrictive. SMF itself (just the fiber) is intrinsically lower in cost than MMF but devices that connect to it are very much more expensive.

4.5.6 Outdoor Cables

A typical outdoor (buried) cable is shown in cross section in Fig. 4.9d. It consists of six gel-filled loose tubes supported by other cable elements designed to provide strength, mechanical protection, and protection against the ingress of water. Note that in addition to the gel filling in the tubes carrying the fibers there is gel surrounding these tubes within the cable. The illustrated cable has six fibers per tube for a total fiber count of 36. This general cable geometry is used with up to 12 tubes supporting as many as 8 fibers each for a total of 96 fibers in the cable. If an electrical power supply is needed, copper wires can replace the fibers in one or more of the tubes. The central strength member is often made from a hard plastic material rather than steel. Indeed the whole cable is often constructed from nonmetallic materials. The outer layer of armoring wire is a customer option which is only added to cable destined for use in places where there is a significant danger of damage from the environment. The use of stainless steel is unusual (but optional) as this is a high-cost material. Ordinary steel is much lower in cost but will rust if the outer sheath is penetrated and water gets in. In some locations (such as tropical areas) it is customary to add an outer nylon covering to prevent attack by termites. An alternative loose tube construction is shown in Fig. 4.9e. Here the cable is formed from an extruded plastic member in the shape of a gear-wheel. Individual primary coated fibers are carried in the indentations (channels) around the outside. In the illustrated case there are six channels for fibers and up to eight fibers may be carried in each. In some geometries up to twenty or so channels are used with as few as one fiber per channel. Outer elements of the cable are the same as for the regular loose tube construction. This is a simpler and thus lower cost way of making the cable. Aerial cables are designed to be supported from towers and there is a significant problem with stress. One alternative is to have a very strong central strengthening wire. However, if this is done there is often a lot of crushing pressure on the cable at the points of support. The design shown in Fig. 4.9f shows a typical cable designed for overhead installation. It is not very different from underground cable except it is enclosed in a common sheath with a strong separate support wire. The weight of the fiber cable is supported evenly all the way along its length and stresses are minimized. The basic construction shown in Fig. 4.9g was used extensively in the US in the early 1980s for medium distance communication (up to 10 km or so) with MMF. A flat cable containing 12 fibers is constructed by sandwiching the fibers between two layers of Mylar tape with a glue (to hold the fibers in place) in between. The flat cable is only about 5 mm in width. A stack of 12 of these flat cables forms a square cross section which is then embedded within a strong protective cable structure. The key to this construction was the use of a metallic connector which terminated all 12 fibers of one flat cable in one operation. While this worked reasonably well for MMF it didn't have sufficient precision for SMFs and the system is no longer used (Dutton 1998).

4.5.7 Undersea Cables

As shown in Fig. 4.9h, undersea cables are significantly different from other types of cable. They usually have a very small number of fibers (between perhaps 4 and 20). This contrasts with terrestrial cables which often have up to 100 fibers in them. At great depths water is under very high pressure and such cables have to prevent water ingress. For this reason all spaces in the cable are filled with very dense plastic or polymer material except for the cavity immediately around the fibers themselves. This is a gel-filled tube as in terrestrial cable construction. The copper sheath shown in the figure is an unbroken tube intended to help keep water out. In contrast to what we might expect, under the sea is not a completely safe place. Ship's anchors and fishing trawlers can do significant damage to undersea optical cables. Thus the undersea operators commonly make a distinction between "shallow" and "deep" water. Water is considered "deep" if the bottom is more than 1,000 m from the surface. Deep water is a relatively safe place to situate a cable and in this environment it is typically laid on the sea floor without special armoring. Thus cable operators plan the cable route to maximize the amount of cable laid in deep water. In shallow water it is common to dig a trench in the sea floor and bury the cable. Also, cables laid in shallow water are typically heavily armored (Dutton 1998).

4.6 Summary

Optical fiber has become the transmission medium of choice for most communications. Its high speed and long-distance capability make it the most cost effective communications medium for most applications. While higher performance and lower cost components are in continual development, it has also become critical to train competent personnel to design, install, and maintain state-of-the-art fiber-optic networks. Light rays that impinge fiber at angles greater than the acceptance angle are refracted into the fiber, but are not guided for a long distance, as they are not totally reflected at the core-cladding interface, but rather are partially refracted into the cladding. Major phenomena that have an effect on light propagation in optical fibers include dispersion, nonlinear interactions, effect of loss, and effects gain and noise. Modes are fields that maintain the same transverse distribution and polarization at all distances along the waveguide axis. They can be regarded as transverse resonances of the fields in the waveguide. Optical fibers are manufactured in three main types: multimode step-index, multimode graded-index, and single mode. The difference between them is in the way light travels along the fiber. Light is transmitted (with very low loss) down the fiber by reflection from the mirror boundary between the core and the cladding, following principle of TIR. Perhaps the most important characteristic is that the fiber will bend around corners to a radius of only a few centimeters without any loss of the light.

Optical fibers are long and flexible kinds of optical waveguides. They are essentially based either on some glass or on polymers (plastic optical fibers). PCF is a new class of optical fiber based on the properties of photonic crystals, although the base materials used for PCF are still glass or polymers. Because of its ability to confine light in hollow cores or with confinement characteristics not possible in conventional optical fiber, PCF is now finding applications in fiber-optic communications, fiber lasers, nonlinear devices, high-power transmission, highly sensitive gas sensors, and other areas. In addition, optical nano-fibers including nano-engineered fibers and optical fiber nanowires have attracted increasing interest because of their tremendous potential in a wide range of fields ranging from telecommunications to chemistry, biology, lasers, and sensors.

In manufacturing fiber, major challenges include: Purifying the materials (especially silica) to an extreme degree; levels of heavy metal impurity of one in a billion are quite sufficient to have an adverse effect on the performance of the fiber; and achieving extreme precision in fiber dimensions and tolerances. The size of the core, its position and the outer dimension of the cladding must be precise to tolerances of 1 μm or less. Refractive indices must also be precise especially in the case of graded-index fibers.

To make practical use of fiber for communication the fiber needs to be enclosed in a cable. Fiber cables vary widely in their characteristics due to the differing environments in which they are installed and requirements they must fulfill. Fiber cables are made to suit the application they are to perform and there are hundreds, perhaps thousands of types. The objective of the cable is to protect the installed fiber from anything that may damage it.

References

Aggarwal ID, Sanghera JS (2002) Development and applications of chalcogenide glass optical fibers at NRL. J Optolectron Adv Mater 4(3):665–678

Argyros A (2009) Microstructured polymer optical fibers. Journal of Lightwave Technology 27(11):1571–1579

Birks TA et al (1997) Endlessly single-mode photonic crystal fiber. Opt Lett 22:961–963

Brambilla G (2010) Optical fibre nanowires and micowires: a review. Journal of Optics 12:1–19

Carlie N, Petit L, Richardon K (2009) Engineering of glasses for advanced optical fiber applications. J Eng Fiber Fabr 4(4):21–29

Corning (2013) Fiber-optic technology. http://www.imedea.uib.es/~salvador/coms_optiques/addicional/Corning/fiber_optic.pdf. Accessed 21 Feb 2013

Digonnet MJF (2001) Rare-earth-doped fiber lasers and amplifiers, 2nd edn. CRC Press, Boca Raton, FL

Dutton HJR (1998) Understand optical communications. IBM, Research triangle park

Gambling WA (2000) The rise and rise of optical fibers. IEEE J Sel Top Quantum Electron 6(6):1084–1093

Labadie L, Wallner O (2009) Mid-infrared guided optics: a perspective for astronomical instruments. http://www.Arxiv.org/pdf/0902.0785. Accessed 12 May 2010

Lewis G (1990) Selection of engineering materials. Prentice Hall, Englewood Cliffs, NJ, pp 421–426

Lezal D (2003) Chalcogenide glasses—survey and progress. J Optoelectron Adv Mater 5(1):23–34

Li MJ et al (2010) Nano-engineered optical fibers and applications. In: Optical fiber communication conference (OFC), San Diego, California, March 21, 2010

Litchinitser N, Iakhnine V (2011) Optical waveguides: numerical modeling. http://optical-waveguides-modeling.net. Accessed 09 Feb 2012

Oh K, Paek UC (2012) Silica optical fiber technology for devices and components: design, fabrication, and international standards. Wiley, Hoboken

Paschotta R (2012) Rare-earth-doped fibers. http://www.rp-photonics.com/rare_earth_doped_fibers.html. RP Photonics Consulting GmbH. Accessed 26 Dec 2012

Peters K (2011) Polymer optical fiber sensors—a review. Smart Mater Struct 20:1–17

Rayscience (2013) Hard polymer cladding optical fiber (HPCF). http://www.oelabs.com/images/goodspdf/201109/HardPolymerCladdingOpticalFiber(HPCF).pdf. Accessed 12 Feb 2013

Scribd (2012) http://www.scribd.com/doc/40088106/1/CHAPTER-III-OPTICAL-FIBER-FOR-COMMUNICATIONS. Accessed 09 Dec 2012

Wikipedia (2012) Modal dispersion. http://en.wikipedia.org/wiki/Modal_dispersion. Accessed 21 Dec 2012

Chapter 5
Semiconductor Waveguides

Abstract Semiconductor optical waveguides are a very important part of modern integrated optoelectronic systems, especially for electrically active devices. Applications range from semiconductor lasers, optical filters, switches, modulators, isolators, and photodetectors. Semiconductor waveguides have many advantages especially for use in slow light applications. They offer a significant enhancement of interaction length that, to first order, scales with the possible delay. With a tight confinement of the optical mode, the required optical power can be drastically reduced while the mode overlap with the active material is strongly enhanced. The use of semiconductor material is of particular interest since it offers compactness and enables for monolithic integration into optoelectronic devices using well established processing techniques. Furthermore, semiconductors are attractive since the operating wavelength, to a large extend, can be designed while performing with bandwidths in the GHz regime that is well suited for communication signals. Based on III–V, II–VI, or IV–VI group elements, two semiconductors with different refractive indices are generally synthesized for fabrication of optical waveguides. They must have different band gaps but same lattice constant. An attractive feature of the binary compounds is that they can be combined or alloyed to form ternary or quaternary compounds, or mixed crystals, for varying the band gap continuously and monotonically together with the variation of band structure, electronic, and optical properties. The formation of ternary and quaternary compounds of varying band gaps also enables the formation of heterojunctions, which have become essential for the design of high performance electronic and optoelectronic devices. This chapter will give a brief review about fundamental theory, semiconductor materials, and fabrication technologies of various semiconductor waveguides.

X.C. Tong, *Advanced Materials for Integrated Optical Waveguides*, Springer Series in Advanced Microelectronics 46, DOI 10.1007/978-3-319-01550-7_5,

5.1 Fundamental Theory

5.1.1 Crystal Structure

Atoms in solid-state materials are located at energetically most favorable sites. The locations depend on the characteristics of the atoms, such as their size and their electric structure. Generally, the ordering of the atoms can be monocrystalline, polycrystalline or amorphous. For example, III–V compound semiconductor materials are made by epitaxy with monocrystalline structure.

When materials crystallize, the atoms are arranged into a crystal lattice. The crystal lattice represents a set of points in space which form a periodic structure. The smallest repetitive structure in any crystal lattice is called the primitive unit cell. The whole crystal can be formed by repeating primitive unit cells. Primitive vectors are linearly independent and connect the lattice points of crystal lattice. Primitive vectors can be chosen by numerous ways and, therefore, primitive vectors are not unique. Primitive vectors have a useful property that all the lattice points $\mathbf{R}$ can be derived from one lattice point $\mathbf{R}_0$ when some linear combination of primitive vectors is added to it (Jussila 2010):

$$\mathbf{R} = \mathbf{R}_0 + m_1\mathbf{a}_1 + m_2\mathbf{a}_2 + m_3\mathbf{a}_3 \tag{5.1}$$

where m_1, m_2 and m_3 are integers, and $\mathbf{a}_1$, $\mathbf{a}_2$ and $\mathbf{a}_3$ are the primitive vectors. As a result of this, the whole crystal can be formed by choosing all combinations of the integer m_1, m_2 and m_3.

To illustrate this type of crystal lattice, a conventional unit cell of a face centered cubic (FCC) lattice is shown in Fig. 5.1a. Atoms in FCC lattice are located at every corner and at the center of every face of a cube. The zincblende lattice, for instance, is composed of two separate FCC sublattices one of which is comprised of group III atoms and the other of group V atoms. The dimensions of the two FCC-lattices are equal but the origin of the other lattice is displaced from the other by a vector (a/4; a/4; a/4) along the body diagonal. Figure 5.1b illustrates the conventional unit cell of a zincblende crystal lattice; the binary III–V compound semiconductors InAs, InP, GaAs, and GaP all are zincblende crystal, but with different lattice constant (Vurgaftman et al. 2001).

The crystal structure of a solid restricts the movement of carriers. In a semiconductor material, the outer electrons of the atoms are delocalized over the entire crystal, with the periodicity of the crystal structure limiting their movement. For certain electron energy, the carrier is allowed to move in one direction, but its motion in a different direction is restricted as a result of destructive interaction from the atomic lattice. This dependence of the electron energy on the momentum of the carrier results in a structure of energy bands where the carrier can exist (Hoogland 2013).

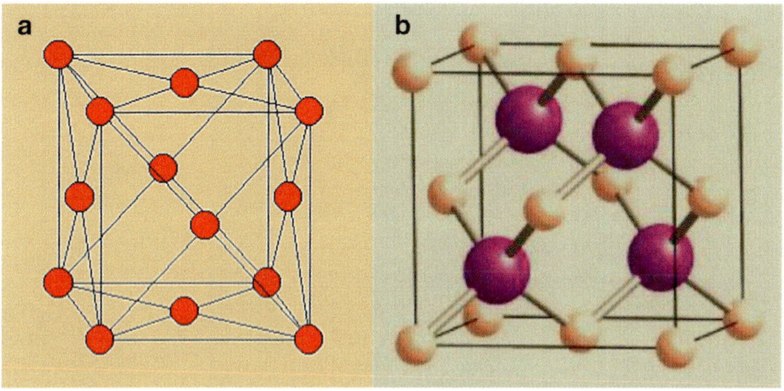

Fig. 5.1 Illustration of some crystal structures: (**a**) conventional unit cell of FCC lattice; (**b**) conventional unit cell of the zincblende lattice

5.1.2 Energy Band Structure

The energy band structure is used to explain the electronic structure of the material, which is derived from quantum mechanics. The foundation of quantum mechanics is the Schrödinger equation, represented in time independent form as (Jussila 2010),

$$\left(-\frac{h^2}{2m}\nabla^2 + V(\mathbf{r})\right)\psi(\mathbf{r}) = E\psi(\mathbf{r}) \tag{5.2}$$

where ∇ is the gradient, h is the Planck's constant, and m is the mass of the particle. $V(\mathbf{r})$ denotes the potential energy experienced by the particle. Typically in solids, potential energy is comprised of interactions between the particle and all the atoms in solids. In semiconductor crystals, the potential energy $V(\mathbf{r})$ is periodic and expressed as (Jussila 2010)

$$V(\mathbf{r}) = V(\mathbf{r} + \mathbf{a}) \tag{5.3}$$

where **a** is the primitive vector of the material. Periodicity of the atoms makes the energy states of crystals periodic in E–**k** space, where **k** is the electron wave vector. In addition, the semiconductor crystals have some energy states which are forbidden for every **k**-vector. The relation between the electron wave vector **k** and the energy E is called the dispersion relation, which allowed energy states in E–**k** space form continuous energy bands. The two most important energy bands in semiconductors are the valence band and the conduction band. The valence band is defined as the highest occupied energy band, and the conduction band as the lowest empty energy band at the temperature of 0 K. Band gap of the material is

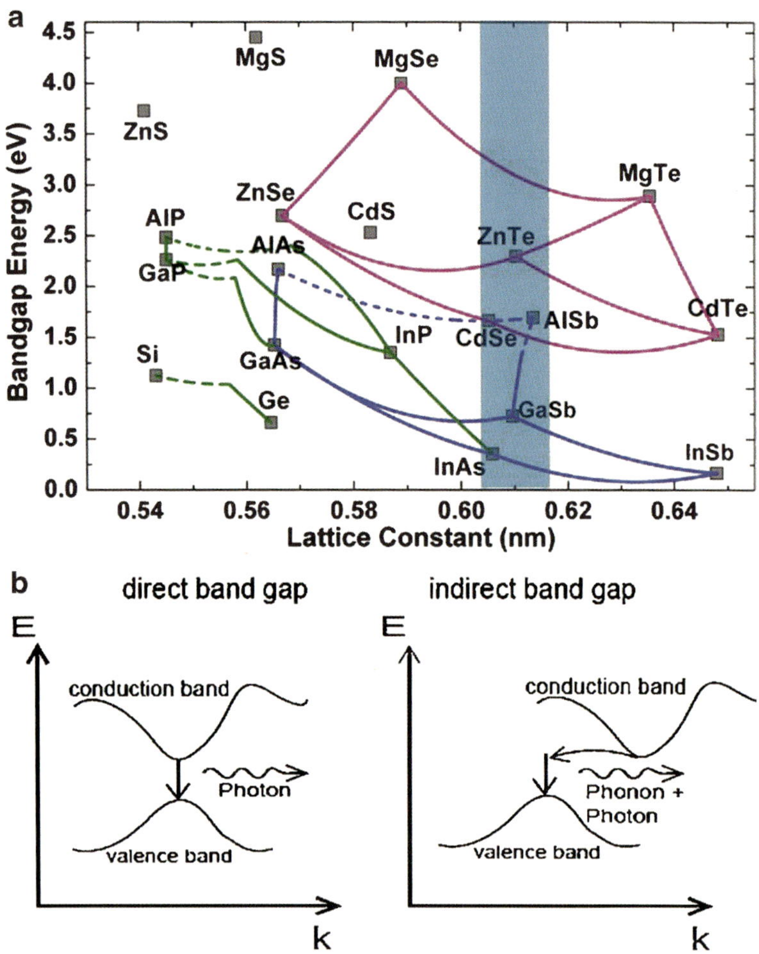

Fig. 5.2 Illustration of energy band structures: (**a**) band gap energy of III–V and II–VI semiconductors; (**b**) energy band structures and photon generation in direct and indirect band gap semiconductors

defined as the energy difference between the conduction band minimum and the valence band maximum. Figure 5.2a presents the band gap energy of III–V and II–VI compound semiconductors. All optical processes occurring in different materials follow the conservation laws of energy and momentum, and can be described with E–$\mathbf{k}$ diagrams. Semiconductors distinguish themselves from other materials by having an energy gap between the highest occupied energy band (the valence band) and the lowest unoccupied energy band (the conduction band). The band gap of a semiconductor is measured between the top of the valence band and the bottom of the conduction band and is called direct if these extrema coincide in momentum (e.g., GaAs) and is called indirect if they do not (Si), as shown in Fig. 5.2b. In direct band gap materials, the conduction band minimum value is located at the same **k**-vector value as the valence band maximum value. The operation of optoelectronic devices that emit or amplify light is based on the phenomenon of either luminescence or stimulated emission. In these phenomena, an electron is relaxed from the conduction band to the valence band, thus generating a photon whose energy equals to the difference of these energy states. In indirect band gap materials, these phenomena are inefficient because the energy transition between the conduction band and the valence band requires a change in momentum which usually occurs via an interaction with a phonon. Therefore, optical-emitting transitions are most likely to occur in direct band gap materials, which are more commonly used in optoelectronic devices (Jussila 2010).

In the real world, the semiconductor crystal is finite, which puts restrictions on the allowed wave vectors within the energy bands, resulting in a set of discrete states available in each band with a wave vector separation determined by the size of the semiconductor (Hoogland 2013).

5.1.3 III–V Compound Semiconductors

III–V compound semiconductors are composed of group III and group V atoms. Binary compounds include InP, InAs, GaAs, and GaN. A fraction of the lattice sites in a binary semiconductor is replaced by other elements to form ternary (e.g., InGaAs, AlGaN) or quaternary (e.g., InGaAsP, InAlGaAs) compounds. For example, $Al_xGa_{1-x}As$ is made by replacing a fraction x of Ga atoms by Al atoms. Band gap varies with x as $E_g(x) = 1.424 + 1.247x$, when $0 < x < 0.45$. Quaternary compound $In_{1-x}Ga_xAs_yP_{1-y}$ is useful in the wavelength range of 1.1–1.6 μm. For matching lattice constant to InP substrate, $x/y = 0.45$. Band gap varies with y as $E_g(y) = 1.35–0.72y + 0.12y^2$. III–V compound semiconductors provide the possibility to engineer the band gap and the lattice constant by changing the composition of the different elements. Figure 5.3 shows the relation between the crystal lattice constant, the band gap energy and corresponding emission wavelength of different III–V compound semiconductors (Jussila 2010).

III–V semiconductors can provide significant advantages over other materials such as $LiNO_3$ or glass for many optical applications, which may include (Deri and Kapon 1991): (a) the potential for monolithic integration with other optoelectronic (e.g., lasers, detectors) and electronic (transistor) devices to enhance the functionality of transmitter or receiver circuits using additional guided-wave signal processing; (b) suitability for high-speed low-drive voltage modulators and switches, with further advances from multiple quantum-well (MQW) technology; and (c) the availability of sophisticated, controllable fabrication processes (e.g., epitaxial growth, selective etching) and simple methods for endfacet preparation (scribe-and-cleave versus cut and polish). For these reasons, interest in III–V semiconductor guided-wave devices continues to grow, particularly as single-mode optical waveguides have increasingly accepted for communications transmission applications.

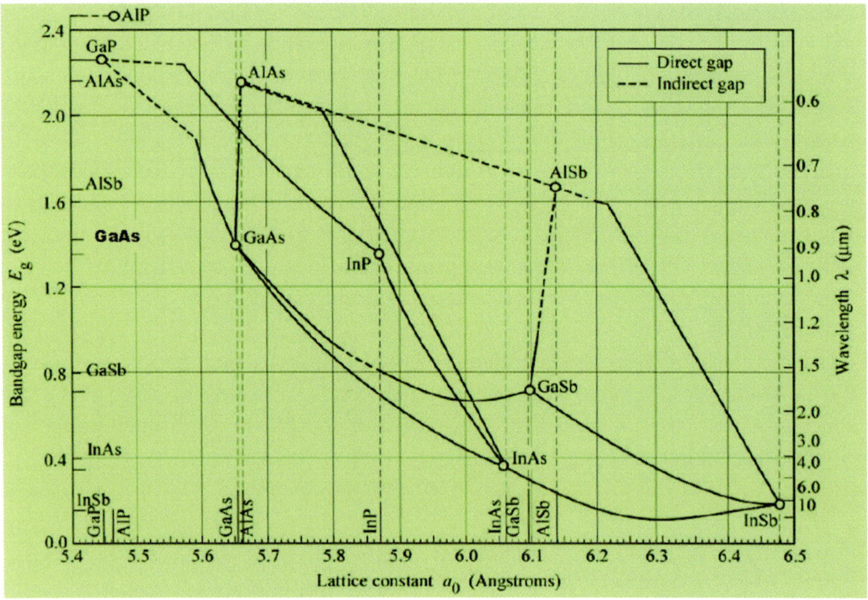

Fig. 5.3 Band gap and the lattice constant of different III–V compound semiconductor materials. The *lines* drawn in the image describe the lattice constant and the band gap of different ternary alloys. For instance, the band gap and the lattice constant of $In_{(1-x)}Ga_xAs$ is located on the *line* which is drawn between the binary compounds GaAs and InAs (Jussila 2010)

5.1.4 Quantum Structure

Quantum structures confine the charge carriers in one, two or three dimensions to such a small size that quantum mechanical effects are observed, as shown in Fig. 5.4. The structure which confines the charge carriers in one dimension is called a quantum well (QW). QWs can be fabricated by growing a layer of a small band gap material sandwiched between large band gap materials. Electrons in the small band gap material experience a potential energy well when moving along the growth direction. The structure which confines the charge carriers in two spatial dimensions is called quantum wire (QR). Quantum dots (QDs) confine the charge carriers in all three spatial dimensions. Fabrication of QDs can be performed by different methods, all of which result in a structure where a small band gap material is located inside a large band gap material (Jussila 2010).

As a practical matter, the states in a macro-size chunk of semiconductor are so closely spaced that the conduction and valence bands appear to be continua. The number of these states per unit volume in an energy band is described by the density of states. Because no available states exist within the band gap, the density of states is measured from the bottom of the conduction band and the top of the valence band from which it is, for a bulk semiconductor, a continuously increasing function with carrier energy (Fig. 5.4a). The magnitude is determined by the effective mass of the

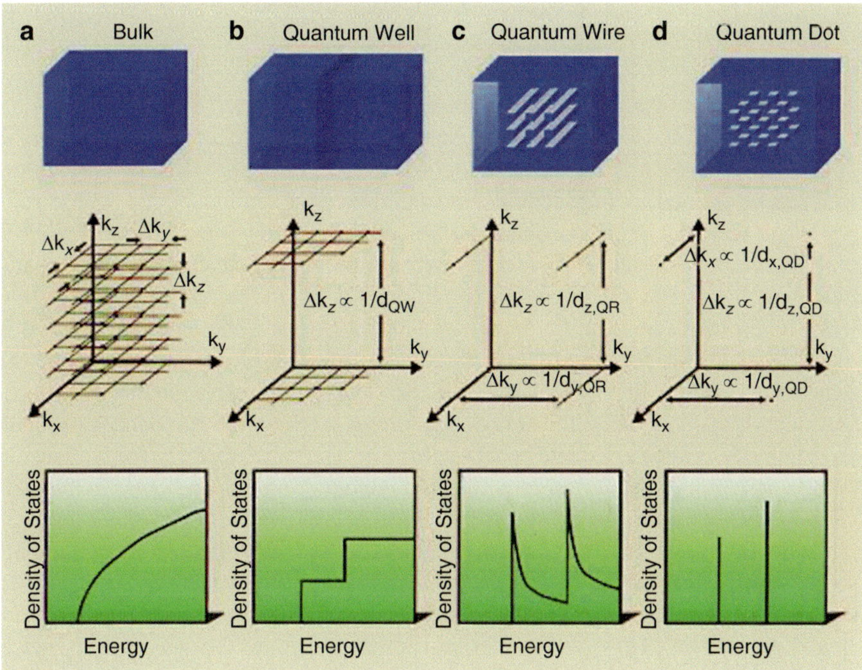

Fig. 5.4 Schematic illustrations of the active region (*top*), allowed states in momentum space (*middle*) and density of states (*bottom*) for confinement in no dimensions (i.e., bulk material) (**a**), in one dimension (i.e., a quantum well) (**b**), in two dimensions (i.e., a quantum wire) (**c**) and in three dimensions (i.e., a quantum dot) (**d**). *d* dimension(s) of quantum structure, *k* momentum, *QW* quantum well, *QR* quantum wire, *QD* quantum dot (Hoogland 2013)

carrier, which is a measure of the degree of curvature of the energy band around the extrema—the larger the curvature, the smaller the effective mass of the carrier; the larger the wave vector separation between states, the smaller the density of states. In semiconductors, electrons can be excited thermally to the conduction band, leaving holes in the valence band. Alternatively, electrons can be actively excited to the conduction band, for example, by absorbing a photon with energy larger than the semiconductor band gap. The excited carriers will thermalize very quickly to the bottom of the conduction band (electrons) and to the top of the valence band (holes) by losing energy through carrier–carrier and carrier–phonon scattering. The probability of finding an excited carrier at certain energy in a semiconductor at thermal equilibrium is described by temperature-dependent Fermi–Dirac statistics. The overall carrier distribution in a semiconductor thus is a function of both the density of states and of the temperature-dependent Fermi–Dirac distribution. In a bulk semiconductor, this distribution is widespread, with a small number of carriers right near the edge of the conduction and valence bands and with a peak that is situated above (below) the conduction (valence) band, whose position is temperature dependent. Because the emission spectrum depends on the distribution and number of available transitions, a bulk semiconductor exhibits a temperature-sensitive emission spectrum, which is undesirable for stable lasing operation. Moreover, the temperature-dependent carrier distribution also makes the laser threshold temperature dependent. These unfavorable optical properties of bulk semiconductors can be overcome by confining carriers to dimensions approximating the de Broglie wavelength of the carriers. A quantum well can be formed by introducing a very thin layer of semiconductor whose band gap is less than that of the surrounding semiconductor (Fig. 5.4b). Charge carriers in this potential well can survive only when their energies correspond to wave functions that satisfy the boundary conditions for a standing wave. Because these wave functions depend on the quantum-well dimensions, the energies allowed in the well—and, hence, the effective band gap—are tunable by adjusting the well dimensions. When the size of the well becomes comparable to the de Broglie wavelength of the carrier, the energy separation between the resonances becomes large enough to form a discrete energy spectrum. The small width of the well results in the restriction of the carrier movement in the direction perpendicular to the well, causing the discrete energy spectrum. Because the width of the quantum well typically is several orders of magnitude smaller than the dimensions of the bulk semiconductor crystal, the wave vector separation between allowed states in the direction of the quantum well is greatly enlarged, compared with the bulk case (Fig. 5.4a, middle). However, carrier movement parallel to the well is unrestricted, just as in bulk material, thus yielding a dense distribution of states in plane parallel to the well (Fig. 5.4b, middle panel). Therefore, at each allowed quantum-confined level, the overall density of states is determined by the wave vectors parallel to the well, thus yielding a two-dimensional density of states: At each allowed energy level, a sharp increase of the density states occurs, as shown in the bottom panel of Fig. 5.4b. This shape results in a less temperature-sensitive carrier distribution than in bulk material, with the peak emission wavelength more stable around the quantum-confined transition energies. In addition, because of the increased number of carriers near the transition energy, the emitted light spectrum has a narrower bandwidth, the peak material gain is increased, and the transparency threshold is decreased. All these improvements in the optical properties arise from quantum confinement of carriers in only one dimension. When taking the degree of confinement to the maximum by creating a quantum dot, the carriers are restricted in their motion in all three dimensions, eliminating the possibility of bulk-like carrier motion. This leads to a truly discrete density of states exhibiting delta functions at the allowed energy states (Fig. 5.4d). Such shape is analogous to atomic energy levels, and therefore quantum dots also are known as artificial atoms. The number of charge carriers accumulated at the working transition is enlarged greatly at the expense of the higher energy parasitic states of the continuum. Therefore, 3D quantum confinement ensures not only temperature insensitivity of the emission wavelength but also the narrowest emission and gain bandwidths, as well as the lowest transparency threshold. An excited electron and hole can interact with each other through Coulomb attractions, creating a quasiparticle called an exciton. However, the binding energy required to create one is much smaller than the thermal energy; therefore, excitons have a very low probability of existing in a bulk semiconductor at room temperature. In quantum-confined systems, and particularly in a quantum dot, the excited electron and hole are forced to exist very close to each other, causing large Coulomb interactions between them. In this case, the binding energy and oscillator strengths are increased, and the excitons can exist at room temperature. The degree of quantum confinement is determined by the interaction length over which the bond between an electron and a hole extends in an exciton, compared with the size of the quantum dot. In weak quantum confinement, the interaction length is shorter than the dimensions of the quantum dot, resulting in very closely spaced quantum-confined energy states. In this case, the quantum dot seems more bulk-like because the electron and hole can separate beyond the exciton interaction length, thus breaking up the exciton and making the transition energy independent of the quantum dot size. In the strong quantum confinement regime, the quantum dot size is smaller than the electron hole interaction length, resulting in widely spaced energy states. The quantum dot sizes can be so small that the energy spacing between the allowed states in the conduction and valence bands is large enough to cause a so-called phonon bottleneck. In most semiconductors, excited electrons can relax down to the bottom of the conduction band by losing energy through carrier–carrier and carrier–phonon interactions. However, a phonon bottleneck occurs when the spacing between energy states is much larger than these interaction energies—as is the case in strong quantum confinement. In this case, the probability for a carrier in a higher energetic state to lose its energy to a lower energetic state through these interactions becomes very low. As a result, electrons cannot relax to the bottom of the conduction band, and so quantum dots with strong quantum confinement should not be expected to emit light from the normal transition across the band gap. However, this phonon bottleneck has not been observed experimentally. This could be explained by the fact that fabricated quantum dots still are imperfect 3D quantum-confined systems (Hoogland 2013).

Quantum structures provide the possibility to engineer the energy band structure of the material. The dispersion relation of the conduction band is presented for bulk semiconductors and different quantum structures. The similar results are also valid for the valence band with the only difference then occurring in the effective mass of the electron, which is replaced by the effective mass of the hole in the valence band. The dispersion relation of the bulk material is assumed to be parabolic near the band edge (i.e., near the Γ-point, $\mathbf{k} = 0$). Hence (Jussila 2010),

$$E_{\text{bulk}}(\mathbf{k}) = E_g + \frac{h^2}{2m_e}\mathbf{k}^2 \tag{5.4}$$

where the E_g is the band gap of bulk material, and m_e the elective mass of electron. The zero of the energy is set at the valence band maximum. The dispersion relation of a QW with infinitely high barriers gives the following energy states (Jussila 2010):

$$E_{\text{QW}}(\mathbf{k}) = E_g + \frac{\pi^2 h^2 n^2}{2m_e L^2} + \frac{h^2}{2m_e}\mathbf{k}_{\parallel}^2 \tag{5.5}$$

where $\mathbf{k}_{\|}$ is the wave vector component in the quantum-well plane, L the width of the quantum well, and n any integer. The solution is valid only near the energy band edge. The conduction band minimum ($n = 1$) for the QW is located at a different energy than for the bulk material, and the width of the quantum well affects the allowed energy states. Therefore, the energy band position of the QW structure can be engineered by varying the thickness of the QW.

For a spherical QD structure with infinitely high barriers the allowed energies are (Jussila 2010):

$$E_{\mathrm{QD}}(\mathbf{k}) = E_{\mathrm{g}} + \frac{\hbar^2}{2m_{\mathrm{e}}}\left(\frac{k_{n,l}}{R}\right)^2 \tag{5.6}$$

where $k_{n,l}$ is the nth zero of lth Bessel function and R the radius of the quantum dot. The radius of the QD affects the energy states, so again the energy bands can be engineered by varying the size of the quantum structure. In addition to this, the energy states of QDs are completely discrete.

The height of the potential barrier was set to infinite to obtain previous results. In real semiconductor QW and QD systems this infinite potential well approximation is not true; and in fact, the height of the potential well also affects the allowed energy states of real quantum structures. However, the results obtained using this approximation provide insight into the nature of the energy states and are not far from the ones obtained by the more advanced numerical techniques.

Quantum structures provide attractive means to engineer the energy band structure of the material. In addition to this, quantum structures possess one extra advantage which is their density of state (DOS) function. DOS function explains how electrons can occupy the energy states of the structure. For instance, if the DOS function has a value of zero for some energy, then no electron can occupy that energy state. Figure 5.4 illustrates the DOS functions of different quantum structures. The DOS function in bulk materials is proportional to the square root of the energy. In quantum structures, the charge carriers are more strictly confined to specific energy states. QWs have a step-function-like DOS function and QDs have a delta-function-like DOS function. The delta-function-like DOS function is the main reason why QDs have been of great interest in recent years. This type of DOS function ensures that the charge carriers are strictly confined to the energy states described in (5.6). Therefore, QD structures enable the fabrication of components that, in principle, operate extremely efficiently. Semiconductor QDs can be fabricated by the following techniques (Jussila 2010).

Initially techniques used for the fabrication of QDs and quantum wires included lithography and etching steps. The etching step can result in degradation of the crystal structure of the material. Therefore alternative method called in situ growth of coherent (i.e., dislocation free) QDs has been used. This self-assembled technique is based on the Stranski–Krastanow growth mode of thin films. The term self-assembly refers to the behavior of the adsorbed atoms to form islands without any external interactions. In this technique, the self-assembly emerges due to the crystal lattice mismatch between the two different materials; and as a result of this, QDs are then fabricated from these islands. As explained in the previous section, QDs emit luminescence at discrete energies because of their DOS function. However, the luminescence spectrum in real QD systems has not fully realized this promise and QD structures can have relatively broad luminescence spectra compared to QWs. This broadening originates from the fabrication processes of the QDs. For instance, the broadening in self-assembled QDs comes from the size distribution of the islands fabricated by the coherent Stranski–Krastanow growth mode. In practice, the growth conditions in the self-assembled technique are optimized so that the fabricated QDs are as homogenous as possible. Still, QDs with different sizes always exist in real QD structures. As seen from (5.6), the size of the QD affects the energy states significantly. In addition to this, the luminescence intensity of the QD energy states has usually been lower compared to QW structures, for example. First, this was attributed to the reduced material quality resulting in the fabrication process of the QDs. However, this poor luminescence can result from the relaxation processes of the charge carriers. This phenomenon is the phonon bottleneck of charge carriers, as discussed earlier. In order to observe luminescence from QDs, charge carriers first have to relax close to the QD energy state minima. In general, the charge carriers can relax in the energy band minima with many relaxation mechanisms. The most usual mechanisms are via Coulomb interaction and Auger process, and the relaxation induced by an interaction with a phonon. To explain the inefficiency of the QD luminescence it has been suggested that the relaxation of charge carriers is a significantly slower process in QDs than in QW and bulk materials due to the more discrete energy states of QDs. Scattering occurs mainly via Auger and Coulombic processes. The relaxation processes via the phonons are more improbable in QDs since only few phonons have suitable energies for interaction (Jussila 2010).

5.1.5 Superlattice Heterostructure

Superlattice is a periodic structure of layers of two (or more) materials. Typically, the thickness of one layer is several nanometers. It can also refer to a lower-dimensional structure such as an array of quantum dots or quantum wires. If the superlattice is made of two semiconductor materials with different band gap, each quantum wells sets up a new set of selection rules which affects the conditions for charges to flow through the structure. The two different semiconductor materials are deposited alternately on each other to form a periodic structure in the growth direction. Since the first proposal of synthetic artificial superlattices in 1970, great advances in the physics of such ultra-fine semiconductors, presently called quantum structures, have been made. The concept of quantum confinement has led to the observation of quantum size effects in isolated quantum-well heterostructures and is closely related to superlattices through the tunneling phenomena. Therefore, these two ideas are often discussed on the same physical basis, but each field has its own intrigue and different physics useful for applications in many electric and optical devices (GoldenMap 2013).

Semiconductor superlattices exhibit the resonant tunneling of carriers under the applied bias. This tunneling results from the overlay of the electron waves in the bound states of the periodic potential wells. This overlay can either occur among the bound states at the same energy level, or it can occur when the bound states become degenerate among the excited energy levels (in the case of sequential tunneling). The tunneling process was reported in the case of Zener diodes and Tunnel diodes in 1950s, whereas in the case of superlattices, it became apparent after the significant advancements in the growth procedures, such as molecular-beam epitaxy (MBE). The use of this procedure, along with its in situ analysis, a precise control over the thickness and composition of thin semiconductor layers can be achieved. The carrier transport measurements performed in GaAs–AlGaAs-based double barrier structure, and GaAs–AlAs-based superlattice structure showed the tunneling at specific bias amplitudes which coincide with the quantum-confined states (Younis 2010).

The superlattice miniband structures depend on the heterostructure type, called type I, type II and type III. For the type I heterostructures the bottom of the conduction band and the top of the valence sub-band are formed in the same semiconductor layer. In the type II the conduction and valence sub-bands are staggered in both real and reciprocal space, so that electrons and holes are confined in different layers. Type III superlattice involves semimetal material, one example is HgTe/CdTe superlattice. Although the bottom of the conduction sub-band and the top of the valence sub-band are formed in the same semiconductor layer in type III superlattice, which is similar with Type I superlattice, the band gap of type III superlattice can be continuously adjusted from semiconductor to zero band gap material and to semimetal with negative band gap. There also exists a class of quasiperiodic superlattices named after Fibonacci. A Fibonacci superlattice can be viewed as a one-dimensional quasicrystal, where either electron hopping transfer or on-site energy takes two values arranged in a Fibonacci sequence (GoldenMap 2013).

Semiconductor materials, which are actually used to fabricate the superlattice structures, may be divided by the element groups, IV, III–V and II–VI. While the group III–V semiconductors have been extensively studied, group IV heterostructures such as the Si_xGe_{1-x} system are much more difficult to realize because of the large lattice mismatch. Nevertheless, the strain modification of the sub-band structures is interesting in these quantum structures and has attracted much attention. So far mostly the III–V compound semiconductors represented by the GaAs/Al_xGa_{1-x}As heterostructures have been investigated. In particular, one distinguished merit of the GaAs/AlAs system is that the difference in lattice constant between GaAs and AlAs is very small. In addition, the difference of their thermal expansion coefficient is also small. Thus, the strain remaining at room temperature can be minimized after cooling down from the higher epitaxial growth temperatures. The first compositional superlattice was realized using the GaAs/Al_xGa_{1-x}As material system. Superlattices can be produced using various techniques, but the most common are MBE and sputtering. With these methods, layers can be produced with thicknesses of only a few atomic spacing. In these highly sophisticated growth technologies chemical compositions and thicknesses of a material are controlled precisely under high vacuum. These material structures are grown by atomic/molecular interactions with the crystalline surfaces; and thin layers can be easily grown with a thickness of 2.8 Å, i.e., a monolayer (ML). In these heterostructures, where the layer thicknesses are comparable to the electron/hole wavelengths, the interlayer carrier interactions are apparent. These interactions of carriers with the potential energy barriers result in resonant tunneling. Resonant tunneling is a quantum effect in which the carrier transport through the barrier layers is transparent, which results in the formation of sub/mini bands in the superlattice region. The MBE technology as a means of fabricating semiconductor superlattices is of primary importance. In addition to the MBE technology, metal-organic chemical vapor deposition (MOCVD) has also contributed to the development of superconductor superlattices, which are composed of quaternary III–V compound semiconductors like InGaAsP alloys. Moreover, a combination of gas-source handling and ultrahigh vacuum (UHV) technologies such as metal-organic molecules as source materials, is becoming popular as well as gas-source MBE using hybrid gases such as arsine (AsH_3) and phosphine (PH_3) (GoldenMap 2013).

The motion of charge carriers in a superlattice is different from that in the individual layers: mobility of charge carriers can be enhanced, which is beneficial for high-frequency devices, and specific optical properties are used in semiconductor lasers. If an external bias is applied to a conductor, such as a metal or a semiconductor, typically an electrical current is generated. The magnitude of this current is determined by the band structure of the material, scattering processes, the applied field strength, as well as the equilibrium carrier distribution of the conductor. A particular case of superlattices called superstripes are made of superconducting units separated by spacers. In each miniband the superconducting order parameter, called the superconducting gap, takes different values producing a multi-gap, or two-gap or multiband superconductivity. Besides, artificial two-dimensional crystals can also be viewed as a 2D/2D case (2D modulation of a 2D system) and other combinations are experimentally available: an array of quantum wires (1D/2D) or 3D/3D photonic crystals (GoldenMap 2013).

5.2 Semiconductor Materials and Fabrication Process for Waveguides

Semiconductor waveguides are generally made from elements of III–V, II–VI, or IV–VI and so on. Historically, the III–V compounds have been the first and mostly used for optoelectronic device applications, such as GaAs and InP. Compared to Si and Ge, GaAs and InP have high electron mobility and velocities, properties that are extremely important for the development of high speed electronic devices. Their direct band gaps and therefore their high radiative efficiency make them important for optoelectronic materials. An attractive feature of the binary compounds is that they can be combined or alloyed to form ternary or quaternary compounds, or mixed crystals. These compounds are made up of three or four group III and group V atoms and by choosing different binary compounds, it is possible to select different band gaps and therefore varying emission energies for light sources. A wide range of compound semiconductors can be exploited to access a wide range of output wavelengths. However, by alloying it is possible to vary the band gap continuously and monotonically and together with it the band structure, electronic, and optical properties. The formation of ternary and quaternary compounds of varying band gaps also enables the formation of heterojunctions, which have become essential for the design of high performance electronic and optoelectronic devices. Among the common GaAs and InP-based ternary and quaternary compounds, the properties of $Al_xGa_{1-x}As$ have been most thoroughly investigated. Besides, there are $In_xGa_yAl_{1-x-y}As$ and $In_{1-x}Ga_xAs_yP_{1-y}$ that usually grown on InP substrates. The quaternary compounds have emerged as being extremely important as sources for optical fiber communication, since their band gaps correspond to the special window in which silica fibers have their lowest loss (at 1.55 μm) and minimal dispersion (at 1.3 μm). GaAs/AlGaAs waveguides provide interesting optical properties for nonlinear optical signal processing. In addition to the band-gap engineering capability, GaAs/AlGaAs has excellent electro-optic properties. The optical Kerr coefficient of GaAs is approximately 4 times larger than that of crystalline silicon, which can relax fabrication processes, and facilitate coupling to the fiber (Simatupang 2009).

Moreover, porous silicon waveguides provide an interesting alternative to conventional silicon-on-insulator (SOI) structures that could offer advantages for silicon interconnects modulators and components. Porous silicon is a nanoscale composite material with an effective refractive index between that of air and silicon that can easily be fabricated in multilayer structures (Apiratikul 2009).

With integrated optics, high-index contrast semiconductor waveguides provide exciting new opportunities to both study basic nonlinear optics in these materials as well as investigate new photonic devices such as Raman lasers and chemical sensors. Unlike a standard optical fiber (characterized by a core-cladding index difference of 0.01) in which most laser-induced fluorescence or Raman scattering will be lost out of the fundamental waveguide mode, in a high-index-contrast waveguide (characterized by a core-cladding index difference as high as 2.5), most of the emitted Stokes-shifted light will be captured into a propagating mode. This property, coupled with the long interaction lengths provided by waveguides, make integrated high-index contrast semiconductor waveguides extremely efficient sources for optical experiments in which a pump laser induces signal light at a Stokes-shifted wavelength. In addition, the strong evanescent fields inherent in air-clad submicron waveguides make these systems intriguing candidates for chemical sensing based on fluorescence or Raman scattering from adsorbed analyte species (NRL 2013).

5.2.1 Silicon Waveguides

Silicon is an especially useful material for photonic components because it is transparent at the infrared wavelengths at which optical communication systems operate. Therefore, silicon waveguides have been primarily operated at wavelengths in the near-infrared (NIR), typically around 1.4–1.6 μm. This is convenient due to the large number of commercial optical components available in this regime (Spott et al. 2010). The silicon is usually patterned with sub-micrometer precision, into microphotonic components. These operate in the infrared, most commonly at the 1.55 μm wavelength used by most fiber-optic telecommunication systems. For instance, the silicon can typically lie on top of a layer of silica in what is known as SOI. That is, silicon photonic devices can be made using existing semiconductor fabrication techniques, and because silicon is already used as the substrate for most integrated circuits, it is possible to create hybrid devices in which the optical and electronic components are integrated onto a single microchip (Ganguly and Rahul 2011). Moreover, silicon waveguides have been investigated to provide faster data transfer both between and within microchips.

The propagation of light through silicon devices is governed by a range of nonlinear optical phenomena including the Kerr effect, the Raman effect, two photon absorption (TPA) and interactions between photons and free charge carriers. The presence of nonlinearity is of fundamental importance, as it enables light to interact with light, thus permitting applications such as wavelength conversion and all-optical signal routing, in addition to the passive transmission of light. Silicon waveguides are also of great academic interest, due to their ability to support exotic nonlinear optical phenomena such as soliton propagation (Zhang et al. 2007).

As silicon is transparent to infrared light with wavelengths above about 1.1 μm, and also has a very high refractive index, of about 3.5, the tight optical confinement provides by this high index allows for microscopic optical waveguides, which may have cross-sectional dimensions of only a few hundred nanometers. This is substantially less than the wavelength of the light itself, and is analogous to a subwavelength-diameter optical fiber. Single-mode propagation can be achieved, thus (like single-mode optical fiber) eliminating the problem of modal dispersion. The strong dielectric boundary effects that result from this tight confinement substantially alter the optical dispersion relation. By selecting the waveguide geometry, it possible to tailor the dispersion to have desired properties, which is of crucial importance to applications requiring ultrashort pulses (Dekker et al. 2008). In particular, the group velocity dispersion (GVD) (that is, the extent to which group velocity varies with wavelength) can be closely controlled. In bulk silicon at 1.55 μm, the GVD is normal in that pulses with longer wavelengths travel with higher group velocity than those with shorter wavelength. By selecting suitable waveguide geometry, however, it is possible to reverse this, and achieve anomalous GVD, in which pulses with shorter wavelengths travel faster. Anomalous dispersion is significant, as it is a prerequisite for soliton propagation, and modulational instability (Agrawal 1995). In order for the silicon photonic components to remain optically independent from the bulk silicon of the wafer on which they are fabricated, it is necessary to have a layer of intervening material. This is usually silica, which has a much lower refractive index (of about 1.44 in the wavelength region of interest), and thus light at the silicon–silica interface will (like light at the silicon–air interface) undergo total internal reflection, and remain in the silicon. This construct is so called SOI as mentioned earlier. It is named after the technology of SOI in electronics, whereby components are built upon a layer of insulator in order to reduce parasitic capacitance and so improve performance (Celler and Cristoloveanu 2003). Silicon has a focusing Kerr nonlinearity, in that the refractive index increases with optical intensity. This effect is not especially strong in bulk silicon, but it can be greatly enhanced by using a silicon waveguide to concentrate light into a very small cross-sectional area. This allows nonlinear optical effects to be seen at low powers. The nonlinearity can be enhanced further by using a slot waveguide, in which the high refractive index of the silicon is used to confine light into a central region filled with a strongly nonlinear polymer (Koos et al. 2007). Kerr nonlinearity underlies a wide variety of optical phenomena. One example is four wave mixing, which has been applied in silicon to realize both optical parametric amplification and parametric wavelength conversion. Kerr nonlinearity can also cause modulational instability, in which it reinforces deviations from an optical waveform, leading to the generation of spectral-sidebands and the eventual breakup of the waveform into a train of pulses (Nicolae et al. 2006). Another example is soliton propagation. The evolution of light through silicon waveguides can be approximated with a cubic nonlinear Schrödinger equation, which is notable for admitting sech-like soliton solutions. These optical solitons (which are also known in optical fiber) result from a balance between self phase modulation (which causes the leading edge of the pulse to be red-shifted and the trailing edge blue-shifted) and anomalous GVD (Dekker et al. 2008; Agrawal 1995).

Silicon exhibits TPA, in which a pair of photons can act to excite an electron–hole pair. This process is related to the Kerr effect, and by analogy with complex refractive index, can be thought of as the imaginary part of a complex Kerr nonlinearity. At the 1.55 μm telecommunication wavelength, this imaginary part is approximately 10 % of the real part. The influence of TPA is highly disruptive, as it both wastes light, and generates unwanted heat. It can be mitigated, however, either by switching to longer wavelengths (at which the TPA to Kerr ratio drops), or by using slot waveguides (in which the internal nonlinear material has a lower TPA to

Kerr ratio). Alternatively, the energy lost through TPA can be partially recovered by extracting it from the generated charge carriers (Dekker et al. 2008; Bristow et al. 2007). The free charge carriers within silicon can both absorb photons and change its refractive index. This is particularly significant at high intensities and for long durations, due to the carrier concentration being built up by TPA. The influence of free charge carriers is often (but not always) unwanted, and various means have been proposed to remove them. One such scheme is to implant the silicon with helium in order to enhance carrier recombination. A suitable choice of geometry can also be used to reduce the carrier lifetime. Rib waveguides (in which the waveguides consist of thicker regions in a wider layer of silicon) enhance both the carrier recombination at the silica–silicon interface and the diffusion of carriers from the waveguide core. A more advanced scheme for carrier removal is to integrate the waveguide into the intrinsic region of a PIN diode, which is reverse biased so that the carriers are attracted away from the waveguide core. A more sophisticated scheme still, is to use the diode as part of a circuit in which voltage and current are out of phase, thus allowing power to be extracted from the waveguide. The source of this power is the light lost to TPA, and so by recovering some of it, the net loss (and the rate at which heat is generated) can be reduced (Dimitropoulos et al. 2005; Jones et al. 2005). Free charge carrier effects can also be used constructively, in order to modulate the light. Silicon exhibits the Raman effect, in which a photon is exchanged for a photon with a slightly different energy, corresponding to an excitation or a relaxation of the material. Silicon's Raman transition is dominated by a single, very narrow frequency peak, which is problematic for broadband phenomena such as Raman amplification, but is beneficial for narrowband devices such as Raman lasers. Consequently, all-silicon Raman lasers have been fabricated (Ganguly and Rahul 2011).

Silicon optical interconnects require the integration of a range of technologies. Firstly, an on-chip laser source is required. One such device is the hybrid silicon laser, in which the silicon is bonded to a different semiconductor (such as indium phosphide) which acts as the lasing medium. Another possibility is the all-silicon Raman laser, in which the silicon itself acts as the lasing medium. There must also be a means to modulate the light, thus causing it to carry data in the form of optical pulses. One such technique is to control the density of free charge carriers, which alter the optical properties of the waveguide. Modulators have been constructed where the light passes through the intrinsic region of a PIN diode, into which carriers can be injected or removed by altering the polarity of an applied voltage. By using an optical ring resonator with a built in PIN diode, data transmission rates of 18 Gbit/s have been achieved. By constructing devices where the electrical signal co-moves with the light, data rates of 30 Gbit/s have been achieved. By constructing devices where the multiple wavelengths are used to scale the modulation capacity 50 Gbit/s has been achieved. After passage through a silicon waveguide to a different chip (or region of the same chip) the light must be detected, converting the data back into electronic form. Detectors based on metal–semiconductor junctions (with germanium as the semiconductor) have been integrated into silicon waveguides. Moreover, silicon–germanium avalanche photodiodes capable of operating at 40 Gbit/s have been fabricated. Complete transceivers have been developed and successfully commercialized in the form of active optical cables (Narasimha et al. 2008).

In addition, the choice of silicon as a substrate for the fabrication of optical waveguides follows naturally from its domination of microelectronics. It is important to note that no single material possesses the optimum properties for each individual device found in an IC, but silicon contains a base material from which all the required devices can be fabricated. This flexibility is also important in the fabrication of silicon-based, planar light wave circuits (PLCs). Many of the optical properties of silicon would suggest it to be an ideal material for PLC fabrication. It is virtually transparent to the important telecom wavelengths around 1,550 nm. Further, silicon dioxide (SiO_2) shares its chemical composition with glass fiber, providing a degree of compatibility. Silicon has a relatively high refractive index around 3.5, compared to that for fiber (around 1.5) which allows the fabrication of submicron dimensional waveguides in the SOI system. SOI's large refractive index contrast imposes design constraints on single-mode rib waveguides; however, it can also be considered an advantage. It makes possible core sizes much smaller than 1 μm, which in turn makes possible very compact integrated optical devices with waveguide bending radii as small as a few microns. Silicon layer thicknesses below about 250 nm are required for slab waveguides to be unconditionally single mode. When two-dimensional waveguides are patterned in such thin silicon, a guide with a width of a few hundred nanometers is typically defined by etching away the surrounding silicon all the way down to the SiO_2 substrate and then depositing a layer of SiO_2 or a transparent polymer to embed the silicon. The result is a photonic wire surrounded on all sides by a low-index medium. The small dimensions of photonic wires are conceptually very attractive for on-chip optical interconnects and integrated optical devices with small footprints. However, they are very demanding in terms of fabrication technology. Conventional optical lithography (with illumination wavelengths of 300 nm or more) cannot provide the required resolution. Electron-beam lithography provides the necessary accuracy, but it is not well suited for mass production because it is a serial process. For this, nanophotonic waveguides using the available tools of advanced CMOS fabrication based on deep ultraviolet (UV) lithography have been fabricated (Knights and Jessop 2007).

On the other hand, porous silicon offers an interesting alternative to conventional silicon structures that could offer advantages for silicon interconnects, modulators and components. Porous silicon is fabricated by electrochemical etching of crystalline silicon, which produces a nanoporous skeleton comprised of silicon and air. The typical pore size formed in this process ranges from 5 to 100 nm, depending on the etching chemistry and substrate doping. Because the pores are smaller than the optical wavelength, porous silicon behaves like an effective medium with refractive index between that of air and silicon. The refractive index depends on the porosity, which can be controlled by adjusting the electrochemical current density. Complex multilayer porous structures can be fabricated by varying the current density during fabrication. Optoelectronic devices based on porous silicon including waveguides, LEDs, photodetectors, passive optical filters, microcavities, sensors and optical switches have been developed. In addition to the flexibility of a controllable refractive index, porous silicon has a number of unique properties that make it attractive for modulators and sensors. It is electrically conductive in the direction perpendicular to the surface, while being optically transparent for signals polarized parallel to the surface, which could facilitate optoelectronic connectivity. In addition, the large internal surface area of porous silicon increases the surface recombination rate, which dramatically reduces the free-carrier lifetime (Apiratikul 2009).

There has been a great deal of success in attempts to monolithically fabricate the vast majority of components required in an optical system such as waveguides, modulators, (de)multiplexers, and detectors on a silicon substrate. However, the outstanding limitation of silicon in the photonics arena is its indirect band gap that prevents the straightforward formation of efficient optical sources. Indeed, the search for silicon-based optical source forms the greatest challenge to the widespread dominance of silicon in optoelectronics. Although not yet dominant as a material for optoelectronic fabrication, it is difficult to imagine highly integrated devices of any kind not based on silicon technology. The next few decades will likely witness the increased migration of silicon photonics from the research laboratory to the manufacturing facility with benefit to areas as diverse as microelectronics, telecommunications, and the biological, environmental, and medical sensor industries (Knights and Jessop 2007).

5.2.2 *Gallium Arsenide Waveguides*

GaAs/AlGaAs waveguides provide interesting optical properties for nonlinear optical signal processing. The optical Kerr coefficient (nonlinear refractive index) n_2 of GaAs is approximately 4 times larger than that of crystalline silicon and three orders of magnitude larger than that of standard optical fiber. Table 5.1 compares parameters such as propagation loss (α), optical Kerr coefficient (n_2) and nonlinear parameter (γ) of some of the nonlinear waveguides that have been proposed for optical signal processing. One of the parameters used to evaluate the efficiency of third-order-nonlinearity-based waveguide devices is the nonlinear parameter γ, which depends not only on material nonlinearity (n^2) but also on the geometry of the waveguide. γ is defined as (Apiratikul 2009).

$$\gamma = \frac{2\pi n_2}{\lambda A_{\text{eff}}} \tag{5.7}$$

where λ is the signal wavelength, and A_{eff} is the effective area of the waveguide. For nanoscale semiconductor waveguides, this nonlinear factor (γ) has to be redefined to take into account the tensor nature of the third-order nonlinear susceptibility, the direction of propagation relative to the crystal axes and the vector nature of the electromagnetic mode, including nonzero z-components of the electric field. The

Table 5.1 Comparison among parameters in nonlinear waveguide devices (Apiratikul 2009)

Waveguide devices	Propagation loss (dB/cm)	Optical Kerr coefficient (cm^2/W)	Effective area of the waveguide (μm^2)	Nonlinear factor (1/(Wm))
Standard fiber (SMF-28)	5×10^{-7}	2.2×10^{-16}	80	1.1×10^{-3}
Bismuth-oxide fiber	3×10^{-2}	8.2×10^{-15}	3	1.1
Chalcogenide fiber taper	1.7	1.1×10^{-13}	0.64	68
Chalcogenide waveguide	0.2	3×10^{-14}	7.2	1.7
Organic χ^3 in slot waveguide	23	8.2×10^{-15}	0.1	104
Silicon nanowire	3.6	4.5×10^{-14}	0.14	130
AlGaAs nanowire	82.5	1.5×10^{-13}	0.3	202
GaAs ridge waveguide	4–6	2.9×10^{-13}	1.8	64

nonlinear efficiency could be improved by using a material with large nonlinearity or by reducing the guided-mode area of the waveguides. In addition to the nonlinearity parameter (λ), another figure of merit that has been used to compare the nonlinearity in different materials and waveguides is FOM_{2PA}. FOM_{2PA} takes into account the TPA loss (β_{2PA}) in addition to the nonlinearity (Apiratikul 2009)

$$FOM_{2PA} = \frac{n_2}{\lambda \beta_{2PA}} \tag{5.8}$$

The GaAs/AlGaAs material system could be designed to achieve a large FOM_{2PA}. This material system allows for band-gap engineering: by adding Al into AlGaAs, the energy band gap of AlGaAs can be altered. For example, the band gap energy of AlGaAs can be designed such that the photon energy of the optical signal at communication wavelengths falls below the half-band gap, thus avoiding TPA, while maintaining a large optical Kerr coefficient. GaAs waveguides could also allow for integration of nonlinear optical devices with optically active devices such as lasers, amplifiers, detectors and modulators. A variety of GaAs/AlGaAs waveguide devices have been developed, including directional couplers, microring resonators and photonic crystal cavities. In addition, GaAs and AlGaAs nanowire waveguides have been investigated. For example, a cross-section as small as 0.2 μm^2 was achieved with the linear propagation loss of approximately 50 dB/cm, which is still too high for practical nonlinear applications; GaAs/AlGaAs waveguides with an effective area of 0.4 μm^2 were demonstrated with propagation loss as low as 0.9 dB/cm at 1,550 nm (Shin et al. 2009).

Figure 5.5a shows a cross-section of the material layer of GaAs/AlGaAs waveguide structure. Its epistructure was grown by MBE on a n-GaAs substrate. The structure comprises a 0.8-μm-thick GaAs guiding layer surrounding by a 1-μm-thick $Al_{0.2}Ga_{0.8}As$ upper cladding and a 3.5-μm-thick $Al_{0.2}Ga_{0.8}As$ lower cladding. The layer thicknesses were to ensure single-mode operation and to minimize leakage loss to the substrate. This structure was used both for making a P–I–N waveguide photodetector, and for a passive waveguide. Both parts of the top and bottom cladding layers, which are close to the core, are left undoped in order to help suppress free-carrier absorption (FCA) loss. A 2.4-μm wide ridge waveguide was fabricated in a single photolithography step. First, a 2,000-nm positive photoresist (PR1-2000A) was spin on a small cleaved sample (typically 1.5 cm × 1.5 cm). Next, the waveguide patterns were exposed using UV contact photolithography with a clear field mask. The waveguides were then etched to a depth of 1.8 μm via inductively coupled plasma (ICP) etching with gas mixtures of BCl_3 and Cl_2. For this process, the BCl_3 and Cl_2 flow rates were controlled to be 12.5 and 2.5 sccm, respectively, with a chamber pressure of 1 mTorr at a substrate temperature of 25± °C. Two RF sources (70 and 500 mW) were used to generate high-density plasma and to introduce a self-biased electrical field, respectively. With two RF power sources, the ICP system has the flexibility to control ion density and ion energy independently, unlike a conventional reactive ion etching system. These etching parameters are optimized to create a vertical etch profile with the smooth sidewalls required for a low-loss waveguide. For these parameters, the ICP etch rate was estimated to be 300–350 nm/min with a GaAs:photoresist selectivity of 3:1. Figure 5.5b depicts a cross-sectional schematic illustration and a scanning electron micrograph of a GaAs/AlGaAs waveguide. The calculated TE mode contours were superposed on the cross-section in Fig. 5.5b with the effective area A_{eff} estimated to be 1.8 μm^2, based on the calculated mode. During the ICP etching process, a thin layer of photoresist byproducts containing Cl–C–H polymers is deposited on the top and sidewalls of the waveguides. These residual polymers are difficult to remove with standard solvent cleaning using acetone, methanol and isopropanol. The sample is dipped into H_3PO_4 at room temperature for 2 min immediately after the ICP etch, followed by O_2 plasma cleaning in a reactive ion etching (RIE) system for 5 min and solvent-based (NMP) stripping at 70± C for 15 min to remove the residue. The sample substrate is then thinned down to 120 μm to facilitate accurate cleaving. After that, the sample is scribed using high-power laser scriber setup and is cleaved on the cleave stage on both facets to a length of 4.5 mm. Finally, a silicon nitride antireflection (Si_3N_4-AR) coating was deposited on both facets of the waveguide by high-density chemical vapor deposition (HDCVD). First, ammonia (NH_3) is deposited for 3 min to improve the adhesion between the Si_3N_4 film and GaAs. Then, Si_3N_4 was deposited with gas mixtures of N_2 and SiH_4 with RF power of 500 W under the chamber pressure of 5 mTorr at 300± C. Both N_2 and SiH_4 flow rates were controlled to be 10 sccm. Under this condition, the deposition rate for the horizontal surface was measured to be approximately 34–37 nm/min. The Si_3N_4 film has a refractive index n of 2 at the wavelength of 1,550 nm. For a single layer AR coating, a quarter-wavelength-thick Si_3N_4 film is deposited at both facets of the waveguide. A quarter-wavelength thick film is chosen in order to minimize the normal incidence reflection. For applications operating at the wavelength of 1,550 nm, the target Si_3N_4 thickness is calculated to be $l/4 = 194$ nm. The deposition time is calculated by taking into account that the deposition rate on vertical surfaces, like the waveguide facets, is approximately 50 % slower than on horizontal surfaces. After AR-coating, the sample is mounted to the submount and is ready to be tested (Apiratikul 2009).

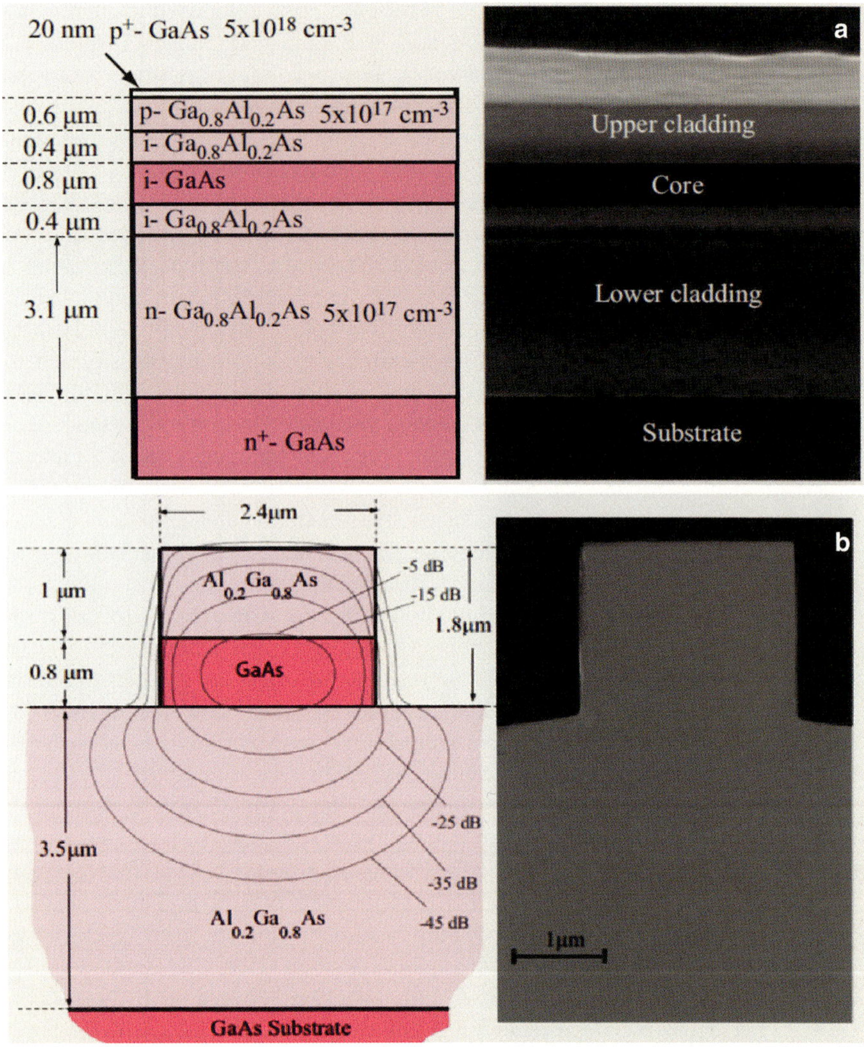

Fig. 5.5 Schematic cross-section and scanning electron micrograph of material structure for a GaAs/AlGaAs optical waveguide (Apiratikul 2009): (**a**) Cross-section of a GaAs/AlGaAs waveguide. (**b**) Schematic illustration with calculated fundamental TE mode profile (contours labeled in dB relative to peak value)

5.2.3 InAs Quantum Dots

As shown in Fig. 5.6, InAs QD structures can be grown on 500 µm-thick semi-insulating (100) InP substrates (Apiratikul 2009): (a) 100-nm thick InP buffer layer was grown on the substrate with a V/III ratio of 220, a growth temperature of 640 °C, and a growth time of 450 s. Then, InAs island layer was grown on the InP

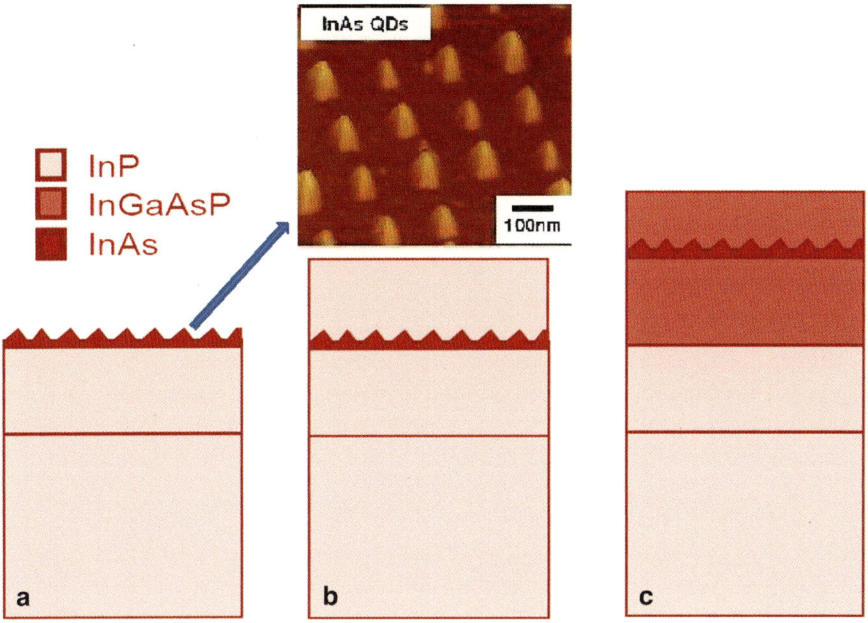

Fig. 5.6 Cross-section of the fabricated InAs quantum dot samples (Apiratikul 2009): (**a**) InAs/InP island structure, (**b**) InAs/InP QD structure, and (**c**) InAs/InGaAsP/InP QD structure

buffer layer with a V/III ratio of 12. The growth temperature and growth time of this layer were changed between 420 and 550 °C, and between 1.25 and 8.25 s, respectively. The schematic structure of these samples is shown in Fig. 5.6a. (b) 100-nm thick InP buffer layer was grown on the substrate with the same growth parameters as in (a). Then, an InAs layer was grown on the InP buffer layer at the temperature of 520 °C. The growth time of the layer was between 1.25 and 10.6 s. After this, the InAs layer was capped with a 50-nm thick InP layer. The InP capping layer was grown at the same temperature as the InAs layer. This was performed to avoid unnecessary temperature ramping which was evaluated to cause more harm than the relatively low growth temperature of InP. The schematic structure of these samples is shown in Fig. 5.6b. (c) 100-nm thick InP buffer layer was again grown with the same growth parameters as in (a) and (b). Then, a 100-nm thick $In_xGa_{1-x}As_yP_{1-y}$ barrier layer was grown on the InP barrier layer with a V/III ratio of 120, a growth temperature of 640 °C and a growth time of 350 s. The aim was to grow the InGaAsP layer as lattice-matched to InP as possible. After this, the InAs QD layer was grown on InGaAsP at 520 °C with a V/III ratio of 12 and a growth time of 2 s. Then, this InAs layer was capped with a 50-nm thick $In_xGa_{1-x}As_yP_{1-y}$ layer. This layer was grown at the same temperature as the InAs layer. The schematic structure of these samples is shown in Fig. 5.6c. Self-assembled growth of InAs islands on InP can be examined by fabricating two series of samples (structure of Fig. 5.6a). The nominal thickness and growth temperature of the InAs layer can be varied in these series. In the first series, the nominal thickness of the InAs layer can vary from one monolayer (ML) to six MLs. The growth rate of InAs is higher at higher temperatures. In addition, the cooling down process also affects the island formation (Apiratikul 2009).

5.3 Quantum-Well Technology

The key to realize on-chip integrated optical devices is working with a low-loss, highly nonlinear material system that allows the creation of both active and passive photonic devices. With quantum-well technology, III–V group semiconductors, such as GaAs and its alloys, can be used to create low-loss waveguides with very high second and third-order nonlinearities, or direct band gaps that enable the fabrication of bright lasing sources. For example, AlGaAs-based heterostructures such as superlattices and MQWs allow for tailoring the linear and nonlinear optical and electronic properties of the devices to suit the given application requirements. GaAs:AlGaAs MQW structures can efficiently confine free carriers in the lower-energy gap layers, and hence enable the fabrication of efficient laser sources. Superlattices allow control over the devices' band gap and tailoring it to the required application. This band structure is characterized by the formation of minibands of allowed energy levels that are dependent on the composition of the superlattice constituents and the thickness of the barriers and wells. The band gap (the absorption edge) change directly alters the macroscopic refractive index through the Kramers–Kronig relations. The change in band gap alters the magnitude of the tensor elements of the second and third-order nonlinear susceptibilities (Sigal 2010).

5.3.1 Characterization of Quantum Well

The GaAs quantum well (~20 nm) with AlGaAs barriers and claddings, grown using MOCVD, was first used to achieve the room temperature injection laser action, which had the threshold current densities of ~3,000 A/cm^2. Later on, much work has been reported in quantum-well laser structure improvement. The graded index separate confinement heterostructure (GRIN-SCH) symmetric waveguide laser, grown using MBE, reported the threshold current density of 500 A/cm^2, and further improved it down to 160 A/cm^2. The GRIN-SCH structure, not only can achieve very low threshold current density but also the selection of GRIN layers can result in any desired far-field pattern. The minimum threshold current density can usually be achieved in short devices with a large number of quantum wells, and long devices with a small number of quantum wells. This is explained by the higher mirror loss for a shorter cavity, which requires a higher gain factor for the laser action to occur, and is provided by an increased number of quantum wells. The excitation wavelength can be varied as a function of decreasing GaAs quantum-well width, bounded by 80 Å $Al_xGa_{1-x}As$ barriers. For instance, the pulsed excitation was observed in 10 and 20 GaAs quantum-well material systems, and a red shift of about 20 nm was reported. In a different material system which had 20 GaAs quantum wells (25 Å) bounded by $Al_{0.31}Ga_{0.69}As$ (40 Å) barriers, indicated coupling between the quantum confinements among the quantum wells. The excitation in this case also observed a red shift of 18 nm which was attributed to reabsorption (Younis 2010).

In addition, suspended quantum-well heterostructures have focused on lasing in photonic crystals, microcantilever photodetectors, and lateral band structure deformation. The successful fabrication of a suspended quantum-well waveguide in the III–V material system depends critically on the epitaxial strain, band structure, and etches selectivity of the different materials. With a proper choice of materials, alloy fractions, and layer thicknesses within the heterostructure, a suspended MQW waveguide can be fabricated to allow low-loss propagation in the optical L-band, just 125 nm below band gap. Such a design allows strong electro-optic effects in the near band edge regime due to the quantum-confined Stark effect, without adding significant interband absorption loss. InGaAs quantum wells with InGaAsP barriers grown on InP have been designed to have a band edge near 1,500 nm. Both InGaAs and InGaAsP (with a phosphorous content below about 60 %) are also known to have a high resistance to a hydrochloric acid (HCl) etch, which is used to selectively remove an InAlAs sacrificial layer. By growing either the InGaAs or InGaAsP layers with a small amount of tensile strain, the suspended waveguide will not crack or buckle upon release from the sacrificial layer. This 590 nm thick waveguide layer is grown on a 1.6 μm lattice-matched $In_{0.52}Al_{0.48}As$ sacrificial layer and a 50 nm thick InGaAs etch stop layer. The waveguides are patterned and etched using electron-beam lithography followed by a chlorine-based inductively ICP etch and then released by selectively etching the sacrificial layer, resulting in a MQW core surrounded by an air cladding. The waveguides are attached using tethers spaced at regular intervals along the waveguide length. The tethers vary in width from 2 to 4 μm while the waveguide width ranges from 1 to 6 μm on a given sample. The waveguide lengths range between 1.2 and 2.4 mm, and the tether spacing along the waveguide ranges between 120 and 575 μm. An example of such a waveguide is shown in Fig. 5.7. Figure 5.7a shows a fabricated waveguide 2 μm wide along with two tether pairs of the same width. Figure 5.7b shows a close-up of one of the facets of this waveguide. Suspended MQW waveguides can exhibit a propagation loss that is approximately 4 dB/cm at wavelengths of only 125–150 nm below band gap. Besides, tether losses can be as low as 0.15 dB per tether pair, permitting their use in devices that require multiple tethers in an active area. Epitaxial tensile strain ensures that the waveguides remain flat upon release. These waveguides, when fabricated with the MQW between P-doped and N-doped layers, should exhibit extremely high electro-optic coefficients due to the strong overlap between the small optical mode and the electrical bias field. Coupled or asymmetric MQWs should increase the electro-optic coefficients even further, enabling extremely low-drive voltage ($V\pi$) modulators in lengths of 1 mm or less. Forward biased P–I–N MQWs with band edges near 1,550 nm could be used as electroabsorptive

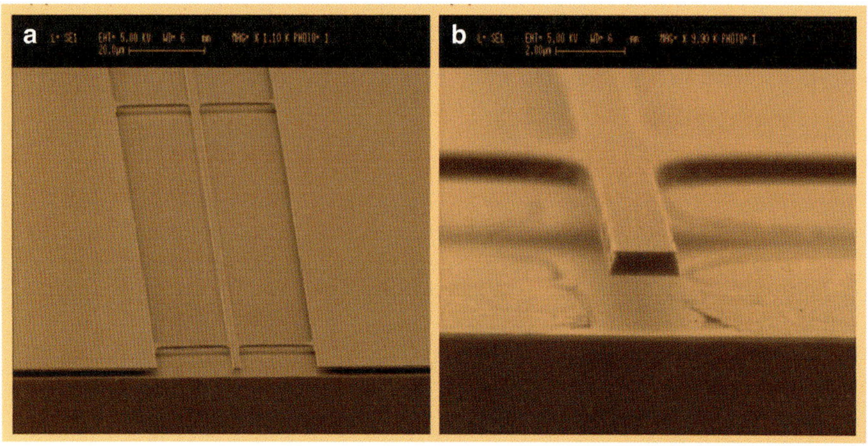

Fig. 5.7 An example of MQW waveguide (Stievater et al. 2008): (**a**) fabricated waveguide 2 μm wide along with two tether pairs of the same width; (**b**) a close-up of one of the facets of this waveguide

modulators, lasers, or amplifiers that could be integrated with suspended waveguide switches or sensors. In addition, since the waveguides can be surrounded by a low-index chemoselective coating after the sacrificial etch, the waveguides can act as chemical sensors (Stievater et al. 2008).

5.3.2 Quantum-Well Intermixing

The properties of MQWs and superlattices are highly desirable for implementing all-optical integrated devices. In order to implement a cost-effective multifunctional optical device on-chip a direct control over the optical properties of the material, and specifically the band gap, is necessary. Quantum-well intermixing (QWI) is a post-growth fabrication technique that allows modifying the band gap of a MQW or superlattice structure on the same wafer, and thus allowing the integration of smaller-band gap lasers with larger-band gap, lower-loss linear and nonlinear waveguides. QWI, in the simplest terms, is induced disorder in a MQW structure that happens due to the diffusion of MQW constituents between the barriers and wells. Even a small amount of such diffusion breaks the structure symmetry and changes the band gap of the structure back to its bulk value. The disordering, in turn, can be achieved by means of a mechanism called impurity-free vacancy disordering (IFVD), where no foreign impurities are introduced into the wafer and losses are minimized. Disorder is instead achieved by shallow As^+ ion implantation, which creates defects in the lattice. The implantation is followed by rapid thermal annealing (RTA) which enables the defects to diffuse across the superlattice and render the material as bulk. The controlled ion implantation and subsequent RTA allows patterning the wafer before the etching process, making appropriate modifications to the wafer's band gap (and linear and nonlinear properties) to suit the application at hand. Such fabrication process has proven to introduce lower scattering losses than the Etch-and-Regrowth process for shallow-contrast waveguides (Younis 2010).

5.3.3 *Micromachining*

Micromachining technologies have the potential to dramatically enhance the functionality and performance of integrated photonic devices. For example, integrated switches based on deflection of a suspended waveguide have been demonstrated in SOI and gallium arsenide (GaAs) systems and in indium phosphide (InP)-based systems. However, the optically active nature of III–V semiconductors has yet to be fully exploited in these micromachined waveguide devices; Lasers, optical gain segments, or electro-optic modulators can be integrated with micromachined III–V waveguides. Not only would the use of electro-optically active heterostructures eliminate the need for off-chip photonic gain or modulation components but the electro-optic properties could also be enhanced by the unique optical properties of suspended semiconductor waveguides, which include high-index contrast and strong evanescent fields. The high-index contrast inherent in suspended waveguides enables small cross-sectional mode areas, which in turn enhance nonlinear optical and electro-optical properties. In addition, the evanescent fields in the air surrounding the waveguide suggest uses in sensing that could also take advantage of integrated gain segments (Stievater et al. 2008).

5.4 Doped Semiconductor Waveguides

Many lanthanide elements have played an important role in various optoelectronic and photonic applications, ranging from emitting elements in solid-state lasers (for example the Nd:YAG laser) and in phosphors for color lamps and displays (for example Eu and Tb) to optical waveguides (using Er or Pr). These rare earth (RE) elements have a partially filled inner (4*f*) shell shielded from its surroundings by completely filled outer (5*s* and 5*p*) orbitals. Due to this shielding, the intra 4*f* shell transitions result in very sharp optical emissions at wavelengths from the ultraviolet to the infrared (IR). The wavelengths of these emission lines are determined by the energy of the transition between 4*f* states of the RE and are relatively independent of the host material. However, the host material does have a very strong effect on the radiative transition probability, in other words on the photoemission intensity. In general, RE-doped conventional semiconductors (Si, GaAs, etc.) have exhibited limited photoemission at room temperature due to low RE solubility and severe temperature quenching. The thermal quenching in Er-doped semiconductors decreases with increasing band gap. Therefore, wide-band gap semiconductors (WBGS) are attractive hosts for RE elements. For instance, GaN is a WBGS that is intensely investigated for (intrinsic) optical and electronic applications. Er-doping of GaN was shown to produce strong near-IR 1.5-m emission suitable for fiber-optic telecommunications from the lowest excited state. Photoemission from higher excited RE states in GaN can cover the entire visible spectrum: light emission in the green (from Er at 537/558 nm), red (Pr at 650 nm, Eu at 621 nm), and blue (Tm at 477 nm). Emission in the NIR is also obtained at 801 nm from Tm, at 1,000 and 1,540 nm from Er, and at 956, 1,303, and 1,914 nm from Pr. The RE doping of GaN (as well as AlN and GaN) can be accomplished by ion implantation or by in situ doping during growth. Ion implantation has the advantage of a simple process and can control the doping concentration independent of the growth conditions. However, the implantation doping approach also has disadvantages: introduction of damage in the GaN (which cannot be completely removed by annealing, especially at the higher concentrations used in ELDs), and the inability to uniformly dope thick layers because of the limited penetration range of the heavy RE ions (Steckl et al. 2002).

5.4.1 In Situ RE Doping

In situ RE doping requires a good understanding and control of the overall growth process and of the interaction of the RE flux with the main V/III fluxes. However, once that is established, good control of the RE concentration over many orders of magnitude can be obtained and very uniform doping can be accomplished. Furthermore, in situ doping does not suffer from the damage effects of ion implantation and, hence, results in more efficient emission. For instance, REs can in situ introduced during growth of the GaN layer by MBE. RE doping of GaN represents an interesting alternative to semiconductor alloying (GaN/InN/AlN) for visible light-emission applications and has the additional attractive aspect of strong IR emission for telecommunications and other applications. Theoretically, RE intra-4*f* atomic transitions are parity forbidden by the Laporte selection rule. RE ions incorporated into a partially ionic solid favor substitutional occupation of the cation site. In addition to RE incorporation, the cation site must provide an uneven ligand crystal field in order to relax the selection rule and increase the probability of intra-4*f* transitions. These 4*f*–4*f* transitions, however, are still not fully allowed, resulting in excited state lifetimes of ~1 ms for RE-doped systems. In wurtzitic GaN, which has a significant component of ionic bonding, the RE ions have strong optical activity levels, since they are generally substitutionally located on the Ga sublattice where the lack of inversion symmetry produces strong ligand fields, thereby increasing the 4*f*–4*f* transition probability. These substitutional RE dopants are, therefore, likely to be the optically active RE centers observed in GaN:RE. The specific 4*f* transitions and associated emission wavelengths in GaN doped with Eu, Er, and Tm are shown in Fig. 5.8a. The dominant transitions producing visible

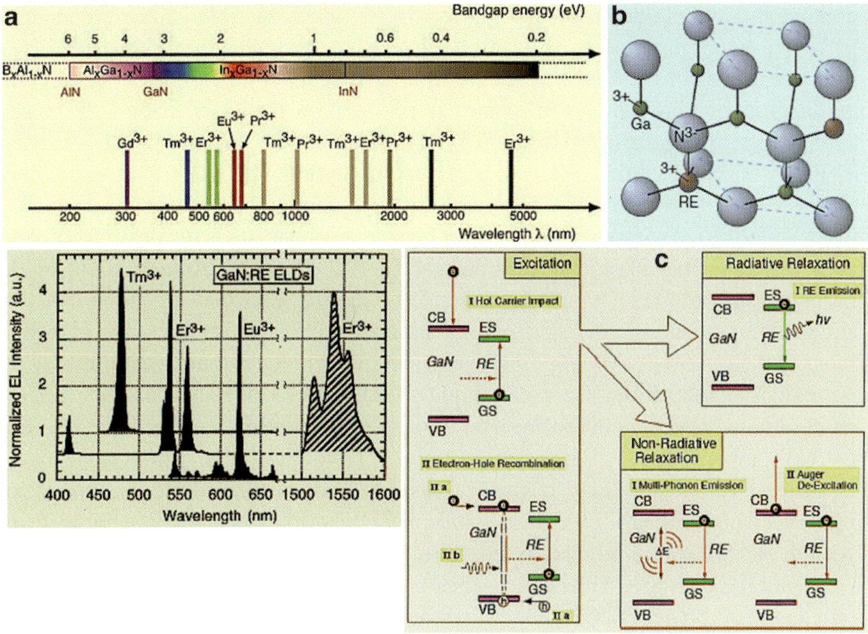

Fig. 5.8 RE-doped GaN photoemission (Steckl et al. 2002): (**a**) emission spectra from Tm-, Er-, and Eu-doped GaN ELDs and emission wavelengths from selected transitions in rare earth ions and associated band gap energies of III-N compound semiconductors; (**b**) bonding model for RE-doped GaN, with the RE ion located substitutionally on the Ga sublattice; (**c**) mechanisms for RE dopant excitation and relaxation in the GaN host

emission are indicated with thicker lines. A simple model of the GaN:RE crystal structure is shown in Fig. 5.8b. A strongly bonded GaN lattice, in conjunction with substitutional incorporation allows unusually high RE doping concentrations (up to 3–5 at.%), while preserving the optical activation of RE dopants. By comparison, the use of RE-doped II–VI semiconductors as emitters and phosphors suffers from a more weakly bonded lattice, and substitutional location of the RE ions on the two cation sites, which generates additional defects due to lack of charge neutrality. Rutherford back scattering (RBS) channeling analysis confirms that a great majority 95 % of the Er ions occupy substitutional sites on the Ga sublattice even at relatively high concentrations of 1 at.%. The Er–N bond has been measured by EXAFS analysis to be 2.17, versus a Ga–N bond length of 1.95. This unusually short Er-nearest neighbor bond length in GaN is thought to be due to two major factors (Steckl et al. 2002): (a) the low fourfold coordination, compared for example to a 12-fold coordination in ErSi; (b) a more polar bond for Er–N than for Ga–N (due to electronegativity differences), which helps to energetically compensate for the Ga–Er size mismatch. Light emission from GaN:RE has been demonstrated via photoluminescence (PL), cathodoluminescence (CL), and EL. The dominant mechanisms for excitation and subsequent relaxation of RE dopants in GaN are depicted in Fig. 5.8c. In PL, electron–hole pairs are generated by above band-gap photon absorption, charge carrier generation is provided by a high-energy electron beam in CL, and in EL, carrier injection occurs by the application of bias voltage to electrical contacts on the GaN layer. The charge carrier energy is transferred to the RE dopants by impact excitation of hot carriers or as a result of nearby electron–hole recombination. The REs then experience either nonradiative relaxation (through multiphonon emission and/or Auger electron excitation) or radiative relaxation. The latter mechanism is the desired outcome as it results in the photoemission utilized in ELDs. The relative strength of the radiative relaxation mechanism is a complex function of the GaN crystalline quality and the RE concentration. For example, increasing the crystallinity of the GaN host reduces the concentration of point defects, and thus reduces the opportunity for RE nonradiative relaxation. However, high crystallinity is normally achieved under conditions which prevent the incorporation of the optimum RE concentration into the GaN layer. Further, for the case of above-band gap photo-pumping of highly crystalline RE-doped GaN, intrinsic (band-to-band) electron–hole recombination provides a high probability radiative intrinsic relaxation mechanism in competition with the RE radiative relaxation (Steckl et al. 2002).

In situ RE doping of GaN has resulted in the successful fabrication of ELDs with red, green, and blue (RGB) color emissions using Eu, Er, and Tm, respectively. The most crucial factors for obtaining and improving EL emissions include RE concentration, growth temperature, and stoichiometry (i.e., V/III ratio) of the host material. With regard to concentration, the EL emission increases more or less linearly with RE concentration at low levels. On the other hand, at high concentration levels, an emission quenching effect occurs due to an increase in RE–RE energy transfer, rather than radiative relaxation. Hence, there is an optimum RE concentration for which maximum emission is observed. Most phosphor systems exhibit this concentration quenching phenomenon. For example, ZnS:Mn exhibits its luminance maximum at 0.5–1.0 wt% of Mn. Growth temperature and Ga flux (V/III ratio) affect the RE emission through the properties of the GaN host (such as crystallinity and resistivity) and directly through their influence on RE incorporation. These three parameters, RE flux (determined by the RE cell temperature), growth (substrate) temperature, and Ga flux (determined by the Ga cell temperature) are externally controlled and adjusted during GaN:RE MBE growth. In general, all Er-doped host materials exhibit a practical limit in Er concentration beyond which the optical emission begins to decrease. The ultimate doping limit is the RE solid solubility in the host material. Concentrations which are introduced in a nonequilibrium process and which are larger than the solid solubility can result in the formation of a second phase material which contains some (or all) of the RE concentration. This precipitation is accompanied by the concomitant degradation of optical and electronic properties of the host material. A practical limit, which is usually reached first, has to do with the concentration quenching effect. For example, the solubility limit of Er in Si is $\sim 10^{18}/\mathrm{cm}^3$, while the PL intensity at 1.54 μm was to saturate at even lower concentration of $\sim 5 \times 10^{17}/\mathrm{cm}^3$, which represents an Er atomic percentage of only $\sim 10^{-3}$ at.%. Maximum Er concentration is ~5 × $10^{18}/cm^3$ in GaAs and ~4 × $10^{19}/cm^3$ in Al_2O_3. Er concentration in GaN ranges from 0.025 to 10 at.% by varying the Er cell temperature from 740 to 980 °C. The Er concentration follows an exponential dependence on the Er cell temperature. An Arrhenius-like thermal activation energy estimated from this curve was 2.9 eV, which gives an excellent agreement with the activation energy of 3.0 eV for the Er vapor pressure in this temperature range. As the RE ion concentration is increased, the average distance between ions is reduced proportional to the cube root of the RE concentration. When RE ions are located sufficiently close to each other, the excitation residing in one ion can migrate to a neighboring ion of the same species as a result of resonant energy transfer. This process is known as cross-relaxation. The energy migration process increases the possibility that the excitation is dissipated at nonradiative sites, resulting in saturation or a decrease in optical emission. The optimum Er concentration is 1 at.%. The corresponding Er–Er spacing is estimated to be ~15 Å. On another hand, an MBE growth temperature in the neighborhood of 600 °C produces the optimum ELD emission intensity. In addition, RE optical emission from GaN films is a strong function of the ratio of the Ga- and N-bearing fluxes (V/III) during growth. The V/III ratio can be modified by changing either component, but is typically adjusted by controlling the Ga flux. The Ga flux is critical to GaN crystalline quality and good crystalline GaN is usually grown under slightly Ga-rich growth conditions. GaN crystallinity and intrinsic luminescence are strong functions of Ga flux during growth. Therefore, it is clear that the RE-related emission, which is dependent on the crystallinity of the GaN host, will also be strongly affected by the Ga flux. The optimum growth condition for Er optical activity is found under slightly N-rich flux near the stoichiometric region. Under these growth conditions, the resulting GaN crystallinity is high enough for efficient Er excitation but the competition from intrinsic carrier recombination is not yet very strong. Furthermore, the higher resistivity of GaN grown under slightly N-rich conditions is favorable for supporting high electric fields for hot carrier generation. GaN:Er is of special interest since it has been shown to be relatively immune to the thermal quenching seen in other Er-doped semiconductors and has the ability to incorporate significant concentration of RE ions without precipitation and without quenching the photoluminescence or electroluminescence intensity. These properties, in conjunction with the fact that GaN is a semiconductor and can provide charge carrier excitation of the Er ions (unlike insulators such as SiO:Er), meet the key requirements for an electrically pumped waveguide amplifier. GaN optical channel waveguides can be designed for use with the commercial material using the effective index method. GaN refractive index values of 2.35 and 2.25 at 632.8 nm and 1.5 μm. The fixed values of the width and film thickness could be calculated using the effective indices for the etched region and the unetched region. An etch depth of ~0.3 μm was chosen since the resulting structure would have an appropriate effective index difference and support only a few modes, limiting the effect of higher order lossy modes. An application area of GaN:RE technology is infrared optical telecommunications at 1.3 and 1.5 μm. The versatility and robust nature of RE-doped GaN technology holds great promise for optoelectronic and photonic applications (Steckl et al. 2002).

5.4.2 Erbium-Doped Semiconductor Waveguides

Erbium-doped semiconductor waveguide amplifiers (EDWAs) have been extensively explored due to their compact integrated size and the performance advantages inherited from Erbium-doped fiber amplifiers (EDFAs), namely low noise figure, negligible polarization dependence, good temperature stability and the absence of interchannel crosstalk to achieve performance comparable to EDFAs, a high doping level is required to compensate for the small optical transition cross-section of erbium (Er) as well as the waveguide length limit of a few centimeter to tens of centimeter, making essential the selection of Er-host materials which can avoid deleterious high concentration effects. There are many similarities in the fundamental basics of lasers and optical amplifiers, where the stimulated emission in both is achieved by maintaining a suitable degree of population inversion between an excited state and the ground state through either electrical or optical pumping. Thus, an Er-doped III–V oxide-based monolithically integrated waveguide optical amplifier can be readily modified for photon generation as either an incoherent amplified spontaneous emission (ASE) light source or an Er-doped waveguide laser (EDWL). An ASE source is basically an EDWA with no signal input, while an EDWL requires feedback through the incorporation of a resonant cavity achieved through high reflectance cleaved facets and/or deposited mirror coatings that are commonly used in III–V semiconductor diode lasers. With the growing field of fiber lasers and broadband ASE sources for sensing and other purposes, many applications might also benefit from this more compact, monolithically integrated planar waveguide implementation (Huang 2006).

Among rare earths, erbium belongs to the group of lanthanides, and is extensively studied due to the fortuitous coincidence of the near 1.55 μm optical transitions of Er with the minimum attenuation window of standard silica optical fibers. The optical properties of interest occur for trivalently ionized Er ions, Er^{3+}. These have the electronic configuration of Xe in common (1*s*22*s*22*p*63*s*23*p*63*d*104*s*24*p*64*d*105*s*25*p*6) with additional electrons added to the 6*s*-, 5*d*- and 4*f*-shells, respectively. The trivalently ionized Er ions have an electronic configuration of [Xe] 4*f*11, where the partially filled 4*f* shell is shielded by the outer 5*s*25*p*6 closed shells of the Xe configuration. Accordingly, the position of energy levels and related optical transitions are barely dependent on the surrounding host material into which the ions are incorporated. The different possible electronic configurations of 4*f* electrons introduce different energy states due to spin–spin and spin–orbit interactions. Erbium ions can be incorporated into a variety of host materials. In general, materials with ionic structures such as oxides, fluoride glasses or ionic crystals are better Er hosts than covalent semiconductors, offering a higher Er solubility in the solid phase for minimal clustering and segregation. Silica-based glass is the most widely used Er host due to its application in optical fibers. However, due to the small solubility of rare earth ions in silica (the suggested maximum Er concentration in silica for optimal amplifier performance is under 100 ppm), other glass materials have been developed such as alumina, sodalime and phosphosilicate glass. Phosphate-based glasses are known to have a higher Er solubility, having been shown to accommodate several weight percent concentrations of rare earth oxides without clustering in rare earth-based bulk lasers. Aluminum oxides can be added to glass matrixes as network modifiers for both higher Er solubility and broadband emission. Rare earth concentrations of 2 % have been claimed without clustering in a host glass consisting 8.5 mol% of Al_2O_3. In addition to the solubility, the luminescence linewidth can be substantially affected by the host composition and the local environment where Er^{3+} ions locate. Besides the broadening by the Stark-splitting of energy levels induced by the local crystal field, the emission linewidth also contains the contributions from a homogeneous and inhomogeneous effect. For a transition between two energy states of an Er^{3+} ion, the homogeneous broadening arises from the lifetime of the states and depends on both radiative and nonradiative processes. The shorter the lifetime is, the broader the state. The inhomogeneous broadening are due to the multiplicity of sites and environments available to the Er ions. In glasses, both broadening can be quite large, as compared to crystal, and is very important for optical amplifiers. A broader emission linewidth covers more signal channels in dense wavelength division multiplexing (DWDM) systems, or provides a wider wavelength tuning range for EDWLs. Besides its effect as a network modifier, Al_2O_3 is known among various oxide glasses for its larger inhomogeneous broadening. The wet oxides of InAlP and AlGaAs usually contain a large amount of Al_2O_3 (and phosphate as in InAlP oxides), showing a broader emission linewidth and higher Er solubility. Most importantly, grown on GaAs, these host materials provide a unique feasibility for monolithic integration with active devices such as pump lasers, an advantage most glass materials don't have (Huang 2006).

In order to achieve very compact integration such as a long spiral of waveguide occupying a small area, a large core/cladding index contrast is needed to increase the optical confinement and reduce the bending loss. Semiconductor/oxide hybrid waveguide structures have been designed to combine the high refractive indices of semiconductors with a suitably positioned Er-doped oxide active layer. In this structure, the bottom cladding layer is a semiconductor layer with a lower refractive index, and the active Er-doped oxide layer is grown and implanted on top of a high refractive index semiconductor core layer. These two layers of oxide/semiconductor can be viewed as forming a composite waveguide core, where the active Er ions are placed along the top edge instead of near the center of the core. With most of the Er emission then guided through an underlying transparent semiconductor core through what is essentially a waveguide coupling mechanism (the emission from Er^{3+} ions can be stimulated via evanescence field and still be guided within the semiconductor core), this offset of the Er-active layer may also serve to reduce the undesirable excited state absorption (ESA). ESA occurs where an excited Er^{3+} ion absorbs a photon emitted at 1.53 μm from other Er^{3+} ions, boosting itself to a higher 4I9/2 energy level. To reduce scattering loss from the higher index contrast oxide/air interface, an optional top oxide cladding (SiO_2, Al_2O_3, SiN_x, etc.) can be deposited after Er-implantation. Due to the large refractive index contrast between the semiconductor and oxide layers, very strong optical confinement can be achieved in very compact structures employing thinner waveguiding layers. In addition, the higher lateral index contrast of ridge waveguide structures enables smaller bend radius curved waveguides desirable for fabricating long spiral-coiled waveguides in monolithically integrated EDWA applications. Moreover, the epitaxially smooth interface possible between semiconductor layers may also help to further reduce the propagation loss compared to a fully oxidized heterostructure waveguide. As with semiconductor heterostructures, such composite semiconductor/oxide hybrid structures provide tremendous design flexibility through the variation of many parameters such as Al composition, and the order, number and thickness of layers. More advanced designs of this type will require simultaneous consideration of an optimization of waveguide optical mode confinement, the multilayer vertical reflectance properties and their influence on the VIP cavity, etc. As an example, in order to maintain a suitable waveguide structure while increasing the overall volume of the Er-doped InAlP native oxide active medium as required for high-power applications. An alternative design uses a multilayer structure containing buried Er-doped AlGaAs and InAlP oxides sandwiched between high-index, oxidation-resistant semiconductor layers (such as InGaP which has been used in MOSFET structures as an oxidation barrier layer), a composite core layer with a higher average effective index can be realized. To provide enough Er penetration in this core layer, Er-doping via multiple high-energy Er-implants is required, considering the thickness of the core and the heavy mass densities of both native oxides and un-oxidized semiconductors. Building on a GaAs platform, various schemes can be envisioned to monolithically integrate pump or signal lasers with Er-doped planar waveguides. Due to the high Er concentration required to reach a high gain per unit length, cooperative up-conversion interactions between the closely spaced Er ions can reduce the pumping efficiency such that a rather high pump power is required to obtain a net gain. The DBR mirror structure is realized by epitaxial growth of alternative quarter-wave optical thickness high (GaAs) and low (AlGaAs) index semiconductor layers. Appropriate P and N-type dopants are incorporated above and below the InGaAs quantum well active region to form the required P–N junction diode for electrical injection. An electroluminescent structure with only one bottom DBR mirror, sometimes referred to as a "half-VCSEL," is not itself a new concept, being utilized in early VCSEL devices having deposited top mirrors and, more recently, in high-power external cavity VCSEL devices. The reflectance of a top gold metal mirror need not match the high reflectance typically found in multilayer DBR reflectors because, with the use of Yb co-doping, even broad luminescence below lasing threshold can be effectively absorbed to pump the Er ions in the EDWA core. Lateral lasing in the quantum-well layer will naturally be prevented by use of spiral-coiled device geometry. Considering that the high-index Er-doped waveguiding structure does not extend down into the semiconductor active layer, the weak gain guiding will be insufficient to bend the light with low loss, and absorption in the un-pumped regions will attenuate light leaving the pumped regions. Further lateral losses in the QW layer can be introduced by use of diamond dicing saw cuts or etched Q-spoiling trenches where needed around the waveguide structure (Huang 2006).

5.5 Semiconductor Nanomaterials for Waveguides

Semiconductor nanowires have been developed to improve the stability of a wurtzite crystalline phase of standard III–V semiconductors such as InP and GaAs, which in their bulk forms always adopt a zincblende crystal structure. This dual stability results in the unanticipated challenge of producing single crystals during the growth process, as well as in necessitating characterization of these new forms of otherwise well-known semiconductors in terms of basic bulk material properties. While wurtzite–zincblende polytypism is known for III-nitride semiconductors, among the III-arsenides and III-phosphides it is unique to the nanoscale under ambient temperature and pressure conditions. For understanding exhibited linear and nonlinear optical properties, modeling the behavior of, or designing devices upon the nanostructures in which the wurtzite modification appears, the bulk material characterization is of both fundamental and applied interest. The uniaxial nature of wurtzite GaAs makes it naturally birefringent, distinguishing it from the normal cubic zincblende form of GaAs, which is spatially isotropic. Wurtzite crystals are consequently interesting as nonlinear media, since their birefringence makes phase-matching conditions possible to maximize nonlinear wave mixing processes, the simplest case of which is second harmonic generation (SHG). The exhibited SHG intensity has polarization dependences that originate from the form of the electrical susceptibility tensor. This is intimately related to the form of the Raman tensor, which of course is itself defined by a derivate of the susceptibility with respect to the atomic positions, as both are reliant upon the crystalline symmetry (Crankshaw 2009).

Semiconductor nanowires have attracted great interest both for basic studies of one-dimensional physics and as building blocks for nanoscale devices. Many III–V and II–VI material systems have been developed with various growth modes, such as laser-assisted catalytic growth and chemical beam epitaxy, as well as some catalyst-free methods such as selective-area epitaxy. The vapor–liquid–solid (VLS) growth mechanism is a common mode of synthesis, in which gas-phase reactants saturate a liquid metal catalyst and precipitate underneath to form a crystalline solid. When grown on [111]-oriented substrates, nanowires tend to align vertically, such that they are unclad, free-standing structures with a lateral dimension determined primarily by the size of the catalyst metal droplet. The multitude of parameters involved in any synthesis method has been optimized, including the VLS growth of InP nanowires. One striking feature observed in the nanowire growth is their ability to crystallize in the wurtzite structure, much like the GaAs nanoneedles. Although bulk InP always occurs in a zincblende crystal structure, both zincblende and wurtzite crystal phases have proven to be stable forms for nanowires made of InP and other typically zincblende III–V materials, indeed sometimes within the same nanowire structure. This presents the unexpected challenge of synthesizing nanowires with a uniform, single crystallinity, as well as the unexpected opportunity to investigate basic material properties of the new crystal forms. For example, a substantial blue-shift of the emission energy of

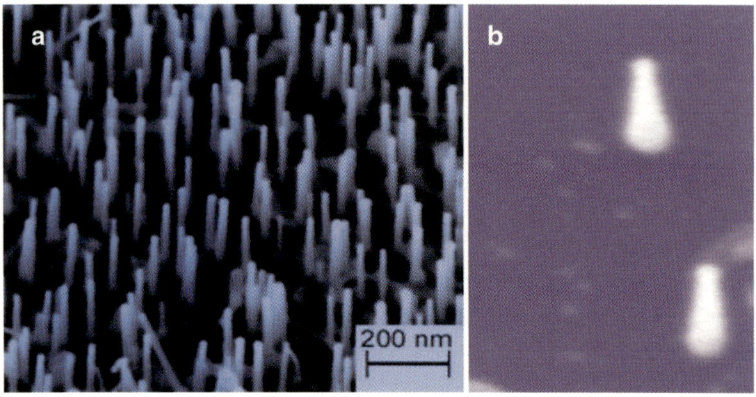

Fig. 5.9 InP nanowires (Crankshaw 2009): (**a**) quantum-confined, having an average diameter of ~16 nm; (**b**) the incorporation of an axial heterostructure within the nanowire (~300 nm in length, and ~20 nm in diameter)

wurtzite InP compared to zincblende InP have been investigated. As shown in Fig. 5.9a, InP nanowires are quantum-confined, having an average diameter of ~16 nm. InP nanowires were grown on (111) silicon substrates via the VLS growth mechanism in a MOCVD reactor. The incorporation of an axial heterostructure within the nanowire (NW)—an $InAs_xP_{1-x}$ section with InP barriers—which is rather unique in that the heterostructures in this material system are at NW diameters small enough to show quantum confinement. For this sample, the wires are ~300 nm in length, and ~20 nm in diameter, as shown in the scanning electron micrograph in Fig. 5.9b. With higher As fractions to push the wavelength into the 1.2–1.55 μm range, these structures could be useful for silicon photonics and optical telecommunications applications (Crankshaw 2009).

5.6 Summary

Semiconductor waveguides are generally made from elements of III–V, II–VI, or IV–VI, which have been mostly used for optoelectronic device applications, such as GaAs and InP. Compared to Si and Ge, GaAs and InP have high electron mobility and velocities, properties that are extremely important for the development of high speed electronic devices. Their direct band gaps and therefore their high radiative efficiency make them important for optoelectronic materials. An attractive feature of the binary compounds is that they can be combined or alloyed to form ternary or quaternary compounds, or mixed crystals. These compounds are made up of three or four group III and group V atoms and by choosing different binary compounds, it is possible to select different band gaps and therefore varying emission energies for light sources. A wide range of compound semiconductors can be exploited to access a wide range of output wavelengths. However, by alloying it is possible to vary the band gap continuously and monotonically and together with it the band structure, electronic, and optical properties. The formation of ternary and quaternary compounds of varying band gaps also enables the formation of heterojunctions, which have become essential for the design of high performance electronic and optoelectronic devices.

Silicon waveguides have been primarily operated at wavelengths in the NIR, typically around 1.4–1.6 μm. This has been convenient due to the large number of commercial optical components available in this regime. The silicon is usually patterned with sub-micrometer precision, into microphotonic components. These operate in the infrared, most commonly at the 1.55 μm wavelength used by most fiber-optic telecommunication systems. The silicon typically lies on top of a layer of silica known as SOI. Moreover, porous silicon waveguides provide an interesting alternative to conventional SOI structures that could offer advantages for silicon interconnect waveguides, modulators and components. Porous silicon is a nanoscale composite material with an effective refractive index between that of air and silicon that can easily be fabricated in multilayer structures. GaAs/AlGaAs waveguides provide interesting optical properties for nonlinear optical signal processing. The optical Kerr coefficient of GaAs is approximately 4 times larger than that of crystalline silicon and three orders of magnitude larger than that of standard optical fiber. The key to realize on-chip integrated optical devices is working with a low-loss, highly nonlinear material system that allows the creation of both active and passive photonic devices. With quantum-well technology, III–V group semiconductors, such as GaAs and its alloys, can be used to create low-loss waveguides with very high second and third-order nonlinearities, or direct band gaps that enable the fabrication of bright lasing sources.

Many lanthanide elements have played an important role in various optoelectronic and photonic applications, ranging from emitting elements in solid-state lasers and in phosphors for color lamps and displays to optical waveguides. These rare earth (RE) elements have a partially filled inner ($4f$) shell shielded from its surroundings by completely filled outer ($5s$ and $5p$) orbitals. Due to this shielding, the intra $4f$ shell transitions result in very sharp optical emissions at wavelengths from the ultraviolet to the infrared (IR). In general, RE-doped conventional semiconductors (Si, GaAs, etc.) have exhibited limited photoemission at room temperature due low RE solubility and severe temperature quenching. The thermal quenching in Er-doped semiconductors decreases with increasing band gap. Therefore, WBGS are attractive hosts for RE elements.

Semiconductor nanowires have attracted great interest both for basic studies of one-dimensional physics and as building blocks for nanoscale devices. Many III–V and II–VI material systems have been developed with various growth modes, such as laser-assisted catalytic growth and chemical beam epitaxy, as well as some catalyst-free methods such as selective-area epitaxy.

References

Agrawal GP (1995) Nonlinear fiber optics, 2nd edn. Academic, San Diego

Apiratikul P (2009) Semiconductor waveguides for nonlinear optical signal processing. PhD dissertation, University of Maryland, College Park

Bristow AD et al (2007) Two-photon absorption and Kerr coefficients of silicon for 850–2,200 nmi (4,100 km). Appl Phys Lett 90(2):191104

Celler GK, Cristoloveanu S (2003) Frontiers of silicon-on-insulator. J Appl Phys 93(9):4955

Crankshaw SM (2009) Spectroscopic characterization of III-V semiconductor nanomaterials. PhD dissertation, University of California, Berkeley

Dekker R, Usechak N, Först M, Driessen A (2008) Ultrafast nonlinear all-optical processes in silicon-on-insulator waveguides. J Phys D 40:R249–R271

Deri RJ, Kapon E (1991) Low-loss III-V semiconductor optical waveguides. IEEE J Quantum Electron 27(3):626–640

Dimitropoulos D et al (2005) Lifetime of photogenerated carriers in silicon-on-insulator rib waveguides. Appl Phys Lett 86:071115

Ganguly P, Rahul S (2011) Silicon photonics: a solution for ultrahigh speed data transfer. ACEEE Int J Commun 2(3):26–31

GoldenMap (2013) Superlattice. http://en.goldenmap.com/Superlattice#. Accessed 25 Feb 2013

Hoogland S (2013) The fuss about quantum dots. http://www.photonics.com/Article.aspx?AID=31908. Accessed 23 Feb 2013

Huang M (2006) Erbium-doped wet oxides of AlGaAs and InAlP on GaAs substrates for optoelectronic integration. PhD dissertation, University of Notre Dame, Notre Dame

Jones R et al (2005) Net continuous wave optical gain in a low loss silicon-on-insulator waveguide by stimulated Raman scattering. Opt Express 13(2):519–525

Jussila H (2010) Fabrication of indium arsenide quantum dot structure for semiconductor optical amplifiers. Master thesis, AALTO University, Aalto-yliopisto Teknillinen Korkeakoulu, Espoo

Knights AP, Jessop PE (2007) Silicon waveguides for integrated optics. In: Calvo ML, Lakshminarayanan V (eds) Optical waveguides: from theory to applied technologies. CRC Press, Boca Raton, pp 231–270

Koos C et al (2007) Nonlinear silicon-on-insulator waveguides for all-optical signal processing. Opt Express 15(10):5976–5990

Narasimha A et al (2008) A 40-Gb/s QSFP optoelectronic transceiver in a 0.13 μm CMOS silicon-on-insulator technology. In: Proceedings of the optical fiber communication conference (OFC), OMK7

Nicolae CP et al (2006) Modulation instability in silicon photonic nanowires. Opt Lett 31 (24):3609

NRL (2013) Nonlinear optics in semiconductor waveguides. http://www.nrl.navy.mil/photonics/5654/nonlinearOpticsWaveguides.php. Accessed 26 Feb 2013

Shin J, Chang YC, Dagli N (2009) Propagation loss study of very compact GaAs/AlGaAs substrate removed waveguides. Opt Express 17(5):3390–3395

Sigal I (2010) Three wave mixing in periodically quantum-well-intermixed GaAs: AlGaAs superlattices: modeling, optimization, and parametric generation. Master thesis, University of Toronto, Toronto

Simatupang JW (2009) Vertical InGaAsP/InP taper FP-LD for injection-locking applications. Master thesis, National Taiwan University of Science and Technology, Taipei

Spott A et al (2010) Silicon waveguides and ring resonators at 5.5 μm. Appl Phys Lett 97:213501 (1–3)

Steckl AJ et al (2002) Rare-earth-doped GaN: growth, properties, and fabrication of electroluminescent devices. IEEE J Sel Top Quantum Electron 8(4):749–766

Stievater TH et al (2008) Low-loss suspended quantum well waveguides. Opt Express 16 (4):2621–2627

Vurgaftman I et al (2001) Band parameters for III-V compound semiconductors and their alloys. J Appl Phys 89(11):5815–5875
Younis U (2010) Monolithic integration for nonlinear optical frequency conversion in semiconductor waveguides. PhD dissertation, University of Glasgow, Scotland
Zhang J et al (2007) Optical solitons in a silicon waveguide. Opt Express 15(12):7682–7688

Chapter 6
Silicon-on-Insulator Waveguides

Abstract While developed for the needs of microelectronics, the silicon-on-insulator (SOI) wafers are excellent substrates for optical waveguides. SOI is a kind of structures formed by a thin layer of crystalline silicon (Si) on an insulating layer, which typically is silicon dioxide (SiO_2). SOI optical waveguides possess unique optical properties due to the high transparency of silicon in the infrared spectrum and the large refractive index difference between silicon (guiding layer or core, $n = 3.45$) and SiO_2 (insulator layer or cladding, $n = 1.46$). This high difference in indices of refraction strongly confines the electromagnetic field into the silicon layer. The widely used SOI waveguides may take the form of a channel waveguide, ridge waveguide, photonic-crystal waveguide, or slot waveguide. The photonic-crystal waveguide is an exceptional option for making SOI waveguides. The refractive indices of different areas of the cladding can be flexibly engineered by varying the diameter of the holes and the lattice constants. These excellent optical properties, as well as compatibility with silicon complimentary-metal-oxide semiconductor (CMOS) integrated technology, enable low-cost and dense optoelectronic integrated circuits. In fact, SOI material has become a main platform for both photonics and VLSI CMOS electronics, with fully compatible processing procedures. This chapter will give a brief review about the principle design, materials selection, and fabrication process of the SOI waveguides.

6.1 Silicon Photonics

Silicon photonics represents a combination of existing silicon manufacturing capability with the advantages of optical interconnects, which claims the implementation of an efficient and low loss platform at low cost. The resulting dual-function technology seems to be a viable approach to extending the limits imposed by conventional metal interconnects to high speed communications in modern computing systems, while the development of this technology opens the possibility of new applications in many different areas (Huante-Ceron 2011).

X.C. Tong, *Advanced Materials for Integrated Optical Waveguides*, Springer Series in Advanced Microelectronics 46, DOI 10.1007/978-3-319-01550-7_6,

Silicon has excellent material properties important in photonics devices, such as high thermal conductivity, high optical damage threshold, a high quality and stable oxide, a low-loss wavelength window extending from 1.1 μm to nearly 7 μm, and high third-order optical nonlinearities. In addition, silicon wafers have the lowest cost and the highest crystal quality of any semiconductor material. Nevertheless, silicon has some basic material drawbacks. It is an inefficient light generator; the electro-optic effect is nonexistent and there is no conventional photodetection in the region 1.3–1.6 μm (communication band). Silicon photonic components employ the optical properties of crystalline silicon, a group IV dielectric material. The impetus for silicon photonics came from optoelectronics integration of optics and electronics on the same chip, since optoelectronic integrated circuits (OEIC) exhibit better performances compared with separated optical and electrical chips. Silicon-based optoelectronic systems have the potential to incorporate with silicon electronics and to achieve high integration densities (Deng 2005).

With the advances in fabricating silicon electronic devices, it is now possible to fabricate silicon photonic devices using the same complimentary-metal-oxide semiconductor (CMOS) technology, especially with the help of the mature silicon-on-insulator (SOI) wafer technique and nano-scale photolithography. The fact that silicon photonics is truly CMOS compatible, and that silicon is transparent in the wide spectral regions extending from near to mid-infrared, make it very promising for making passive and active optoelectronic components (Yin 2009). The significant importance of silicon in optoelectronics comes from the following factors (Deng 2005): (a) SOI material has become a main platform for both photonics and VLSI CMOS electronics, with fully compatible processing procedures; (b) Si-based OEIC have economic advantages; (c) compared with III-V semiconductors, silicon has better crystal perfection, better native oxide, and superior thermal and mechanical properties; (d) SOI CMOS circuits have reduced parasitics and latch-up that enable high speed at low power; and (e) strong optical-confinement of SOI waveguides enables low-loss and compact device-size.

Comparing with fused silica, silicon is promising for making nonlinear optical devices. Silicon is transparent in the spectral region beyond 1.1 mm up to 6 mm. The refractive index of silicon (around 3.5) is much larger than that of fused silica ($n = 1.45$). This implies a much stronger light confinement inside SOI waveguides, which is beneficial for both nonlinear light interactions and for controlling the size of optical devices. The nonlinear refractive index of silicon is about 200 times larger than that of silica. The Raman gain coefficient of silicon is about 3,000 times larger than that of silica, and is strongly polarization dependent. However, comparing with silica, silicon has some additional complications, such as two-photon absorption (TPA), free-carrier absorption (FCA), and free-carrier-induced change in the refractive index. Also, the polarization properties of silicon are different from silica because of its lattice structure (Yin 2009).

In practice, most optoelectronic devices have employed III-V semiconductor materials such as AlGaAs/GaAs and InGaAsP/InP, because the indirect band gap and low carrier mobility in silicon were once obvious obstacles to optoelectronic applications. However, there are now methods to surmount these problems.

Modulation doping enhances higher mobilities; optical and electrical properties can be transformed by adopting heterostructures or by tetragonal distortion of the lattice due to coherent strain. Emitting, guiding, detecting, modulating, and switching infrared light have been realized in silicon. Waveguide structures have been demonstrated in $Si/Si_{1-x}Ge_x$, silicon-on-silicon (SOS), SOI and silicon-germanium-on-silicon (SGOS) materials. Therefore it is reasonable to foresee more silicon-based photonic components since the performance of Si-based OEIC could eventually surpass that of III-V OEIC in some areas. Additionally, silicon photonics is expected to influence other areas of technology like the sensing industry, biomedical applications, telecommunications, and consumer electronics. The biggest challenge for silicon photonics is the manufacture of efficient and low-cost information processing components. Ultimately, the biggest limitation on the success of the technology might not be the laws of physics but those of economics. Therefore, silicon photonics is developed within the protocols of a manufacturing technology based on compatibility with the mature Complementary-Metal-Oxide Semiconductor (CMOS) process and the availability of high-quality SOI (Deng 2005).

6.2 Silicon-on-Insulator Materials

SOI is most commonly described as a thin layer of silicon (typically hundreds of Å to a few microns thick) on top of an insulating material with an underlying bulk silicon substrate. The most common insulating material currently is silicon dioxide. Early SOI materials, such as silicon-on-sapphire (SOS), were mainly used in niche markets including space exploration and high temperature environments. In the late 1970s, separation by implanted oxygen (SIMOX) was developed and has become the most mature of all commercially available SOI materials. Since 1990s advanced techniques have been developed for fabrication of SOI materials, such as the Smart-cut® and Nanocleave® methods (Saavedra 2004). There are many sophisticated techniques to fabricate low-defect SOI wafers. Other methods include silicon nitride, and bond-and-etch-back SOI (BESOI). These methods enable SOI wafers to be widely used for commercial high-speed CMOS and DRAM chips. Moreover, SOI has been applied to many guided-wave optical devices and circuits, such as high-speed modulators, photodetectors, directional couplers, Mach–Zehnder Interferometer (MZI) switches, Corner-Mirrors and T-Branches, Y-Branches, Bragg gratings, star couplers, multi-mode interference (MMI) coupler, and arrayed-waveguide gratings (AWG).

6.2.1 Silicon-on-Silica

The most mature of all SOI materials have been the separation by implantation oxygen (SIMOX) process. It is basically the implantation of high doses ($>10^{18}$ cm^{-2}) of oxygen ions below the surface of a silicon wafer. In order to prevent an amorphous silicon over-layer from forming during implantation the wafer is kept at high temperatures (600 °C) during implantation. The thickness of the silicon over-layer is determined by the energy used for oxygen implantation since that determines the depth of the SiO_2. Due to the high oxygen doses, the peak concentration of the implantation profile saturates, thus producing a uniform concentration that results in a complete buried SiO_2 layer. After implantation the wafer is annealed approximately at 1,300 °C for several hours. This annealing process removes implantation-related defects from the silicon over-layer and further defines the buried SiO_2 layer. The silicon over-layer thickness may be increased subsequent to ion implantation via epitaxial growth (Gili de Villasante 2010).

Oxygen ion implantation was first used for the synthesis of silicon oxide in the late 1960s. However, it was not until the late 1970s that the SIMOX process was actually developed.

In order to form a buried insulator using conventional ion implantation took over 2.5 days to implant a dose of 1.2×10^{18} cm^{-2} with a beam current of 100 mA. The development of a high current oxygen implanter significantly reduced the processing time devoted to the implant step. The quality of the surface Si layer was enhanced by the use of lower doses and higher post-implant annealing temperatures. The schematic SIMOX process and its advantages and disadvantages are shown in Fig. 6.1a. The SIMOX process basically consists of three steps (Saavedra 2004): (a) high dose oxygen implantation at an elevated temperature (>500 °C). This is done in order to prevent complete amorphization of the surface Si layer. Creation of an amorphous surface Si layer would be disastrous because it would be impossible to recrystallize off an amorphous buried oxide (BOX). Typical implant energies and doses would be 180 keV, 1.8×10^{18} cm^{-2} for standard SIMOX (180 nm surface Si/400 nm BOX), or 30 keV, 1×10^{17} cm^{-2} for low dose SIMOX (57 nm surface Si/47 nm BOX). (b) High temperature annealing step at 1,300 °C for 6 h. The annealing step is necessary in order to synthesize a box shaped buried SiO_x layer. (c) Polishing in order to remove oxidation at the surface as a result of the high temperature annealing. Despite the high temperature annealing, the BOX properties vary from thermally grown SiO_2 due to the presence of silicon islands, stoichiometry, increased oxygen concentration in surface Si, etc. For this reason, other methods, such as the Smart-cut® and Nanocleave® processes have been developed. In addition, the internal oxidation of Si (ITOX) process have been used to improve the dielectric properties of the surface Si/BOX interface in SIMOX materials, as well as eliminate many of the silicon islands.

While the SOI is with the development of the SIMOX process, higher quality SOI materials could be made by utilizing thermal oxidation followed by wafer bonding. Early attempts, such as bonded and etch-back SOI (BESOI), could not overcome the problem of material waste. Initially, two silicon wafers are thermally oxidized. The two oxidized surfaces come into contact at room temperature to form a chemical bond which is subsequently strengthened by a thermal process at a temperature of approximately 1,100 °C. The thin silicon over-layer is formed by controllably removing one of the bonded wafers, typically, via Chemical Mechanical Polishing (CMP). In the early 1990s, the Smart-cut® process was invented. The schematic Smart-cut® process

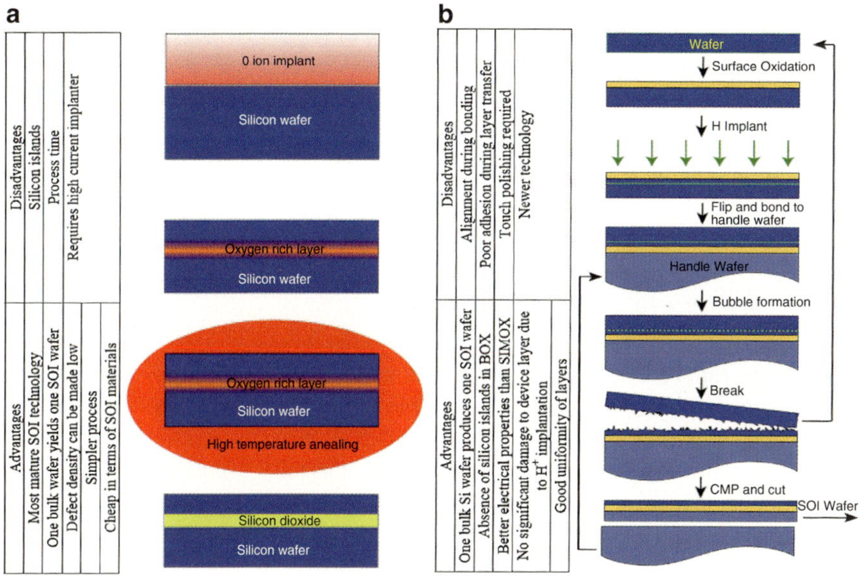

Fig. 6.1 Fabrication of SOI platform (Saavedra 2004; Gili de Villasante 2010): (**a**) SIMOX process and its advantages and disadvantages; (**b**) Smart-cut process and its advantages and disadvantages

and its advantages and disadvantages are shown in Fig. 6.1b. This technique combines steps of both SIMOX and BESOI procedures (Saavedra 2004; Gili de Villasante 2010): It utilizes two wafers, but results in one final SOI wafer and another bulk silicon wafer, which may be reused. The first step involves thermal oxidation of one wafer, which will later provide the BOX of the SOI wafer. Next, hydrogen ion implantation is performed through the BOX to the surface silicon thickness desired. The implant energy can be tailored to dictate the thickness of the surface Si layer, while the dose is around $5 \times 10^{16}\,cm^{-2}$. The depth of the implantation peak depends on the energy used but generally is of the order of hundreds of nanometers to a few microns. Silicon lattice bonds are weakened as a result of the damage induced by the implantation coupled with the presence of hydrogen. The two wafers are then bonded together at room temperature via van der Waals forces at the two surfaces. A low temperature anneal (~600–700 °C) is then performed in order for the implanted hydrogen to coalesce into micro bubbles. As the micro bubbles grow, the pressure inside them increases until they fracture resulting in the splitting of the SOI wafer from the recyclable wafer. At room temperature, a second wafer (with or without a thermally oxidized surface) is brought into contact with the implanted wafer, and the two weakly bond together. A second anneal is performed at 1,100 °C for 2 h to strengthen the bond at the bonded interface. A final polishing step is required in order to smooth the surface after splitting. Usually a fine CMP is used to reduce the surface roughness of the silicon over-layer. If required, the silicon over-layer thickness can be increased by epitaxial growth. The Smart-cut® process provides significant flexibility in final wafer dimension, coupled with high quality material. It is currently the most popular method for SOI formation (Celler and Cristoloveanu 2003).

For optical wavelengths in the range of 1.3–1.55 μm and silicon guiding layer thicknesses of a few microns, the thickness of the BOX layer must be approximately 1 μm in order to prevent the evanescent field of the propagating signal from penetrating into the silicon substrate.

6.2.2 *Silicon-on-Sapphire*

Silicon-on-sapphire (SOS) is made with a hetero-epitaxial process for integrated circuit manufacturing that consists of a thin layer (typically thinner than 0.6 μm) of silicon grown on a sapphire (Al_2O_3) wafer. Typically, high-purity artificially grown sapphire crystals are used. The silicon is usually deposited by the decomposition of silane gas (SiH_4) on heated sapphire substrates. The advantage of sapphire is that it is an excellent electrical insulator, preventing stray currents caused by radiation from spreading to nearby circuit elements. SOS faced early challenges in commercial manufacturing because of difficulties in fabricating the very small transistors used in modern high-density applications. This is because the SOS process results in the formation of dislocations, twinning and stacking faults from crystal lattice disparities between the sapphire and silicon. Additionally, there is some aluminum, a P-type dopant, contamination from the substrate in the silicon closest to the interface. SOS was first used in aerospace and military applications because of its inherent resistance to radiation. More recently, advancements in SOS processing and design have been made, allowing SOS to be commercialized in high volume for high-performance radio-frequency (RF) applications (Wikipedia 2013).

Optical waveguides have been made using the silicon-on-sapphire (SOS) materials system, as shown in Fig. 6.2, SOS is particularly desirable for this application because of the lack of a high index substrate, which eliminates the issue of substrate leakage. The resistivity of the silicon for the wafers used is about 100 Ω-cm, suggesting that optical loss due to free carriers can be minimal. Mid-infrared (MIR) guides at longer wavelengths could also be built using free-standing silicon guides or germanium-on-silicon. SOS, in particular, has the advantage of offering the ability to build high-confinement, fully etched waveguides from 1.1 μm all the way to around 6.2 μm—over two octaves of bandwidth, including the telecommunications region, while maintaining electronics compatibility. While SOI has been used to build waveguides at 2.2 μm, substrate leakage will become an increasingly large problem at longer wavelengths. Another problem is that silica becomes extremely lossy at wavelengths longer than 4 μm. The SOS waveguides have been developed to overcome these problems. The waveguides can be fabricated using standard semiconductor processing techniques. Epitaxial

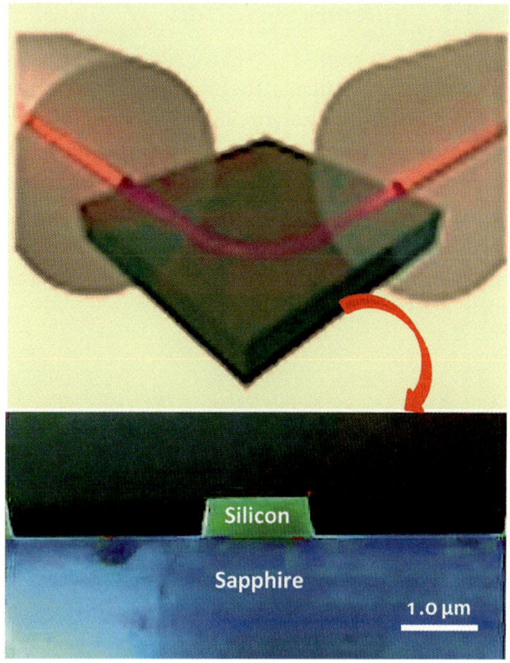

Fig. 6.2 Illustration of silicon-on-sapphire (SOS) waveguide (Baehr-Jones et al. 2010)

silicon-on-sapphire wafers (100 mm diameter) are used as starting material. Patterning is accomplished using an electron-beam lithography system and HSQ resist on a wafer fragment. The resist is developed and the chip is etched using CF_4 plasma. Resist is first removed using just a wet resist remover, although some resist residue appeared to remain on the surface of the waveguides due to incomplete stripping. Further processing then involved cleaning the chip with a piranha etch, which appeared to improve waveguide loss dramatically. It is likely that the piranha etch removed some remnants of the resist. The chip is finally cleaved through the 8 μm wide waveguide segment, leaving an optical quality edge. With such fabricated low-loss waveguides, it will be possible to build integrated MIR lasers and detectors using techniques such as wafer bonding and selective-area growth, and to construct a wide variety of further devices within the silicon platform. It is also possible to build high-confinement integrated nonlinear optical devices, such as integrated OPOs and difference frequency generators, because these wavelengths are so long that more than several photons are required to reach the silicon band gap energy. Another interesting opportunity emerges from the limits of lithography: A 20 nm trench represents a small fraction of a wavelength at 1.55 μm, but represents a significantly smaller fraction for light at 4.5 μm; MIR waveguides may be an ideal playground for exploring ideas in ultrasubwavelength photonics. In the long run, integrated MIR optical systems could be much smaller and cheaper than contemporary systems (Baehr-Jones et al. 2010).

6.2.3 Silicon-on-Nitride

The silicon-on-nitride (SON) waveguide exhibits an ultrabroad band (4,200 nm), low chromatic dispersion in the MIR wavelength region from 2,430 to 6,630 nm. It has two zero-dispersion wavelengths within the span. Even at 6, the nonlinear coefficients of the SON waveguides are still comparable with the ones of integrated waveguides around 1,550 nm, which is widely used for octave-spanning nonlinear process. This enables a potential nonlinear optical platform for broadband signal processing across the over-one-octave MIR bandwidth (Yue et al. 2012).

6.2.4 Other Perspective Materials

Si-based waveguide structures are constructed mainly in the shape of ridges (ribs), strips (photonic wires), rings, or disks: N+/N Si, P+/P Si, silica (with a doped core) upon silicon, SOI, silicon oxynitride on SiO_2 on Si, Si_3N_4 on SiO_2 on Si, nanocrystalline Si within SiO_2 on Si, polycrystalline Si on Si, nanocrystalline Si in SiO_2 on Si, β-SiC on SiO_2 on Si (SiCOI and SICOI), silicon-on-sapphire, low porosity porous-Si upon high-porosity porous-Si, amorphous hydrogenated Si on Si, anti-reflecting resonant optical waveguides upon Si, amorphous hydrogenated SiC on ZnO on Si, low-index core (SiO_2 or air) "hollow" guides with photonic-crystal cladding, double SOI ($Si/SiO_2/Si/SiO_2/Si$), SiGe upon Si, SiGeC upon Si, α-SiC:H/α-Si:H/Si, silicon line-defect in a Si-membrane photonic-band gap lattice, self-collimated waveguide in a dispersion-engineered Si-membrane photonic crystal, electro-optic polymer on low density polymer on Si, $BaTiO_3$ on MgO on Si, $LiNbO_3$ epitaxy on Si, metal-dot plasmonic array on Si, silicon upon a buried layer of silicide ($CoSi_2$ or $NiSi_2$) on silicon, SiGe/Si surface plasmon waveguide gold-coated on top and bottom, SiGe/Si multi-quantum well stack grown strain balanced upon an SiGe buffer on Si, deep, nanometer-diameter holes in a dielectric that are coated on the inside with Si or Ge. The perspective waveguide structures may include (Soref 2005): (1) layered heterostructures such as GeSn/Ge/SiGeSn on Si, or SiGeSn/GeSn/SiGeSn on Si, which can be asymmetrically strained or (preferably) coherently strained strain-balanced alloy-heterostructure waveguides wherein the waveguide core alloy has a higher index than that of the buffer-on-Si or of the Si substrate; (2) a layered multi-quantum well (MQW) alloy heterostructure of GeSn/Ge on SiGeSn/Si or SiGeSn/GeSn on SiGeSn/Si; (3) quantum-confined SOI having a top Si layer less than 10 nm thick; (4) plasmonic waveguides with a metal dot or metal strip array buried beneath the surface of an Si wafer; (5) tensile-strained-silicon-on-insulator (s-SOI); (6) silicon-germanium-on-insulator (SGOI); (7) germanium-on-insulator

(GOI); (8) silicon-on-epitaxial-insulator (SOEI) such as crystal CeO_2 or HfO_2 grown epitaxially upon a silicon wafer with a layer of c-Si then grown upon the oxide layer; (9) N-layer SOI offering N independent Si levels for guided light transmission together with "optical vias" for vertical interlevel optical communication; and (10) a 3-dimensional Si photonic crystal offering sub-surface self-collimated waveguides and interlevel communications within a 3-dimensional silicon photonic crystal.

Plasmon optics is a new metallo-dielectric technology in which undoped silicon can serve as the "insulator" or "dielectric" in a composite photonic structure. Generally, photonic materials can be categorized by the sign of their dielectric constant ε and the sign of their permeability μ. A metal has $(-\varepsilon, +\mu)$ and is "singly negative," whereas silicon $(+\varepsilon, +\mu)$ is doubly positive. In the future, artificial or man-made materials with $(-\varepsilon, -\mu)$ and $(+\varepsilon, -\mu)$ will likely be synthesized. The double and triple combinations of these four materials-types (composite photonic structures that include silicon), such as M-I-M or I-M-I will prove to have unique and valuable optical properties. Microphotonics refers to micrometer-scale devices. As to the wavelength scale, the minimum width of a silicon microphotonic strip waveguide (fundamental mode only) is approximately λ/n where λ is the wavelength in vacuum and n is the core refractive index, while the interguide separation in a coupler is approximately $\lambda/2n$, the minimum feature size in microphotonics. Photonic components with nanometer-scale features—nanophotonics—can be fabricated in several ways; for example, using quantum wires or quantum dots. A photonic-crystal device is the prime example of nanophotonics and here the 2D dielectric lattice of holes or posts (or a 3D porous dielectric crystal) has a periodicity of order λ/n with a pore diameter ~λ/n. The photonic crystal feature sizes are thus comparable to the microstrip sizes. The Si-based optical-network chip of the future will include integration of individual microphotonic, nanophotonic, and plasmon-optic components along with composite or hybrid components. In other words, a particular component may unite the micro, nano, plasmo techniques in some combination. A variety of interconnected photonic components comprising the network will be feasible. This micro/nano/plasma "convergence" is an emerging trend (Soref 2005).

6.3 Silicon-on-Insulator Technology

The need for a faster and more efficient CMOS device and integrated photonic circuits has led to the insurgence of SOI. As shown in Fig. 6.3 (Chen 2012), SOI technology has relied heavily on the ability of ion beams to create shallow, abrupt as-implanted doping profiles in bulk Si. Implant thermal processing must be performed in order for the dopant atoms to occupy substitutional lattice sites and contribute electrically. This needs to deal with fundamentals of ion implantation, atomistic diffusion, and segregation in the proximity of interfaces.

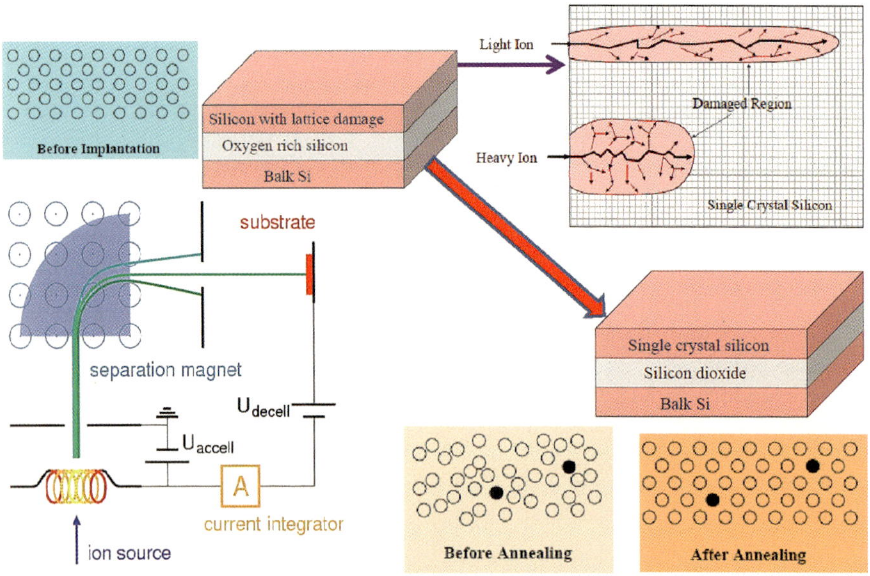

Fig. 6.3 Illustration of SOI formation through oxygen-ion implantation and high temperature annealing for a precipitation of silicon dioxide (Chen 2012)

6.3.1 Ion Implantation and Damage Recovery

Ion implantation was first proposed as a method for fabricating doped regions within semiconductor devices. Ever since, it has been the preferred technique for introducing dopants into silicon in controlled amounts. It offers a number of advantages over gas and solid source diffusion, such as easy introduction of desired impurity into target, good mass-charge separation, accurate dose control, not confined to surface, not limited by solid solubility, reproducibility of impurity profiles, lower process temperature, and ability to tailor doping profiles. Ion implantation is a very versatile process that has also been used for introducing gettering layers, synthesis of compounds, and surface modification of metals and polymers. Unfortunately, the main drawback of using ion implantation is resulting damage to the target, consisting mainly of point defects. For silicon, this damage can lead to the deleterious effects of transient enhanced diffusion (TED), dopant-defect clustering, as well as leakage current within the depletion region of transistors (Saavedra 2004).

The process of ion implantation is a highly nonequilibrium one. Physical, chemical, and structural changes may occur when the ions become embedded in the target material. A number of elastic and inelastic effects also result, leaving the silicon in a damaged, metastable state.

Lattice atoms may be displaced from their equilibrium positions when the displacement energy (15 eV for Si) is exceeded. As an energetic ion comes to rest, it may undergo a number of collisions with atoms in the Si lattice, which leads to the production of a damage or collision cascade. This damage cascade consists of interstitials, vacancies, amorphous regions, ionized atoms, etc. These defects produced in the "as-implanted" state are typically referred to as primary defects. The density of a damage cascade depends significantly on the ion mass and stopping or energy-loss mechanism. The two primary stopping mechanisms are nuclear and electronic stopping. Nuclear stopping is characterized by a significant transfer of energy between the nuclei of the energetic ion and the nuclei of a lattice atom. This results in a very dense damage cascade. An ion that undergoes electronic stopping is decelerated by interacting with the electron cloud surrounding a lattice atom. Thus, the energy losses in electronic stopping are much less per stopping event and the collision cascade is less dense. In general, nuclear stopping is observed at lower implant energies and heavier ions (e.g., As, Sb), while electronic stopping occurs at higher energies and for low mass ions (e.g., B, H). Damage cascades are also affected by the ability of the ion to channel, which is related to the ion mass, as well as the crystal orientation relative to the incident ion beam. Channeling refers to the phenomenon whereby an ion is able to traverse great distances into the crystal by moving through the interstices present in the lattice. For this reason, {100} Si wafers are typically oriented relative to the beam direction with a tilt of 7° in the [110] direction followed by a rotation of 22° around the [100] direction. This aids in reducing channeling and improving the reproducibility of implanted profiles. Thin screen oxides and pre-amorphization are also common procedures for controlling channeling (Saavedra 2004).

A large increase in the excess interstitial population occurs after ion implantation. Secondary defects are formed during subsequent thermal processing and are nearly exclusively extrinsic, or interstitial, in nature. Post-implant annealing is always required since the majority of implanted dopant ions are not on substitutional lattice sites. These implanted dopants do not contribute electrically and are considered inactive. Upon annealing, the damaged silicon lattice tends toward a more equilibrium state. Excess interstitials undergo a number of evolutionary processes in order to reduce the free energy associated with the silicon lattice. These processes may be broadly classified into recombination and clustering. Recombination occurs when the strain field surrounding an interstitial interacts with that of a vacancy resulting in a mutual attraction and annihilation. Frenkel pairs, an interstitial and vacancy pair around a host lattice site, may recombine during implantation or annealing at temperatures <600 °C. Clustering of excess interstitials often results in the formation of submicroscopic interstitial clusters (SMICs), {311} defects, and dislocation loops in order to reduce the Gibbs free energy of the system. While the final annealed state may not be an entirely equilibrium one, it is desired that the final state does not significantly change with time at the temperature required for device operation. The location of the majority of excess interstitials depends on whether the implanted dose is sufficient to produce enough disorder in the silicon lattice (~10 %) to create an amorphous layer. This dictates where the secondary defects form relative to the surface. For non-amorphizing implants the majority of excess interstitials lie around the projected range of the implant. In the case of amorphization, the excess interstitials are found just beyond the amorphous-crystalline interface after regrowth of the amorphous layer has commenced. Secondary defects are particularly important because they drive the phenomenon of TED. They do this by maintaining an interstitial supersaturation until their eventual dissolution. Extended defects, such as {311} and loops, may also act as sources of leakage current in devices. This occurs because the strain field of the dislocation behaves as a mid-band gap recombination center (Kim et al. 1997; Saavedra 2004).

One kind of secondary defect is SMICs. SMICs, after the interstitial point defect, are the basis from which the microscopic extended defects evolve, although a structural transformation appears necessary. TED can occur in the absence of extended defects, providing a basis for the existence of SMICs. SMICs are either the precursors of the {311} or they compete as sinks for the self-interstitials. For ion-implanted Si, the rod-like {311} defect is commonly observed. Their elongation results due to the fact that self-interstitials can be added along the {110} cross-section without introducing dangling bonds. For this reason, these defects do not significantly change in width of ~4 nm. Below a threshold dose, however, {311} defects do not form and only SMICs might be present. This threshold appears to be between 7×10^{12} and 1×10^{13} cm^{-2} Si^{+} doses. These doses are easily reached in modern IC fabrication, so {311} defects will nearly always form. However, the misconception that {311} defects are the sole source of interstitials should be avoided. {311} defects can be made to dissolve upon annealing above 700 °C. Release of interstitials from {311} defects has been shown to follow an exponential relationship according to (Saavedra 2004)

$$\mathrm{Si_I} = \mathrm{Si}(0)\mathrm{e}^{-t/\tau} \tag{6.1}$$

where $\mathrm{Si_I}$ is the planar density of interstitials trapped in {311} facets, Si(0) is the pre-exponential, t is the anneal time, and τ is the time constant for dissolution. This time constant obeys first-order reaction kinetics to yield an activation energy for dissolution via the Arrhenius relationship (Saavedra 2004)

$$\tau = \tau(0)\mathrm{e}^{-E_\mathrm{a}/kT} \tag{6.2}$$

where $\tau(0)$ is the pre-exponential, E_a the activation energy for {311} dissolution, k Boltzmann's constant, and T the temperature in Kelvin. The value of E_a for {311} defects has been determined to be approximately 3.7 eV. {311} defects do not necessarily have to dissolve, though. {311} defects are the source of dislocation loops for non-amorphizing implants, which can either dissolve or undergo an unfaulting reaction to form dislocation loops. Dislocation loops can be stable at moderately high temperatures (750–850 °C) for hours. Another drawback of these defects is their ability to provide leakage current paths and degrade carrier lifetimes when lying across a junction. This is due to the introduction of localized energy levels that sit near the middle of the band gap in Si. On the other hand, a great advantage can be gained by using dislocation loops to getter out metallic impurities. By introducing a dislocation loop band well below the device junctions, any metallic impurities lying in the active regions will be attracted to the strain field introduced by the loop band. These defects always form under amorphizing implant conditions, assuming the implant energy is not ultralow energy, as well as non-amorphizing conditions when the dose is marginally high. Faulted Frank loops and perfect elongated loops are the two common dislocation loops observed in ion-implanted Si. They are both two-dimensional precipitates placed in between adjacent {111} planes of Si, but have different burger's vectors. The faulted loops have $b = a/3$ <111>, while the perfect loops have $b = a/2$ <110>. The planar density of interstitials is believed to be approximately the same, $1.566 \times 10^{15}\ \mathrm{cm}^{-2}$. Loops that form as a result of amorphizing implants are termed end-of-range (EOR), but bear no relation to the faulted or perfect loop; both types are observed as EOR loops. Compared to {311} defects, dislocation loops are much more thermally stable. They exhibit activation energy for dissolution in range of 5 eV. Another key difference is their ability to Ostwald ripen, allowing a larger dislocation loop to absorb the interstitials from a smaller one. Thus, the larger dislocation loops are more stable. The average size of {311} defects to increase as annealing proceeded. {311} should acquire an equilibrium shape at long times, but the aspect ratio should be constant during ripening. Dissolution of {311} depends on the ability of interstitials to hop off the ends of the defects, thus smaller {311} dissolve faster than larger ones, but not necessarily because they are more stable. Similarly, interstitials can only be added to the {311} by attaching to the ends of the defect, whereas in dislocation loops the interstitials may attach anywhere along the edge (Saavedra 2004).

6.3.2 *Dopant Diffusion in Bulk Silicon*

Dopants migrate through the Si lattice by interaction with point defects. Most dopants are dominated by diffusion of either interstitials or vacancies. B, P, and Ga, diffuse mainly through interaction with interstitials, while Sb is nearly a pure vacancy diffuser. Arsenic, on the other hand, diffuses by interactions with both types of point defects. Although point defects are efficiently created at room temperature, it does not approach the large supersaturation that can be created using ion implantation or other processes such as oxidation or nitridation. Under this case, a near equilibrium formulation for dopant diffusion can be introduced based on Fick's Laws of diffusion. Under these intrinsic and dilute dopant concentrations the flux of dopant A is expressed as (Saavedra 2004)

$$-J_{\mathrm{A}} = d_{\mathrm{AV}}\frac{\partial C_{\mathrm{AV}}}{\partial x} + d_{\mathrm{AI}}\frac{\partial C_{\mathrm{AI}}}{\partial x} \tag{6.3}$$

where d_{AV} and d_{AI} are the diffusivities associated with the particular defect complex, C_{AV} and C_{AI} are the concentrations of the particular complex, and x is the one-dimensional distance of interest. This is a form of Fick's first law of diffusion which states that a flux of the impurity will occur in the presence of a concentration gradient. It can be shown that Fick's second law can be applied to near equilibrium such that (Saavedra 2004)

$$\frac{\partial C_{\mathrm{A}}}{\partial t} = D_{\mathrm{A}} \frac{\partial^2 C_{\mathrm{A}}}{\partial x^2} \tag{6.4}$$

where C_{A} is the concentration of the dopant, t is time, and D_{A} is the equilibrium diffusivity of the dopant defined to be the sum of the equilibrium diffusivities of the A_{V} and A_{I} complexes. Basically, the change in concentration with time within a volume element is dependent on the difference in the flux of the impurity entering and leaving the volume element. Unfortunately, these formulations are highly idealized situations and are not applicable to the processes that take place during IC fabrication. This requires the use of equations that take into account the local point defect populations.

Nonequilibrium/enhanced diffusion can be used for dopants to diffuse rapidly under a supersaturation of point defects. TED is a type of enhanced diffusion that proceeds for a specified amount of time, as long as secondary defects are able to store interstitials. A number of processes that commonly occur during IC fabrication significantly alter the point defect populations, for instance, damage created by ion implantation can easily do this. Oxidation of the silicon surface is another method. Only a submonolayer of Si is required to form a monolayer of SiO_2 (ratio of ~1:2.25). This results in the injection of interstitials due to the net volume expansion difference between Si and SiO_2. Oxidation enhanced diffusion (OED) of B and P is observed, while oxidation retarded diffusion (ORD) is observed for Sb. The opposite phenomena occurs as a result of nitridation (NED) and silicidation (SED). NED is believed to occur as interstitials are swept toward the surface, where they react to form Si_xN_y leaving behind an excess of vacancies. In SED, vacancies are generated at the interface as silicon atoms are removed to react at the silicidemetal interface. Under these conditions, B and P diffusion is retarded, whereas Sb and As are enhanced (Saavedra 2004).

TED of B has a number of interesting characteristics and is affected by a number of implant and anneal parameters, including dose, energy, time, temperature, ramp rate, etc. Implant energy and dose tend to increase the amount of observed TED. However, saturation in the amount of TED occurs as the dose increases above approximately 1×10^{15} cm^{-2}. This is attributed to stable dislocation loop formations that act as strong sinks for excess interstitials. TED can be reduced by annealing at higher temperatures for shorter times, though. For this reason there has been considerable effort aimed at alternative annealing technologies such as rapid thermal annealing (RTA), laser thermal annealing (LTA), and FLASH lamp annealing. Each of these has specific advantages and disadvantages, but in general the goal is to obtain shallow junction depths by reducing TED, while at the same time enabling high dopant activation (Saavedra 2004).

Equilibrium solid solubility of boron is well above 1×10^{20} cm^{-3} at temperatures of 850 °C and higher. However, these levels of dopant activation are not obtained after annealing of ion-implanted boron in silicon, even after TED has ended. The reason for this is attributed to the formation of boron interstitial clusters (BICs) due to a high interstitial supersaturation. These defects consist of one or more boron atoms bound with one or more self-interstitials. In general, BICs are immobile and electrically inactive for the most part, although active BICs are believed to exist. Complete dopant activation cannot be obtained until they dissolve; this requires long periods at high temperatures. The boron concentration threshold for BIC formation appears to be between 1×10^{18} and 1×10^{19} cm^{-3}. The amount of clustering depends strongly on the separation between the region of high boron concentration and peak of the interstitial concentration. Thermal activation energy for BIC dissolution has been experimentally determined to be between 3.0 and 3.2 eV. In general, BICs with a high interstitial content (e.g., BI_2, B_3I_3, B_4I_4) form at early times when the interstitial supersaturation is greatest. As annealing proceeds, the BICs emit interstitials that can contribute to TED leading to BICs with a lower interstitial content. Therefore, the most stable configurations are those where $m < n$ for a B_nI_m cluster. When the BICs completely dissociate the immobile peak then dissolve out, but only long after TED has ended (Saavedra 2004).

After ion implantation, overall, interstitials may undergo a number of evolutionary stages that determine the effect they have on the diffusion of dopants. The presence of interfaces is unavoidable in microelectronic processing and their interactions with point defects and dopants must be accounted for in order for process simulators to provide accurate results. Several methods are currently available for fabricating SOI substrates, but the most prevalent are the Smart-cut and SIMOX methods, as discussed earlier. It is clear that the type and age of a SOI substrate may significantly alter the process physics and must be accounted for when fabricating SOI waveguides.

6.4 Silicon-on-Insulator Waveguide Structures

The use of optical waveguides is the most promising candidate to solve the challenges imposed by electrical interconnection, both for off-chip and possibly on-chip applications. The band gap of silicon (~1.1 eV) is such that the material is transparent to wavelengths commonly used for optical transport (around 1.3–1.6 μm). Standard CMOS processing techniques can be used to sculpt optical waveguides onto the silicon surface. Similar to an optical fiber, these optical waveguides can be used to confine and direct light as it passes through the silicon. Due to the wavelengths typically used for optical transmission and silicon's high index of refraction, the feature sizes needed for processing silicon waveguides are on the order of 0.5–1 μm. The typical configuration of some SOI waveguides is shown in Fig. 6.4 (Kishima 2007; Milosevic et al. 2009). A silicon dioxide layer is grown under the surface of the silicon wafer. The top silicon layer is then used as the guiding layer. The purpose of the BOX layer is to act as the lower cladding layer, and hence prevent the field associated with the optical modes from penetrating the silicon substrate below.

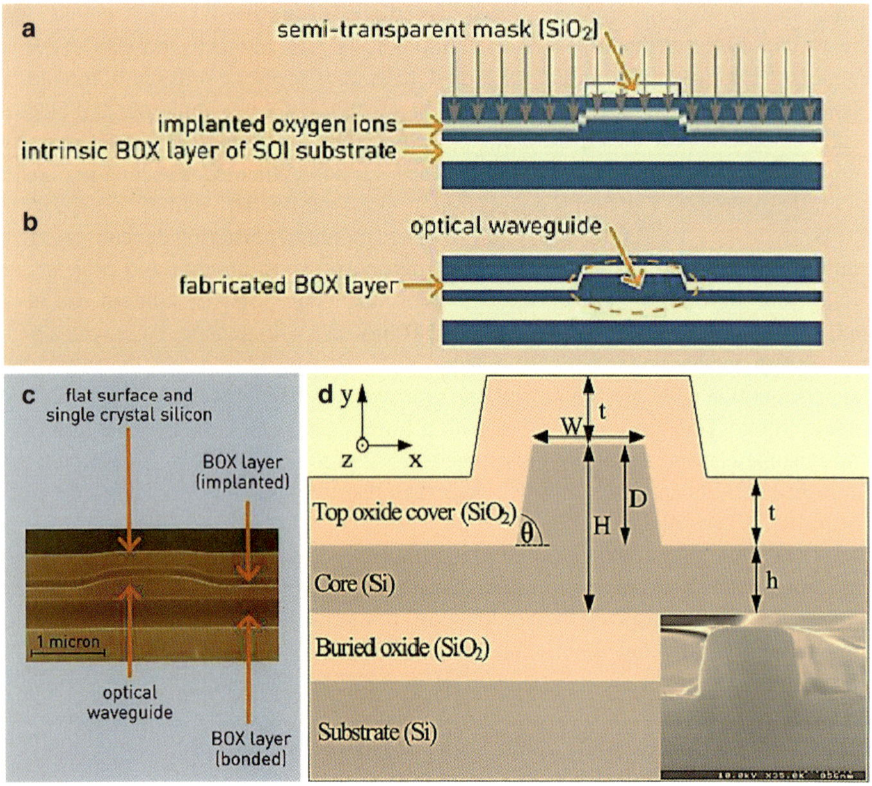

Fig. 6.4 Typical configuration of SOI waveguides (Kishima 2007; Milosevic et al. 2009): (**a**) oxygen-ion implantation through a mask into a bonded SOI wafer to form double-SOI waveguides; (**b**) fabricated BOX layer through precipitation of silicon oxide during high temperature annealing; (**c**) double-SOI waveguides (cross sectional view); and (**d**) rib waveguides

Therefore as long as the oxide is thicker than the evanescent fields associated with the modes it will be satisfactory. In some cases, the waveguide is transformed from an asymmetrical waveguide into a symmetrical one by the addition of a surface oxide layer. In practice, however, the refractive indices of both air ($n = 1$) and silicon dioxide ($n \sim 1.5$) are so different from that of silicon ($n \sim 3.5$) that the two configurations are very similar. For many applications, two-dimensional confinement is required. This is achieved in silicon by etching a two-dimensional waveguide. There are two main possible structures that can be etched in the silicon layer: (a) Rib or ridge waveguide structure when the dimensions of the core width range from 1 to 4 μm; and (b) strip waveguide structure with the dimensions of the core width to be in the order of hundreds of nanometers (Gili de Villasante 2010).

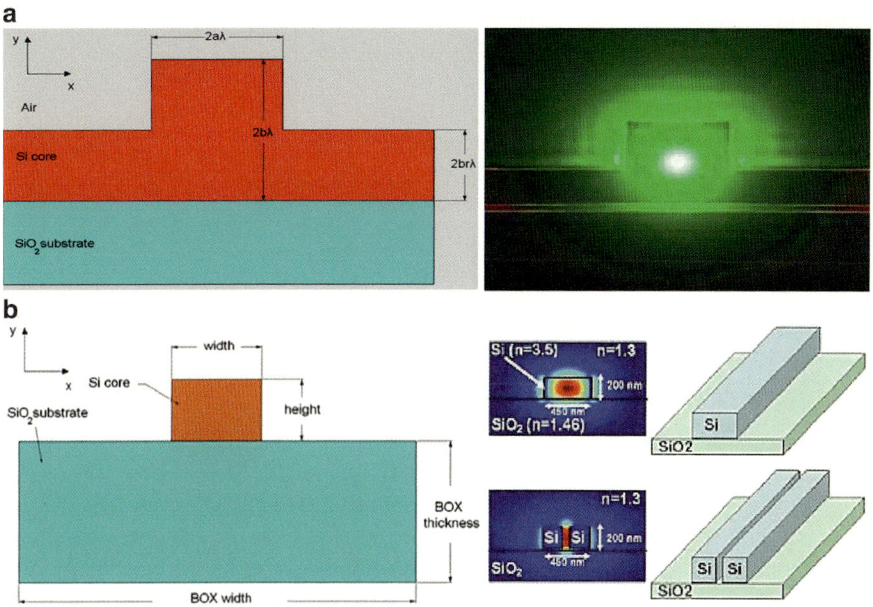

Fig. 6.5 Typical SOI waveguide structures (Gili de Villasante 2010): (**a**) large single-mode rib waveguides; and (**b**) strip nano-waveguides

6.4.1 Large Single-Mode Waveguides

Rib (or ridge) waveguides are composed of a core and two lateral slab regions etched on the silicon layer. The lateral confinement is achieved by means of the leakage of the higher-order modes in these lateral slab regions. The cross-section of a rib waveguide is shown in Fig. 6.5a.

The core has a width of $2a\lambda$ and a height of $2b\lambda$, while the lateral slab regions have a height of $2br\lambda$, and the parameter r is defined as the ratio of slab height to rib height, or outer-inner ratio.

This structure can be single mode when the next condition is true (in the limit of large b) (Gili de Villasante 2010):

$$\frac{a}{b} \leq c + \frac{r}{\sqrt{(1-r^2)}} \quad (\text{for } 0.5 \leq r \leq 1) \tag{6.5}$$

where a/b is the ratio between width and height, and r is the ratio of etch thickness and waveguide height, and c a constant to be equal to –0.05 (Pogossian et al. 1998). For $0.5 \leq r < 1$, higher-order modes in the vertical direction will be cut off because the higher-order modes in the central rib section will be coupled to the fundamental mode of the slab section, which becomes leaky for $r \geq 0.5$. This happens at $r \geq 0.5$ because the effective index of the fundamental slab mode becomes higher than the effective index of any higher-order vertical mode in the central rib region. The second-order modes in the vertical direction have a double-peaked intensity distribution along the vertical axis. For $r \geq 0.5$, one of the two peaks (the lobe near the bottom of the guide) will couple out into the fundamental slab mode of the rib side regions. This lateral leakage ensures that the first-order modes will not propagate. That is why the condition $0.5 \leq r < 1.0$ guarantees single-mode operation of the rib in the vertical direction. In this type of structure, it has to be assured that the SiO_2 layer thickness is thick enough. The dimensions of the waveguide (width, height, and slab height) are critical to define the single-mode region and have to be accurately designed.

6.4.2 *Strip Nano-Waveguides*

The perspective trend toward miniaturization is bringing the waveguide to dimensions in the order of several nanometers. In this case the strip structure is used, in order to achieve higher light confinement. Figure 6.5b shows the structure of strip nano-waveguides. The core width and height define the single-mode region, but the SiO_2 layer width and thickness have to be determined to obtain good coupling conditions, because the input field is launched in the cladding and then couples to the core. The single-mode condition is the following (Wang and Ho 2008):

$$H < 2.109\mathrm{e}^{-11.93w} + 0.7365\mathrm{e}^{-1.956w} \tag{6.6}$$

For a single polarization mode (TE or TM) the condition is modified to (Gili de Villasante 2010):

$$H < 63.13\mathrm{e}^{-23.92w} + 0.2684\mathrm{e}^{-0.85776w} \tag{6.7}$$

Many different designs and experimental approaches have been developed in order to find the optimal strip waveguide structure that provides minimal propagation losses (Gili de Villasante 2010).

6.5 Fabrication Techniques of SOI Waveguides

6.5.1 *Wafer Fabrication*

As discussed earlier, the SOI wafer is composed of a uniform layer of SiO_2, which is sandwiched between a thicker (hundreds of microns) silicon substrate and a thinner surface layer of crystalline silicon. This structure forms a waveguide structure. The first step to producing the waveguides is obtaining the SOI wafer structure from a silicon wafer. There are three most relevant methods.

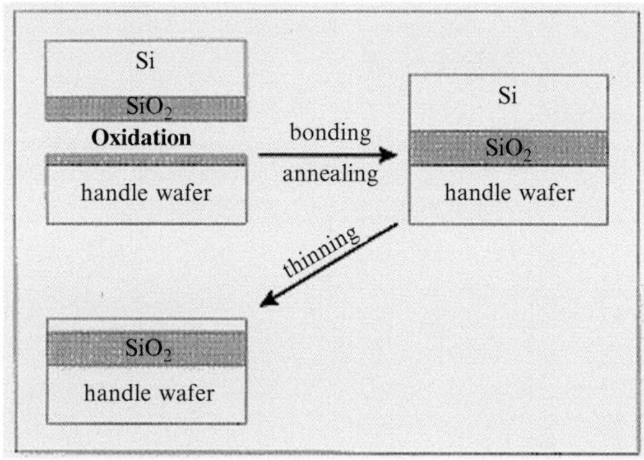

Fig. 6.6 Principle of the bond-and-etch-back approach (BESOI) (PlÖßl and Kräuter 2000): oxidation of two wafers to be bonded; formation of the chemical bond through annealing; and thinning of one of the wafers through etching

1. Separation by implanted oxygen (SIMOX)
   In this fabrication method, a large number of oxygen ions are implanted under the silicon surface to form the SiO_2 layer. The oxygen ions are implanted into crystalline silicon at an energy level of up to 200 keV. This energy subsequently determines the depth of the SiO_2 and hence the thickness of the silicon over-layers. The silicon substrate is maintained at a temperature of approximately 600 °C during implantation, to avoid the formation of an unwanted amorphous silicon over-layer. Several parameters that are critical for the waveguide performance, such as the crystal dislocations which could influence the optical propagation properties and interface micro-roughness have been improved through the years, so SIMOX has proved the most popular method for the fabrication of large volumes of SOI material (Gili de Villasante 2010).
2. Bond-and-Etch-back SOI (BESOI)
   This method takes advantage of the fact that bringing two hydrophilic surfaces (such as SiO_2) into intimate contact can result in the formation of a very strong bond. In general, the wafers are brought into contact at room temperature, at which point an initial bond is formed. The bond strength is increased to that of bulk material via subsequent thermal processing to temperatures as high as 1,100 °C. The production of BESOI has three steps as shown in Fig. 6.6 (Gili de Villasante 2010; PlÖßl and Kräuter 2000): oxidation of two wafers to be bonded; formation of the chemical bond through annealing; and thinning of one of the wafers through etching.
3. Smart-cut Process
   Smart-cut process combines advantages of steps from the SIMOX and BESOI processes. First, a wafer is oxidized to create the BOX layer. Then, hydrogen ions are implanted at a well-controlled depth, creating a "smart-cut." This wafer is then bonded to a clean silicon wafer. The substrate of the first wafer can now be separated along the "smart-cut" interface and then annealed and polished. The process is shown schematically in Fig. 6.7 (Bruel et al. 1997; Ma et al. 2006).

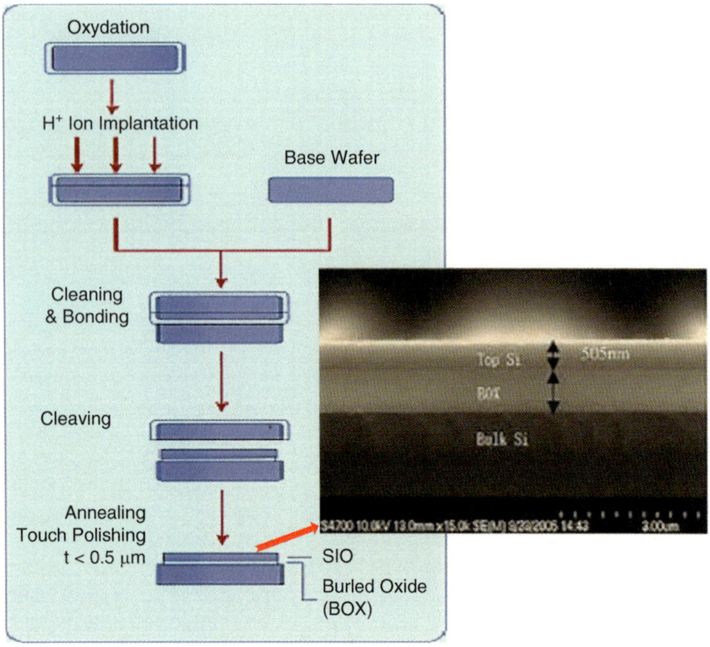

Fig. 6.7 Illustration of the Smart-Cut process (Ma et al. 2006): oxidized wafer is implanted with a high dose of hydrogen; a second wafer is bonded to the first as in the BESOI process; and then split/cleave the implanted wafer

The flexibility, high quality, and efficient use of silicon offered by this process makes it an excellent platform for the fabrication of low-cost SOI waveguides.

4. Deposition of silica optical layers
As the SIMOX and Smart-Cut processes do not usually provide enough silica overlay thickness, an extra thickness has to be grown on the wafer to achieve the required thickness. This is done by depositing a solid film on the surface of a silicon wafer (Gili de Villasante 2010).

Two deposition techniques have been mainly used, including flame hydrolysis deposition (FHD) and plasma-enhanced chemical vapor deposition (PECVD). For both FHD and PECVD techniques, the refractive indices of the resulting films can be controlled by varying gas flows; adding small amounts of a dopant (usually Ge) is the favored method. A high temperature annealing step is commonly required to improve the structures of the layers and the optical propagation. For the top cladding over the etched waveguide, boron and phosphorus dopants are also added to improve the flowing behavior of the silica during the temperature annealing, resulting in lower intrinsic stress and better planarization. The control of intrinsic stress build-up in these optical layers is often critical for waveguide devices. FHD is based on the process of fabrication of optical fiber preforms. For doping, it uses $SiCl_4$ vapors as well as $GeCl_4$, PCl_3, and BCl_3 burned in an oxygen/hydrogen torch, resulting in soot particles deposited on a rotating substrate. The technique has given good results for thick layers, up to 100 μm. The soot requires a consolidation step at 1,000–1,300 °C. The PECVD technique involves a vacuum system in which radio-frequency plasma is generated between parallel plates with a gas mixture typically of SiH_4 and N_2O. Carrier gases (He, N_2, or Ar) and dopant gases (GeH_4, PH_3, BH_4, or CF_4) may also be added. The basic reaction can be described by (Lamontagne 2007):

$$SiH_4(\text{gas}) + 4N_2O(\text{gas}) \rightarrow SiO_2(\text{solid}) + 2H_2O(\text{gas}) + 4N_2(\text{gas}) \qquad (6.8)$$

The deposition usually takes place at a temperature of 250–350 °C. PECVD layers usually require thermal annealing (~1,000 °C) to reduce the concentrations of Si–H and N–H bonds, and their associated absorption lines near 1,480 and 1,510 nm close to the C band (1,525–1,565 nm). The doped P and/or B cladding layers usually require multiple deposition and anneal steps to optimize glass flow. The advantages of high-contrast optical waveguides brought increased attention to PECVD SiN_x and SiO_xN_y layers. They provide the possibility of higher integration (Lamontagne 2007).

6.5.2 *Waveguide Fabrication*

The general fabrication process silicon-on-insulator waveguides is described in Fig. 6.8. The silicon etching can be wet or dry, but dry etching is regarded as the most suitable solution to meet the requirement for flexible process capability, tight tolerances, and reproducible production. It uses chemically reactive plasma to remove the silicon of the upper layer. The plasma is generated under in the vacuum by an electromagnetic field. High-energy ions from the plasma attack the wafer surface and react with it. Reactive Ion Etching (RIE) is an example of dry etching that is used for etching effectively vertical walls (Zimmermann et al. 1998). The asymmetrical waveguide may be converted to a symmetrical one with the deposition or growth of a SiO_2 upper layer (Gili de Villasante 2010).

6.5.2.1 Critical Dimension Control

A very important issue in the fabrication of waveguides is the control of the critical dimensions, that is, with which resolution of the dimensions that rule the performance of the waveguide can be fabricated. These critical dimensions include rib width (w), rib wall angle (θ), slab height (D), and silicon thickness (h), for a general rib waveguide as shown in Fig. 6.4d. The silicon overlay thickness is determined by the process used to define the SOI wafer waveguiding layer. In SIMOX, this accuracy is 3 nm, and in BESOI it is of 500 nm. The rib width and the rib wall angle (θ) depend on the photolithography and removal process and the etching process. The slab height is

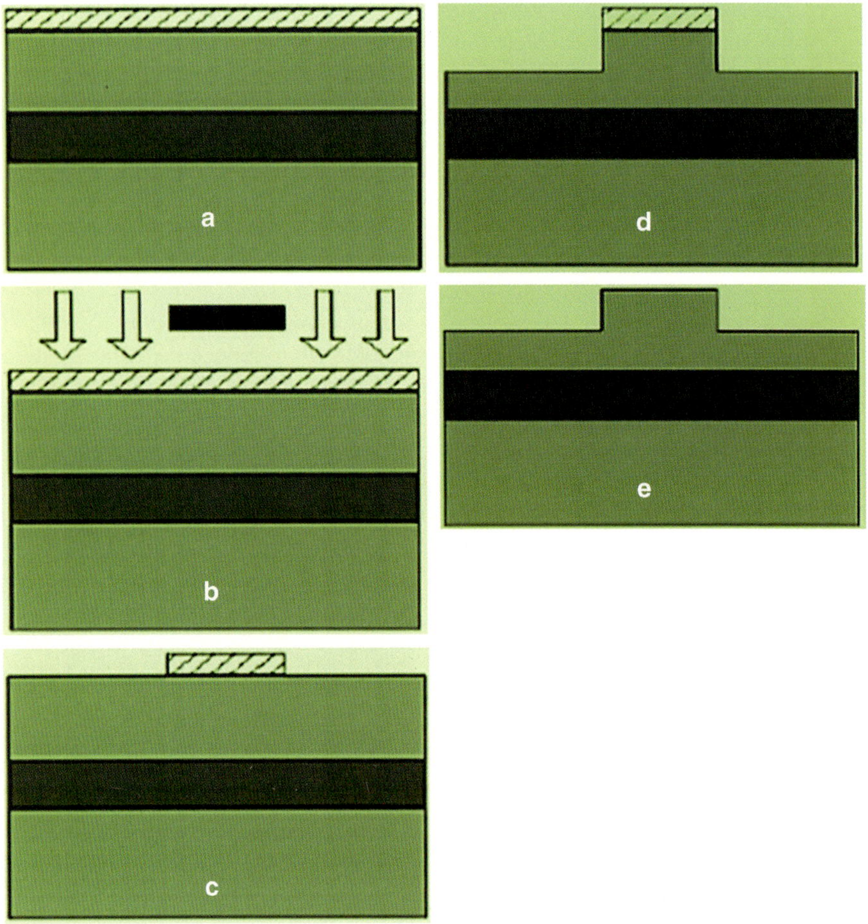

Fig. 6.8 Fabrication procedure of SOI waveguides (Gili de Villasante 2010): (**a**) firstly, a photoresist is applied all over the surface of the SOI wafer, which has been previously cleaned and dried; (**b**) the wafer is then aligned with the mask with submicron precision and exposed to UV light, so the photosensitive elements of the resist are activated. This process is so called photolithography; (**c**) the exposed parts are removed with the exposition of the wafer to a developing solution; (**d**) the silicon upper layer is etched to the desired thickness; and (**d**) the remaining photoresist is removed

determined by the silicon etch process. The minimum critical dimension control (according to the International Technology Roadmap for Semiconductors) is 10 nm, with minimum feature sizes lower than 100 nm. These characteristics should allow the production of submicron waveguides as well, provided that some critical features in this field are improved, such as minimum SiO_2 thickness, and interface and sidewall roughness (Lee et al. 2000).

6.5.2.2 Grating Fabrication

Grating couplers are structures that are much more complicated to produce than straight waveguides, because the dimensions are much smaller and require higher accuracy. The process of fabrication is essentially the same that has already been exposed for waveguides. First a photoresist is applied on the surface of the SOI wafer to define the shape of the element (lithography), and then this shape is etched on the silicon upper layer. There are a couple of lithography and etching techniques that can be used for making this kind of structures (Gili de Villasante 2010):

1. Optical lithography
   This technique uses light to image a pattern on the photoresist. The resolution depends mainly on the used wavelength. The highest resolutions can be achieved for the so-called deep UV lithography (DUV), which uses $\lambda = 248$ or 193 nm, and are <200 nm.
2. Electron-beam lithography
   Instead of using photons, electrons can also be used to expose the photoresist. The resolution can be very high, around 10 nm. Two other advantages of electron-beam lithography are the large depth of field compared to optical lithography and very low levels of contamination damage. High resolution optical lithography suffers from small depth of field, which makes it more difficult to process nonideal wafers from warping and already patterned (topography) wafers.

   The two main types of electron-beam lithography are direct writing and projection. Direct writing systems use different spot shapes and writing modes but generally yield very low throughputs. In this technique, a focused electron-beam scans all the surface of the photoresist and writes the pattern directly on it, so no mask is needed. This system is highly flexible and the spot-size of the beam can be as small as 5 nm. However, it is quite slow and not suitable for mass production. Projection systems use the specialized mask. During the exposure process, back-scattered and secondary electrons result in proximity effects that complicate the patterning of dense features.
3. Interference lithography
   In this technique, two coherent light beams coming from a laser are combined, resulting in a periodic intensity pattern. The angle between the two beams determines the grating period. Grating periods of about 250 nm can be defined.
4. Imprinting lithography
   Imprinting lithography involves a mechanical contact between a mold and a soft (polymer) material to transfer a pattern from the mold to the sample. Among the different versions of imprint lithography are nanocasting, displacement printing, and chemical imprinting. The three basic versions are nanoimprinting, soft-lithography, and step-flash lithography. They allow very high resolution ($\sim$10 nm) and three-dimensional patterns. They can pattern flat or curved and small or large (8–12 inches) substrates. The mask (template mold) is usually fabricated using electron-beam lithography. Nanoimprint lithography involves a

solid mold, for example, silicon or nickel. The process requires heating the surface polymer over its glass transition temperature, then embossing the pattern with the mold. Soft-lithography involves the transfer of self-assembled monolayers using a flexible template. A transparent mold of quartz is brought into contact with a sample surface previously coated with a monomer. Before removing the mold, the patterned monomer is exposed to UV light through the mold and the residual monomer is etched away (where pressed) before transferring the pattern onto the sample. This technique does not require hot and hard embossing, and is particularly suitable for good overlay accuracy. None of these imprint techniques aims for very high throughputs. One challenge facing imprint lithography is the development of good anti-adhesion layers required for efficient and high yield release of templates (Lamontagne 2007).

5. Dry etching
   As in waveguides, a dry etching process such as RIE can be used. Different etching depths can be achieved by controlling the etch depth, and the error in this depth can be under 5 nm if the fabrication environment is well controlled.
6. Focused Ion-Beam etching (FIB)
   The technique used in FIB etching is similar to the electron-beam lithography technique, but heavy ions are used instead of electrons to etch the silicon directly. This process can be quite slow, but high accuracy can be achieved.

6.6 Thallium-Doped SOI Rib Waveguides

The optical absorption of thallium-doped silicon at the 1,550 nm wavelength results from the excitation of an electron from the valence band into a neutral thallium center located 0.246 eV above the valence band. This transition changes the charge state of the center from neutral to negative. This optically excited transition has an optical cross-section of approximately 2.6×10^{-17} cm^2 for photons with an energy equivalent to 1,550 nm wavelength. The energy required to optically excite an electron from the thallium center into the conduction band is 0.87 eV, which is higher than the energy associated with 1,550 nm photons. The absorption coefficient strength for 1,550 nm has been found to be proportional to the overlap integral for the optical waveguide mode and the concentration of unoccupied (neutrally charged) thallium centers, defined as (Huante-Ceron 2011)

$$\alpha_{\mathrm{d}} = \sigma_{\mathrm{p}}^{\mathrm{opt}} \iint_{x,y} \Phi(x,y)\left[N_{\mathrm{T}}(x,y) - N_{\mathrm{T}}^{-}(x,y)\right]dxdy \tag{6.9}$$

where $\sigma_{\mathrm{p}}^{\mathrm{opt}}$ is the cross-section for optical absorption at 1,550 nm wavelength; $\Phi(x, y)$ is the normalized (unit power) optical waveguide mode profile; N_{T} (x, y) is the thallium concentration profile and N_T^- is the ionized (negatively charged) thallium center profile. Thallium is a group-III impurity, which makes it a P-type dopant in silicon. Since thallium is a deep-acceptor-level, not all the impurity centers are ionized at room temperature, where this ionization fraction depends on the concentration of background doping, thallium doping, capture rate, emission rate, and temperature. The fraction of neutrally-charged to the total concentration of thallium centers for a constant background concentration can be increased by increasing the thallium concentration and, for a constant concentration of thallium, the fraction can be decreased as the background doping concentration is increased.

Silicon rib waveguides, for instance, can be fabricated using a SOI wafer with a guiding layer of 2.5 μm on a 0.375 μm thick BOX. Approximately 2×2 cm^2 silicon wafer was cleaved, cleaned using the RCA cleaning procedure, and then annealed in a dry-oxygen environment for 30 min to grow 100 nm thick oxide on the silicon over-layer. Prior to the photoresist (PR) deposition, the SOI sample was heated at 110 °C for 5 min to remove moisture. The samples was subjected to standard photolithography so that a waveguide pattern could be transferred to the PR, and etched into the oxide using a 10 % buffered HF. After the PR was removed, the waveguide pattern was transferred to the silicon via wet etching (using a 40 % KOH + IPA solution), where the thermal oxide layer was used as an etching mask. The etching depth in silicon was chosen to ensure single-mode propagation. For a waveguide width of 4 μm, no more than 0.25 μm etching depth is required to ensure single-mode propagation. 4 μm thick PR was used as a masking layer during ion implantation of thallium with an energy level of 1 MeV for a dose of 7.4×10^{13} or 1.2×10^{14} cm^{-2}. After the PR was removed from the sample, it was annealed at 1,100 °C in a dry-oxygen environment for 90 min in order to diffuse and activate the implanted thallium. The sample was subjected to subsequent (unmasked) phosphorus implantation at an energy level of 385 keV for doses ranging from 3×10^{12} to 1×10^{14} cm^{-2} and was further annealed at 1,100 °C in a dry-oxygen environment for 90 min. This range of phosphorus doses provided co-doping of the thallium such that the thallium mediated optical absorption. The implantation energies, annealing temperatures, and annealing times were chosen in order to position the resulting thallium and phosphorus concentration profiles coincidentally. In addition, their peak concentration was required to be located as close to the waveguide center as possible. Phosphorus implantation contributes to FCA along the whole waveguide length, while thallium centers contribute to the absorption only in the unmasked window. This design thus allows the separation of absorption due to free carriers and due to thallium mediated absorption. The final sample preparation step was the definition of optical quality end-facets using a chip thinning and edge cleaving process. Using a 1,550 nm wavelength laser, optical absorption measurements were performed for all the waveguide on the sample. This was done by butt-coupling light into the rib waveguide through a tapered optical fiber. Light was collected by an objective lens located at the opposite end of the waveguide. The collimated light was propagated through free space and measured using an InGaAs photodetector. Total measured loss, in dB, can be summarized by the following relationship (Huante-Ceron 2011):

$$\alpha_T = \alpha_d l_w + \alpha_i L + \zeta \tag{6.10}$$

where α_d is the absorption coefficient of thallium centers, in dB/cm, defined by (6.9), l_W is the length of the thallium implantation window, α_i is the intrinsic absorption coefficient, which includes free carriers absorption induced by phosphorus doping, L is the waveguide length and ζ is the coupling loss.

For high phosphorus concentration, full compensation of thallium centers is expected. In an already implanted material with a heavy ion, and annealing in an oxidation environment phosphorus diffusion is enhanced and therefore a lower concentration of phosphorus (coincident with thallium) is induced. However, for higher phosphorus concentrations it may be necessary to consider the presence of other effects induced by high doping concentration in order to explain the residual absorption of thallium. Qualitatively, it is know that the diffusivity coefficient depends on the size of the impurity atom and on the concentration of the diffused impurity. For example, the elastic interaction between a point defect and a dopant atom depends on the mismatch in size, so that a large dopant atom may prefer to migrate with a vacancy, while a small dopant atom may prefer to migrate with an interstitial. Furthermore, it is known that phosphorus diffusion coefficient is enhanced when the surface of silicon is oxidized. i.e., Phosphorus diffusion is enhanced by injecting self-interstitials, resulting from the oxidation, which recombine with vacancies in the bulk, reducing the vacancy concentration and possibly retarding thallium diffusion (Huante-Ceron 2011).

6.7 Indium-Doped SOI Rib Waveguides

Indium is a group-III impurity, which makes it a P-type dopant in silicon with a single acceptor level at 0.157 eV above the valence band. The optical excitation of an electron from the valence band to the neutrally-charged indium level has an optical cross-section, of approximately 1.7×10^{17} cm^2 for a 1,550 nm wavelength. For indium the capture cross-section for holes is approximately 8×10^{-15} cm^2 and for electrons 2×10^{-22} cm^2. The maximum indium doping that can be achieved is limited by its solid solubility in silicon which is approximately 1.5×10^{18} cm^{-3} at room temperature (Huante-Ceron 2011).

Indium has been chosen to be used as the deep level impurity for the optical modulator. Periodic P-N junctions were made by implanting indium and phosphorus to form the P-region and N-region respectively. The device operates by injecting (forward-bias) or removing (reversed-bias) free carriers in order to modulate the propagating signal. The presence of the deep levels and the configuration of the P-N junction perpendicular to the propagation direction of a SOI optical waveguide mode are intended to enhance the change in refractive index. There is strong variation in absorption on 1,550 nm wavelength via deep-level (indium, thallium) doping through the variation of background N-type doping and absorption, and therefore the efficiency of the modulator compared to that of equivalent dimensions which rely on free-carrier absorption effects alone (Knights et al. 2011).

Ideally, a silicon optical modulator should have high modulation speed and modulation depth, small footprint, low losses, large optical bandwidths, low power consumption and must also be CMOS compatible. However, these requirements may impose contradictory requirements on the device design and compromises are often sought. For instance, silicon rib waveguide structure with periodically arranged P-N junctions that are formed on the rib waveguide by implanting indium as the P-type and phosphorus as the N-type dopant. The P-region, in the periodically arranged P-N junctions, can be formed by only implanting indium, hence free-charge carriers (free-holes) and neutrally charged concentration of deep-levels are responsible for the optical absorption in that particular region. The N-region is formed by masked implantation of phosphorus in the regions already doped with indium so that the phosphorus over-compensates the indium. Therefore, the optical absorption in the N-region will be proportional to the availability of neutrally-charged deep-levels (which for this region is always negligible) and the free-charge carrier concentration (the dominant effect) (Huante-Ceron 2011).

Commonly, the P-N junctions are formed parallel to the beam propagation direction in such a way that the depletion region coincides with the maximum intensity of the waveguide mode, and the orientation of the P-N junction can be either horizontal or vertical with respect to the SOI wafer plane. In order to increase the optical mode and the depletion region overlap, for example, a modulator structure has been proposed as shown in Fig. 6.9a. In this structure the P-N junctions are periodically located along the waveguide, oriented perpendicular to the light propagation direction and vertically to the wafer plane, as shown in Fig. 6.9b. With this configuration the whole waveguide mode profile interacts with the depletion region instead of only a fraction of it (Fig. 6.9c). Since the optical mode passes through the successive depletion regions, the total optical modulation can be calculated by multiplying the overlap ratio between the depletion region and optical waveguide mode, depletion region width and the total number of P-N junctions along the waveguide (Li et al. 2009; Huante-Ceron 2011).

6.8 SOI Waveguide Applications

6.8.1 Type of SOI Waveguides

The widely used SOI waveguides may take the form of a channel waveguide, ridge waveguide, photonic-crystal waveguide, or slot waveguide, as shown in Fig. 6.10 (Lipson 2005). Channel waveguides have been used to reduce the sidewall roughness resulting in reduced propagation losses. However for a waveguide with

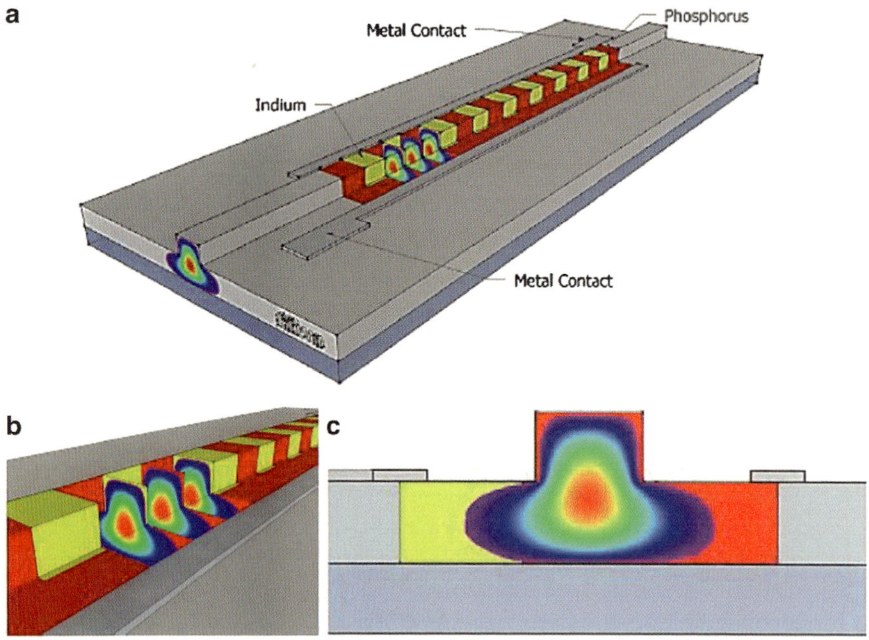

Fig. 6.9 Design of indium-doped SOI waveguides (Li et al. 2009; Huante-Ceron 2011): (**a**) waveguide with periodic P-N junction; (**b**) close-up of the interaction of optical mode and the P-N junctions; and (**c**) optical mode and P-N junction overlap

thickness and width of 1 mm or more the number of modes will be large. For single-mode operation in an SOI waveguide, for example at a wavelength of 1.55 mm, its dimensions need to be less than or close to 240 nm. The rib waveguides are a little bit different. A large rib waveguide can be single-moded if its aspect ratio satisfies the (6.5). For a typical ridge waveguide where $r = 0.5$, the waveguide is single-mode as long as $a = b < 0.96$. A reasonable set of dimensions will be $b = 400$ nm, $a = 385$ nm. There are some advantages of using rib waveguides: (a) free carriers dissipate faster into the wings of the waveguide, which reduces the free-carrier effects; and (b) a P-I-N structure with metal contacts is easily fabricated using this structure.

A photonic-crystal waveguide is another option for making SOI waveguides. The refractive indices of the left and right sides of the cladding in Fig. 6.10c can be flexibly engineered by varying the diameter of the holes and the lattice constants. The disadvantage is a high propagation loss due to the roughness of sidewalls. In addition, the slot waveguide has been proposed and fabricated as another candidate for silicon photonics. In a slot waveguide, an ultra-thin low-refractive-index layer is sandwiched between two high-refractive-index layers, as shown in Fig. 6.11 (Anderson et al. 2006). In such a waveguide, the TE mode is strongly confined in the low-index slot region. This structure has been

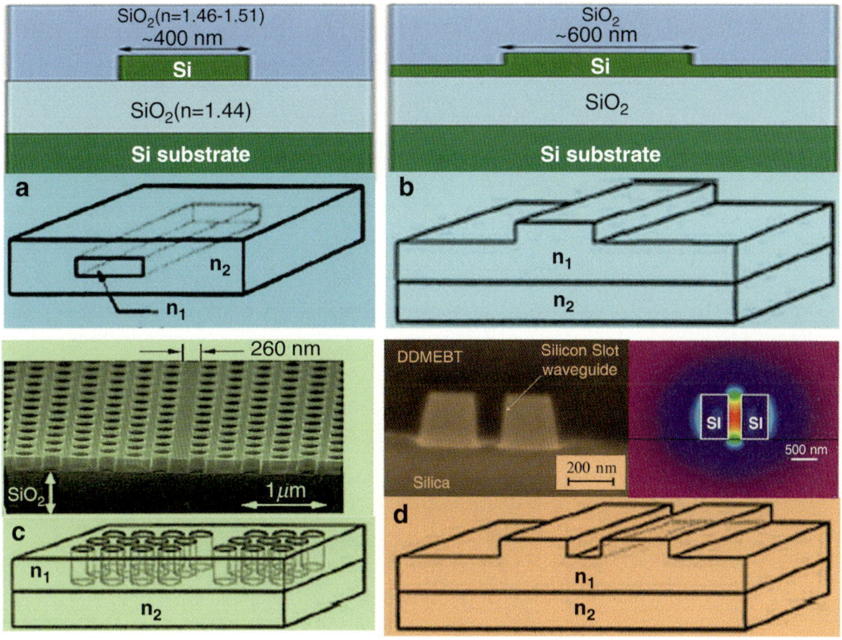

Fig. 6.10 Four configurations for making SOI waveguides (Lipson 2005): (**a**) channel waveguides; (**b**) rib waveguides; (**c**) photonic-crystal waveguides; and (**d**) slot waveguides

applied in light-emission and bio-sensing areas. For example, a multi-slot waveguide is built with alternating thin Si and SiO_2 layers, and light can be confined in the lower-index SiO_2 layers. This structure of slot waveguides combines the advantages of both silicon and silica to explore the promising nonlinear interactions of light in the waveguides, such as rare-earth material doping to provide optical gain. Moreover, a two-dimensional photonic-crystal slot microcavity sensor can be built for virus-sized particle detection. The resonance of the cavity shifts when a particle is captured. Comparing with normal photonic-crystal cavities, the slot cavity structure enhances the sensitivity exponentially with the size of particles (Yin 2009).

6.8.2 Low-Loss SOI Waveguides

SOI waveguides have great potential for use as nonlinear optical devices. However, two manufacturing challenges are high propagation losses and high fiber-to-waveguide coupling losses. The main origin of the propagation loss in a silicon submicron waveguide is scattering at its sidewalls. By reducing the sidewall roughness, scattering losses can be reduced significantly. Lithography and etching are the

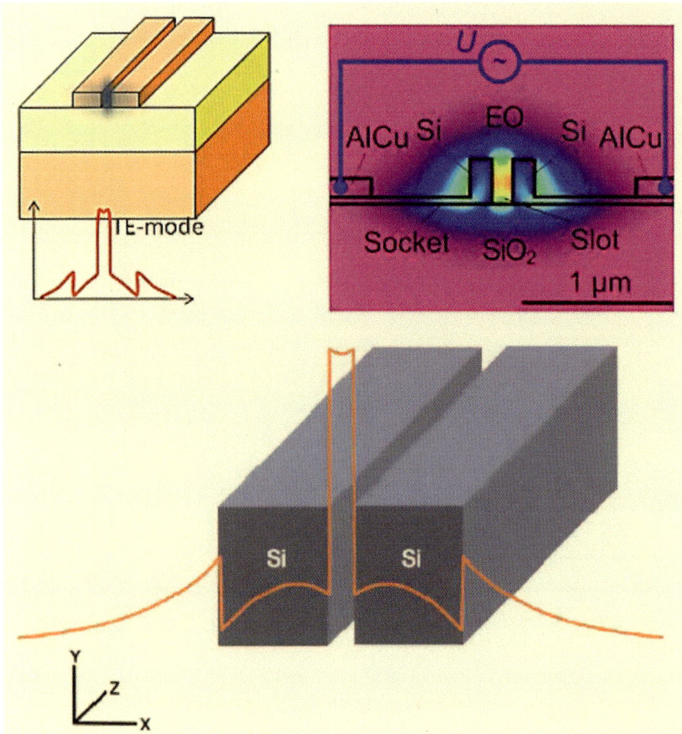

Fig. 6.11 Schematic of a slot SOI waveguide. The slot is filled with air. The TE mode is strongly confined in the low-index slot region (Anderson et al. 2006)

keys to making a smooth silicon pattern. To do this, electron-beam lithography and plasma etching have been employed for the silicon submicron-size waveguides. For instance, SOI waveguide with a core of 400 × 200 nm was successfully made with a relatively low propagation loss of 2.8 dB/cm (Tsuchizawa et al. 2005). An oxidation procedure can further improve the smoothness of sidewalls. For example, losses in channel waveguides typically range from 0.2 to 5 dB/cm. Losses in photonic-crystal and slot waveguides are even higher. Rib waveguides have typically lower losses of less than 0.1 dB/cm due to their larger dimensions (Lipson 2005). The most difficult problem with silicon waveguides is how to connect them to external circuits, where optical fibers are generally used. An SOI waveguide is typically very small, and its core is about 500 nm wide. On the other hand, the fundamental mode of single-mode fiber is about 9 mm in diameter. Moreover, the air gap between the facets causes interference and puts some wavelength dependent characteristics on top of the original spectra. The coupling element can be either a microscope objective lens or a tapered fiber lens. The latter is widely used nowadays for convenience. The waveguide itself can be either un-tapered, or tapered with normal tapering or inverse tapering. The advantage of inverse tapering is that the mode index at the narrow tip is much smaller than that of bulk silicon and the effective mode area is larger than that of the silicon region, both of which offer better mode matching and more efficient coupling. The disadvantages are that the coupling is very sensitive to environmental vibrations, and also a bigger portion of the optical field is propagating along the sidewalls so that scattering loss from the sidewalls is huge. Low-loss, low-reflection, mode-field converter can be used to solve the coupling problem even better. The converter has a silicon adiabatic taper that gradually becomes thinner toward the end, and a second low-index waveguide covers the taper. Coupling losses can be made as low as 0.5 dB per connection with this approach (Yin 2009).

6.8.3 Linear Applications

Using low-loss SOI waveguides, passive devices including wavelength filters, MMI couplers, and WDM filters have been developed, which can be in the configuration of Mach–Zehnder interferometers, Bragg gratings, ring resonators and photonic-crystal waveguides. Other than passive devices, active devices have also been developed to control the flow of light, including modulators and switches. The basic idea is to change the transmission properties of the device through a change in the refractive index. Modulation of the refractive index of silicon can be done using either the thermo-optic effect or the electro-optic effect. However, the thermo-optic effect is relatively slow and can only be used for up to 1-MHz modulation frequencies (Lipson 2005). For high-speed modulation, electro-optic devices are required. However, pure crystalline silicon does not exhibit a linear electro-optic effect. An effective mechanism for changing the refractive index in silicon at a high speed is the free-carrier plasma-dispersion effect. The refractive index of silicon varies with the free-carrier concentration, which can be manipulated through injecting or depleting carriers by applying an electric field to the device. However, FCA is a disadvantage associated with this technique. For instance, an MOS configuration embedded in silicon waveguides is used in a Mach–Zehnder geometry and able to demonstrate high-frequency modulation with 3 dB bandwidth of 20 GHz and data transmission up to 30 Gb/s (Liu et al. 2007). This performance is limited by the free-carrier's lifetime. The high speed is the result of a unique device design with a traveling-wave drive scheme allowing electrical and optical signal co-propagation along the waveguide. The traveling-wave electrode which is based on a coplanar waveguide was designed to match the velocity for both optical and electrical signals. The speed was pushed even further up to 40 Gb/s by optimizing the device packaging (Yin 2009).

Although a lot of optical devices have been developed with the silicon platform, a room-temperature, electrically pumped, silicon laser is still the missing piece for monolithic integration because silicon has a poor stimulated emission cross-section due to its indirect band gap structure. There have been many efforts toward the goal of silicon lasers. Silicon light-emitting diodes have also been demonstrated. Another effort was carried out on the hybrid AlGaInAs-silicon evanescent laser.

The device is comprised of a multiple-quantum-well structure bonded to a silicon waveguide on the SOI wafer. The optical mode overlaps both the III-V material and the silicon waveguide so that the mode can obtain electrically pumped gain from the III-V region while being guided by the silicon waveguide region. The device showed room temperature continuous-wave (CW) lasing with 65 mA threshold, 1.8 mW output power and overall differential efficiency of 12.7 %. Meanwhile, mode-locked silicon evanescent lasers have also been demonstrated. Besides silicon-based light sources, an integrated germanium-on-silicon detector has also been developed, which can operate at the speed of 40 Gb/s (Yin et al. 2007).

6.8.4 Nonlinear Applications

Silicon is also promising for making nonlinear optical devices, because of its high Kerr and Raman nonlinearities. Nonlinear applications in SOI waveguides are based on self-phase modulation (SPM), cross-phase modulation (XPM), stimulated Raman scattering (SRS), and four wave mixing (FWM). In addition, TPA, FCA and free-carrier-induced dispersion (FCD) should also be taken into consideration in silicon. SPM is a very efficient process in SOI waveguides. SPM induced spectral broadening is found to be significant at coupled peak powers of even a few tens of milli-watts. XPM was used to demonstrate strong modulation instability in silicon waveguides in the pump-probe configuration. The result showed modulation instability gain spectrum that is two to three orders of magnitude larger than that achieved in optical fibers. XPM in a silicon MZI was shown to work as an optical modulator. The dependence of XPM on walk-off was also observed experimentally. Raman amplification in an SOI waveguide is promising because the Raman gain peak is about 3,000 times stronger than in silica fibers. Using SRS, many nonlinear optical functions have been demonstrated, including Raman amplification, optical modulation, and wavelength conversion. Net Raman gain was observed, while Silicon Raman lasers and a cascaded silicon Raman laser have also been demonstrated. They provide the ability to generate coherent light in wavelength regions that are not easily accessible with other conventional types of lasers. FWM is promising for all-optical signal processing. Applications of FWM include wavelength conversion, parametric amplification, and photon-pair generation. Net gain from FWM is not promising when CW pumps are used to pump a signal in SOI waveguide because of the effect of FCA. However, net gain in the case of pulsed pumps is achievable. For instance, highly tunable optical parametric oscillation is achieved using silicon micro-resonators. The pump was chosen to beyond 2.2 mm to get around the issues of TPA and the consequent free-carrier effects. The carrier-induced plasma-dispersion effect makes use of an electric field applied across the waveguide. An alternative way to generate free carriers is through photon-absorption process using optical pumps. This is faster than using external electrical field because the free carriers are generated locally.

Applications based on this effect include carrier-induced optical bistability in silicon ring resonators, all-optical switching, optical modulation, wavelength conversion, and silicon photonic memory (Yin 2009).

6.9 Summary

With the advances in fabricating silicon electronic devices, it is now possible to fabricate silicon photonic devices using the same CMOS technology, especially with the help of the mature SOI wafer technique and nano-scale photolithography. SOI material has become a main platform for both photonics and VLSI CMOS electronics, with fully compatible processing procedures; Si-based OEIC have economic advantages; Compared with III-V semiconductors, silicon has better crystal perfection, better native oxide, and superior thermal and mechanical properties; SOI CMOS circuits can reduce parasitics and latch-up that enable high speed at low power; and strong optical-confinement of SOI waveguides enables low-loss and compact device-size.

SOI is usually described as a thin layer of silicon (typically hundreds of Å to a few microns thick) on top of an insulating material with an underlying bulk silicon substrate. The most common insulating material currently is silicon dioxide. SOI technology has relied heavily on the ability of ion beams to create shallow, abrupt as-implanted doping profiles in bulk Si. Implant thermal processing must be performed in order for the dopant atoms to occupy substitutional lattice sites and contribute electrically. This needs to deal with fundamentals of ion implantation, atomistic diffusion, and segregation in the proximity of interfaces.

The widely used SOI waveguides may take the form of a channel waveguide, ridge waveguide, photonic-crystal waveguide, or slot waveguide. Channel waveguides have been used to reduce the sidewall roughness resulting in reduced propagation losses. However for a waveguide with thickness and width of 1 mm or more the number of modes will be large. The advantages of using rib waveguides include free carriers dissipate faster into the wings of the waveguide, which reduces the free-carrier effects; and a P-I-N structure with metal contacts is easily fabricated using this structure. A photonic-crystal waveguide is a promising option for making SOI waveguides. In addition, the slot waveguide has been proposed and fabricated as another candidate for silicon photonics. In a slot waveguide, an ultra-thin low-refractive-index layer is sandwiched between two high-refractive-index layers. In such a waveguide, the TE mode is strongly confined in the low-index slot region. This structure has been applied in light-emission and bio-sensing areas.

Using low-loss SOI waveguides, passive devices including wavelength filters, MMI couplers, and WDM filters have been developed, which can be in the configuration of Mach–Zehnder interferometers, Bragg gratings, ring resonators and photonic-crystal waveguides. Other than passive devices, active devices have also been developed to control the flow of light, including modulators and switches.

References

Anderson PA, Schmidt BS, Lipson M (2006) High confinement in silicon slot waveguides with sharp bends. Opt Express 14:9197–9202

Baehr-Jones T et al (2010) Silicon-on-sapphire integrated waveguides for the mid-infrared. Opt Express 18(12):12127–12135

Bruel M, Aspar B, Auberton-Hervé A-J (1997) Smart-cut: a new silicon on insulator material technology based on hydrogen implantation and wafer bonding. Jpn J Appl Phys 36:1636–1641

Celler GK, Cristoloveanu S (2003) Frontiers of silicon-on-insulator. J Appl Phys 93(9):4955–4978

Chen GS (2012) Chapter 8 Ion implantation. http://www.me.ntut.edu.tw/introduction/teacher/lu/IC%20fabrication_GA/IC_ch08.pdf. Accessed 21 Feb 2013

Deng H (2005) Design and characterization of silicon-on-insulator passive polarization convert with finite-element analysis. PhD dissertation, University of Waterloo, Ontario, Canada

Gili de Villasante O (2010) Design and simulation of vertical grating coupler for photonic integrated system-in-package. Master thesis, Technische Universitat Berlin, Berlin

Huante-Ceron E (2011) Optical modulation by controlling the charge state of deep impurity levels. PhD dissertation, McMaster University, Hamilton, Ontario

Kim et al (1997) Extended Si{311}defects. Phys Rev B 55(24):16186–16197

Kishima K (2007) Double-SOI waveguide: the communication pathway beneath the surface. http://www.advancedsubstratenews.com/2007/10/double-soi-waveguidethe-communication-pathway-beneath-the-surface/. Accessed 23 Feb 2010

Knights A et al (2011) Optical modulators employing charge state control of deep levels. WIPO Patent Application WO/2011/011393. http://www.sumobrain.com/patents/wipo/Optical-modulators-employing-charge-state/WO2011011393.html. Accessed 16 Feb 2012

Lamontagne B (2007) Enabling fabrication technologies for planar waveguides devices. In: Calvo ML, Lakshminarayanan V (eds) Optical waveguides: form theory to applied technologies. CRC Press, Boca Raton. pp 271–307

Lee KK, Lim DR, Luan H-C, Agarwal A, Foresi J, Kimerling LC (2000) Effect of size and roughness on light transmission in a Si/SiO_2 waveguide: experiments and model. Appl Phys Lett 77(11):1617–1619

Li ZY et al (2009) Silicon waveguide modulator base on carrier depletion in periodically interleaved PN-junction. Opt Express 17(18):15947–15958

Lipson M (2005) Guiding, modulating, and emitting light on silicon-changes and opportunities. J Lightwave Technol 23(12):4222–4238

Liu A et al (2007) High-speed optical modulation based on carrier depletion in a silicon waveguide. Opt Express 15:660–668

Ma X et al (2006) A high-quality SOI structure fabricated by low-temperature technology with B^+/H^+ co-implantation and plasma bonding. Semicond Sci Technol 21:959

Milosevic MM et al (2009) Rib waveguides for mid-infrared silicon photonics. J Opt Soc Am B 26(9):1760–1766

PlÖßl A, Kräuter G (2000) Silicon-on-insulator: materials aspects and applications. Solid State Electron 44(5):775–782

Pogossian SP, Vescan L, Vonsovici A (1998) The single-mode condition for semiconductor rib waveguides with large cross section. J Lightwave Technol 16(10):1851–1853

Saavedra AF (2004) Interfacial phenomena in ion implanted silicon-on-insulator materials. PhD dissertation, University of Florida

Soref R (2005) Silicon photonics technology: past, present and future. Optoelectronic integration on silicon II, edited by Joel A. Kubby, Ghassan E. Jabbour. In: Proceedings of SPIE, vol 5730. pp 19–28

Tsuchizawa T et al (2005) Microphotonics devices based on silicon microfabrication technology. IEEE J Sel Top Quantum Electron 11:232–240

Wang Q, Ho ST (2008) Modal analysis of silicon photonic wire and TM pass nanowaveguide polarizer. PhotonicsGlobal@Singapore, IPGC 2008. IEEE, pp 1–4

Wikipedia (2013) Silicon on sapphire. http://en.wikipedia.org/wiki/Silicon_on_sapphire. Accessed 16 Feb 2013

Yin L (2009) Study of nonlinear optical effects in silicon waveguides. PhD dissertation, University of Rochester, Rochester, New York

Yin T et al (2007) 31 GHz Ge n-i-p waveguide photodetectors on Silicon-on-Insulator substrate. Opt Express 15:13965–13971

Yue Y et al (2012) Silicon-on-nitride waveguide with ultralow dispersion over an octave-spaning mid-infrared wavelength range. IEEE Photonics J 4(1):126–132

Zimmermann L, Schnarrenberger M, Mitze T, Bruns J, Petermann K (1998) Study of reactive ion etched (RIE) facets of silicon-on-insulator (SOI) rip waveguides. In: IEEE international SOI conference proceedings, October 1998, pp 15–16

Chapter 7
Glass Waveguides

Abstract Optical waveguide components have been made with various glasses for commercial photonic devices, such as phase-arrays, and Mach–Zehnder interferometer switches. The devices are based on planar optical waveguides, in which light is confined to substrate-surface channels and routed onto the chip. Silica, SiON, fluoroaluminates, chalcogenides and doped glasses are important glasses for making optical waveguide devices. Bulk silica (SiO_2) and silica-on-silicon (SiO_2/Si) are by far the most common materials used to manufacture planar light wave circuits (PLCs), due to their refractive-index match with silica-based optical fiber. Doped glass and silica waveguide-based PLC technology is a promising way to integrate optical devices and thus reduce the receiver size and cost. Glass/silica waveguide PLCs typically consist of a planar arrangement of glass waveguides with a higher index of refraction buried in glass all on a silicon or glass substrate. Glass/silica waveguide PLCs have successfully used in optical fiber communications because of their reliability, low insertion loss, ease of coupling to optical fibers, integration capability, and ability to produce optical filters with high accuracy. This chapter will provide an overview about glass-based waveguide devices, including materials selection, fabrication process and applications.

7.1 Glass Structure and Composition

Glass is a solid material that has the atomic structure of a liquid. Atoms in glass form a three-dimensional connected structure without a long-range periodic order. The following four rules should be fulfilled for an oxide to form glass (Lehky 2006): (a) each oxygen must be linked to maximum two cations; (b) the number of oxygen atoms around any cation must be small; (c) the cation polyhedra must share corners only (not edges or faces), to form a three-dimensional network; and (d) at least three corners must be shared. Depending on geometric (coordination number) or bond strength considerations, three types of oxides can be discerned: network or glass formers, intermediates and modifiers. Network formers (e.g., B_2O_3, SiO_2,

X.C. Tong, *Advanced Materials for Integrated Optical Waveguides*, Springer Series in Advanced Microelectronics 46, DOI 10.1007/978-3-319-01550-7_7,

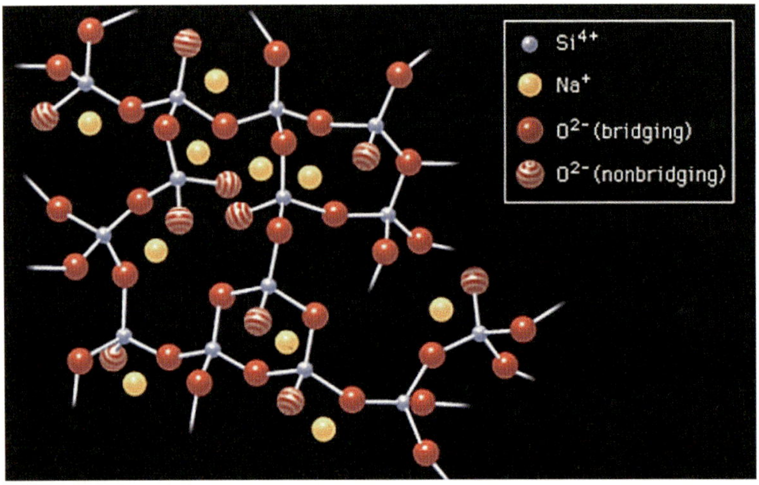

Fig. 7.1 The irregular arrangement of ions in a sodium silicate glass. The building blocks of the glass network are the network-forming cations (e.g., Si^{4+}). The four positive charges of the Si^{4+} ion form the SiO_4 tetrahedra, connected to each other at the corners. An oxygen atom that connects two tetrahedra is called bridging oxygen—an oxygen linked to only one silicon atom is a non-bridging oxygen. The remaining negative charge is satisfied by bonding to a network-modifying cation (e.g., Na+), which occupies an interstice adjacent to the SiO_4 tetrahedron (Rudolph 2004; Lehky 2006)

GeO_2, P_2O_5, As_2O_5, and Ta_2O_5) have cation-oxygen bond strength greater than 355 kJ/mol. The only fluoride fulfilling the rules for network formation is BeF_2. In multicomponent glasses, oxides with lower bond strengths do not become part of the network (e.g., Na_2O, K_2O, CaO, BaO, MgO, Li_2O, Rb_2O, and Cs_2O). They are called network modifiers and are distributed throughout the holes in the network. These network modifiers are generally added to the glass to change its properties (softening point, fluidity, resistivity, thermal expansion coefficient, and chemical durability, etc.). Finally, intermediates may or may not become part of the network (e.g., BeO, ZnO, ZrO_2, TiO_2), but they cannot form a glass network on their own.

Figure 7.1 shows the structure of silicate glass (Rudolph 2004). The basic structural unit is the silicon-oxygen tetrahedron (SiO_4) in which a silicon atom is connected to four surrounding oxygen atoms. The oxygen atoms shared between two tetrahedral are called bridging oxygen (BO). Those not shared are referred to as non-bridging oxygen (NBO). The remaining negative charge of the NBO is satisfied by bonding (ionic bonding) to a network-modifying cation, in this case a monovalent sodium ion (Na^+), which occupies an interstice adjacent to the SiO_4 tetrahedron. In pure vitreous silica (SiO_2), nearly all oxygen atoms are bridging (Lehky 2006).

Most properties of glass are sensitive to its chemical composition and its atomic structure. In oxide glasses, the specific composition-structure-property relationships are based upon the following factors: (a) the coordination number of the network-forming (glass-forming) ion; (b) the connectivity of the structure, determined by the concentration of NBO, which, in turn, depends on the concentration and nature of network-modifying ions; (c) the openness of the structure, also determined by the concentration of the network-modifying ions; and (d) the mobility of the network-modifying ions. As an example, tetrahedrally connected networks, such as those formed by silicates, are more viscous than the triangularly connected networks formed by borates. In silicates, the addition of network-modifying alkali ions raises the concentration of NBOs, and the resulting lowered connectivity leads to a lowering of viscosity. Networks in which the interstitial spaces are less filled with network-modifying ions possess lower density and allow greater permeation of gases through them. Since alkali ions are the most mobile species through interstices of oxide glasses, a higher alkali concentration will lower the chemical durability and electrical resistivity of the glass. Glass types often encountered in optical waveguide fabrication are listed as below (Lehky 2006):

1. Soda-lime glasses: Nearly 90 % of all the glass produced is soda-lime glass. In addition to Na_2O (soda—actuating as a flux), CaO (lime—stabilizing the silica) and SiO_2, these glasses may contain MgO, Al_2O_3, BaO or K_2O and various colorants. Soda-lime glass is sometimes also referred to as crown glass and has a low optical dispersion. Alumina (Al_2O_3) increases the durability of the glass (maximum of viscosity for equal soda and alumina amount) and the glass withstands higher temperatures (aluminosilicate glasses). Furthermore, if the molar ratio $R_2O{:}Al_2O_3$ or $RO{:}Al_2O_3$ is unity, then the aluminosilicate glass has a silica-like structure, where almost all oxygens are bridging oxygens. MgO is added to prevent devitrification. If PbO is added, lead glasses are obtained, which have a lower melting point, a long working range and a high refractive index (lead crystal and optical glass). Lead glass is also called flint glass and is commonly used to make lenses and prisms. It absorbs most of the UV light.
2. Borosilicate glasses: Borosilicate glasses are easily melted, low expansion commercial glasses. The replacement of the alkali by boric oxide (B_2O_3) results in a lower expansion glass, because of the fluxing action of the boron. This is due to the presence of planar three-coordinate boron that weaken the silicate network at high temperature. Alumino-borosilicate glasses are obtained, if alumina (Al_2O_3) is added to the borosilicate glass. These glasses are characterized by a high stability of the vitreous state. That is of special importance in the process of transition-metal ion doping, which tends to initiate crystallization. The commonly used fabrication methods for flat glass substrate manufacturing include the float and the draw process. In the float process, the glass melt is poured onto a "bed" of molten tin. The glass therefore floats on the tin, which leads to flawless optical quality surfaces, without the need of surface post-processing. However, it is possible that tin diffuses into the glass surfaces as an impurity. In the draw process, the glass melt is poured through a slot (slot-drawn) or over a fusion pipe (fusion drawn) and subsequently drawn into a uniform sheet of glass. In the case of optical or special "custom-made" glass fabrication, the glass is melted into a block of very limited dimensions. The glass

substrates are then cut from the block and mechanically polished to achieve optical quality surfaces. Therefore, apart from possible fluctuations in composition and manufacturing, such substrates are much more expensive (Lehky 2006).

3. Fluoroaluminates, chalcogenides and doped glasses: Fluoroaluminates, chalcogenides and doped glasses are important glasses for making optical waveguide devices, such as 1.3 μm planar amplifiers and nonlinear switches. However, they are difficult to fabricate as thin films because of their tendency to crystallize and volatility. Reproducing the stoichiometry of multicomponent glasses in planar waveguide form has proved difficult by conventional techniques, such as sputtering, due to the different vapor pressures of each element. Pulsed laser deposition (PLD) can overcome this problem because of the high vaporization rates provided by excitation at ultraviolet wavelengths, but are often troubled by poor surface topography. By dissolving the bulk material in a solvent, spin or dip coating can be used to retain the waveguide composition for a limited range of materials such as sol–gel glasses. At temperatures above the liquidus temperature, several soft glass families, such as the fluorides, sulphides and phosphates, have viscosities comparable to those of spin coating solutions. A neodymium-doped fluoroaluminate (AlF) glass, with a viscosity of 0.1 poise at 1,000 °C, has successfully been spin coated into waveguide geometry directly from its liquid phase. Laser operation around 1,050 nm has been observed in a 20 μm thick, 0.5 % Nd-doped glass waveguide fabricated using this technique. Performance is consistent with propagation losses of 0.5 dB/cm, which have been independently measured by imaging the scattered light from the waveguide surface (Harwood 1998).

7.2 Silica Glass Waveguides

The most prominent feature of silica waveguides is their simple and well-defined waveguide structure. This allows photonics component manufacturers to produce multi-beam or multistage interference devices such as arrayed-waveguide gratings (AWGs) and lattice-form programmable dispersion equalizers. A variety of passive planar lightwave circuits (PLCs), such as N × N star couplers, N × N AWG multiplexers, and thermo-optic matrix switches have been developed (Pham 2004).

7.2.1 Material Processing Technology

Bulk silica (SiO_2) and silica-on-silicon (SiO_2/Si) are by far the most common materials used to manufacture PLCs, due to their refractive-index match with silica-based optical fiber. Two major types of deposition processes are used today: chemical vapor deposition (CVD) and flame hydrolysis deposition (FHD). The CVD approach, which is a modification of standard semiconductor-processing

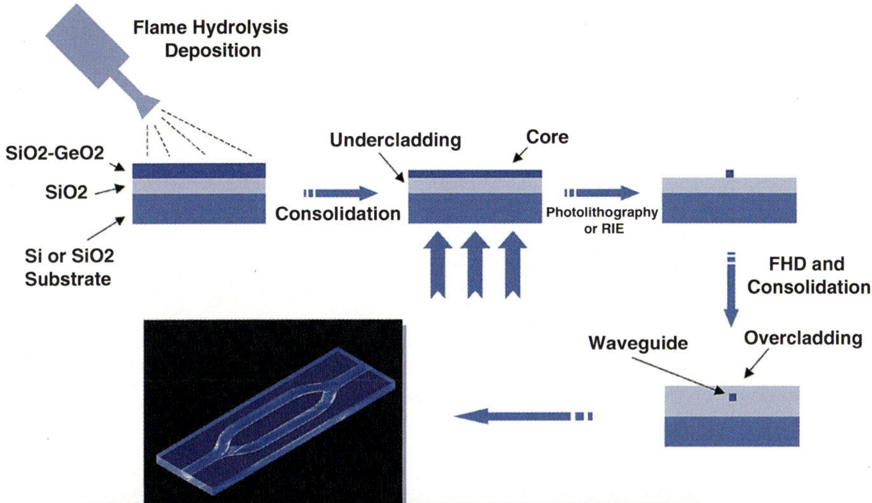

Characteristics	Waveguide Low-Δ	Medium-Δ	High-Δ	Superhigh-Δ
Index difference (%)	0.3	0.45	0.75	1.5 – 2.0
Core size (μm)	8×8	7×7	6×6	4.5×4.5 – 3×3
Loss (dB/cm)	<0.01	0.02	0.04	0.07
Coupling loss (dB/point)	<0.1	0.1	0.4	2.0
Bending radius (mm)	25	15	5	2

Fig. 7.2 Illustration of silica-based planar waveguide fabrication technique (Pham 2004)

techniques, is compatible with both clean-room processes and high-volume wafer production (in which more than 100 wafers can be simultaneously loaded in deposition chambers). A raised refractive index is created in the core-guiding region by the addition of phosphorus, germanium, or both, during the deposition process. Photolithography technology and reactive ion etching (RIE) are then used to pattern the waveguides.

The FHD deposition process is different from the CVD approach. Glass precursor chemicals are introduced into a hydrogen/oxygen torch in which a flame hydrolyzes the chemicals to form the appropriate glass composition, as shown in Fig. 7.2. Glass particles (roughly 0.1 μm in diameter) are deposited onto the substrate (SiO_2 or Si) in a thick, porous and fluffy layer. Finally, this fragile structure is placed in a furnace and heated to consolidate the porous layer into a solid, clear glass layer free of bubbles or any other defects. The waveguides are then patterned using standard photolithography and RIE. From a production standpoint, the use of silica waveguides has a number of significant advantages. An index of refraction roughly equal to that of optical fiber minimizes losses at the fiber-chip interface. The CVD process is also mature and allows low-cost volume production using a wafer-processing technology similar to the one developed for the semiconductor industry. The large dimensions of AWG devices (typically 1 × 3 cm) make it imperative that optimal clean-room conditions be maintained in order to avoid incorporation of unwanted particles. In that respect, CVD processing has an immense advantage. The main drawback of silica waveguides is the limited range of refractive-index differences, which ultimately restricts the reduction in size of individual devices (Pham 2004).

As discussed in previous chapters, silicon and silicon oxynitride (SiON) can also be used to form the waveguide material. A key benefit is the use of advanced process technologies originally developed for the semiconductor industry. Automated assembly techniques can be used to integrate passive devices with active components such as lasers or photodetectors, creating low-cost integrated devices. Furthermore, very-high-contrast waveguides can be made from these materials. This allows tight-bend radii and hence the possibility for compact integrated devices. However, due to the large refractive-index difference at the waveguide-fiber interface, propagation losses and coupling efficiencies constitute challenges for products based on silicon.

7.2.2 Refractive Index Profiling of Planar Waveguides

The refractive index profile (RIP) of the waveguide core plays an important role in characterizing the properties of the planar waveguide. It allows the determination of the waveguide's numerical aperture (NA) and of the number of modes propagating within the light guide core, while defining intermodal and/or profile dispersion caused by the waveguide itself. Furthermore, since the impulse response and, consequently, the information-carrying capacity of the waveguide is RIP-dependent, it is essential for PLC manufacturers to produce controlled waveguide profiles with great accuracy. Accurate knowledge of the RIP allows IO designers to reduce optical waveguide device manufacturing costs through tight control of the planar waveguide fabrication processes during the R&D phase (Pham 2004).

7.2.3 Silica Waveguide Devices

PLCs using silica-based optical waveguides are fabricated on silicon or silica substrate by a combination of FHD and RIE, as shown Fig. 7.2. Fine glass particles are produced in the oxyhydrogen flame and deposited on the host substrate (Si or SiO_2). After undercladding and core glass layers are deposited, the wafer is heated to high temperature for consolidation. The circuit pattern is fabricated by means of photolithography and RIE. Finally, core ridge structures are covered with an over-cladding layer and consolidated again. Since the typical bending radius R of a silica waveguide is between 2 and 25 mm, the chip size of the large-scale integrated circuit (IC) becomes several square centimeters. Therefore, propagation loss reduction and uniformity of refractive indices and core geometries throughout the wafer are essential. Propagation loss of 0.1 dB/cm is obtained in a 2-m long waveguide

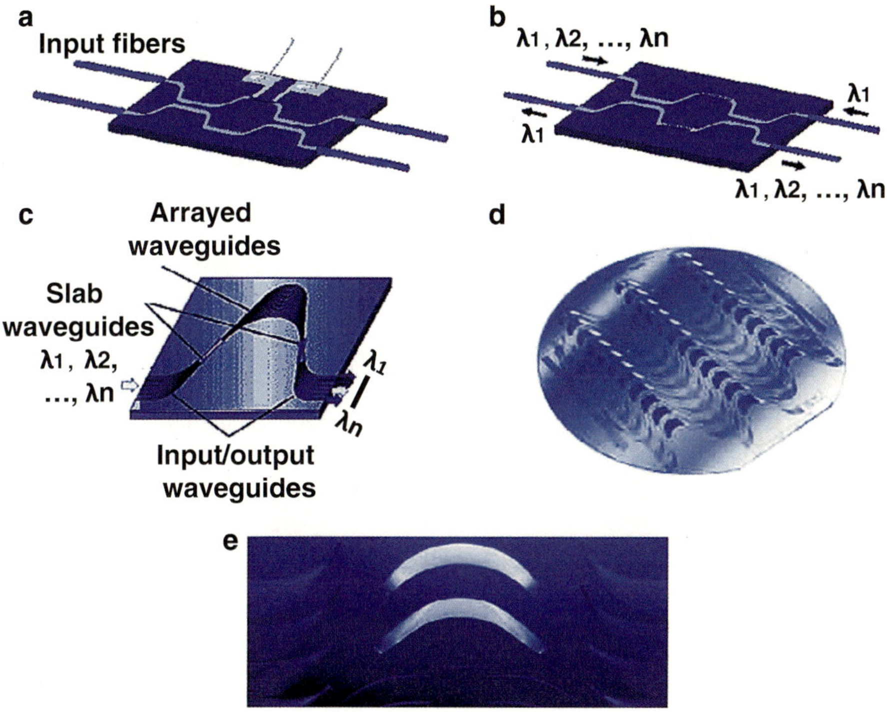

Fig. 7.3 Examples of silica-based waveguides (Pham 2004): (**a**) electro-optic 2 × 2 switch; (**b**) optical add-drop multiplexer; (**c**) optical AWGs; (**d**) SiO_2 AWGs on Si wafer; and (**e**) detailed view of two SiO_2 phasars

with $\Delta = 2\ \%$ index difference ($R = 2$ mm) and loss of 0.035 dB/cm is obtained in a 1.6-m long waveguide with $\Delta = 0.75\ \%$ index difference ($R = 5$ mm). For instance, the propagation losses of low-Δ and medium-Δ waveguides are about 0.01 dB/cm, while high-Δ and super high-Δ waveguide losses are about 0.04 and 0.07 dB/cm, respectively. The low-Δ waveguides are superior to the high-Δ waveguides in terms of fiber coupling losses with standard single-mode fibers. On the other hand, the minimum bending radii for high-Δ waveguides are much smaller than those for low-Δ waveguides. As a result, high-Δ waveguides are indispensable in constructing highly integrated and large-scale optical circuits such as N × N star couplers, AWG multiplexers and dispersion equalizers. Different types of devices can be manufactured to suit various applications. Some examples of silica waveguide-based IO devices include (Pham 2004): (a) N × N star coupler used in high-speed, multiple-access optical networks; (b) AWGs used to increase the aggregated transmission capacity of single-strand optical fiber; (c) Flat spectral response AWGs; (d) Uniform-loss and cyclic-frequency (ULCF) AWGs; (d) Athermal AWGs; (e) Phase error compensated (PEC) AWGs; (f) Optical add-drop multiplexers; (g) N × N matrix switches; and (h) Lattice-form programmable filters. Figure 7.3 shows some of these samples.

The reduction of power consumption is the key for actual deployment of these devices. For instance, silica waveguide TOs, in general, consume significant electrical power and limit the integration density due to thermal management issues. Thermal isolation grooves or of hybrid integration with microelectromechanical systems or polymers show possible directions to improve the thermal management. There are, of course, other PLC technologies to consider besides silica waveguides, such as semiconductors (GaAs, InP), polymers, silicon oxynitride (SiON), lithium niobate ($LiNbO_3$), and silicon-on-insulator, and each will undoubtedly find its own niche, but it is difficult to believe that silica waveguide PLCs will be completely replaced by another technology any time soon because of silica's extreme robustness, very low loss, and near perfect match to optical fiber, which is also made of silica. More specifically, silica waveguides have the following advantages (Doerr 2006): (a) Low propagation loss; (b) The same refractive index and mode field diameter as conventional single-mode fiber (low coupling loss and reflection); (c) Excellent physical and chemical stability (they can withstand end face polishing for fiber coupling and have high long-term reliability); (d) They can be fabricated on an inexpensive large size Si wafer (low cost and large scale); and (e) They offer easy phase controllability with a thin film heater via the thermo-optic effect.

Integrated silica-based waveguides have historically been niches of the analog/digital (A/D) and sensor fiber-optic applications. Other applications may include (Pham 2004): (a) The largest of these is, perhaps, the telecommunications industry, which uses IO devices for multigigabit bidirectional communication data transmission, signal splitting and loop distribution. (b) CATV, where IO modules will be used for external modulation in fiber optic-based signal distribution systems. In both telecommunications and CATV, IO devices enable signal transmission at higher data rates over longer distances. (c) Instrumentation with fiber-optic gyroscopes as the major application.

7.3 Silicon Oxynitride Waveguides

Silicon oxynitride (SiO_xN_y or SiON for short) has been increasingly applied in various integrated optical devices. Application of this material has been mainly motivated by its excellent optical properties, such as low absorption losses in the visible and near infrared wavelength range. Moreover, the refractive index of SiON thin films can be easily adjusted over a large range, i.e., between 1.45 (SiO_2)–2.0 (Si_3N_4). This large flexibility in choosing the refractive index adds to the attractiveness of this material for designers of integrated optical devices. In addition, standard silicon substrates can be used and reliable techniques that are well compatible with standard silicon integrated circuit processing, are available for controlled deposition of homogeneous, high quality thin films. This opens a potential route toward low-cost mass production (de Ridder et al. 1998).

7.3.1 Material Processing Technology

Silicon oxynitride thin films can be deposited using various technologies. The most extensively applied technologies for the deposition of SiON waveguiding layers are: plasma enhanced chemical vapor deposition (PECVD) and low pressure chemical vapor deposition (LPCVD).

PECVD layers can be deposited from 2 % SiH_4 –N_2, N_2O, and NH_3, applying a parallel-plate Electrotech 210 PECVD machine with a low frequency RF-source (187.5 kHz), for instance. The LPCVD layers can be grown from SiH_2Cl_2, O_2, and NH_3, applying a Tempress hot-wall LPCVD reactor, for example. For the production of SiON layers with specific parameters, the LPCVD process are mostly used for depositing waveguides in the higher refractive index range ($n \geq 1.7$), while the PECVD process is most suitable for controlled deposition of layers with refractive indices below 1.7. In the respective refractive index ranges, both processes have an excellent uniformity of the layer thickness and the refractive index and a good run-to-run reproducibility of both these parameters. For some of the most extensively applied refractive indices the layer deposition can be further optimized by introducing proper technological procedures. For both PECVD and LPCVD grown layers, the optical loss of slab-type waveguides is below 0.2 dB/cm in the visible range of light. At wavelengths around 1,500 nm, the optical losses are considerably increased, due to absorption caused by vibrational overtones of the N–H and Si–H bonds. The absorption loss of as-deposited layers depends on the deposition process and the composition (O/N-ratio) of the SiON thin film. For layers with comparable composition, the absorption loss of LPCVD grown layers will be significantly lower than for PECVD layers. For example, in LPCVD silicon nitride the total hydrogen content (i.e., N–H Si–H) is approximately 4 at%, while in PECVD nitride this hydrogen content is about 22 at%. When increasing the O/N-ratio, the total hydrogen content of the layers will be decreased. For example, in silicon oxynitride layers with a refractive index around 1.49, a total hydrogen content of 4 at% has been measured. However, the remaining absorption loss can be reduced by post deposition heat treatment (annealing) at temperatures up to 1,150 °C. In that way, the optical loss of low-index PECVD SiON can be reduced to less than 0.2 dB/cm at 1,550-nm wavelength, corresponding to a decrease of the hydrogen content to less than 1 at% (de Ridder et al. 1998).

7.3.2 SiON Waveguide Design and Fabrication

A standardized, SiON-based waveguide structure for application in telecommunication devices, for example, tunable wavelength division multiplexed (WDM) add–drop wavelength filters, operating at 1,550-nm wavelength has been developed. The waveguide structure has to fulfill the following demands for commercialization: low insertion loss, polarization-independent operation, high integration density, and

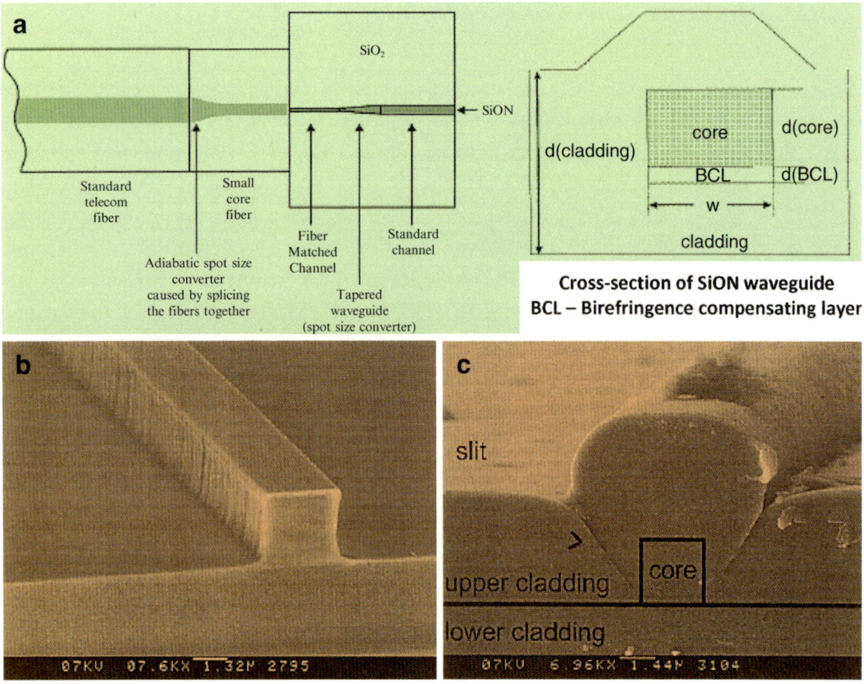

Fig. 7.4 SiON waveguide structures (de Ridder et al. 1998): (**a**) fiber-to-chip coupling; (**b**) RIE etching profile; and (**c**) cross section of the growth profile of the upper cladding

the accessibility of reliable low-cost fabrication technologies. Moreover, requirements for polarization dependence can be inferred from specifications like Telcordia GR1209. Other requirements may include monomode waveguide operation; bending radius (<1 mm) at bending loss (<0.01 dB/90°); fiber-to-chip coupling loss (<1 dB/facet) and standard single-mode fiber with 9-μm core diameter; waveguide loss (<0.2 dB/cm); and polarization dependence of channel or the material birefringence $\Delta n_{\mathrm{eff}} = n_{\mathrm{eff,\ TM}} - n_{\mathrm{eff,\ TM}} < 5 \times 10^{-5}$, where NTM and NTE are the refractive-index components, which are relevant for TM and TE polarization, respectively (Worhoff et al. 2007). In addition, the relevant properties can be determined by simulation on various structures having different refractive indices of the waveguiding layer and different shapes of the channel structure. Furthermore, the influence of a nonideal waveguiding structure, due to the tolerances in the applied technological fabrication processes, should be taken into account. The structure that is shown in Fig. 7.4, has found to be very promising for application in telecommunication devices (de Ridder et al. 1998).

The birefringence compensating layer (BCL) has been introduced for geometrically compensating the material birefringence introduced polarization dependence. The TE-field distribution is much stronger influenced by the very thin BCL than is the TM field, leading to a shape-anisotropy that has a sign opposite to the stress-induced birefringence. The BCL will cancel out the stress birefringence ($\Delta n_{\mathrm{eff}} = 1.7 \times 10^{-3}$) at a thickness $d_{\mathrm{BCL}} = 26$ nm. The TE effective refractive index of this standardized waveguiding channel is calculated to be $n_{eff,\ TE} = 1.47$ with a residual waveguide birefringence($\Delta n_{eff} = 2 \times 10^{-6}$). The mismatch loss can be reduced by expanding the modal field in the waveguiding channel, for example, by horizontal tapering resulting in a decrease of the channel width to, as is shown in Fig. 7.4a. When tapering the channel width to 1 m, the mode mismatch loss will be below the demanded value, even at a slight fiber displacement (de Ridder et al. 1998).

The waveguiding structure can be fabricated on thermally oxidized Si wafers. The cladding and the core layers are PECVD-deposited from 2 % SiH_4 –N_2 and N_2O, applying a substrate temperature of 300 °C, a pressure of 650 m-torr. For definition of the waveguiding channels, a 200-nm-thick chromium masking layer has been sputtered on top of the core layer. The waveguide pattern is defined in the chromium layer with standard lithography, applying S1805 photoresist, and a wet Cr-etching step. Then, the waveguiding channels are etched in an electrotech 240 RIE machine applying an SF plasma, at 10-m-torr pressure, 75-W plasma power and cooling of the lower electrode to 10 °C. The etching profile is highly anisotropic, as can be seen in the SEM photograph in Fig. 7.4b. After removal of the chromium mask by wet etching, the upper cladding layer (PECVD SiO_2) is deposited. A cross section of the growth profile of the upper cladding is shown in Fig. 7.4c. Although the step coverage of the cladding deposition turned out to be excellent, a nano-slit remains between the two growth planes. This nano-slit can hamper the overgrowth with for instance, thin metal layers which is essential when introducing electrodes on top of the upper cladding. The profile of the upper cladding layer can be planarized, by reflow, which would be realized at acceptable temperatures (1,150 °C) when doping the PECVD cladding material with softeners (de Ridder et al. 1998).

7.3.3 SiON Waveguide Devices

1. Optical wavelength filter
   In optical WDM networks, an optical fiber carries signals at a number of different wavelengths. Add–drop multiplexers are important elements in such networks as they are required for extracting a signal in a single-wavelength channel from the fiber and substituting it with another signal, while leaving the other channels undisturbed. A very elegant and popular type of wavelength (de)multiplexer is the phased array. This type of multiplexer cannot be easily and quickly tuned. However, at the expense of additional complexity, a discrete tuning functionality can be obtained by combining a phased array with a switching matrix. Another class of wavelength (de)multiplexers is based on cascading Mach–Zehnder interferometers (MZI's) having unequal branches (Fig. 7.5a). In order to obtain also add–drop functionality, a cascading configuration of the MZI's has been used as shown in Fig. 7.5b. This configuration shows a binary filter operation implying only logarithmic growth of the number of filter elements with the number of optical wavelength channels. The periodic nature of the MZI-response is exploited by halving the number of wavelength channels at each stage. By tuning an MZI

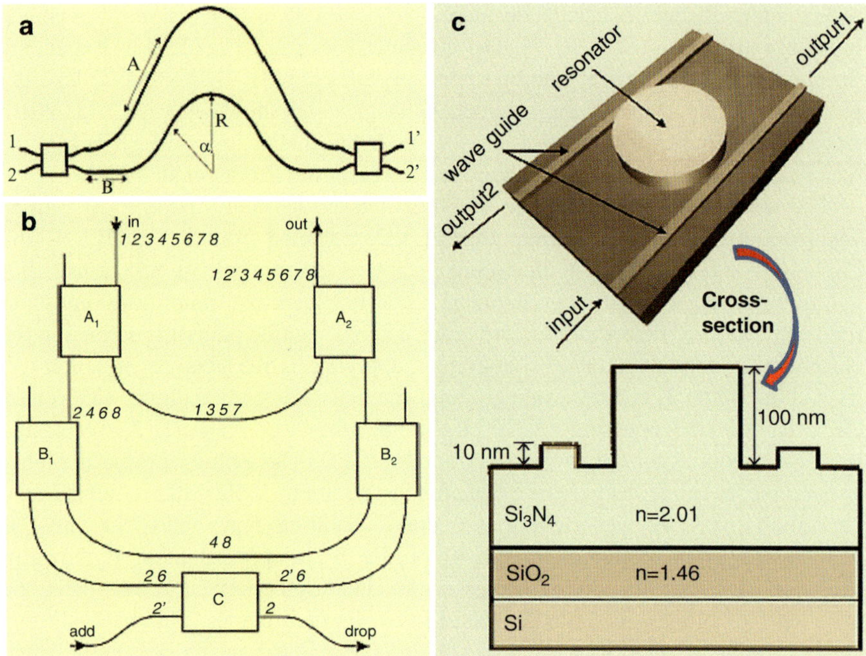

Fig. 7.5 Illustration of MZI and resonator devices (de Ridder et al. 1998): (**a**) schematic diagram of a MZI wavelength filter. All bends in the interferometer branches have the same bending radius R and the same total subtended angle of are α, so that the path length difference is given by 2(A–B). The rectangular structures are 3-dB MMI-couplers. (**b**) Eight-channel (numbered 1…8, equal wavelength spacing) add–drop multiplexer. Each rectangle stands for an MZI as shown in (**a**) (three different types A, B, C having FSR's in proportion 1:2:4, respectively). This example shows adding and dropping channel 2 and (**c**) cylindrical microcavity resonator in a channel-dropping filter configuration

over half a free spectral range (FSR), the outputs are reversed. Due to mirror symmetry, new information at the "add" port at the same wavelength as the dropped signal will be combined with all other channels through the MZI's C, B and A, if B and A are identical to B and A, respectively. For instance, thermo-optically tunable MZI filter elements have designed with having 2-nm FSR at 1,550 nm (for 1-nm channel spacing) of the type shown in Fig. 7.5a, having a $3 \times 3\ \mu m^2$ core but without the BCL. The path-length difference of both branches was 820 m, the bend radii were conservatively chosen to be 7 mm and the MMI-couplers were 35 m wide by 920 m long. The angle, subtended by the bends was optimized in such a way that, under the constraint of the output channels being in line with the input channels, the total MZI filters had minimum dimensions for a given path-length difference. The heaters (nonoptimized: 200-nm-thick gold, 20 m wide by 4.5 mm long) for actuating the thermo-optical tuning mechanism were deposited on top of the 5-m-thick SiO_2 cladding layer, above one of the MZI branches.

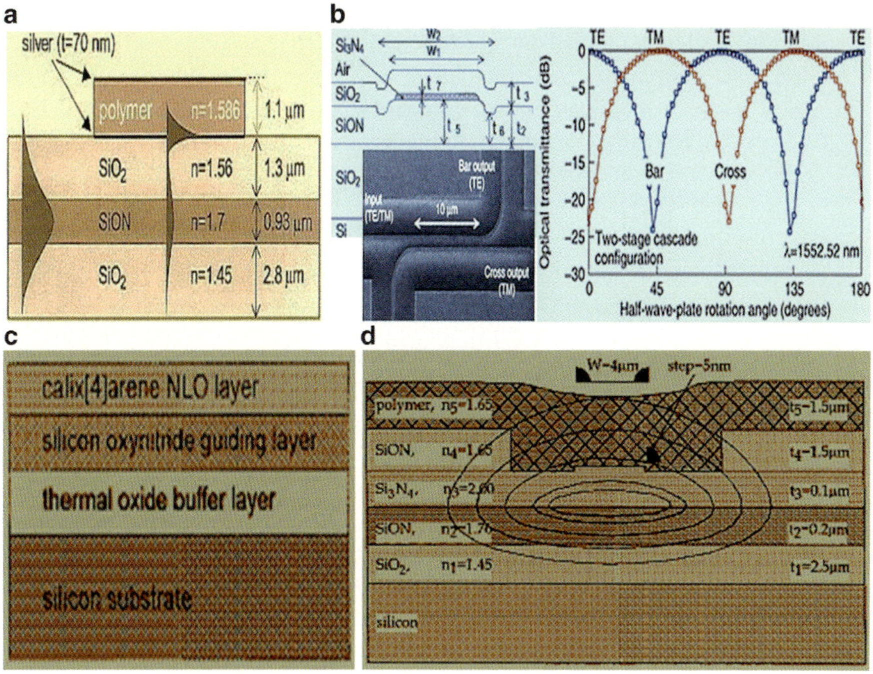

Fig. 7.6 Some examples of SION-based waveguide structures (de Ridder et al. 1998): (**a**) electrooptical modulator based on resonant coupling between a low-loss SiON waveguide and a lossy surface plasmon waveguide; (**b**) polarization splitter and splitter characteristics for a two-stage cascade configuration; (**c**) slab-type waveguiding structure for noncritically phase-matched SHG; and (**d**) cross section of nonlinear waveguide, which can be defined locally in an otherwise linear waveguiding structure by etching a window into the top SiON cladding (n_4) and applying a nonlinear polymer on top of it

The crosstalk was −20 dB for the cross-ports (1-2′ and 2-1′ in Fig. 7.5a) and −10 dB for the bar ports (1-1′ and 2-2′). The latter result could be explained by asymmetry introduced by nonperfect MMI's. A heating power of 0.8 W was required for tuning the device over a full FSR, giving a tuning sensitivity of 0.4 W/nm (de Ridder et al. 1998).

2. Microresonator
   Microcavity resonators have an interesting application as channel-dropping filters in WDM systems. For instance, microcavities with radii down to 23 m, coupled to straight channel waveguides as shown in Fig. 7.5c, have been fabricated (de Ridder et al. 1998).
3. Electro-optic modulator
   In an electro-optic modulator, a SiON waveguide has designed to obtain phase matching at $\lambda = 1{,}523$ nm between a guided mode in this waveguide and a lossy surface Plasmon mode bound to a thin silver film with an electro-optic polymer cladding. The waveguide structure is shown in Fig. 7.6a. By applying a voltage to the two silver electrodes, the resulting electric field changes the refractive

index of the polymer, which in turn changes the deviation from phase matching between the mode guided by the low-loss SiON waveguide and a high-loss surface plasmon mode, and hence the power transmitted by the system. In this type of resonant coupling devices, the exact control of refractive index and layer thickness is crucial to device performance. An extinction ratio of more than 15 dB is feasible with an on-chip excess loss with respect to a straight waveguide of circa 3 dB (de Ridder et al. 1998).

4. Polarization splitter
   An adiabatic TE/TM mode splitter, shown in Fig. 7.6b, has been designed and fabricated, by making use of shape birefringence in SiON-based waveguiding channels. It is based on the mode-sorting effect in an asymmetric Y-junction. The design calls for a waveguiding material with a refractive index between the low one ($n = 1.45$) of the SiO_2 cladding and the high one ($n = 1.98$) of an asymmetrically placed Si_3N_4 layer which induces the shape birefringence. In this case SiON with $n = 1.55$ fits the requirements. In addition, fine-tuning of the SiON index allowed to optimize the design with respect to sensitivity to the technological tolerances. A crosstalk of -11 dB for both TE and TM has been measured, and a best value of -17 dB for TE. The excess loss was 1 dB compared to a straight channel waveguide (de Ridder et al. 1998).
5. Second harmonic generator
   LPCVD grown SiON layers have been applied in second harmonic generating (SHG) devices based on phase-match between the TM0 and the TM1 modes. The applied slab-type structure is schematically shown in Fig. 7.6c. The refractive index and layer thickness of the silicon oxynitride layer have been optimized with respect to the following device requirements, which are important for efficient SHG (de Ridder et al. 1998): (a) maximization of the field overlap of the fundamental TM0 and second harmonic TM1 mode in the optically nonlinear (NLO) layer; and (b) minimization of the phase mismatch between the TM0 and the TM1 modes over the device length of 10 mm by minimizing the sensitivity of the designed structure toward technologically critical parameters. This approach was shown to be successful. In an optimized layer structure, applying an SiON layer with a refractive index of $n{-}1.689$, blue light at a wavelength of 481.6 nm has been generated with an efficiency of 1.8×10^{-3} %W^{-1}.
6. Nonlinear MZI
   An MZI having a nonlinear optical material in one of its branches may act as an opto-optical switch. One of the problems to be solved in that kind of device is to integrate the linear (passive) and nonlinear waveguides. A solution for the nonlinear waveguide is shown in Fig. 7.6d. SiON technology gives the opportunity to design the waveguide in such a way that a relatively large field extends into the nonlinear polymer. A seamless integration with the passive waveguide is possible by choosing the refractive index n_4 of the top SiON cladding identical to that of the polymer (the nonlinear waveguide is made by etching a window into the top cladding and applying the polymer on top). An all-optical phase-shift of 0.5 rad. has been measured, using 100 ps pulses at 1,053 nm wavelength.

The peak intensity of the pulse in the input waveguide channel was 11 W (de Ridder et al. 1998).

Silicon oxynitride with low optical losses can be made with good uniformity and reproducibility of refractive index (n = 1.45–2.0) and layer thickness. Therefore, the material can be applied to a great variety of devices. On the one hand, general-purpose waveguides with low birefringence, good fiber-chip coupling efficiency and bend radius down to 2 mm can be made. This type of waveguide can be very useful for passive or thermo-optic devices. On the other hand special-purpose waveguides have been designed and used, which demonstrate the usefulness of both the high contrast features like polarization splitter, and microresonator, and the possibility of index-matching as EO-modulator, SHG-generator, and nonlinear Mach–Zehnder (de Ridder et al. 1998).

7.4 Ion-Exchanged Glass Waveguides

The fabrication of optical waveguides in glass by ion exchange was first achieved in 1972 using a melt containing thallium ions. The Tl^+–Na^+ system was problematic due not only to the mild toxicity of Tl^+ but also to the large index change (~0.1), which causes difficulty in repeatedly producing single-mode waveguides. Subsequently, waveguides using a melt containing silver ions were produced and become the most common process. Other dopant ions include Cs^+, Rb^+, K^+, and Li^+. The creation of integrated optical devices in glass offers several obvious benefits over other technologies: Intrinsic absorption is very low in the near infrared region of the spectrum. Coupling losses to optical fiber are minimized due to the similarity in refractive index. In addition, glasses are amorphous, meaning that they exhibit no intrinsic material birefringence, unlike crystalline semiconductors. This is not to say that birefringence is not an issue in glass waveguides—both the shape of the waveguide and the stresses that arise during fabrication contribute to birefringence, but with proper design, these can be balanced against each other to produce single-mode devices with very low polarization dependence (West 2004; Honkanen et al. 2006).

In addition to ion exchange, other processes exist through which glass waveguides have been fabricated. Most involve the deposition of thin glass films such as CVD, FHD, and sol–gel coating, followed by RIE to define the device geometry, and subsequent deposition of the over-cladding. The multiple deposition steps and etching make these methods costly and time consuming. The benefits of ion exchange over competing glass-based technologies are numerous. Ion-exchanged waveguides possess many desirable characteristics (Honkanen et al. 2006): (a) It is very easy to make waveguides exhibiting very low propagation losses of less than 0.1 dB/cm. (b) The increase in refractive index between the waveguide and the substrate is relatively small, allowing for easy fabrication of single-mode waveguides, and excellent mode matching to single-mode fiber can be achieved through a technique of waveguide burial. And (c) ion-exchanged waveguides can exhibit very low birefringence across a broad range of waveguide widths. This is critical for certain resonant devices and devices that contain both single- and multimode waveguides. The process is cost effective, requiring no complicated material growth steps following the photolithography that is common to all processes. In addition, it is tolerant to imperfections in the photolithography—edge roughness in the mask that defines the waveguide geometry is rendered less damaging due to the diffusive nature of the ion exchange.

As a result, ion exchange has the potential for the fabrication of low-cost, low-loss and thermally stable optical waveguides. The main motivations for the use of ion exchange in glass for optical interconnects are the following (Lehky 2006):

1. Glass is the material of predilection for optics. It provides minimal optical attenuation and is rugged against a diversity of atmospheric, thermal and mechanical strains. Glass is commercially available in a multitude of forms and compositions. It is important to find a substrate, which allows the fabrication of low-loss waveguides. Here, aspects like glass composition, production methods, available substrate size (i.e., >50 cm side for board-to-board interconnects) and price have to be considered.
2. Ion-exchange, almost exclusively from salt melts, is a technology which was successfully used in integrated optics (especially for telecommunications and sensor applications) to produce low-loss, fiber-compatible optical devices at low costs. Ion exchange from metal films (silver and possibly copper) may provide a scalable and industrially compatible planar process for optical waveguide fabrication. The use of copper for optical waveguide fabrication is further interesting as it is the key material in printed circuit board manufacturing and could benefit from the large available manufacturing and patterning knowhow, while simultaneously providing for process and material compatibility.

The applications of glass-based waveguide devices for integrated optics extend far beyond the telecommunications industry. One rapidly emerging area is the design of integrated optical sensors, in which a guided wave interacts with the environment, causing a perturbation of the optical field. For sensor applications, glass-based devices have a distinct advantage over those fabricated in semiconductors or inorganic crystals such as $LiNbO_3$. Regardless of whether the sensor operates on refractometry or absorption, sensitivity is proportional to the overlap of the guided mode with a sensitized superstrate. Depending on the measurand, this layer is commonly based on organic compounds having an index of around 1.4–1.6, or an aqueous solution with an index near 1.33 (West 2004). These indices are very close to that of glass ($n \sim 1.5$). The small dielectric barrier between the glass and the superstrate results in a large mode overlap, maximizing the sensitivity. In contrast, the higher indices of $LiNbO_3$ ($n = 2 \cdot 2$ at 633 nm) and semiconductors ($n > 3$) greatly reduce the influence of the superstrate. Of all glass waveguide technologies, ion exchange holds the most potential for sensor applications due to its low cost. Unlike telecommunications devices, many sensors are designed to be disposable. With the introduction of the erbium doped fiber amplifier (EDFA) and the resulting importance of WDM telecommunication systems, there has been a recent emphasis on the development of glass-based devices that operate in the 1.55 μm wavelength range. Such devices include *y*-branches, Mach–Zehnder interferometers, and ring resonator gyroscopes. The observation of ultraviolet photosensitivity in some glasses used for ion exchange has supported the successful production of grating-based devices. Furthermore, the success in forming waveguides by ion exchange in quantum dot doped glass shows promise for nonlinear device applications (Honkanen et al. 2006).

The ion exchange process has been utilized for the production of active devices as well. The earliest such devices were splitters that included an amplifying region to compensate for the splitting loss. One tremendous benefit of ion exchange over semiconductor-based processes is the presence of hybrid substrates, glass wafers that consist of active and passive sections that have been joined together in the same plane. In semiconductor processes, fabricating such a waveguide usually requires regrowth techniques. Ion-exchanged waveguide amplifiers have been extensively investigated. However, they have achieved little practical success. Two major factors contribute to this. First, the short length of waveguide devices requires an extremely high gain per unit length in order to compete with fiber amplifiers. This necessitates a high concentration of rare-earth ions within the glass, which eventually leads to a decrease in pumping efficiency due to cooperative upconversion. Second, unlike rare-earth doped fiber, which has radial concentration and index profiles that are optimized for pump/signal/gain medium overlap, waveguides that are ion exchanged into a homogeneously doped substrate exhibit a relatively poor overlap, decreasing the gain. Waveguide amplifiers do have an advantage over their fiber-based counterparts, nonetheless, in that the passive elements (pump–signal combiner, tap coupler for power monitoring, etc.) can be integrated onto the same substrate as the amplifier. Ion-exchanged waveguide lasers, however, are becoming increasingly popular. High reflectivity dielectric mirrors or Bragg gratings can be used to provide a net round-trip gain in ion-exchanged Yb^{3+}/Er^{3+}-doped waveguides of a few centimeters in length, leading to single longitudinal mode lasing. Alternatively, the feedback may be provided by a grating etched directly into the active substrate (Honkanen et al. 2006).

7.4.1 The Ion-Exchange Techniques

A wide range of optical waveguide properties can be achieved by ion-exchange. These waveguides include both strongly and weakly guiding single-mode waveguides, large multimode waveguides, surface guides as well as buried waveguides.

1. Thermal ion-exchange
   During ion-exchange, the network-modifying ions are not very tightly bound to the silicate network. They are ionically bound to the non-bridging oxygen atoms.

The mobility D of these ions follows an Arrhenius-type temperature dependence, i.e., $D(T) = De^{\left(-E_A/R\right)}$ with E_A the activation energy of the diffusion. Thus, at appropriate temperatures, a concentration gradient of similar ions at the glass surface will generate the movement of these network-modifier ions. The exchange of mobile charged species occurs on a one-to-one basis to preserve glass neutrality and the rate of ion exchange is controlled by the slower mobile ions. As the non-bridging oxygen atoms are covalently bound to the silicon atoms and tightly fixed to the network, their movement in the glass is impeded (Lehky 2006).

When a glass substrate with the network-modifying ion A^+ (e.g., Na^+) is immersed into a molten salt with a cation B^+ (e.g., Ag^+, K^+, Tl^+,...), at the glass-melt interface the concentration of both ions abruptly drops to zero. To recover an equilibrium situation, thermal agitation at the interface causes B^+-ions to replace A^+-ions and vice versa. In other words, the cations diffuse away from the interface. The mobility of the A^+ cations in the molten salt being higher, they move rapidly away from the interface, whereas the B^+ cations slowly move into the glass, creating a thin layer near the glass surface. The concentration of exchanged B^+ cations is highest at the glass surface and decreases monotonically inside the glass. This process is accelerated at higher temperatures, because of the increase of ion mobility with temperature, but also because the glass structure relaxes with increasing temperature (decrease of the viscosity of the glass substrate). Table 7.1 gives a few relevant parameters for the commonly used ions for ion-exchange processes: ionic radius, electronic polarizability, commonly used salts and the corresponding melting or decomposition point and also the induced refractive index change, birefringence and propagation loss after ion-exchange. The one-dimensional diffusion equation of thermal ion exchange is given by (Lehky 2006):

$$\frac{\partial N_B}{\partial t} \frac{\partial}{\partial x}\left(D\frac{\partial N_B}{\partial x}\right) \tag{7.1}$$

$$D = \frac{D_A\, D_B}{N_A\, D_A + N_B\, D_B} m \tag{7.1a}$$

$$N_A = \frac{C_A}{C_A + C_B} \tag{7.1b}$$

$$N_B = \frac{C_B}{C_A + C_B} \tag{7.1c}$$

where N_A (N_B) is the mole fraction of the A^+ (B^+) ion in the glass, C_A and C_B are the concentrations of A^+ and B^+ ions, respectively, in the glass, D the interdiffusion coefficient, D_A (D_B) the self-diffusion coefficients of the A^+ (B^+) ion, and $m = d\,(\ln a_A)/d\,(\ln N_A)$ is related to the concentration derivative of the A^+ ion activity a_A in the glass [111] and is equal to unity for an ideal system. For the initial and boundary conditions: $C(x,0) = 0$, $C(\infty, t) = 0$ and $C(0, t) = C_{B0}$,

Table 7.1 Typical parameters for the commonly used ions in ion-exchange processes (Lehky 2006)

Ion	Ionic radius [Å]	Electronic polarizability α [Å^3] [104] ([105])	Salt	Melting (decomp.) point [°C]	Refractive index change Δn	Birefringence $\delta n \times 10^4$	Propagation loss [dB/cm]	Comments
Li^+	0.65	0.03 (0.029)	$LiNO_3$	264 (600)	0.012	2	>1	Tensile stresses and surface crystallization
Na^+	0.95	0.41 (0.24)	$NaNO_3$	307 (380)	−0.02–0.002	–	–	–
K^+	1.33	1.33 (0.9)	KNO_3	334 (400)	0.009	2–20	<0.2	Compressive stresses
Rb^+	1.49	1.98 (1.4)	$RbNO_3$	310 (–)	0.015–0.02	2–10	>1	Expensive
Cs^+	1.65	3.34 (2.44)	$CsNO_3$	414 (–)	0.04	2	0.1	Slow diffusion
Ag^+	1.26	2.4 (1.89)	$AgNO_3$	212 (444)	0.1–0.22	0–10	<0.2	Low thermal stability and photo-stability
Tl^+	1.49	5.2 (3.88)	$TlNO_3$	206 (430)	0.1–0.2	0–50	<0.1	High toxicity
Cu^+	0.96	1.73 (1.22)	CuCl	430 (–)	0.015–0.06	–	–	Risk of coloration (Cu^{2+})

where C_{B0} is the concentration of B^+ ions in the glass at the surface, the solution of (7.1) can be expressed as (Lehky 2006):

$$C_B(x,t) = C_{B0}\mathrm{erfc}\left(\frac{x}{2\sqrt{Dt}}\right) \tag{7.2}$$

where $\mathrm{erfc}(z) = \frac{2}{\sqrt{\pi}}\int_z^{\infty} \mathrm{e}^{-t^2}$ is the complementary error function.

2. Field-assisted ion-exchange
   If an electric field is applied across the glass substrate, in contrast to the previous case, both A^+ and B^+ ions move in the same direction in the glass, driven by the potential difference. For each ion B^+ coming in from one side, an ion A^+ comes out on the other side. The mobile ions become the charge carriers for the current flowing through the glass. Therefore, the integrated current represents the total number of ions introduced into the glass during the exchange and can be used to monitor the process. The field-assisted exchange is a thermal as well as an electric field-dependent. By adjusting the temperature and driving voltage, the relative contributions of the temperature and the electric field can be varied and thus the ion concentration profile is further controlled. The concentration profile can be written as follows (Lehky 2006):

$$C_B(x,t) = \frac{C_{B0}}{2}\left[\mathrm{erfc}\left(\frac{x-\mu Et}{2\sqrt{Dt}}\right) + \exp\left(\frac{\mu Ex}{D}\right)\mathrm{erfc}\left(\frac{x+\mu Et}{2\sqrt{Dt}}\right)\right] \tag{7.3}$$

where μ is the mobility of the (slower) B^+ ions in glass and is connected to the diffusion constant through the Einstein relation $D = \mu\, kT/q$, where q is the electrical charge of a particle, k is Boltzmann's constant, and T is the absolute temperature. For large electric fields the second term of (7.3) can be neglected.

Figure 7.7 shows the standard processing steps used for waveguide production by thermal and field-assisted ion exchange from molten salts (Honkanen et al. 2006). A diffusion barrier mask defining the waveguides is structured by photolithography. This mask usually consists of Al or Ti. Anodized Al (where the Al is oxidized on the surface) shows improved resistance against corrosion from the melts and prevents deposition of silver or other ions (e.g., trivalent indium in fluoride glasses) beneath the edge of the mask during ion-exchange. In a field-assisted ion-exchange process, the anode (+) is the molten salt containing the exchanging ions (Ag^+, K^+, Tl^+, ...). The cathode (−) can be a molten salt (usually $NaNO_3$), acting as a sink for the out coming sodium ions, or a metal film electrode (usually Al or Au). When metal films are used as a cathode, care has to be taken that the out coming sodium ions can be reduced and incorporated to the metal in order to avoid the accumulation of sodium ions at the glass surface, leading to the build-up of an electric field counteracting the externally applied field. After the first ion-exchange additional treatments can be performed with the aim of burying or modifying the concentration profile of the exchanged ion.

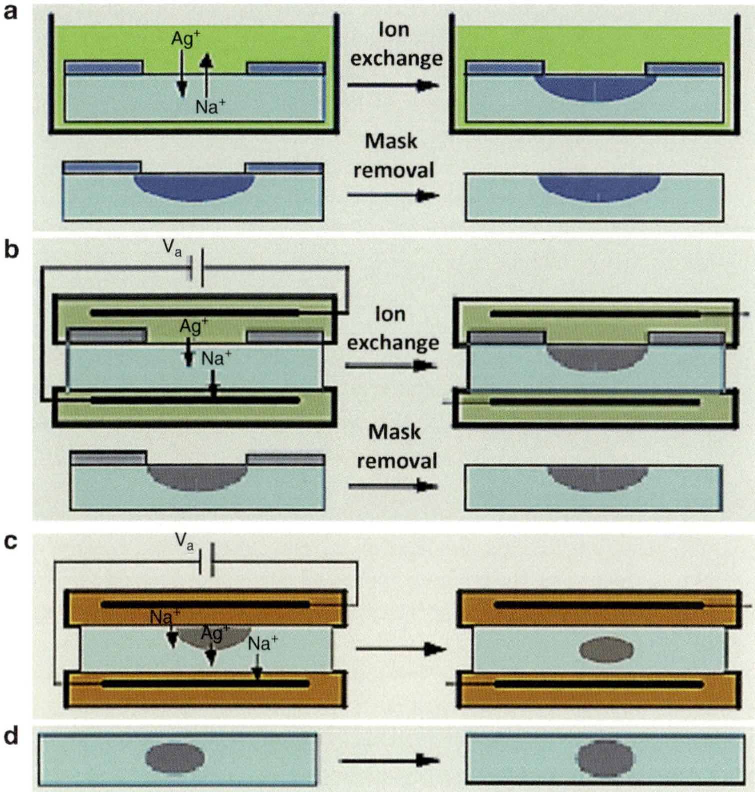

Fig. 7.7 Common processing configurations for ion exchange (Honkanen et al. 2006): (**a**) thermal exchange from a molten salt; (**b**) field assisted exchange from a molten salt; and (**c**) field assisted burial; and (**d**) thermal annealing

The application of an electric field during the ion exchange introduces a second ion-motion mechanism, also referred to as electro-migration. As a consequence the ion-exchange process will proceed much faster and, depending on the mobilities of the respective ions, a more step-like concentration profile can be achieved. However, with ion exchange from molten salts, it is a technological challenge to electrically insulate the anode from the cathode, at the generally high melting and exchange temperatures of the commonly used salts. Therefore, such a process is impractical and will not be useful for scale-up or even mass production. To circumvent this insulation issue, for example, a method is disclosed, where the mask is removed after the first ion exchange (e.g., with $AgNO_3$), then a "dry" electric field step is applied to increase the depth of the exchanged region, and a second, burying ion-exchange step, which decreases the refractive index at the surface, with e.g., a $NaNO_3$ melt, is performed. The benefit of removing the mask for the second step is that the profile broadening of the waveguide can be diminished, because Ag^+ and Na^+ ions move together in parallel, driven by the electric field. The concentration profile, and therefore also the RIP, can be shaped by a subsequent annealing step. However, it should be mentioned, that all these additional steps, which do not introduce new ions into the glass, will result in a decrease of the peak concentration. As a method to improve the lateral confinement of the waveguide in the first exchange step, so-called guard bands can be used. These closely spaced mask openings provide the desired field confinement into the depth of the glass substrate, thus giving a preferred direction for the exchanging ion movement (Lehky 2006).

3. Film ion-exchange technique
   An alternative to molten salt ion exchange is given by the metal film (mainly silver or copper) ion-exchange technique. This process is represented schematically in Fig. 7.8a. The fundamental difference between a molten salt and a metallic film is that the metallic film does not contain ions. Therefore, an electrochemical reaction (oxidation) is necessary to convert the metal atoms into ions: $B \rightarrow B^{+} + e^{-}$. The whole system can be considered as an electrochemical process occurring in a solid-state cell, where at the anode (+) the metal film is oxidized and at the cathode (−) the sodium ions are reduced to metallic sodium, while the glass acts like a solid electrolyte. Thus, the applied voltage during the process must overcome the chemical potential barrier between anode and cathode to activate the exchange process. No appreciable thermal diffusion of ions occurs without applied electric field. Another significant difference compared to a molten salt ion exchange is the limited amount of metal in a thin film. While only a fraction of ions in a salt melt participates in ion exchange, a metal thin film can be used practically completely. This has a substantial influence on the profile of the waveguide (e.g., uniformity in depth). The reaction kinetics of this oxidation is the result of several cascaded processes. A thin metal-oxide layer (e.g., Ag_2O) is formed at the metal surface, whenever it is exposed to an oxidizing medium. Oxygen is adsorbed on the surface of this oxide layer and captures an electron, which can tunnel through from the metal and converts the oxygen to O^{-}. Thereby, a strong inherent potential is established (Mott-Potential) between the metal and the adsorbed oxygen, with an electric field in the order of ~10^6 V/mm. This strong electric field causes the weakly bound metal cations to move through the metal-oxide, whereas the anions remain fixed. Every metal ion, which escapes from the metal surface as result of a defect, is impeded to recombine with the metal due the strong electric field and forms an interstitial cation in the oxide layer and finally recombines with the adsorbed O^{-} to form more oxide. The followed cation migration process (diffusion and electro-migration) depletes this interface region (oxide region) from cations and promotes further oxidation of the metal, which would otherwise be hindered by a too thick oxide layer, and a consequently weaker electric field generated by the Mott-potential. Figure 7.8b shows typical electrode structures for metal film ion-exchange. An additional metallic layer (e.g., Au) is often deposited over the film (anode), sometimes with a diffusion barrier layer (e.g., Ni) between the film and the cover layer. This cover layer improves the electrical contact and prevents oxidation and flaking of the

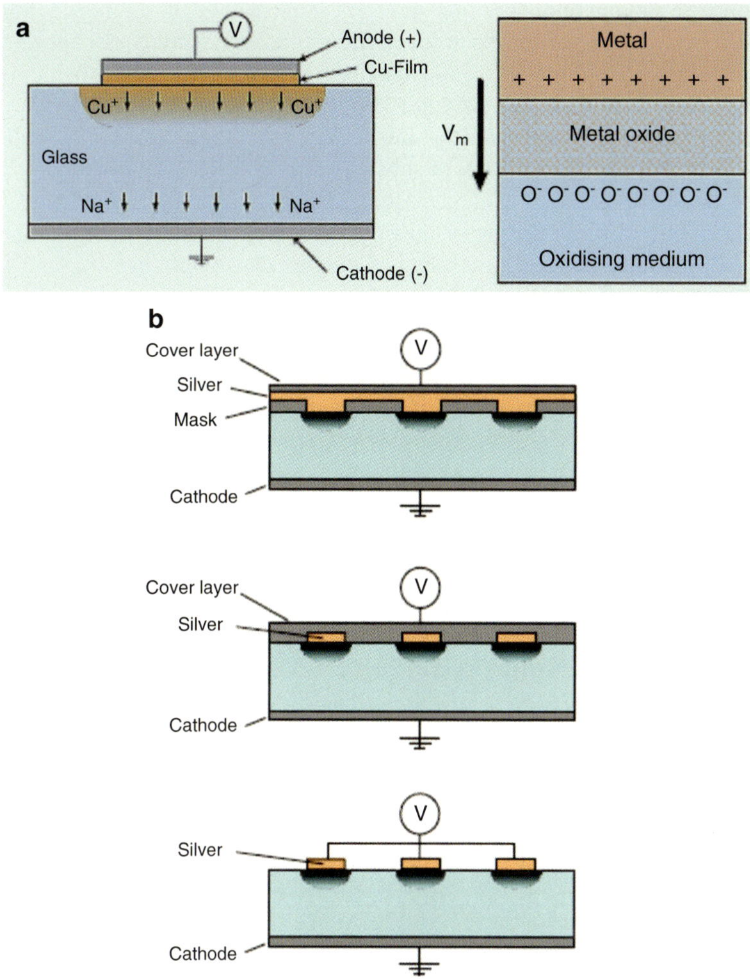

Fig. 7.8 Schematic representations of film ion-exchange process and typical electrode structures (Lehky 2006): (**a**) copper film ion-exchange process, where at the anode the metal is oxidized and at the cathode the sodium ions are reduced. The glass acts as a solid electrolyte. As this process necessarily needs an applied electrical field (to overcome the chemical potential barrier), the two exchanging ions move into the same direction in the glass; and (**b**) Typical electrode structures in dry silver-film ion-exchange. A cover layer is often used to improve the electrical contact and also prevents flaking of the film to be exchanged. One important advantage of the film ion-exchange technique is that it is a dry process and does not need sophisticated sample holders, as they are needed in molten salt ion exchange. The process is thus easier to accomplish and well suited for mass production

film to be exchanged. Several metals (e.g., Al, Au, Ag) can be used as back electrode (cathode), however they tend to deteriorate with time due to an accumulation of sodium metal (reduction of Na^+-ions at the cathode: $Na^+ + e^- \rightarrow Na$). In the case of a gold cathode, the sodium and gold form an alloy, which allows the sodium to diffuse in it and form stable phases of varying composition. When masks are used for the process, there is more flexibility in choosing mask material than in molten salt ion-exchange processes, which are often corrosive. Common mask materials are Al and TiW, but Cr also effectively prevents silver-film migration and presents the interesting possibility to directly utilize the integrated circuit mask plates as substrates in the ion-exchange process. Metal film ion exchange could also be performed from evaporated metal layers, which were electro-diffused during or directly subsequent to the evaporation process. During the field-assisted ion exchange, the flux density of silver j_{Ag} and sodium ions j_{Na} in one dimension can be written as (Lehky 2006):

$$j_{\mathrm{Ag}} = D_{\mathrm{Ag}}\left(\frac{eEC_{\mathrm{Ag}}}{fkT}\right) - D_{\mathrm{Ag}}\left(\frac{\partial C_{\mathrm{Ag}}}{\partial x}\right) \tag{7.4}$$

$$j_{\mathrm{Na}} = D_{\mathrm{Na}}\left(\frac{eEC_{\mathrm{Na}}}{fkT}\right) - D_{\mathrm{Na}}\left(\frac{\partial C_{\mathrm{Na}}}{\partial x}\right) \tag{7.4a}$$

where e is the electric charge, E is the applied electric field, T is the exchange temperature and k is the Boltzmann constant. The self-diffusion coefficients D_i are assumed to be constant with respect to the position x and the concentration C_i (not considering mixed alkali effects). The factor $f = D_i/D_i(\sigma_i)$, also called Haven ratio, relates the tracer diffusion coefficient D_i to the conductivity diffusion coefficient $D_i(\sigma_i)$ of the particle i and is supposed to be the same for both of the ions. The conditions for equilibrium (electrical neutrality) inside the glass are (Lehky 2006):

$$C_{\mathrm{Ag}} + C_{\mathrm{Na}} = C_0 \tag{7.5}$$

$$j_{\mathrm{Ag}} + j_{\mathrm{Na}} = j_0 \tag{7.5a}$$

where C_0 is the initial sodium concentration in the glass and $J_0 = I/F$, where I is the used ion current density and F the Faraday constant. Using (7.4), (7.4a) and (7.5), (7.5a) the electric field can be written as (Lehky 2006):

$$E = \frac{\left(\frac{fkT}{e}\right)\left[Mj_0 + D_{\mathrm{Ag}}(M-1)\left(\frac{\partial C_{\mathrm{Ag}}}{\partial x}\right)\right]}{D_{\mathrm{Ag}}\left[C_0 + (M-1)C_{\mathrm{Ag}}\right]} \tag{7.6}$$

where M is the ratio of the self-diffusion coefficients of the ions ($M = D_{\mathrm{Ag}}/D_{\mathrm{Na}}$). Finally, using the continuity condition for the silver ions (Lehky 2006)

$$\frac{\partial C_{\mathrm{Ag}}}{\partial t} + \frac{\partial j_{\mathrm{Ag}}}{\partial x} = 0 \tag{7.7}$$

leads to a second-order nonlinear differential equation, which can be solved analytically only if a profile with stationary shape can be assumed ($M < 1$). The solution of the stationary state is (i.e., for constant ion-exchange current) (Lehky 2006)

$$\frac{C_{\mathrm{Ag}}}{C_0} = \frac{1}{1 + \exp\left(\frac{j_0(1-M)}{D_{\mathrm{Ag}}C_0}\left(x - \frac{j_0 t}{C_0}\right)\right)} \tag{7.8}$$

In metal film ion exchange the thermal contribution is usually negligible and the surface concentration of the exchanged ions is mainly determined by the applied electric field strength and the duration of the ion-exchange process. Even with relatively high electric fields it takes a certain time to reach the maximum surface concentration. This is in contrast to the molten salt ion-exchange, where the maximum surface concentration, corresponding to the total replacement of B^+ ions (in the melt) for A^+ ions (in the glass), is reached very rapidly due to diffusion. The time needed to drive the whole metal film into the glass is given by (Lehky 2006)

$$t = d\rho_{\mathrm{Ag}}F/\left(M_{\mathrm{Ag}}I\right) \tag{7.9}$$

where d is the film thickness, ρ_{Ag} is the density of silver, F is the Faraday constant, M_{Ag} is the molecular weight of silver and I is the used ion current density. The ion exchange time t depends strongly on temperature (because of the strong temperature dependence of the ion mobility σ), but a drastic reduction in current can be observed when the silver film becomes depleted. Rearranging (7.9) yields (Lehky 2006):

$$d = t\left(\frac{I}{F}\right)\left(\frac{M_{\mathrm{Ag}}}{\rho_{\mathrm{Ag}}}\right) = \frac{tj_0}{C_0} \tag{7.10}$$

Comparing (7.10) with (7.8), it can be seen that if the metal film is totally driven into the glass, the translational term of the concentration profile only depends on the original film thickness d. Finally, because the ion-exchange time t is determined by the silver becoming exhausted (which can be monitored by current measurement), the thickness of the film d can be used as a process controlling parameter, instead of time and temperature. In the case of channel waveguide fabrication, the electric field distribution must be known.

7.4.2 Optical Property of Ion-Exchanged Waveguides

The presence of Ag^+ ions locally perturbs the index via three physical mechanisms: ionic size, ionic polarisability, and induced stress. The stress-induced index change is relatively small in Ag^+–Na^+ exchange and is complicated to take accurately into account. The RIP of the waveguide cross section can be expressed as (Honkanen et al. 2006)

$$n(x,y,\lambda) = n_{\text{sub}}(\lambda) + \Delta n_{\text{o}}(\lambda) C_{\text{Ag}}(x,y) \tag{7.11}$$

where $n_{\text{sub}}(\lambda)$ is the substrate index before ion exchange and $\Delta n_{\text{o}}(\lambda)$ is the increase in refractive index resulting from $C_{\text{Ag}} = 1$, determined experimentally. The electric field E_n of each scalar mode supported by the waveguide is found by solving the Helmholtz equation (Honkanen et al. 2006)

$$\left(\nabla^2 + k^2\right) E_n = \beta_n^2 E_n \tag{7.12}$$

where $k = k_{\text{o}} n(x,y) = 2\pi n(x,y)/\lambda$ is the wave number and β_n is the propagation constant of the n_{th} mode.

7.4.3 Ion-Exchange Systems in Glass Waveguides

Some exchanging systems in glass waveguides have been developed, which are capable of providing relatively high refractive index changes ($\Delta n > 0.02$) and large penetration depths ($x \gg 10$ μm), for a reasonable exchange time, thus allowing the fabrication of waveguides with high numerical aperture N_{A} (multimode waveguides). In Table 7.1 the most frequently used monovalent ions (Li^+, Na^+, K^+, Rb^+, Cs^+, Ag^+, Tl^+ and Cu^+) and some of their properties are listed. Divalent ions (Be^{2+}, Mg^{2+}, Ca^{2+}, …) exhibit only a very low mobility and would require long diffusion times or exchange temperatures above the transformation temperature (T_{g}) of the glass. Lithium ions produce only a small index change ($\Delta n \approx 0.012$). Additionally, when a small ion such as Li^+ replaces a larger ion of the glass such as Na^+, the substrate network will collapse around the smaller ion to produce a more densely packed structure. This change is accompanied of significant tensile stresses in the exchanged layer, which in turn induce high birefringence δ_n. To reduce the high tensile stresses induced by the Li^+-exchange it is necessary to work at very high temperatures (usually above T_{g}, i.e., 540–580 °C) allowing the relaxation of the glass matrix. However, it is not possible to work with applied fields, because sealants do not resist at these high temperatures. Further problems consist in the reaction of the melts with the masking materials (namely Al) creating aluminum-silica compounds, which generate optical losses (combinations of Al and Ti or Cr seem to reduce these reactions). Finally, the glasses must be resistant against recrystallization (i.e., glasses containing Al_2O_3 and ZrO_2). Rubidium exchange, apart from its high price, produces a similar refractive index change to the one of potassium ($\Delta n \approx 0.015$–0.02) and is thus not suitable for multimode waveguides. Cesium ions have the largest ion radius of the listed ions and show also the lowest mobility. Due to their high polarizability, the maximum refractive index change is about 0.04. Their low mobility requires an applied electrical field to reduce the diffusion time drastically. The potassium ion produces only a small refractive index change ($\Delta n \approx 0.01$) in sodium containing glasses and is thus mainly used for single-mode applications. Due to the large ionic radius of the potassium, these ions cause important compressive stresses in the exchanged layer. This property is often used for the strengthening of glasses, also known as ion-stuffing. However, the compressive stresses induce a rather high birefringence δ_n. Interestingly, in the case of some float glasses (the glass is poured onto molten tin during the fabrication to achieve flawless optical surfaces), the presence of Sn-ions as impurities resulting from this float process already induce a small layer with increased refractive index allowing for single-mode operation (Lehky 2006).

1. Thallium ion-exchange
   Thallium has the largest electronic polarizability and therefore yields a large refractive index change ($\Delta n \approx 0.1$) through thallium ion-exchange. Thus, numerical apertures $N_{\mathrm{A}} > 0.5$, with $N_{\mathrm{A}} = \sqrt{n_{\mathrm{core}}^2 - n_{\mathrm{cladding}}^2}$ are possible. The main drawback of this ion is the high toxicity of the thallium salts. The refractive index change can be adjusted from 0.0001 to 0.1 by using diluted salt melts (usually $NaNO_3$/KNO_3_$TlNO_3$—salt melts). Buried waveguides are obtained by performing a second ion-exchange step, thus reducing surface scattering losses. With this process, propagation losses of less than 0.1 dB/cm have been achieved. In addition, Tl_2O has been introduced into the glass by the ion-exchange, decreases the viscosity of the glass and augments the local ion exchange rate. This effect is responsible for the generation of a step-like RIP in KF_3 glass, without electric field-assistance (Lehky 2006).
2. Silver ion-exchange
   Silver ion exchange is the most explored ion-exchange technique. The Ag^+-ions from the $AgNO_3$ melt exchange very well with the Na^+-ions of the glass, even at low temperatures (typically 220–300 °C). The refractive index change Δn is about 0.1 for soda-lime glasses and can reach 0.22 for TiF_6 phosphate glass for a thermal ion exchange at 410 °C. A reduced refractive index increase can be achieved by using diluted salt melts. Through electrolytic release of silver ions from a silver rod, precise control of the silver concentration in the Ag^+–Na^+ melt throughout the process can be obtained, thus allowing the fabrication of reproducible waveguides (reproducibility of ~10^{-4} in propagation constant). The main drawback of the silver exchange is the silver reduction, which occurs at higher exchange temperatures and is supported by reducing ions like Fe^{2+}, Sb^{3+} and As^{3+} (acting as electron donors) present in the glass as impurities. The presence of NBOs in the glass, which act as electron donors, much more promote the reduction of silver ions than the only marginal presence of impurities in the glass matrix. Due to the low solubility of the elementary silver, precipitation of silver as submicroscopic crystals occurs. These silver clusters are responsible for the yellow coloring and the increase of the absorption losses of the waveguides. The losses depend strongly on the wavelength used and have a maximum slightly above 400 nm, where the resonance for the precipitated small silver particles occurs. The loss increase due to the colloidal silver is much smaller at wavelengths >800 nm. The absorption losses vary from less than 0.1 dB/cm for

BK7 glass (where the waveguides were annealed at 500 °C in air after the ion-exchange) to a few dB/cm for soda-lime glasses as well as Pyrex glass (a borosilicate glass not optimized with respect to impurities). Additionally, the annealing as well as the burying of the waveguides usually reduces the propagation losses significantly. To lower the process temperature and therefore diminish silver reduction effects, eutectic melts can be used. Waveguide burying, which considerably reduces optical losses (originating mainly from surface scattering) as well as the polarization dependence of the waveguides, can be obtained by a second ion exchange in a $NaNO_3$ melt as well as an eutectic melt of KNO_3 and $NaNO_3$. Because of the lower melting point of the eutectic melts, profile broadening of the waveguides due to thermal diffusion is reduced. By the application of an electric field, the ion-exchange time can be diminished and the exchange depth increased. An electric field can also be used to bury the waveguide. The use of anodized masks can improve resistance against corrosion from the melts and also prevents deposition of silver, therefore diminishing optical losses. Finally, the refractive index at the glass surface induced by silver ion exchange from melts shows a dependence on the melt concentration which furthermore varies for different glasses. The ion exchange depth is increased when the intermediate oxide evolves from ionic to more covalent bonding; that is from compositions containing CaO and MgO to compositions containing only ZnO as the intermediate oxide. Thus, the more covalent character of the glass makes its structure more open for the penetrating particles (i.e., the exchanging ions). The ionic character of the modifiers (Ca^{2+} and Mg^{2+}) means quite strong electrostatic forces between the non-bridging oxygen (NBO) and the neighboring Na^+ that impede the free movement of the sodium ion throughout the glass matrix, thus slowing down the exchange of the Na^+ for another ion. Therefore, for the choice of an adequate substrate for ion-exchange, a glass will be chosen, which ideally is free of MgO and CaO. Also, the number of non-bridging oxygens (NBO) shall be as low as possible. Ion exchange from thin silver films has been used ion exchange of various glasses. The advantages of ion exchange from thin silver films over the molten salt ion exchange can be summarized as (Lehky 2006): (a) Dry process suitable for mass production—no sophisticated leak proof and electrically insulated sample holders necessary allowing high temperature operation (as it is the case for ion exchange from melts); (b) Pure materials deposited by evaporation, sputtering and the like; (c) Large range of exchange temperatures—as a field-assisted process it allows ion exchange at low temperatures (not limited by melting and/or decomposition temperatures of salts); (d) Ion exchange is initiated only when applying the electric field—no ion exchange during heating up (negligible thermal diffusion from the source); (e) Charge controlled process—current monitoring allows for intrinsic process control; and (f) Uniform ion concentration profile due to the finite ion source.

The maximum refractive index change Δn is similar to molten salt ion exchange because the same ions are exchanged. However, there is still a dependence of the surface refractive index on the applied electric field. It is possible to fabricate multimode waveguides using relatively small fields and short diffusion times. In multimode silver-film ion-exchanged waveguides, losses below 0.1 dB/cm for slab and 0.2 dB/cm for channel waveguides have achieved. Multimode waveguides (10 guided modes) made with soda-lime glass (14.31 wt% of Na_2O, with $T = 275\,°C, E = 15$ V/mm and $t = 30$ min) presented 0.5 dB/cm while single-mode waveguides showed 1 dB/cm absorption losses. For the fabrication of multimode waveguides it is convenient to drive the ions quickly into the glass and perform a subsequent thermal or field assisted step to change the shape of the RIP (Lehky 2006).

3. Copper ion-exchange
An alternative to silver ion exchange is the Cu^+–Na^+-exchange. Molten salt ion exchange and ion exchange from thin solid copper films have all been implemented. The introduction of copper into the glass was initially achieved by melting glasses adding copper-oxides. At the high melt temperature the Cu^+ state is thermodynamically more stable than Cu^{2+}. Copper (II) complexes of different kinds can be formed in a given glass depending on its history of forming. The Cu^{2+} ions select their environment in the melt themselves to form their characteristic complexes. The introduction of copper into the glass with the ion-exchange technique gives the copper ions the possibility to create its complexes in the glass, depending on the environment made available by the ions replaced by the copper ions. The increase of the basicity of a glass by changing its composition increases the absorption peak of the cupric state (A glass of less basicity corresponds e.g., to a glass composition containing less NBO). Moreover, the ion exchange allows the introduction of copper into the glass in an amount far in excess of the solubility limit of copper in the glass. For Cu^+–Na^+ ion exchange from molten CuCl salt for glasses of the alkali-alumino-silicate type Na_2O–CaO–Al_2O_3–SiO_2 in the temperature range of 500–650 °C, for instance, copper enters the glass as monovalent Cu^+ ion by ion-exchanging with Na^+ ions. Some of these Cu^+ ions are converted into Cu^{2+} ions, Cu_2O, CuO or Cu colloids after being introduced into the glass (depending on the glass composition and temperature). However, as the diffusion coefficient of divalent cations is very small, their influence in Cu–Na exchange should be negligible. Furthermore, the role of Cu^{2+} in the index increase may be ruled out, since its electronic polarizability is much lower than that of Cu^+ and is comparable to that of Na^+-ions. The remaining Cu^+-ions gradually migrate further into the glass. The increase in Al_2O_3 content of the glass markedly increase the Cu–Na ion-exchange rate, with a maximum for a glass composition with an oxide ratio Al_2O_3/Na_2O or $Al_2O_3/(Na_2O + CaO) \geq 1$. For these compositions non-bridging oxygens are removed in the glass and it is assumed that this absence of NBOs leads to a reduced chance of possible electrostatic interaction between the diffusing cations and the NBOs. The presence of divalent cations like Ca^{2+} and Pb^{2+} in the glass have tendency to suppress the Cu–Na as well as the K–Na ion-exchange. Their slowing down effect on Cu–Na diffusion rate follows the order Mg^{2+}, Ca^{2+} and Ba^{2+}, where Ba^{2+} exhibits the slowest exchange rate. In other words, the exchange rate decreases with decreasing

field strength of the alkaline earth ion. The effect of different types of alkaline earth ions on the inhibition of the Cu–Na ion–exchange may be attributed to the polarization of oxygen ions by alkaline earth ions. The diffusing Cu or Na ions become less mobile by the coulombic interaction with the non-bridging oxygens produced by the addition of alkaline earth oxides to the glass. This coulombic interaction would be smaller for an alkaline earth ion with stronger field strength due to its stronger polarizing power against oxygen ions. The presence of divalent ions (Ca^{2+}) in the exchange melt (already at concentrations of about 0.1 mol%) blocks further K–Na ion-exchange. Additionally, there is a strong influence on the atmosphere (air or N_2) during the ion exchange (from a CuCl melt at 550 °C) on Cu^+ or Cu^{2+} state. Almost 90 % of the exchanged copper was in the Cu^+ state when exchanging under N_2 atmosphere, while only about 40 % of the copper was in the monovalent state when exchanged in air. When exchanging Cu for Li (in Li_2O containing glasses) about 90 % of the copper was in the Cu^+ state despite of being exchanged in air. Good quality multimode waveguides were obtained (T = 530–585 °C, t = 5–20 min) with the Cu–Na ion exchange from molten $CuSO_4$:Na_2SO_4 and $CuSO_4$:K_2SO_4 eutectic baths in soda-lime and BK7 glasses. Ion exchange in the $CuSO_4$:K_2SO_4 bath (competition between Cu–Na and K–Na exchanges) exhibits significant (stress-induced) modal birefringence, similar to the birefringence observed in pure K–Na exchange, where the larger K^+-ions ($r_{K+} = 1.33$ Å) replace the smaller Na^+-ions ($r_{Na+} = 0.95$ Å) in the glass and thus building up planar stress fields. Ion exchange using the $CuSO_4$:Na_2SO_4 bath showed no birefringence. This is attributed to the similarity of the ion radii of the exchanging ions ($r_{Cu+} = 0.96$ Å and $r_{Na+} = 0.95$ Å). The electronic polarizability of Cu^+ is $\alpha(Cu^+) = 1.73$ Å^3 and significantly higher than the one of Na^+, $\alpha(Na^+) = 0.41$ Å^3, contributing to the increase of the refractive index. Copper ion-exchanged waveguides have been fabricated from melts containing Cu^+ or Cu^{2+} in BK7 borosilicate glass and GIL49 low impurity soda-lime glass. The Cu–Na exchange was performed for exchange temperatures from T = 350–500 °C and exchange times from t = 5 min to 21 h, generating a maximal refractive index increase of $\Delta n = 0.069$ and waveguide depths supporting up to 16 modes. Furthermore, ion exchange from a CuCl–$ZnCl_2$ melt was shown to strongly impede the oxidation of Cu^+ to Cu^{2+}. When studying the spontaneous reduction of silver ions from thin films in soda-lime float glasses, the tin present in the float glass surface does not spontaneously reduce the copper to the metallic state (copper colloids). This is because copper ions form much more stable complexes in the glass and are less sensitive to reduction due to impurity ions contained in the glass matrix (Lehky 2006).

Copper film ion exchange into K8 borosilicate glass has been performed. Before the diffusion, the copper film was thermally oxidized at 400 °C for more than 1 h, resulting in a film containing about 90 % CuO. The process parameters were, T = 350–550 °C, E = 20–80 V/mm and t = 3–50 min. The as-exchanged waveguides presented losses of 5–10 dB/cm, which could be reduced to ~0.5 dB/cm by annealing under a reducing atmosphere at 500 °C during 1.5 h. It is believed, that this lowering of the losses is due to the reduction of Cu^{2+} to

Cu^+ in the exchanged layer. Cu^+–Na^+ exchange from thin solid copper films into 1 mm thick soda-lime glass (Corning 2947) has studied. The copper film (1 μm) was evaporated onto the glass and a 0.5 μm thick gold film was deposited onto the copper film (to prevent oxidation and flaking of the copper during the diffusion process) and onto the backside of the glass substrate (cathode). The parameters of the exchange process were, $T = 350$ °C, $E = 30$ V/mm and $t = 15$–100 min. For a waveguide supporting 5 modes (100 min exchange time and ~5 μm depth) the maximum refractive index change at the surface was $\Delta n = 0.058$. The diffusion coefficient of Cu^+-ions in soda-lime glass was $D_{Cu} = 5.3 \times 10^{-16}$ m²/s (at 350 °C). Finally, the copper ion-exchanged waveguides did not show any coloration. Propagation losses (waveguide having three modes at 632.8 nm) was about 1.1 dB/cm were measured by the Fabry-Pérot interferometric method. In addition, the waveguides of copper film ion exchange into Pyrex glass (Corning 7740) featured a refractive index increase of $\Delta n = 0.0165$ and propagation losses of 0.45 and 0.3 dB/cm at wavelengths of 850 and 1,300 nm, respectively (Lehky 2006).

4. Anionic exchange
   In anionic exchange external anions (mainly OH-) are exchanged for anions of the glass, mainly F- of fluorinated glasses. Silicate glasses of the type SiO_2–Al_2O_3–ZnO–Na_2O containing fluorine is effective in reducing the refractive index of the glass. Waveguides in these glasses could be formed by heating in air, which leads to the diffusion of F-anions from the glass surface, causing a decreased fluorine concentration and thus increasing the refractive index of the surface layer. Thus obtained waveguides showed losses of ~0.2 dB/cm at 633 nm. Anionic ion exchange in fluoride glasses, where OH- or even OD- (when using heavy water) is exchanged or the F-anion shown the propagation losses were still very high (>2 dB/cm) mainly caused from the OH-absorption band and its overtones. Exchanged channel waveguides in ZBLAN glasses showed 10 ± 1 dB/cm at 1,550 nm (Lehky 2006).

Overall, ion exchange in glass has been a mature and well-developed optical waveguide fabrication technique, particularly for small-sized, compact integrated optical devices. The performances of such optical devices even exceed the Telcordia (GR 1221) requirements used for telecommunication applications. However, although ion exchange in glass is successfully established as a very low-cost and high-performance technology for passive as well as active integrated optics applications, no attempts were made to adapt and try to scale-up this well-performing technology to optical board (below ~0.5 m) or even optical backplane dimensions (up to 1 m) with the limitation of adequate commercially available glass substrates (controlled composition as well as reproducible production processes) which should be available in board-scale sizes (i.e., 50 cm or more). Of the different ion-exchange processes, the film ion exchange seems to be best suited for optical interconnects applications. This is mainly due to the considerable simplification of the processing setup (no high-temperature leak proof and electrically insulated sample holders needed) as well as the inherent control of the initiation of the ion-exchange (no ions generated without applying an electric field) and ion-migration through proper monitoring. The film ion-exchange presents the properties of an industrially compatible and scalable process, suitable for mass production. Finally, the exchanging ion remains to be chosen. In the case of film ion exchange the possible alternatives are either silver or copper ions. Compared with silver film ion-exchange, copper ions are known to form stable complexes and they are less sensitive to reduction from impurities in the glass and light-induced reduction (photoreduction) than silver ions. The ionic radius of the copper ion is almost identical to the ionic radius of sodium, and thus low stresses are expected. However, copper may be present as Cu^{2+} ion, which can induce absorption around 800 nm. Copper is the key material in printed circuit board manufacturing, which could make copper film ion exchange certainly very interesting for future applications (Lehky 2006).

7.4.4 Applications of Ion-Exchanged Glass Waveguides

1. Add-drop wavelength filter
   In a point-to-point dense wavelength division multiplexing system (DWDM), multiple wavelengths are coupled into optical fiber by a wavelength multiplexer at the transmitter end. At the receiver end, those wavelengths are fanned out by a wavelength demultiplexer. This point-to-point fiber optic network forms the backbone of the DWDM transmission system. Wavelength add/drop filters are required to have access to the individual wavelengths at any point of the system. The difference between a demultiplexer and an add/drop filter is that the add/drop filter not only has the function of demultiplexing but also has the function of wavelength routing or coupling. It can drop (or add) the required one wavelength from (or to) the fiber network without affecting the transmission of other wavelengths. A wavelength channel add/drop filter is a key device in fiber optical networks. The device made with an ion-exchanged waveguide add/drop wavelength filter is shown schematically in Fig. 7.9a. Two buried single-mode waveguides of different widths are brought together adiabatically into a two-mode section. The two-mode section is later separated again into the two single-mode waveguides. Bragg gratings are ultraviolet written at a small angle in the two-mode section. The filter functions as follows (Honkanen et al. 2006): input signals on the narrow waveguide are coupled to the odd mode of the two-mode section. For wavelengths on resonance, a tilted grating breaks the orthogonality of the two modes and reflects the signal to the backwards propagating even mode. This propagates back out of the wide "drop" port. Wavelengths off resonance pass through the Bragg grating and continue on to the narrow output port. Additionally, a signal on resonance at the dropped wavelength can be added through the fourth port. A fabricated device can yield excellent results with a 20 dB extinction ratio and a 0.4 nm 3-dB bandwidth.
2. Er-doped waveguide lasers
   Perhaps one of the most attractive uses of ion-exchanged waveguides is to employ them in an Er-doped glass substrate, with the goal to develop compact

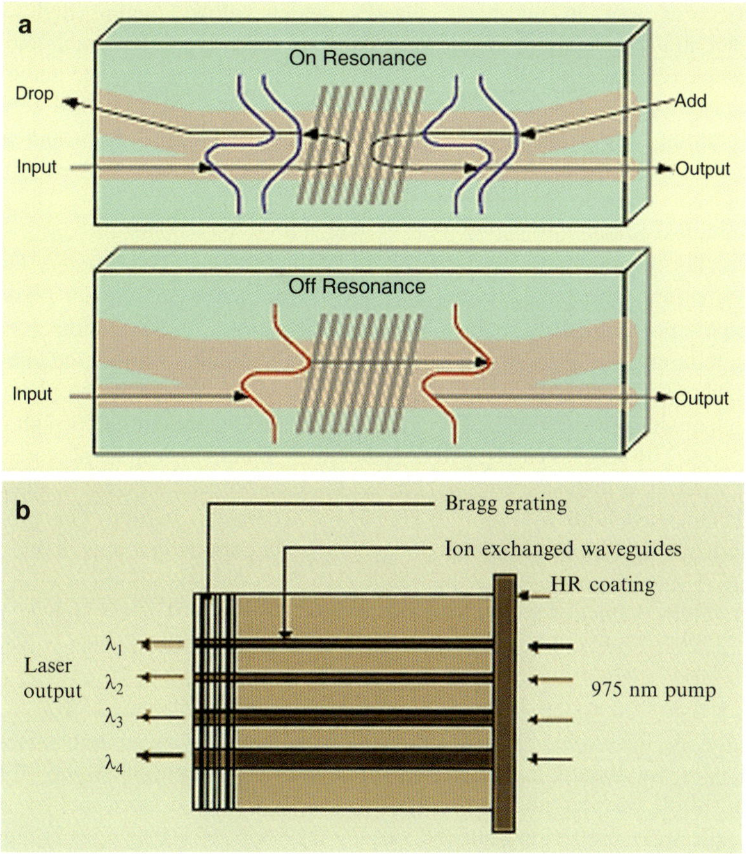

Fig. 7.9 Examples of ion-exchanged glass waveguides (Honkanen et al. 2006): (**a**) schematic illustration of an add/drop waveguide filter; (**b**) DBR waveguide laser array implemented with a surface grating and a dielectric mirror on Ag film ion-exchanged waveguides

glass waveguide amplifiers and lasers for 1,550 nm wavelength region. For example, an Ag film ion exchange technique has used to fabricate waveguide lasers in Er-doped phosphate glass. The first process step introduces Ag^+ ions in the glass just beneath the surface, where they replace Na^+ ions and locally increase the refractive index of the glass, thus forming a waveguide. In contrast to conventional ion exchange processes using molten salts as ion sources a thin silver film is used as an ion source in Ag film ion exchange. The silver ions are released from the film and driven into the glass with an electric field. The second step of the process is a simple thermal annealing to modify the waveguide profile. Compared to the molten salt ion exchange processes the Ag film ion exchange has its unique advantages, especially when a "difficult" substrate such as a phosphate glass is used. Phosphate glasses are preferred for erbium doped waveguide amplifiers (EDWA) and lasers because of the high solubility of rare-earth ions in phosphate glasses, which allows doping high erbium concentrations without significant lifetime reduction. However, phosphate glasses react easily with molten salt baths and etchants used to remove the diffusion mask, resulting in damage to the surface. Also, hybrid substrates containing active and passive regions are required when active and passive devices are integrated monolithically in a single substrate. Molten salt baths can easily damage the joint between the active and passive regions. Therefore a gentle waveguide fabrication process, which does not damage the surface of the glass nor the joint between the active and passive regions, is preferred. This will ensure low propagation losses and allows repeatable subsequent photolithographic processing of the glass, e.g., for fabrication of DBR gratings for waveguide lasers. Ag film ion exchange is a gentle technique without the use of salt melts or strong etchants and causes no damage to the surface of phosphate glass. High quality surface waveguides can be fabricated in rare-earth doped phosphate glasses. The process can be performed at low temperatures and a photoresist can be used as a diffusion mask. The use of a photoresist as a mask makes the channel waveguide patterning as simple and accurate as possible. One process to fabricate Ag film ion-exchanged waveguides in commercially available Schott IOG-1 phosphate glass is doped with very high concentrations of erbium and ytterbium. Typical doping levels are 1.15 wt% Er_2O_3 (1.0×10^{20} ions/cm^3) and 4.73 wt% Yb_2O_3 (4.0×10^{20} ions/cm^3). For single-mode channel waveguides the mask aperture widths in the photoresist range from 2 to 10 μm and the ion exchange is done at ~100 °C for about 1 h with a voltage of ~100 V. After removing the remaining silver film and resist, the glass substrate is annealed at ~230 °C for about 2 h. The propagation losses of the waveguides are ~0.15 dB/cm. Multi wavelength laser arrays can conveniently be fabricated by using a simple surface relief grating and varying the width of waveguides in the array. The wavelength range is about 0.3 nm (less than 50 GHz) due to the low index change resulting from potassium ion exchange. The use of Ag film ion exchange in this type of a DBR waveguide laser array can result in much broader wavelength range (covering six ITU grid wavelengths). Molten salt silver ion-exchanged multiwavelength DFB laser arrays based on the same idea have also been demonstrated. A schematic diagram of the waveguide laser array is shown in Fig. 7.9b. The laser cavity is implemented with a broad band high reflection dielectric mirror on one end and a surface relief grating on the other end. An array of lasers lasing at different wavelengths can be fabricated by changing the width of the waveguides and keeping the period of the grating the same. The effective index of the waveguide depends on the width of the waveguide, which changes the Bragg wavelength. By controlling the width of the waveguides, each channel can be designed to lase at different ITU (International Telecommunication Union) specified wavelengths. The number of wavelengths that can be implemented with a single grating depends on the maximum index change that can be obtained by the fabrication process. Ag film ion exchange has the advantage of producing a larger index change compared to molten salt processes. The surface relief gratings in the lasers were fabricated holographically on the photoresist at the surface of the glass and then transferred into glass by Argon ion milling. For the 5 μm wide waveguide at 1540.2 nm wavelength, for a coupled pump power (980 nm wavelength) of 145 mW, the output power was 11 mW with a threshold of 60 mW and a slope efficiency of 13 %. In addition to the waveguide lasers using a single-mode pump laser diode, anerbium doped planar waveguide laser configuration has been developed, which enables easy pumping using low cost multimode semiconductor lasers. The device was fabricated by Ag film ion exchange in a hybrid phosphate glass having Er-doped and passive regions monolithically integrated in a single substrate. This device using the monolithic hybrid glass enables integration of active and passive ion-exchanged devices in a single substrate. This feature is crucial for the success of active planar waveguide devices for fiber optic communications (Honkanen et al. 2006).

7.5 Sol–gel Glass Waveguides

Though the sol–gel technique as a tool to produce silica glass was reported for the first time in 1846, and it was not until 1930 used to produce optical coatings, while the importance of solution-deposited films for planar waveguides was outlined in 1972. Most of the activity thereafter has been concerned with the development of waveguides based on silica glass films, both for their low loss and for their compatibility with optical fibers. In order to increase the refractive index of pure silica glass and thus allow the realization of waveguides over different substrates, the more complex SiO_2–TiO_2 system has also been widely investigated. SiO_2–TiO_2 thin films are usually been produced starting from solutions of tetraethoxysilane (TEOS) and Ti(n-Obu)4 (titanium butoxide) as precursors of silica and titania, respectively. A typical process procedure includes: the sol is obtained starting from two separate solutions of TEOS and titanium butoxide in ethanol. Addition of water in molar ratio 2:1 to TEOS for the hydrolysis of the precursors and of HCI (0.01:1) for the catalysis of the reaction follows. Acetylacetone (acacH, ≅0.25:1), a complexing agent, is used to further stabilize the solution, reducing the hydrolysis rate of the TiO_2 precursor. Actually, the precursors must be separately soluted, since the hydrolysis rate of TEOS is much lower than that of titanium butoxide and a single-step preparation of the solution could produce the premature precipitation or clusterization of the TiO_2 component of the glass. The solution is then carefully filtered before it is deposited on the substrate by dip coating or spinning under controlled environment (30 ± C, 30 % relative humidity). The environment in which the preparation and deposition of the film is carried out is critical: the presence of dust particles would badly affect the propagation quality of the resulting waveguides, while any change in temperature or external humidity would strongly influence the hydrolysis-gelation processes of the solution, which occur concurrently in a few seconds, even during the dip coating or the spinning deposition. The sintering is then carried out in an oven at 500 ± °C for several hours. Eventually, waveguides showing 0.5 dB/cm average propagation losses are obtained. The maximum thickness of each single layer obtained is about 0.2 μm; thicker films crack during the sintering process due to the high mechanical stress induced by the large shrinkage of the densifying film. Therefore, it is necessary to superimpose several layers in order to realize films having thickness suitable (i.e., 1 μm or more) for integration with other guided-wave components (Righini and Pelli 1997). In order to overcome the mentioned thickness limitation, another route, that of organic-modified silicates (ORMOSILS), has been developed. In this case the precursors are not fully inorganic, but also have some organic components which make the gel network more flexible and less prone to cracking during sintering. Moreover, it is also possible to prepare subhydrolysed starting solutions, having low water content; in this case, the density of the gel is higher and experiences smaller shrinkage during the sintering process, further decreasing the risk of obtaining cracked layers. An example of such an approach is represented by organic–inorganic silica waveguides made with γ-glycidyloxypropyl trimethoxysilane (GPTS)/γ-methacryloxypropyl trimethoxysilane (MPTS) system, often in conjunction with the Zr/methacrylic acid system, obtaining low-loss waveguides with single layer thickness up to 10 μm. If a small amount of photoinitiator (e.g., 0.1 % wt. of IRGAGURE) is added to the components, the resulting film can also be photopolimerized by UV mask photolitography or direct laser writing. The waveguides can also be realized by copolimerisation of Ti(OEt)4 and γ-glycidyloxipropyltrimethoxysilan (GLYMO). The process is completely carried out at temperatures below 80 ± °C, allowing the doping of the film with organic materials which could be destroyed by a high temperature sintering process. The refractive index increase caused by the presence of Ti is low, likely due to the low temperature process and the high organic content of the film; the produced waveguides show rather high loss of 2 dB/cm in the fundamental mode at 514 nm. An intermediate approach is represented by the partial or total substitution of TEOS in the starting solution with methyltrietoxysilane (MTES) or Polydimethylsiloxane (PDMS), leading to more than 1 μm thick single layers of optical quality comparable to that attainable with pure TEOS films (Righini and Pelli 1997).

The production of waveguides via sol–gel technique is appealing also for the possibility of patterning the guiding film by direct laser writing: this is achieved by selective densification due to the local heating produced by laser irradiation. Laser writing would be extremely useful for rapid circuit prototyping and could also lend itself to mass production. In order to obtain the desired heating of the sol–gel film, the film and/or substrate must absorb at the laser operating wavelength or, alternatively, an absorbing metal cladding has to be deposited on the top of the film. Silica films and substrates absorb in the UV and IR spectral region; therefore, suitable laser source candidates are excimer lasers (in the UV) and CO_2 lasers (emitting in the IR at 10.6 μm). Nd:YAG lasers, emitting in the NIR at 1.06 μm, require the use of adsorbing metal claddings. Actually, thanks to their broader availability and simplicity of operation, CO_2 lasers have been preferably employed in the testing of this technique. For example, strip waveguides of reasonably good quality (better sample showing losses slightly lower than 1 dB/cm) have been produced by irradiating films, partially densified at 60–200 °C, with a 50 mW CO_2 laser beam scanning the film at 2 mm/s and focused on the surface by a 50 mm lens; the width of the resulting strip waveguides was found to be about 100 μm. The low spatial resolution of the CO_2 laser, intrinsically due to its 10.6 μm wavelength, limits its application in IO to some specific tasks like the production of highly multimode or tapered structures. The use of excimer or Nd:YAG lasers would allow a more general application (Righini and Pelli 1997).

The large flexibility of the sol–gel process has been exploited to introduce into standard films various materials which can provide the desired functional properties, such as sol–gel films doped with semiconductors, rare earths, nonlinear optical organics, and dye materials. Two main routes have been followed in the development in this class of materials, namely the addition and reaction of the dopants inside the sol–gel starting solution, by chemically controlling the size of the particles, and the impregnation or exposition of a porous sol–gel film to the dopants (pore doping), where the crystal size is limited by the pore size. For instance, guiding sol–gel silica-titania films doped with CdS and PbS microcrystals were fabricated: cadmium and sulphur were added to the solution as cadmium acetate (5 % molar) and thiourea respectively. AcacH was also added to the cadmium acetate solution as stabilizer (CdAc:acacH ≅ 2:1); lead was introduced as lead acetate, together with acetic acid. Films, after dipping, were densified at 300–500 ± °C under nitrogen or air flux. As a matter of fact, other routes which possess large potentialities are constituted by polymer doping, which can produce high third order nonlinearities, as well as by rare earth or dye doping in order to produce light amplification or lasing. A broad variety of dopants have also been investigated in the effort to obtain highly-sensitive and fast-responding integrated optic chemical and/or physical sensors. Nonlinear optical glassy materials produced by sol–gel are particularly promising, but reproducibility and stability are still to be fully achieved (Righini and Pelli 1997).

7.6 Laser-Written Waveguides

High-power femtosecond (fs) lasers systems are being used extensively for laser-material processing of glassy materials in order to fabricate photonic devices. The ultrahigh peak power infrared (IR) radiation generated by many of these systems has been used to induce large index changes in bulk glasses for the fabrication of imbedded waveguides and surface structures. In glass optical fiber, femtosecond laser systems have been used to induce large index changes and fabricate long-period fiber grating structures in a step-and-repeat fashion. The process of induced index change in bulk glasses resulting from femtosecond-IR laser exposure is thought to result from a multiphoton absorption/ionization process resulting in material compaction and/or defect formation depending on the intensity of the exposure. Above the ionization threshold intensity I_{th}, multiphoton ionization (MPI)-induced dielectric breakdown likely results in localized melting and material compaction causing an index change that is permanent up to the glass transition temperature t_g, of the material. The value of I_{th} is material dependent. Below I_{th}, another regime of induced index change is observed that can be erased by annealing with temperatures below the material t_g. In this regime, multiphoton absorption likely results in defect formation similar to that seen for ultraviolet (UV)-induced index changes in photosensitive germanium doped silica glasses. Laser-induced index change is the key feature of most fiber Bragg grating (FBG) manufacturing. Typically FBGs are made by side exposure of the UV-sensitive fiber core to a spatially modulated UV-laser beam. The modulated UV beam is typically created by interfering two or more UV beams using either a bulk interferometer or a zero order-nulled phase mask. For femtosecond systems, interferometric setups for generating beam modulation are nontrivial to align since path lengths of the interfering beams need to be matched to within the spatial location of the femtosecond-pulse, which for a 120-femtosecond pulse would be ~36 μm. The phase mask technique solves the alignment problem since the optical path lengths of the generated beams are automatically matched for an incoming beam that is at normal incidence to the mask. Femtosecond UV-laser inscriptions of fiber gratings using silica phase masks have been performed. Typically, silica-based fibers are placed in close proximity to the mask, and large index changes are generated in suitably UV photosensitive waveguides. A limitation of this technique however is that lower pulse intensities need to be employed since high nonlinear absorption and group velocity dispersion would otherwise occur within the mask. Such nonlinear absorption reduces the amount of light transmitted through the mask that is available to induce an index change in the fiber and simultaneously degrades the phase mask with time. Combining the ultrafast IR laser and the phase mask approach, the efficient fabrication of retroreflecting FBGs in standard telecom and pure silica core single-mode fibers have been performed. The technique has proven to be as simple as the standard UV-laser writing technique but is far more versatile as grating inscription is not limited to materials that are traditionally UV photosensitive such as pure silica, Ge-doped silica, borosilicate glass waveguides, reverse proton exchange $LiNbO_3$ waveguides, sapphire optical fiber, fluoride glass fibers, actively doped silica fibers, phosphate glass fibers, and pure silica photonic crystal fibers and tapers (Mihailov et al. 2008).

Direct-write waveguide fabrication is probably the most widely application of femtosecond laser micromachining in transparent dielectrics. Devices such as buried waveguides, power splitters, couplers, gratings and optical amplifiers have all been demonstrated. Waveguide properties depend critically on the sample material properties and writing laser characteristics. In 1996, it was shown that tightly focused femtosecond infrared laser pulses could create a permanent refractive index modification inside bulk glass materials. Although investigations into understanding the nature of this modification and the conditions that produce it are ongoing, it is widely accepted that the modification process is initiated by the rapid absorption of laser energy through nonlinear excitation mechanisms. The subsequent dissipation of this energy into the lattice causes the modification inside the glass. This result enables the direct-write fabrication of optical devices, active and passive, in a variety of dielectric optical materials including amorphous glasses, crystalline materials and optical polymers simply by moving the glass sample through the focus of a femtosecond laser beam. The material surrounding the focal volume remains largely unaffected by the writing beam passing through it, allowing structures to be written at arbitrary depths and in a three-dimensional fashion. The femtosecond laser direct-write technique has been used to fabricate buried waveguides, power splitters, couplers, gratings and optical amplifiers. These devices are produced using (1) regeneratively amplified Ti:Sapphire laser systems that provide high pulse energies at kHz repetition rates; (2) oscillator-only Ti: Sapphire systems with low energy and high repetition rates (MHz), and (3) high pulse energy ytterbium-doped fiber lasers at high repetition rates (100 kHz-MHz) as well as cavity dumped Yb:KYW laser oscillators. While all of the systems described above are effective at modifying transparent dielectrics significant differences exist between the mechanism underlying the modification and therefore also the strength of the modification, level of damage (if any) and most importantly in terms of waveguides whether the index change is positive or negative. Key parameters which affect the writing properties include the sample translation speed, focussed beam shape, beam polarization, pulse energy, pulse repetition rate, wavelength and pulse duration. Other properties that dictate the type of material modification include, for example, bandgap energy, whether the sample is crystalline or amorphous, thermal characteristics and fracture strength (Huang et al. 2011).

The materials interaction processes at play within the laser focus are strongly dependent on both the material and the laser parameters, and it is common to observe both positive and negative changes in the material refractive index under different laser processing conditions or even within the same interaction region. Fundamental physical processes that occur at the laser focus have been developed with fused silica. In comparison with other optical materials, fused silica can be processed under a wide range of laser pulse frequencies, durations and energies, wavelengths and sample translation speeds. Furthermore, fused silica is easy to obtain in high purity forms by virtue of it being a popular UV optical material. Borosilicate glass has also been extensively studied. The most important property of borosilicate glasses is, however, the response of the glass to cumulative heating resulting from high (>500 kHz) pulse repetition frequency laser exposure. Borosilicate glasses (in contrast to fused silica) have been demonstrated to exhibit controlled growth of the heat affected zone centered at the laser focus within the material thus controlling the dimensions of the written optical waveguide. This facile control of the heat affected zone through judicious selection of the laser processing parameters is an example of how the combination of the correct material and laser processing parameters can be used to great effect in the creation of arbitrary waveguide designs. Fused silica and borosilicate glasses provide an excellent platform in which to create passive optical devices. The solubility of active, rare-earth, ions in these glasses is low and despite the extensive use of rare-earth ion doped silica glasses in optical fiber devices, the relatively low gain-per-unit length value (0.3 dB/cm) of these materials makes it difficult to realize high gain devices in a typical few centimeters long directly written device.

Consequently there has been a great deal of interest in phosphate glass hosts in which tens of percent by weight of rare-earth ions can be held in solution offering a higher gain-per-unit length value (4 dB/cm) without detrimental effects such as ion-clustering. Erbium and Ytterbium co-doped phosphate glass hosts have been successfully laser processed and used to create optical amplifiers and, with the addition of external reflectors, optically pumped waveguide lasers (WGL). Apart from the passive and active glass materials typically used for directly written devices, the femtosecond laser direct-write technique has also been applied to common crystalline materials such as $LiNbO_3$, YAG, LiF and Ti:Sapphire. The dominant material change, when using a low repetition rate femtosecond laser, in most of these materials is a negative refractive index change, however, use of suppressed cladding arrangements or induced stress fields allowed waveguiding regions to be realized. Smooth refractive index change induced by femtosecond laser radiation is likely due to a contribution of factors, such as, color center formation, densification (structural change) and thermal treatment (melting) of the glass. Optical waveguide devices are fabricated in materials using design parameters that give rise to this regime of modification. The femtosecond laser direct-write technique as a technology capable of producing optical waveguide devices inside bulk transparent materials without the need for lithography, etching, a controlled environment or much sample preparation. Specific consideration of the pulse repetition rate and energy, writing beam polarization, sample translation speed, number of fabrication scans, spherical aberration, polishing techniques and material preparation must be taken into account in order to fabricate low-loss positive index guiding waveguide devices in a specific type of glass (Huang et al. 2011).

7.7 Glass Waveguide Lasers

Planar glass waveguide lasers provide some important benefits over widely deployed semiconductor and fiber lasers. The excited state lifetimes of the order of milliseconds in glass materials result in inherently lower noise in waveguide-fiber lasers compared with their semiconductor counterparts having lifetimes of the order of nanoseconds. Higher noise in semiconductor lasers can be associated with their higher spontaneous emission rate. Noise causes undesirable fluctuations in intensity, phase, and center frequency. Phase fluctuations are responsible for the line width broadening while intensity fluctuations result in a reduced signal-to-noise ratio (SNR). Thus, due to the longer lifetimes in glass lasers, significantly narrower line widths can be obtained compared with semiconductor lasers. In glass lasers relaxation oscillations take place at frequency of $\sim$500 kHz whereas in semiconductor lasers relaxation oscillations occur at $\sim$1 GHz making relative intensity noise a serious problem in mode locking applications at high repetition rates (above 1 GHz). Lower spontaneous emission rate in glass lasers has also enabled demonstration of mode locked lasers with low timing jitter. Timing noise is affected both by frequency noise, which can couple to timing noise through group velocity dispersion, and by intensity noise. In contrast to fiber lasers, waveguide lasers realized on glass substrates allow chip-scale integration of other components, such as pump couplers and power dividers, and enable monolithic integration of multiple optical transmitters at different wavelengths. High gain values can be achieved in waveguides in rare-earth doped phosphate glass substrates. Therefore, waveguide lasers can be made short enabling mode locking at the fundamental round-trip frequency instead of higher harmonics utilized in fiber lasers that easily lead to super-mode or pattern noise problems. A drawback in glass-based lasers is that they require optical pumping in contrast to semiconductor lasers that can be pumped electrically. However, Er-Yb-co-doped glass waveguide-fiber lasers benefit from the well-established pump diode lasers developed for the Er-doped fiber amplifiers (Yliniemi 2007).

7.7.1 Short Cavity Er-Yb-Co-doped Ag–Na Ion-Exchanged Waveguide Lasers

In long haul transmission systems, it is beneficial to use the return-to-zero (RZ) pulse format instead of the nonreturn-to-zero (NRZ) pulse format since undesirable nonlinear effects can be more easily avoided with the RZ pulse format. However, the RZ pulse format is more vulnerable against dispersion due to its larger bandwidth, and also configurations providing the RZ pulse format are more costly, thereby making the NRZ pulse format better in short-distance networks. Compact mode locked lasers operating at 1,550 nm region with high repetition rates (10 GHz and above) would provide an ideal, cost effective RZ pulse source for optical transmitters. Good pulse quality of mode locked lasers improves the SNR enabling scaling to higher repetition rates through optical time division multiplexing (OTDM).

Semiconductor Saturable Absorber Mirrors (SESAMs) provide a compact, cost-effective way to realize mode locking (or Q-switching). Passive, self-starting mode locking of a CO_2 laser by a SESAM was first demonstrated in 1974. Since then, SESAMs have been fabricated for wavelengths ranging from the visible to infrared. Pulse repetition rates as high as 160 GHz can be achieved. Pulse durations as short as 6.5 fs have been demonstrated thus approaching the pulse lengths obtained by Kerr lens modelocking. High gain value in short cavity length is a desirable feature in mode locked lasers operating at high pulse frequencies ($>$10 GHz) since the cavity length sets the fundamental limit for pulse repetition rate. High small-signal gain also reduces the risk of Q-switching, decreases the mode locking build-up time and enables shorter pulses at steady state. This is why Er-Yb-co-doped phosphate glass is an ideal gain-providing medium for mode locked applications at 1,550 nm wavelength region. Typical gain values lie between 3 and 4 dB/cm around 1,550 nm wavelengths and in heavily Er-Yb-co-doped phosphate glasses, a gain value as high as 4 dB has been measured from only a 3 mm long sample. In phosphate glass, a pulse repetition rate of 10 GHz is obtained with a cavity length of ~1 cm and by shortening the length to ~2.5 mm, a pulse repetition rate of 40 GHz can be reached.

High repetition rates without Q-switching are difficult to achieve in Er-Ybcodoped glasses because of the small emission cross section of erbium. In spite of this, 40 GHz repetition rates with 4.3-ps-short pulses are obtained with careful cavity and SESAM design optimization. The first mode locked waveguide laser is made with 1 GHz repetition rate, achieved with a 2.2-cm long K^+–Na^+ ion-exchanged waveguide in Er-Yb-co-doped IOG-1 glass. Passive mode locking can be performed by an InGaAs/GaAs multiple-quantum well layer integrated with an AlAs/GaAs thin film stack as a saturable absorber mirror (Yliniemi 2007).

7.7.2 Waveguide DBR Laser with UV-Written Bragg Grating

Fiber optics technology has greatly benefited from the discovery of photosensitivity of silica-based fibers to UV-irradiation, which enables cost- and time-effective fabrication of Bragg gratings into optical fibers. FBGs have made it possible to provide dispersion management, filtering, sensing and wavelength control in fibers, and optical feedback in fiber lasers. They have replaced many bulky optical components traditionally used in wavelength multiplexing. Today, FBGs are an essential part of fiber lasers too. UV-written Bragg gratings are expected to provide a multitude of opportunities in integrated optics, too. So far, Bragg gratings have been used, e.g., in integrated optical add-drop multiplexers in. Use of UV-written Bragg gratings in integrated optical waveguide lasers has awaited because it has turned out to be difficult to write gratings into glass material that provides enough gain in short cavity lengths (phosphate glass). In practice, Bragg gratings have been fabricated by a lithographic process that begins with spinning a resist layer on the sample, then patterning it with standard photolithography and finally etching the Bragg grating into glass. Another possibility is to use thin-film reflectors grown either straight onto the waveguide facets or attached by some mechanism (e.g., by using index matching oil) onto the waveguide facet. Both of these approaches contain multiple steps in the clean room, and are therefore time consuming and expensive processes. Clearly, replacing these reflectors by UV-written Bragg gratings would both simplify the fabrication process and improve the laser efficiency (UV-written volume Bragg gratings are nearly lossless and do not couple pump laser light out from surface waveguides).

It was only as recently as 2004 demonstrated photosensitivity of Er-Yb-co-doped IOG-1 phosphate glass demonstrated. A thin (3 μm-thick) surface-like grating was fabricated operating in the Raman-Nathan regime by irradiating Er-Yb-co-doped IOG-1 glass with KrF excimer laser emitting at 248 nm. UV-irradiation resulted in an index change of 2×10^{-3} in a silver-sodium ion-exchanged sample. Only a small index change of the order of 10^{-5} was observed in a pristine sample. This kind of a surface grating would not work as a mirror in waveguide lasers for two reasons (Yliniemi 2007): (a) surface grating at the pump input side would couple out the pump laser power at 980 nm propagating close to the surface; and (b) a surface grating with a relatively low-index modulation would not provide enough optical feedback for lasing.

Fabrication of a high quality volume grating in Er-Yb-co-doped phosphate glass is a complex process. For typical laser applications, reflectance above 70 % should be obtained. It is possible to write narrow-band high reflectance waveguide gratings into undoped IOG-1 glass. For instance, hybrid glass can be used as a substrate composed of both Er-Yb-co-doped and undoped parts bonded to each other. The Bragg grating is written to the undoped part of the hybrid substrate, and it serves as a wavelength selective cavity mirror as well as the output coupler of the laser. The Er-Yb-co-doped part of the hybrid glass provides the gain required for the laser operation. The Erand Yb-concentrations are 1.0×10^{20} and 6.0×10^{20} ions/cm^3, respectively. The Bragg grating can be written through a phase mask with an ArF pulsed excimer laser. This results in improved spatial coherence but reduced the fluence to 140 mJ/cm^2. The exposure time was 30 min and the laser was run at a frequency of 100 pps. A phase mask with a period of 1,065 nm was designed so that the Bragg grating in IOG-1 glass would have a peak reflectance at 1,535 nm corresponding to the gain maximum of Er-doped glass. The waveguide was fabricated after the grating writing. The transmission minimum occurs at 1534.71 nm (TE) and at 1534.52 nm (TM) with a peak reflectance of 80 %. The grating period (533 nm) calculated from the Bragg law corresponds well with the phase mask period divided by two (the theoretical period produced into glass). A polarization-dependent wavelength shift of 0.2 nm was observed corresponding to a waveguide birefringence of 10^{-4}. This is not surprising for a surface waveguide in which the proximity of the surface causes more form birefringence than in buried waveguides where birefringence is of the order of 10^{-5} or below. A fiber coated with a SiO_2/TiO_2-thin film stack aligned with the waveguide facet in the Er-Yb-co-doped part of the substrate was used as another cavity mirror. Pump power from a fiber-pigtailed semiconductor diode laser emitting at 980 nm was also delivered through this fiber. The thin film stack was designed to have transmittance close to 100 % at the pump wavelength of 980 nm and reflectance close to 100 % at 1,550 nm wavelength region. Since the post-bake annealing could reduce significantly the strength of the UV-written grating, the annealing was performed at 225 °C in three steps. Also the Er-Yb-co-doped waveguide length was shortened between the annealing steps. All available pump power was consumed in a length shorter than the waveguide length in the Er-Yb-co-doped part of the substrate, the rest of the Er-Ybcodoped part of the waveguide being underpumped and adding therefore to losses. Shortening the sample increased pump efficiency resulting in improved output power and reduced threshold power. The threshold power required for lasing steadily decreased with sample annealing and shortening, while the output power and the slope efficiency increased. During annealing, the silver ions diffuse further into glass. Annealing also increases the mode size and smoothens the index profile, therefore reducing coupling and propagation losses. In addition, it improves the mode overlap between the pump and the signal. After the first annealing step, lasing was obtained only when using two pump lasers (the output signal was separated from the pump signal using a 980/1,550 WDM-coupler). After the final annealing step, an output power of ~9.0 mW could be extracted from the sample with a pump power of 200 mW the threshold being ~135 mW. Annealing altogether for 120 min at 225 °C did not reduce the grating strength. The lasing wavelength decreased from 1534.77 to 1534.50 nm with annealing because the effective index of the propagating mode decreases due to smaller maximum refractive index difference, a characteristic feature of diffused waveguides. Diffusion parameters (the self-diffusion coefficient of Ag^+ ions and the ratio of self-diffusion coefficients of Ag^+ and Na^+ ions) in IOG-1 phosphate glass have been determined for the silver-film ion exchange process at temperatures 90 °C (ion exchange) and 230 °C (thermal post-bake). These parameters have significance in designing and modeling waveguide components in IOG-1 glass. Waveguides with negligible birefringence can be fabricated by waveguide burial and thermal post-bake annealing. This approach can be utilized to fabricate waveguide components (such as OADMs) with polarization-independent operation. High quality narrow-band waveguide gratings in undoped phosphate glass have been demonstrated. A single-mode waveguide laser utilizing a UV-written grating as another mirror was demonstrated in hybrid phosphate glass. The waveguide Bragg grating was written with a pulsed ArF excimer laser emitting at 193 nm. A waveguide grating in Er-Ybcodoped phosphate glass with a reflectivity of 15 % has been demonstrated, which is written with a KrF excimer laser emitting at 248 nm. In both cases, the Bragg grating was exposed prior to waveguide fabrication process. It was assumed that silver particles produced during ion exchange prevent UV light from penetrating deep enough into the glass to form a volume grating. Short cavity waveguide lasers with multimode operation have been fabricated. The internal powers obtainable from the 1-cm long Er-Yb-co-doped laser cavities are high enough for either Q-switching or mode locking the laser cavity with a SESAM directly attached to the waveguide facet. These laser cavities have potential to operate as high repetition rate pulsed laser emitters in telecommunication systems. The 1-cm long cavities can produce mode locking at 10 GHz repetition rate. By further reducing the cavity length to 2.5 mm, a 40 GHz repetition rate could be achieved. However, such mode locked laser cavities would require glass substrates with higher doping levels of Er and Yb ions and these kinds of substrates are commercially available. Waveguide Bragg gratings also have potential to improve the pulse quality transmitted by a mode locked laser. Such waveguide gratings could be used to minimize dispersion, and therefore, to obtain pulses with low timing jitter (Yliniemi 2007).

7.8 Summary

Silica, SiON, fluoroaluminates, chalcogenides and doped glasses are important glasses for making optical waveguide devices. Bulk silica (SiO_2) and silica-on-silicon (SiO_2/Si) are by far the most common materials used to manufacture PLCs, due to their refractive-index match with silica-based optical fiber. Two major types of deposition processes have been used: CVD and FHD. The CVD approach, which is a modification of standard semiconductor-processing techniques, is compatible with both clean-room processes and high-volume wafer production. A raised refractive index is created in the core-guiding region by the addition of phosphorus, germanium, or both, during the deposition process.

Silicon oxynitride has been increasingly applied in various integrated optical devices. Application of this material is mainly motivated by its excellent optical properties, such as low absorption losses in the visible and near infrared wavelength range. Moreover, the refractive index of SiON thin films can be easily adjusted over a large range, i.e., between 1.45 (SiO_2)–2.0 (Si_3N_4). This large flexibility in choosing the refractive index adds to the attractiveness of this material for designers of integrated optical devices. In addition, standard silicon substrates can be used and reliable techniques that are well compatible with standard silicon integrated circuit processing, are available for controlled deposition of homogeneous, high quality thin films. This opens a potential route toward low-cost mass production.

Ion-exchanged waveguides possess many desirable characteristics. It is very easy to make waveguides exhibiting very low propagation losses of less than 0.1 dB/cm. The increase in refractive index between the waveguide and the substrate is relatively small, allowing for easy fabrication of single-mode waveguides, and excellent mode matching to single-mode fiber can be achieved through a technique of waveguide burial. Ion-exchanged waveguides exhibit very low birefringence across a broad range of waveguide widths. This is critical for certain resonant devices and devices that contain both single- and multimode waveguides. The process is cost effective, requiring no complicated material growth steps following the photolithography that is common to all processes. In addition, it is tolerant to imperfections in the photolithography—edge roughness in the mask that defines the waveguide geometry is rendered less damaging due to the diffusive nature of the ion exchange.

The large flexibility of the sol–gel can provide the desired functional properties, such as sol–gel films doped with semiconductors, rare earths, nonlinear optical organics, and dye materials. Two main routes have been followed in the development in this class of materials, namely the addition and reaction of the dopants inside the sol–gel starting solution, by chemically controlling the size of the particles, and the impregnation or exposition of a porous sol–gel film to the dopants (pore doping), where the crystal size is limited by the pore size. A broad variety of dopants have also been investigated in the effort to obtain highly-sensitive and fast-responding integrated optic chemical and/or physical sensors. Nonlinear optical glassy materials produced by sol–gel are particularly promising, but reproducibility and stability are still to be fully achieved.

The laser direct-write technique has been used to fabricate buried waveguides, power splitters, couplers, gratings, optical amplifiers and many other optical devices, active and passive, in a variety of dielectric optical materials including amorphous glasses, crystalline materials and optical polymers. The material surrounding the focal volume remains largely unaffected by the writing beam passing through it, allowing structures to be written at arbitrary depths and in a three-dimensional fashion.

Planar glass waveguide lasers provide some important benefits over widely deployed semiconductor and fiber lasers. The excited state lifetimes of the order of milliseconds in glass materials result in inherently lower noise in waveguide-fiber lasers compared with their semiconductor counterparts having lifetimes of the order of nanoseconds. In contrast to fiber lasers, waveguide lasers realized on glass substrates allow chip-scale integration of other components, such as pump couplers and power dividers, and enable monolithic integration of multiple optical transmitters at different wavelengths. High gain values can be achieved in waveguides in rare-earth doped phosphate glass substrates. Therefore, waveguide lasers can be made short enabling mode locking at the fundamental round-trip frequency instead of higher harmonics utilized in fiber lasers that easily lead to super-mode or pattern noise problems. A drawback in glass-based lasers is that they require optical pumping in contrast to semiconductor lasers that can be pumped electrically.

References

de Ridder RM et al (1998) Silicon oxynitride planar waveguiding structures for application in optical communication. IEEE J Sel Top Quantum Electron 4(6):930–937

Doerr CR (2006) Advances in silica planar lightwave circuits. J Light Technol 24(12):4763–4789

Harwood DWJ (1998) Novel fabrication technique for planar glass waveguides. Opt Commun 1:449–450

Honkanen S et al (2006) Recent advances in ion exchanged glass waveguides and devices. Phys Chem Glasses: Eur J Glass Sci Technol B 47(2):110–120

Huang H et al (2011) Femtosecond fiber laser direct writing of optical waveguide in glasses. In: Proc. SPIE 8164, Nanophotonics and Macrophotonics for Space Environments V, 81640B (August 31, 2011), Diego, California. doi:10.1117/12.892942.San

Lehky JM (2006) Optical waveguides with copper film ion-exchange in glass. PhD dissertation, Swiss federal Institute of Technology, Zurich

Mihailov SJ et al (2008) Induced bragg gratings in optical fibers and waveguides using an ultrafast infrared laser and a phase mask. Laser Chemistry. http://www.hindawi.com/archive/2008/416251/. Accessed 26 Feb 2009

Pham A (2004) Planar lightguide circuits: an emerging market for refractive index profile analysis. Application Note 053. http://documents.exfo.com/appnotes/anote053-ang.pdf. Accessed 21 Feb 2009

Righini GC, Pelli S (1997) Sol-fel glass waveguides. J Sol-Gel Sci Technol 8:991–997

Rudolph S (2004) Model structure of glass. http://www.a-m.de/englisch/lexikon/glas-bild1.htm. Accessed 21 Feb 2008

West BR (2004) Modeling and analysis of ion-exchanged photonic devices. PhD dissertation, The University of Arizona, Tuson

Worhoff K et al (2007) Design and application of compact and highly tolerant polarization-independent waveguides. J Light Technol 25(5):1276–1283

Yliniemi S (2007) Studies on passive and active ion-exchanged glass waveguides and devices. PhD dissertation, Helsinki University of Technology, Teknillinen korkeakoulu

Chapter 8
Electro-Optic Waveguides

Abstract Electro-optic waveguides are designed based on changing refractive index of the core layer with an external voltage. Materials used for electro-optic waveguides can be $LiNbO_3$, $LiTaO_3$, $BaTiO_3$, electro-optic polymers, liquid crystal, and strained silicon. The $LiNbO_3$ is a versatile material and used for various applications in guided wave optics, electro-optics, acousto-optics, and nonlinear optics. It has high electro-optic coefficient, optical damage resistance, and low losses. Stoichiometric lithium tantalate is good for applications at UV wavelengths as it is transparent down to 260 nm where most of the standard electro-optical materials, e.g., $LiNbO_3$ or $KNbO_3$, show large light absorption. The optical ferroelectrics in the form of highly transparent thin films are promising materials for communication. $BaTiO_3$ is particularly an attractive candidate for thin film electro-optic modulators due to its large electro-optical coefficients, its high optical transparency and its favorable growth characteristics. Electro-optic (EO) polymers are unique materials having many advantages over inorganic materials for the wide range of applications from optical network components and optical interconnects to millimeter and microwave photonics. They have very large EO coefficients, low optical loss, and low dispersion of refractive index between optical frequencies and millimeter waves, as well as high bandwidth. They are easy to process and have relatively low cost. There is a broad range of EO devices that can be fabricated based on EO polymers having a waveguide structure as the main building block. Liquid-crystal electro-optic waveguide platform has exhibited unprecedented electro-optical phase delays, with very low loss and rapid response time. This technology is based upon a unique liquid-crystal waveguide geometry, which exploits the tremendous electro-optic response of liquid crystals while circumventing historic limitations of liquid crystals. The exceedingly large optical phase delays accessible with this technology enable the design and construction of a new class of previously unrealizable photonic devices. Strained silicon can be another electro-optic material. A significant linear electro-optic effect can be induced in silicon by breaking the crystal symmetry. The strain-induced linear electro-optic effect may be used to remove a bottleneck in modern computers by replacing the electronic bus with a much faster optical alternative. This chapter will

X.C. Tong, *Advanced Materials for Integrated Optical Waveguides*, Springer Series in Advanced Microelectronics 46, DOI 10.1007/978-3-319-01550-7_8,

give a brief review about various types of electro-optic waveguides, and their materials selection, optimum designs, as well as processing technologies depending on applications.

8.1 Physical Effects in Electro-Optic Waveguides

In most designs of optical waveguides with external modulators, the physical effects most commonly used are based on electro-optic, acousto-optic, magneto-optic, and thermo-optic effects. The classic electro-optic (EO) effect is the alteration of the index of refraction due to the modification of the index ellipsoid (or optical indicatrix) by the application of an electric field. If the index change varies linearly with the amplitude of the applied field, the effect is known as the linear electro-optic or Pockel's effect. A change in the refractive indices of the EO material modifies the phase of a propagating optical beam and can be used to produce an optical modulator (Davis 2010).

A voltage V applied to the electrodes placed over or alongside the waveguide, creates an internal electric field of approximate magnitude $|E| \cong V/G$, G being the gap between the electrodes. Since the Pockel's effect is found in crystals without inversion symmetry (noncentrosymmetric), such as lithium niobate, the sign of the induced refractive index change depends on the polarity of the voltage applied to it. On other hand, in centrosymmetric crystals, linear electro-optic effect does not exist, while quadratic electro-optic effect (Kerr effect) is observed where the induced refractive index change is proportional to the square of the applied electric field intensity. The equation of the index ellipsoid in the presence of an applied electric field can be written as (Bhandare 2003)

$$\left(\frac{1}{n_x^2} + r_{1j}E_j\right)x^2 + \left(\frac{1}{n_y^2} + r_{2j}E_j\right)y^2 + \left(\frac{1}{n_z^2} + r_{3j}E_j\right)z^2 + 2yzr_{4j}E_j + 2xzr_{5j}E_j + 2xyr_{6j}E_j = 1 \tag{8.1}$$

where E_j ($j = 1, 2, 3$) is a component of the applied electric field and summation over repeated indices j is assumed. Here 1, 2, and 3 corresponds to the principal dielectric axes x, y, z and n_x, n_y, n_z are the principal refractive indices. This new index ellipsoid reduces to the unperturbed index ellipsoid when $E_j = 0$. In general, principal axes of the perturbed ellipsoid do not coincide with the unperturbed axes (x, y, z). A new set of principal axes can always be found out by a coordinate rotation, which is known as principal-axis transformation of a quadratic form. The linear change in the coefficients of the index ellipsoid due to applied electric field E_j along the principal axes is (Bhandare 2003)

$$(\Delta n)_i = -\frac{n^3}{2}\sum_{j=1}^{j=3} r_{ij}E_j \tag{8.2}$$

where $i = 1, 2, \ldots, 6$ and r_{ij} is the 6×3 electro-optic tensor. For lithium niobate, the coefficients of the electro-optic tensor r_{ij} are in the form (Bhandare 2003)

$$\begin{pmatrix} 0 & -r_{22} & r_{13} \\ 0 & r_{22} & r_{13} \\ 0 & 0 & r_{33} \\ 0 & r_{51} & 0 \\ r_{51} & 0 & 0 \\ -r_{22} & 0 & 0 \end{pmatrix} \tag{8.3}$$

By inserting the electro-optic tensor r_{ij}, the values of n can be written as the elements of a symmetric 3×3 matrix. For lithium niobate (Bhandare 2003)

$$(\Delta n)_{ij} = -\frac{n^3}{2}\begin{pmatrix} -r_{22}E_y + r_{13}E_z & -r_{22}E_x & r_{51}E_x \\ -r_{22}E_x & r_{22}E_y + r_{13}E_z & r_{51}E_y \\ r_{51}E_x & r_{51}E_y & r_{33}E_z \end{pmatrix} \tag{8.4}$$

where n is either the ordinary index n_o or the extraordinary index n_e value. Regarding electro-optic tensor of lithium niobate, $r_{11}, r_{21}, r_{31}, r_{41}$ are all zero. There will be no stretching of the index ellipsoid along the principle axes under applied electrostatic field E_x. r_{43}, r_{53}, r_{63} are all zero. There will be also no stretching of the index ellipsoid along the principle axes under applied electrostatic field E_z. An applied electric field E_y will cause both stretching along the principal axes as well as the rotation of the index ellipsoid cross-section in the Y-Z plane. Perturbation of the index ellipsoid due to electro-optic effect depends on the relative orientation of the polarization state of the input optical signal, the axis of propagation, the crystal cut, and the magnitude and direction or sign of the applied electrostatic field. Utilization of the diagonal elements 11, 22, and 33 of the perturbed refractive index matrix results in an index change and therefore, a phase change, for an incident optical field polarized along the crystallographic x-, y-, and z-axes, respectively. The diagonal elements causes an index change, which is essential for modulators and switches, for the optical field polarized along the crystallographic j-axis, for the given electric field which is applied in the appropriate direction. The off-diagonal elements, on the other hand, represent an electro-optically induced mode mixing or conversion between the otherwise orthogonal polarization components. It represents a rotation of the index ellipsoid that causes a mode coupling which is proportional to the relevant electro-optic coefficient due to the application of electrostatic fields. Utilization of off-diagonal electro-optic elements is necessary to induce polarization change in Ti:$LiNbO_3$ waveguides, i.e., TE–TM mode converters (Bhandare 2003).

Pockel's effect may be the most widely used of the physical effects for constructing temporal optical modulators due to the intrinsic, high speed of the response (sub-picoseconds). However, a variety of other effects can be included under the electro-optic category. These include the Kerr effect (index change varies with the square of the applied field), photorefractive effect (a manifestation of the classic EO-effect on a microscopic scale depending on charge transport properties), and the alteration of polarized light in liquid crystals when an electric field is applied.

Another utilized effect in the EO category is based on electro-absorption, which is a change in the absorption characteristics of a medium in response to an externally applied electric field. The optical index of refraction can be broadly defined as a complex quantity given by (Davis 2010)

$$\widetilde{n} = n + ik \tag{8.5}$$

where n is known as the ordinary index of refraction and k is called the extinction coefficient, which is directly related to the absorption coefficient. Thus, under certain conditions, a change in the real part of the refractive index can be accompanied by a change in the absorption coefficient. Observed primarily in semiconductors, absorption coefficients can be manipulated by the Franz-Keldysh effect, quantum-confined Stark effect, excitonic absorption, or changes of free carrier concentration. The Franz-Keldysh effect is an example of the electro-absorption effect where the application of an external electric field causes the conduction and valence band edges to bend in a way that result in electron tunneling into the forbidden gap, reducing the effective bandgap and thereby shifting the absorption edge to longer wavelength. It can be shown that the average deviation in the bandgap ($\Delta\varepsilon_g$) depends on the applied electric field (E) according to (Davis 2010)

$$\Delta E_g = \frac{3}{2}(m^*)^{-1/3}(qhE)^{2/3} \tag{8.6}$$

where m^* is the electron effective mass, q is the electron charge, and h is the reduced Planck's constant. A modulator can be made for light operating at a wavelength longer than the normal bandgap wavelength. In the absence of an electric field (OFF state), light is transmitted with minimal absorption. Upon application of the electric field (ON state), light is strongly absorbed. This electro-absorption effect is more pronounced in semiconductor multiquantum-well (MQW) structures. When an electric field is appropriately applied to a MQW structure, behavior similar to the Franz-Keldysh effect occurs with additional effects related to exciton absorption, known as the quantum-confined Stark effect. These effects also have high-speed responses (picoseconds). EO modulators, based on purely electro-refraction or electro-absorption, are capable of operating at speeds in the tens of GHz (Davis 2010).

In the acousto-optic (AO) category of modulation mechanisms, sound waves are used to produce mechanical strains that alter the refractive index of an optical medium via the photoelastic effect. Therefore, sound controls or modifies the effect of the medium on an interacting optical beam, and this is known as the acousto-optic effect. The acoustic wave creates a perturbation in the refractive index in the form of the wave. This means the index change will have a periodicity and amplitude proportional to the acoustic intensity such that the medium acts as a phase grating (Bragg cell) to diffract or partially reflect an incident optical beam. The relation between the amplitude of the refractive index change and the acoustic intensity is given by (Davis 2010)

$$\Delta n_0 = \sqrt{\left(\frac{1}{2}\right) M I_s} \tag{8.7}$$

where I_s is the acoustic wave intensity and M is a material parameter representing the effectiveness of sound in altering the refractive index. The quantity M is a figure of merit defined by (Selah and Teich 2007)

$$M = \frac{p^2 n^6}{\rho v_s^3} \tag{8.8}$$

where p is the appropriate component of the photoelastic tensor, n is the refractive index in the absence of sound, ρ is the mass density, and ν_s is the speed of sound in the material. Using an electrically controlled acoustic transducer, the intensity of the reflected light can be varied proportionally to form an analog optical modulator. In general, for temporal-only modulation, each of the acoustic frequency components of the modulated acoustic signal must interact with a focused cone of light incident on an acousto-optic cell near the transducer. It can be shown that the modulation bandwidth is inversely proportional to the transit time of the acoustic wave through the incident optical beam. To increase the bandwidth of an AO modulator the optical beam can be focused to a small diameter. With focused optical beams, typical bandwidths of AO modulators are on the order of tens of MHz (Selah and Teich 2007).

Similar to the EO category, there are many types of magneto-optic (MO) effects. However, the only such effect of real interest for optical modulators is the Faraday effect. Similar to the effect of optically active materials on polarized light, the Faraday effect rotates the plane of polarization of light as a function of the strength of an applied magnetic field. In general, only ferromagnetic materials, such as terbium gallium garnet and yttrium iron garnet (YAG), show an effective response to be useful in practical devices. In these types of materials the Faraday rotation per unit length is specified when the sample material is magnetized to saturation according to (Irvine and Elezzabi 2003)

$$\theta_F = K M_s \cos\gamma \tag{8.8}$$

where M_s is the component of the saturation magnetization parallel to the light propagation direction, K is known as the Kundt constant, and γ is the angle between the light propagation direction and magnetization. The applied magnetic field can be used to change the magnitude or the orientation of the internal magnetization (or both) and thus alter the optical rotation to modulate the light beam. When the direction of light propagation and magnetization are the same, the rotation in degrees of the linear polarization due to an applied magnetic field (H) may be written as (Irvine and Elezzabi 2003)

$$\theta = \chi_{\mathrm{M}} \left(\frac{\theta_{\mathrm{F}}}{4\pi M_{\mathrm{s}}} \right) Hl \tag{8.9}$$

where χ_{M} is the magnetic susceptibility and is the length of travel through the magnetic material. The amount of Faraday rotation is strongly dependent on the wavelength, usually increasing with shorter wavelength. However, many materials with large Faraday rotations also strongly absorb light. Thus, practical MO modulators must be designed and fabricated to simultaneously optimize Faraday rotation and optical transparency. A wideband MO modulator has been constructed in bismuth-substituted YAG material. A bandwidth up to 3 GHz can be achieved for an 800 nm optical beam by driving the device in a nonresonant mode, well below the ferromagnetic resonance frequency of the film (Irvine and Elezzabi 2003).

Modulation employing thermo-optic (TO) effects involves the alteration of the real part of the index of refraction by a change in temperature. A variation in the output intensity of a TO modulator, between a maximum and a minimum, is associated with two working temperature points ($T_{\max}$ and $T_{\min}$) such that a phase shift ($\Delta\phi$) in the input light is produced with a change in temperature ($\Delta T = T_{\max} - T_{\min}$). Given $\phi = 2\pi nl/\lambda$ as the phase factor for a propagating optical beam, the phase shift due to the TO-effect can be derived by differentiating with respect to T such that (Reed and Knights 2004)

$$\Delta\phi = \frac{2\pi nl}{\lambda} \left(\frac{1}{n} \frac{dn}{dT} + \Gamma \right) \Delta T \tag{8.10}$$

where $\Gamma = (1/l)(dl/dT)$ is the material thermal coefficient of expansion, dn/dT is the TO coefficient, n is the index of refraction, λ is the optical wavelength, and l is the propagation distance effected by the temperature change. Thermally activated modulators can be realized with materials with large TO coefficients (10^{-4}/°C), high thermal conductivity, and low specific energy which enable fast cooling rates directly related to the modulation rate. Thermal modulation speeds are generally slow compared to other physical effects utilized in optical modulators. However, TO modulators are developed for both analog and digital applications with speeds ranging from tens of kHz to 1 or 2 MHz (Iodice 2003; Reed and Knights 2004).

Other emerging and specialized modulation mechanisms are being developed in various research areas. Some of these areas include the use of metamaterials, surface plasmons, photonic bandgap structures, and devices based on biological materials. In plasmonics, light can be controlled and manipulated at the nanometer spatial scale by coupling the optical wave with charge-density waves near the surface of a metal-dielectric interface. Plasmonic modulators have been made with sub-wavelength dimensions that combine low optical loss and high mode confinement which show enhancements in characteristic parameters such as sensitivity and extinction ratio. With photonic bandgap materials, optical waves interact with periodic media in a unique way such that spectral bands emerge in which light waves cannot propagate through the medium without severe attenuation. The cut-on and cut-off optical frequencies of the bandgaps can be used as the basis for optical modulators. Optical modulators with photonic crystal materials have been demonstrated utilizing both electrical and thermal tuning of the bandgap. Certain biological materials that exhibit nonlinearities, photochromism, or biochemical affinities have been explored for optical modulation. In addition, biophotonic materials have been integrated into photonic devices to incorporate enhanced sensitivity to changes in properties such as refractive index, absorption, or adsorption. One of the most often used compounds for optical modulation has been bacteriorhodopsin and other amino acids with photochromic groups. Moreover, the use of deoxyribonucleic acid (DNA) as a biopolymer thin film has shown promise in the creation of novel photonic and optoelectronic devices. All of the above nanomaterials and biomaterials are of particular interest as they often have unusual properties that are not easily replicated in conventional organic or inorganic materials. However, in many cases, both materials and devices in nanophotonics and biophotonics are immature, difficult to fabricate, and have not yet achieved the high performance required for optical modulators (Davis 2010).

The discussion of this chapter would focus on electro-optic effects. Actually, electro-optical systems are generally a combination of electronic and electro-optical components such as light emitting diodes, image intensifiers, and lasers. Examples of such devices are the night vision goggles and rifle sights, laser range finders, markers and designators, and forward looking infrared receivers. However, all the electro-optical devices need some source of light to produce an image of an object through illumination, amplification, or thermal imaging. Most of the devices use lasers as the light source because of their low running threshold. The lasers are also associated with electro-optical elements for the production of chaotic carriers. One way to do that consists in embedding a resonator and electro-optical feedback through the Mach-Zehnder modulator and pumps this with semiconductor laser (Nguimdo 2008).

8.2 Electro-Optic Materials and Modulators

8.2.1 Electro-Optic Materials in Photonics

The demand for photonic components and integrated circuits that meet performance criteria as well as economic requirements has stimulated research and development for technologies capable of enabling unique functions. Several material systems, including hybrids, have been pursued for discrete components and multifunctional optical circuits. Currently, no single material system has all desired optical properties, or can achieve every function needed for photonic devices and systems. The key properties of current state-of-the-art photonic material systems are summarized in Table 8.1 (Davis 2010). Commercial production of discrete components typically utilize the highest-performance and most-reliable material systems for the functions they perform.

Among these materials, the developments in engineered organic polymers at the molecular level are enabling passive and active components that are eclipsing what has been possible. Passive devices include photonic components that simply pass optical signals from sources to detectors. Examples of passive photonic components include optical interconnects, couplers and taps, splitters and combiners, and multiplexers. Active devices include all photonic components that alter the wavelength, polarization, or intensity of an optical signal as it propagates along a path. They can be add/drop filters, amplifiers, laser/LED sources, modulators and switches, or photodetectors. When synthesized and processed properly, organic polymers for photonics offer high performance through low optical losses (<1 dB/cm at all communication windows), wide controllability of the refractive index contrast in waveguide devices, environmental stability, absence of diffusion problems, ease of processing, hybridization and synthesis modification (Davis 2010).

8.2.2 *Electro-Optic Modulation in Waveguides*

Optical modulators are devices that can alter the temporal and spatial character of a light beam. In general, the input light going into a modulator can be described by a spatio-amplitude and phase function (Davis 2010):

$$E(x, y, z, t) = A(x, y, t)\mathrm{e}^{-i\varphi(x,y,t)}\ \mathrm{e}^{i(kz-\omega t)} \tag{8.11}$$

where $A(x, y, t)$ is the amplitude of the light wave and the phase information is represented by the expressions in the exponents. The primary function of optical modulators is to convey digital or analog signals over an optical link by modifying the phase φ, frequency ω, or amplitude A (or intensity) of the input light. Modulation devices that perform both spatial and temporal modulation are called a spatial light modulator (SLM). The simplest type modulator is temporal-only and it mimics a time-varying alteration of a measurable property of the light wave.

Figure 8.1 shows a schematic of a typical arrangement of how EO materials can be used to produce a bulk intensity modulator utilizing the Pockel's effect. In general, a birefringent EO material is placed between crossed polarizers (P1 and P2) and an applied electric field is produced by the application of a voltage (V_{app}) to electrode contacts on the top and bottom of the EO material. The electric field changes the extraordinary (n_{e}) and ordinary (n_{o}) indices of refraction by different amounts. When 45° linear polarized light enters the EO material, a voltage-controlled phase shift ($\Delta\varphi$)

Table 8.1 The key properties of current state-of-the-art photonic material systems (Davis 2010)

Material system	Propagation loss (dB/cm)	Refractive index (n)	Birefringence ($n_{TE} - n_{TM}$)	T/O coefficient dn/dT (1/K)	Maximum modulation frequency	Optical functions
Silica (SiO_2)	0.1	1.5	0.0001–0.01	0.00001	1 kHz (T/O)	Amplifiers, polarization controllers, filters, switches, attenuators, chromatic dispersion compensators, PMD compensators, coupler, splitters, combiners, taps, multiplexer, demultiplexer
Er-doped silica	0.7	1.5–1.7	–	–	–	Amplifiers
Silicon (Si)	0.1	3.5	0.0001–0.01	0.0018	1 kHz (T/O)	Amplifiers, modulators, polarization controllers, filters, switches, attenuators
Silicon oxynitride (SiO_xN_y)	0.1	SiO_2:1.5 Si_3N_4:2.0	0.00005–0.0001	0.00001	1 kHz (T/O)	Polarization controllers, filters, switches, attenuators, multiplexer, demultiplexer
Polymers	0.1	1.3–1.7	0.000001–0.01	−1–0.0004	>100 GHz (E/O) 1 MHz (T/O)	Amplifiers, modulators, polarization controllers, filters, switches, attenuators, wavelength converters, chromatic dispersion compensators, PMD compensators, coupler, splitters, combiners, taps
Lithium niobate ($LiNbO_3$)	0.5	2.2	0.01–0.1	0.00001	40 GHz (E/O)	Amplifiers, modulators, polarization controllers, filters, switches, attenuators, wavelength converters, PMD compensators
Indium phosphide (InP)	3	3.1	0.001	0.00008	40 GHz (E/O)	Amplifiers, modulators, filters, switches, attenuators, wavelength converters
Gallium arsenide (GaAs)	0.5	3.4	0.001	0.00025	20 GHz (E/O)	Amplifiers, modulators, filters, switches, attenuators
InP/GaAS	3/0.5	3.1/3.4	–	–	–	Lasers, amplifiers, wavelength converters
InGaAs	0.8	3.3–3.6	–	–	–	Detectors
Garnets (e.g., YIG)	0.5	2.2	0.00001–0.0001	0.00001	1 GHz (M/O)	Isolators, circulators

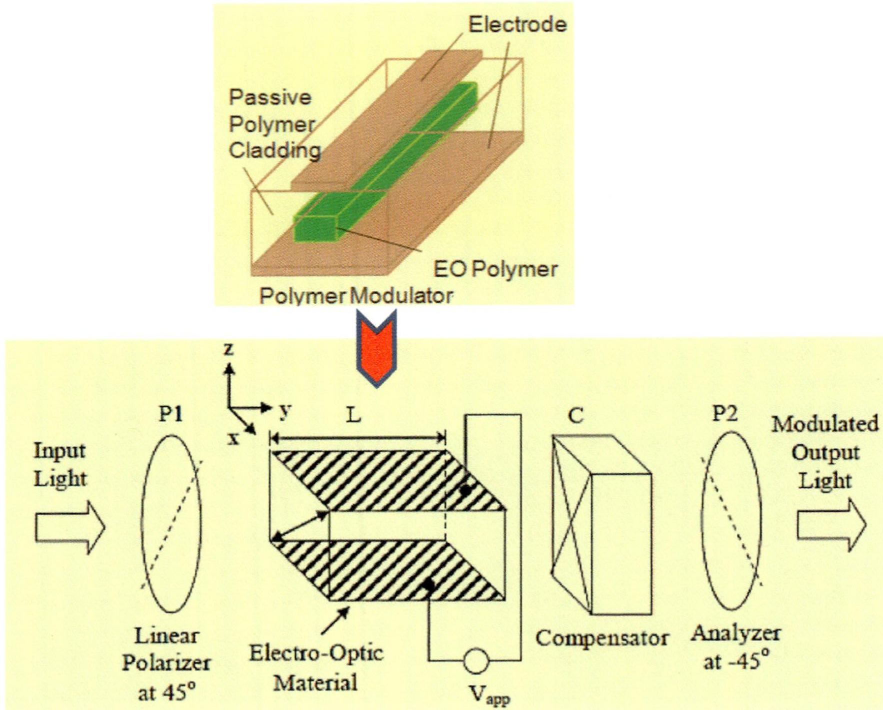

Fig. 8.1 Typical transverse bulk modulator arrangement for intensity modulation with a birefringent electro-optic material between crossed polarizers (Davis 2010)

is accumulated between the orthogonal components of the light wave such that the polarization state of the light is modulated as it exits the EO material. In practice, a phase compensator (*C*) is usually added to compensate for phase factors independent of the applied electric field and to provide a constant phase shift when needed. Thus, at the output of the analyzer (P2) only a fraction of the light intensity is transmitted, according to a sine-squared function of the applied voltage (Davis 2010),

$$T = \frac{I_{\text{out}}}{I_{\text{in}}} = \sin^2\left(\frac{\Delta\varphi}{2}\right) = \sin^2\left(\frac{\pi}{2}\frac{V_m}{V_\pi}\right) \tag{8.12}$$

where V_m is the modulation voltage and V_π is the voltage required to achieve a phase shift between the orthogonal components of the light. Hence, in this arrangement, the EO modulation of the polarization state of light can be used to construct a bulk intensity modulator. Use of bulk modulators is limited in practice due to the size restraints and required electrical drive power needed to efficiently modulate the optical input wave. A fundamental restriction of a bulk EO modulator is that the cross-sectional area of the modulator must be sufficiently large to allow the light beam to pass through. However, the total energy needed to operate the modulator at a maximum frequency equal to its bandwidth is proportional to the total volume of the modulator. Hence, for a fixed interaction length (L), the drive power increases as the cross-sectional area increase. Even for focused optical beams, a lower bound of the cross-sectional area exists due to limitations related to diffraction properties. Thus, in bulk modulators, the need for strong electric fields occupying relatively large volumes leads to high drive powers. To significantly reduce the drive power, EO modulators can be fabricated in optical waveguides. Optical waveguides restrict the cross-sectional area of light to regions with transverse dimensions on the order of the wavelength of light for essentially unbounded distances in the direction of the propagation of light. Drive powers can be reduced by two or three orders of magnitude in waveguides compared to bulk modulators. However, the use of waveguides present additional issues not found in bulk modulators. While bulk modulators need only a single EO active material that must fulfill only the modulator requirements, waveguides require a selection of materials for the core and cladding layers with optical properties that must satisfy optimal waveguide characteristics and, at the same time, have EO activity, electrical, thermal, and mechanical properties that will produce a modulator comparable or better than bulk materials. Single-mode waveguide designs are preferable for modulators since multimode waveguides tend to negate the advantages gained from the reduced cross-sectional area of the optical confinement. Both bulk and waveguide devices require a means to apply electric fields that are restricted to the EO active material regions. However, in waveguides, the optical fields and, in some cases, the applied electric fields are not uniformly confined and may not occupy the same volume. Therefore, a correction factor must be incorporated in the modulation parameters to allow for an overlap integral that accounts for the distributed interaction between the optical field and the applied electric field. The ideal correction factor is equal to one for optical and electric fields coinciding, and less than one for all other cases. Waveguide optical properties also have an effect on the coupling of light to and from the modulator. Light can be coupled into waveguides by direct focusing, end-butt coupling, prism coupling, or using gratings. Unlike coupling in bulk modulators, waveguides require input beam or mode size to closely match the modulator's waveguide mode size to achieve high coupling efficiencies. Coupling efficiency has a direct and significant impact on the insertion loss of the modulator. In practice, waveguide modulators are typically coupled to optical fibers at the input and output via end-butt coupling. Thus, waveguide transverse dimensions are also limited by mode matching conditions for coupling light into and out of the waveguide. The tradeoff between optical coupling and modulation efficiency in an EO waveguide modulator should be considered (Davis 2010).

The benefit of reducing the drive power by using the waveguide configurations is the reduction in the voltage required to achieve a given maximum modulation depth or extinction ratio using the same EO material, as compared to the bulk case. To obtain the high electric fields needed for EO modulation, low voltages can be applied across the electrode gaps on the scale of microns for conventional waveguides and submicrons for emerging nanotechnologies. Low voltage operation simplifies drive power supply requirements. In spite of some challenging tradeoffs associated with optimizing the optical waveguide design, EO waveguide modulators are essential for applications of a variety of systems in order to achieve efficient high-speed modulation. A major goal of EO material and device development is to discover and implement materials with large EO activity and other favorable material properties that would allow for optimum design of waveguide modulators with simultaneous low voltage and high-speed operation, as well as low optical power loss (Davis 2010).

8.2.3 Alternative Electro-Optic Materials

External optical modulators have been made from different classes of materials including inorganics, semiconductors, and polymers. Some examples of the alternative platforms for electro-optic (EO) modulators have been made of titanium-diffused lithium niobate, indium-phosphide multiple quantum wells, gallium arsenide heterostructures, electro-optic polymers as well as sol-gels, silicon, and hybrid systems. Among these alternatives, inorganic lithium niobate is the most common EO modulator material. This material has achieved commercial success not because it is the best with respect to any particular material property but because it has a reasonable combination of many parameters suitable for EO modulation and it is also relatively inexpensive. However, with expanding use and requirements of optical transmission systems, research efforts continue to pursue and implement alterative EO materials with unique and exceptional properties that can enhance external modulator's performance. For instance, a nonlinear optical (NLO) polymer with an electro-optic coefficient larger than lithium niobate is utilized to make an all-polymer waveguide modulator based on the Pockel's effect. In general, the Pockel's effect is only observed in materials that possess a noncentrosymmetric order and originates from the polarizability of the material due to an externally applied electric field. On the macroscopic scale, an induced, nonlinear polarization due to a sufficiently strong, electric field can be described by a power series in the applied field strength such that (Davis 2010)

$$P_i^{\mathrm{NL}} = \chi^{(1)} E_{\mathrm{opt}} + \chi^{(2)} E_{\mathrm{opt}} E_{\mathrm{app}} + R E_{\mathrm{opt}} E_{\mathrm{app}} E_{\mathrm{app}} \tag{8.13}$$

where E_{opt} is the optical field and E_{app} is a low (relative to optical frequencies) frequency electric field due to an applied voltage (V_{app}). The quantity $\chi^{(2)}$ is the second-order, NLO susceptibility and is related to the linear, electro-optic coefficient. The quantity R is the quadratic, electro-optic coefficient. For electro-optic modulation applications employing the Pockel's effect, the most popular and commercially available materials with large, second-order nonlinear susceptibilities are inorganic ferroelectrics such as lithium niobate ($LiNbO_3$), lithium tantalite ($LiTaO_3$), barium titanate ($BaTiO_3$), and potassium niobate ($KNbO_3$). However, research efforts in organic NLO materials, such as poled-polymers and dendrimers, have delivered materials with electro-optic activity that is either competitive with or surpassing inorganics. High-speed, external modulators based on the Pockel's effect are demonstrated with $LiNbO_3$ up to 105 GHz and with polymers up to 150 GHz. Some external modulators have also achieved half-wave voltages less than 1 V. However, other device parameters, link performance, fabrication steps, integration capabilities, and economical factors do not meet simultaneously, the high standards needed for rapidly-growing and advanced system applications. For example, the half-wave voltage of high-speed $LiNbO_3$ modulators is currently still relatively high, ranging between 3 and 7 V; $LiNbO_3$ can be difficult to work with and integrate with electronic semiconductors; long-term stability of polymer materials is still undergoing development and insertion losses of 4–9 dB reduce the performance of system applications. Thus, external modulators with combined stable, broad bandwidth, low operating voltage, and low insertion loss continue to be the motivation of some research efforts in photonic materials and devices for modulator technologies (Davis 2010).

8.3 Lithium Niobate Waveguides

With the progress in the optical interconnect and waveguide systems, a variety of optical components and devices for optical modulation, optical switching, wavelength division multiplexing, and coherent system techniques are required. In many applications, the nonlinearity or the electro-optic constant and the optical damage or the photorefractivity are important. In the optical sensor applications for instance, the He–Ne laser is often used as a light source. In this case, the index instability in the waveguide appears due to the photorefractive effect (optical damage). Therefore, the proton-exchanged waveguides with higher resistance to the photorefractive effect than that in the Ti-diffused waveguides seem to be preferable. The SHG blue laser by using the lithium niobate crystals can be used for the optical information processing such as the high density optical disk. In this case, the index change of lithium niobate crystals due to the photorefractive effect appears mainly by the generated second harmonic beam. On the other hand, the wavelength conversion in the C-band to the L-band becomes important for the large capacity communication. The difference frequency generation (DFG) will be used in such a conversion. The visible wavelength 800 nm may influence on the stability of the devices because the photorefractive effect is induced by the light with such a wavelength when the optical power density is high. The optical damage can be a serious problem also in the communication wavelength when the high-power laser diode is used. Therefore, the optical damage is significant not only in blue wavelength but also in the red, and even at 1,550 nm with high-power (100–300 mW) laser diode for communication. The MgO-doped lithium niobate crystal is effective to reduce the photorefractive effect. The enhancement of the photorefractivity is observed to the applied DC and AC fields independent of the polarity. This means the drifting and the trapping of the photoexcited carrier is induced by the applied electric field (Fujii et al. 2007).

8.3.1 Lithium Niobate Crystal

The lithium niobate is a uniaxial crystal with trigonal 3 m symmetry. The lattice constants for *a* and *c* crystal axes are 5.1483 and 13.8631 Å, respectively. An oblate octahedral structure is the source of large electro-optic constant. The lithum nibate crystal has a large birefringence (the refractive indices for ordinary and extraordinary rays are 2.286 and 2.20, respectively) and it has also a large nonlinearity of the second order (Fujii et al. 2007).

Lithium Niobate ($LiNbO_3$) is a colorless, insoluble with water, and ferroelectric material. Its electro-optical, acousto-optical, and NLO properties attract designers to use it in integrated optics and have been proved as excellent material for optical waveguides manufacturing. The property of high intrinsic modulation bandwidth makes it a suitable candidate for communication technology. Table 8.2 lists various specifications for lithium niobate crystal (Singh et al. 2010). The lithium niobate is a versatile material and used for various applications in guided wave optics, electro-optics, acousto-optics, and nonlinear optics. Its piezoelectric and photoelastic properties have been used for various technical developments. Lithium niobate is used to fabricate integrated waveguides due to its high electro-optic coefficient, optical damage resistance, and low losses. Their Strain-optic effects are used to implement wavelength-tunable polarization converters. The $LiNbO_3$ modulators have been used with bulky diode pumped YAG lasers and cavity RF filters to construct optoelectronic oscillators (OEO's) with ultra-low phase noise capability.

Table 8.2 Typical physical properties of lithium niobate crystal (Singh et al. 2010)

Chemical formula	$LiNbO_3$
Crystal class	Trigonal, 3 m
Solubility in water	None
Color	None
Molecular weight	147.9
Density, g/cm^3 at 293 K	4.644
Transmittance range, nm	350–5,500
Dielectric constant at 100 KHz e_a (perpendicular), e_c (parallel)	85, 29
Melting temperature, K	1,530
Curie temperature, K	1,415
Thermal conductivity, W/(m K) at 300 K	5.6
Bandgap, eV	4.0
Optical damage threshold at 1,064 nm, $t \sim 10$ ns, MW/cm^2	250

The lithium niobate can be used to make QPM structure by reversing the spontaneous polarization under the influence of a sufficiently large electric field. The lithium niobate material has been extensively used to design switches with low loss, but their polarization dependency is the major concern. The choice of LN-based switches is proved as best for high and low speed systems. In addition, $LiNbO_3$ devices have successfully addressed the modulation requirements in digital fiber-optic time-domain-multiplexed (TDM) and wavelength-division-multiplexed (WDM) systems (Singh et al. 2010).

Consequently, properties and applications of Lithium Niobate crystal can be summarized as below (Castaldini 2006; Fujii et al. 2007):

1. Optical nonlinearities
   The optical nonlinearity appears when two input optical frequencies interact with each other (nonlinearity in narrow sense) and when an optical and a radio frequencies interact (electro-optic effect) through the second-order nonlinear polarization. These effects are utilized in optical modulators and also optical frequency converters.
2. Electro-optic modulator
   The modulator is the most important device. Lithium niobate crystal is used mainly in the transverse modulator due to the symmetry of the nonlinear coefficients. The modulation sensitivity of the transverse modulator increases by long crystal, but the response decreases. The traveling type modulator may improve the response, but the phase matching becomes difficult.
3. Other Applications
   The voltage sensor is another application of lithium niobate modulator. The nonlinearity of the crystal is used in the SHG, the optical parametric devices, and also the terahertz parametric oscillator. Another category is the nonvolatile memory. A memory utilizes the photorefractive (optical damage) effect of this crystal.

8.3.2 Fabrication Process of Lithium Niobate Waveguides

The lithium niobate optical waveguides can be used to fabricate the functional devices such as modulators, switches and the various nonlinear devices such as the second harmonic generator and the difference frequency generator. The principle of these devices is the coupling of two waveguides of the *Y*-branching of the waveguide. Typical fabrication processes for the optical waveguides include (Castaldini 2006; Fujii et al. 2007):

1. Ti-indiffusion waveguides
   $LiNbO_3$ waveguides can be fabricated to design switches and modulators using either titanium indiffusion or annealed proton exchange processes. The Ti-indiffused waveguides are fabricated by the indiffusion of Ti-film of 25–80 nm sputtered on the substrate at 1,000–1,100 °C in a crucible of the oxygen or argon atmosphere. The index profile type is Gaussian. In some cases, Mg oxide is doped with $LiNbO_3$ to control optical damages. While the proton-exchanged lithium niobate waveguides are easy to fabricate and can operate with low temperature, Ti-diffused lithium niobate waveguides are useful for various communication, signal processing, and sensor systems. Ti doping in the lithium niobate crystal increases refraction indexes, and which allow both TE and TM modes to propagate along the waveguides, which satisfies desired conditions for optical signal processing.
2. Proton-exchanged waveguides
   The proton-exchanged waveguides are fabricated immersing the substrate into the benzoic acid or pyrophosphoric acid at the temperature 200–250 °C. The index profile type is step-like.

   In order to recover the nonlinearity or the electro-optic constant, the waveguides are usually annealed at 300–350 °C for a few hours. The profile becomes Gaussian after annealing.
3. Liquid epitaxy waveguides
   Five kinds of powders: lithium carbonate, sodium carbonate, niobium oxide, vanadium oxide or boron oxide and MgO are mixed so that the melt composition is 5 mol% MgO doped in the 20–80 mol% mixture of the lithium niobate and the lithium(0.7)-sodium(0.3) vanadate. The lattice constants of the niobate and the tantalite should be matched. This melt is heated at 1,000–1,100 °C in a Pt crucible and cooled down below the saturation temperature at 930–950 °C until the super cooling state. Then the lithium tantalate substrate is touched to the melt for a few minutes rotating at 10–100 rpm. The growth rate is about 1 μm/min. At the room temperature, the residual flux is removed by water and polished. This process may control the thickness of the thin film and the guiding layer.

8.3.3 Erbium-Doped Lithium Niobate Waveguides

Er-doped $LiNbO_3$ is an excellent laser material for integrated optics. It can be easily fabricated in the surface layer of a $LiNbO_3$ substrate by indiffusion of a thin vacuum-deposited Er layer. Afterwards, a single-mode channel waveguide is defined by the standard indiffusion technique of Ti-stripes. If optically pumped by $\lambda = 1{,}480$ nm radiation a wavelength dependent gain of up to 2 dB/cm results. Additional doping by Fe allows defining holographically waveguide gratings of excellent quality. Reflectivities >95 % and a spectral halfwidth of the grating characteristic of <60 pm enabled the development of narrow line width integrated optical DBR-, Distributed Feedback- (DFB-), and coupled DBR-DFB-lasers. In particular, a whole family of Er-doped waveguide lasers of excellent quality have been developed emitting in the wavelength range $1{,}530 \text{ nm} < \lambda < 1{,}603$ nm. Free running lasers of the Fabry Pèrot type, harmonically mode-locked lasers (5 ps/10 GHz), Q-switched lasers (4 ns/1 kHz/1 kW), Distributed Bragg Reflector- (DBR-) lasers, self-frequency doubling devices, and acousto-optically tunable lasers have been developed (Sohler et al. 2005).

The fabrication of erbium-doped waveguide lasers in $LiNbO_3$ consists of a sequence of steps, starting with diffusion doping of the substrate, followed by waveguide and resonator fabrication. Electro-optically and acousto-optically intracavity controlled devices require corresponding electrode structures and waveguides for surface acoustical waves (SAWs), respectively. Small bandwidth resonators for DBR- and DFB-lasers can be developed with etched surface relief or photorefractive gratings; the latter require an additional doping e.g., by Fe (Castaldini 2006; Sohler et al. 2005):

1. Erbium-doping
   The electronic transitions of Er^{3+}:$LiNbO_3$ exploited for lasing around 1,550 nm wavelength represent a quasi-three level system with ground state absorption. Therefore, Er-doping should be restricted to pumped sections of an integrated optical device; otherwise, strong absorption sets in. Such a selective doping cannot be achieved during crystal growth; it can only be achieved by indiffusion of a patterned, vacuum-deposited Er layer. This technique proved to be ideally suited for waveguide laser development avoiding clustering of Er-ions in the $LiNbO_3$ host. Moreover, it is compatible with Ti-indiffusion for waveguide fabrication. Er-diffusion doping of $LiNbO_3$ follows Fick's laws resulting in a nearly Gaussian concentration profile. It allows obtain concentration levels up to the solid solubility limit (about 0.18 at% at 1,060 °C, the waveguide fabrication temperature) without fluorescence quenching; this is a prerequisite for significant optical gain in a short device (up to 2 dB/cm has been demonstrated experimentally). Due to the low diffusivity of the Er^{3+}-ions, high diffusion temperatures up to 1,130 °C (close to the Curie temperature of ferroelectric $LiNbO_3$) and long diffusion times (up to 150 h, depending on crystal orientation) are required. For acousto-optically tunable lasers and for DFB- and DBR-lasers Er is diffused into X-cut wafers, whereas for electro-optically mode-locked and Q-switched lasers it is diffused into

Z-cut wafers. Due to the low ion mobility Er-diffusion doping is always the first step of doped waveguide fabrication (Sohler et al. 2005).

2. Acoustical waveguide fabrication
If an acousto-optically controlled waveguide laser is to be developed, intracavity acousto-optical filter and frequency shifter have to be fabricated in the next step. As Ti-doping changes the elastical properties of the $LiNbO_3$ substrate, it can be used to design waveguides for surface acoustic waves (SAWs). Ti-doped claddings of very high concentration are indiffused (e.g., 160 nm/1,060 °C/24 h) to form an undoped SAW-guiding channel of 100 μm width. Interdigital electrode structures of about 17 μm periodicity, photolithographically defined in vacuum-deposited Al-layers, allow an efficient excitation of SAWs at a frequency of about 170 MHz; they can propagate with losses as low as 0.3 dB/cm. The width of 100 μm of a SAW-waveguide allows incorporating along its axis an optical waveguide, which is considerably narrower. In this way a combined acoustical/optical waveguide can be formed yielding a nearly ideal overlap of acoustical and optical fields. This results in an optimum efficiency of all acousto-optical interactions (Fellers and Davidson 2012).
3. Optical waveguide fabrication
Several technologies such as proton exchange, Ti-indiffusion, or Zn-diffusion can be used for waveguide fabrication in $LiNbO_3$. However, in proton-exchanged Er-doped waveguides, a dramatic fluorescence quenching (radiative lifetime reduction) is observed due to the coupling to OH-phonons. Up to now high quality Er-doped waveguides have only been fabricated by Ti-indiffusion. The resulting channel guides provide single-mode guiding in both, TE- and TM-polarization, which is mandatory for the acousto-optically tunable laser, for instance. The fluorescence lifetime of the upper laser level of 2.7 ms is almost identical to the one of bulk Er:$LiNbO_3$. Single-mode waveguides for wavelengths around 1,550 nm are fabricated by indiffusion (9 h/1,060 °C) of 7 μm wide, 100 nm thick Ti-stripes, photolithographically defined on the surface of the $LiNbO_3$ substrate. For *Q*-switched and mode-locked lasers the waveguides are fabricated on *Z*-cut substrates with propagation along the *X*-axis to utilize the largest electro-optic coefficient for intracavity modulation and to take advantage of the higher optical gain in this crystal orientation. For acousto-optically tunable lasers and for DBR-/DFB-lasers *X*-cut substrates are used to allow efficient surface acoustic wave (SAW) excitation and grating fabrication, respectively. For the former ones propagation is along the *Y*-axis, whereas for the latter ones propagation is along *Z* (polar axis) (Sohler et al. 2005).
4. Fe-doping: If photorefractive gratings are required to form the cavity of DBR- or DFB-lasers, an additional Fe-doping is necessary to increase the photorefractive sensitivity of $LiNbO_3$ in selected sections of a Ti-doped optical waveguide with or without additional Er-doping. In these sections a refractive index grating of sub-μm periodicity can be easily fabricated holographically.
Fe^{2+} acts as a donor for photoionized carriers whereas Fe^{3+} acts as a trap. Charge separation is mainly driven by the photovoltaic effect, which is strongest along the Z-axis. Therefore, optical waveguides for DFB- and DBR-lasers are delineated

along Z. Typically, about 15 mm long sections of a waveguide are sensitized by indiffusion (70 h/1,060 °C) of a vacuum-deposited Fe-layer of about 40 nm thickness leading to a concentration profile of about 40 μm 1/e depth. After indiffusion the sample is annealed in a wet Ar-atmosphere to provide the necessary proton concentration for the fabrication of fixed holographic gratings (Das 2003; Sohler et al. 2005).

5. Holographic grating: Photorefractive gratings are written using a holographic (two beam interference) setup. A "Lloyd" arrangement provides a wavefront splitting and intersection of the expanded collimated beam of an Ar-laser ($\lambda = 488$ nm) on the waveguide surface generating a periodic light pattern. Optically excited electrons move into the dark regions of the periodic light pattern, where they are trapped, generating a periodic space charge field. This periodic field induces a refractive index grating via the electro-optic effect; the resulting Bragg response is measured during holographic illumination by monitoring with an optical spectrum analyzer (OSA) the transmitted amplified spontaneous emission of an Erbium-Doped Fiber Amplifier (EDFA) through the waveguide. However, due to a residual dark conductivity of $LiNbO_3$ the electronic space charge pattern decays with time after the holographic illumination is switched off. This serious problem can be overcome by "fixing" the grating taking advantage of the mobility of protons at elevated temperatures ($\sim$180 °C). The mobile protons compensate the electronic space charge and generate a positive replica. After cooling to room temperature the ionic space charge distribution with the resulting field is "frozen," but the electronic one can be "smoothed" using a constant homogeneous illumination with blue light. In this way stable Bragg gratings can be fabricated; peak reflectivities of the Bragg response of up to 60 % and a bandwidth of 60 pm have been achieved (Das 2003; Sohler et al. 2005).
6. Dielectric mirror deposition: Dielectric mirrors are needed for resonator definition of tunable, mode-locked and Q-switched lasers with a Fabry Perot type cavity. They are comprised of alternating SiO_2 and TiO_2 quarter wave layers directly deposited onto the polished waveguide end faces. O^{2-} ion beam assisted e-beam evaporation is used to fabricate fully oxidized (non-absorbing) layers of high density at low substrate temperatures. Several types of narrow linewidth lasers with optical feedback by photorefractive gratings have been developed: distributed Bragg reflector- (DBR-), distributed feedback- (DFB-), and DBR-/DFB-coupled cavity lasers with single-mode Ti:Er:$LiNbO_3$ waveguide. They have one or two photorefractive gratings in Fe-doped waveguide sections. Two types of DBR-lasers have been developed. One has a cavity consisting of one Bragg-grating, a gain section, and a multilayer dielectric mirror deposited on the opposite waveguide end face. The other DBR-cavity consists of two gratings in Ti:Fe:$LiNbO_3$ waveguide sections on both sides of the Er-doped waveguide. Single-frequency operation could be achieved in the latter case at various wavelengths in the Er-band (1,530 nm $< \lambda <$ 1,575 nm) with up to 1.12 mW output power. The DFB-laser has a thermally fixed photorefractive grating in a Ti:Fe:Er:$LiNbO_3$ waveguide section; it is combined with an integrated optical amplifier on the same substrate. Up to 1.12 mW of output power at

$\lambda = 1531.35$ nm can be emitted by the laser/amplifier combination at a pump power level of 240 mW ($\lambda_P = 1{,}480$ nm). The emission spectrum consists of the two lowest-order DFB-modes of about 3.9 GHz frequency spacing. Moreover, an attractive DBR/DFB coupled cavity laser has been developed. The laser consists of a photorefractive Bragg-grating in the Ti:Fe:Er:$LiNbO_3$ waveguide section (a DFB-laser) close to one end face of the sample, a Ti:Er:$LiNbO_3$ gain section and a broadband multilayer dielectric mirror of high reflectivity on the other end face. Single-frequency operation has been achieved with an output power of up to 8 mW. The optically pumped ($\lambda_P = 1{,}480$ nm, $P = 130$ mW) laser emits up to 8 mW at $\lambda = 1557.2$ nm with a slope efficiency of about 22 %. The single-frequency laser emission wavelength can be thermo-optically and electro-optically tuned by a fraction of a nanometer (Sohler et al. 2005).

Comparably, narrow bandwidth DBR and DFB-lasers increase the potential of $LiNbO_3$ integrated optics significantly; they can be incorporated everywhere in an optical circuit by spatially selective Er-doping. Acousto-optically tunable lasers promise single-frequency emission and mode-hop free continuous tuning. As frequency shifted feedback devices attractive applications become possible such as optical frequency domain ranging. Ring lasers might allow the development of compact optical gyroscopes of high performance. Moreover, if fabricated in a periodically poled substrate the laser can be combined with a nonlinear device in the same waveguide enabling the development of self-frequency doubling lasers or of parametric oscillators with integrated pump (Sohler et al. 2005).

8.4 Lithium Tantalite Waveguides

There are many electro-optical materials which are used at visible or infrared wavelengths, but only a few of them are suitable for applications in the UV. Since laser sources in the UV are becoming available, there is an increasing interest in such low wavelength applications. Using UV light (smaller wavelengths), a higher resolution (proportional to the wavelength λ) can be achieved in applications like holographic interferometry or lithography. For volume holographic data storage, the storage density is increasing proportional to λ^{-3} when going to smaller wavelengths. Another important advantage of using lower wavelengths is the smaller divergence of such light beams. A promising material for applications at UV wavelengths is stoichiometric lithium tantalite ($LiTaO_3$). It is transparent down to 260 nm where most of the standard electro-optical materials, e.g., $LiNbO_3$ or $KNbO_3$, show large light absorption. At 320 nm, congruent $LiNbO_3$ exhibits an absorption constant of about 20 cm^{-1} while in congruent $LiTaO_3$, this value is reached at 275 nm. There is an increasing interest in the use of $LiTaO_3$ for photorefractive and other applications at UV wavelengths. Besides lower absorption edge, stoichiometric $LiTaO_3$ as compared to $LiNbO_3$ shows higher Curie temperature, higher dark resistivity, smaller optical damage in the visible, smaller coercive fields, one order of magnitude longer dark-storage time of holograms and smaller birefringence, which can even be adjusted to zero by controlling the stoichiometry and/or the temperature. The near-stoichiometric lithium tantalite has shown some potential applications in the UV, such as tunable optical filters, nonvolatile holographical data storage, high-speed optical processing with high resolution, light-induced waveguides or frequency doubling into the UV using periodically poled crystals (Juvalta 2008).

Congruent $LiTaO_3$ crystals can be grown by the traditional Czochralski method. These crystals are intrinsically nonstoichiometric owing to a shift of the melt composition from the stoichiometry towards the Ta side. Consequently, to grow stoichiometric crystals, the melt composition has to be kept Li rich. This is achieved by adding stoichiometric powder to the melt at a rate equal to the crystal growth using double crucible Czochralski method. The degree of stoichiometry of the near-stoichiometric crystal can be Li/(Ta + Li) = 49.9 % with a Curie temperature of 684 °C. Below the Curie temperature, the point-group symmetry of $LiTaO_3$ is uniaxial $3m$ with the optical axis, the crystallographic c axis, along the threefold symmetry axis. It has therefore two different refractive indices $n_a = n_b = n_o$ for ordinary and $n_c = n_e$ for extraordinary polarization. According to the $3m$ point-group symmetry, lithium tantalate has eight nonzero elements of the electro-optic tensor in the reduced notation, with only four independent values: $r_{12} = -r_{22} = r_{61}$, $r_{51} = r_{42}$, $r_{13} = r_{23}$ and r_{33}. In NLO applications like harmonic generation or parametric mixing, guiding the light allows maintaining a high intensity over a much longer interaction length leading to larger conversion efficiencies as in bulk materials. There are many techniques for fabricating permanent waveguides including ion indiffusion, ion exchange, ion implantation, laser ablation, photolithography, and epitaxial thin film deposition. By using these techniques, a permanent change of the refractive index in the waveguide region can be obtained. This precludes an easy reconfiguration of the waveguide structure. For instance, several techniques based on light-induced refractive index changes in photorefractive materials have been developed. One of the advantages of such optically induced waveguides is that they can be reconfigured solely by changing the light illumination. For applications such as optical switching, optical routing, or dynamic optical interconnections, fast reconfigurable waveguides are required, which can be easily achieved with the light-induced techniques. Near-stoichiometric and/or Mg-doped $LiTaO_3$ are promising materials for frequency doubling, electro-optics, long-persistent holographic storage, and interband photorefraction, which is transparent down to UV wavelengths near 260 nm. The advantage of this material compared to the as-grown congruent $LiTaO_3$ for the above mentioned applications is the greatly reduced concentration of intrinsic trap levels that are responsible for the conventional photorefractive effect. For example, the fabrication of light-induced waveguides in magnesium doped near-stoichiometric lithium tantalite can be performed via interband photorefractive field screening in a biased crystal. The induced waveguides are reconfigurable in a fast way using controlling light at $\lambda = 257$ nm, with waveguide buildup times of 10 ms at UV intensities of

100 mW/cm^2. A homogeneous external electric field E that is applied to an electro-optic crystal induces a uniform change in the refractive index (Juvalta 2008):

$$\Delta n = -\frac{n^3}{2} r_{\mathrm{eff}} E \tag{8.14}$$

where r_{eff} is the effective electro-optic coefficient for the chosen configuration and n the refractive index for the corresponding wavelength and polarization. If a small portion of such a biased photorefractive crystal is nonuniformly illuminated, free charge carriers are produced in the bright regions, which screen the applied field due to charge transport and trapping in dark regions. As a result one gets a strongly reduced electric field in the bright regions, while in dark regions the field is basically unchanged. Via the electro-optic effect, this field distribution generates the refractive index profile, where the refractive index in the dark regions is lowered by Δn. With a proper field direction and a suitable distribution of the illumination, light-induced waveguide structures can be generated in this way. By changing the illumination the waveguide structure can be reconfigured. By choosing light with photon energy larger than the bandgap energy of the material, charges can be excited directly from one band to the other, i.e., by interband excitation. This process is much more effective in terms of use of the incident photons with respect to a conventional photorefractive effect, where the photoexcitation occurs from dopant or impurity energy levels within the material band gap. In $LiTaO_3$ interband excitation leads to three orders of magnitude faster response compared to conventional trap-level excitation (Dittrich et al. 2004; Juvalta 2008).

8.5 Barium Titanate Waveguides

The optical ferroelectrics in the form of highly transparent thin films are promising materials for communication. Many ferroelectric thin film materials, such as $BaTiO_3$, $KNbO_3$, $LiNbO_3$, (Pb, La)TiO_3, (Na, K)NbO_3, Pb(Zr, Ti)O_3, and (Pb, La) (Zr, Ti)O_3 have been investigated for their optical and waveguiding properties. $BaTiO_3$ (BTO) is particularly an attractive candidate for thin film electro-optic modulators due to its large electro-optical coefficients, its high optical transparency, and its favorable growth characteristics (Petraru 2003).

Barium titanate ($BaTiO_3$) was one of the first ferroelectric materials to be discovered, and also one of the first to be recognized as photorefractive. $BaTiO_3$ is a member of the perovskites family, of which the parent member is the mineral $CaTiO_3$, called perovskite. The perovskite family includes other well-known materials such as $KNbO_3$, $KTaO_3$, $PbTiO_3$, and $SrTiO_3$. The general formula of compound belonging to this family is ABO_3, where A is a monovalent, divalent, or trivalent metal and B a pentavalent, tetravalent, or trivalent element, respectively. From the point of view of practical applications this material is very interesting because it is chemically and mechanically very stable and it exhibits ferroelectric properties at and above room temperature. From the optical applications point of view, $BaTiO_3$ is very interesting because of its high linear and nonlinear electro-optic coefficients. The Curie temperature of $BaTiO_3$ is at 120 °C. When the temperature is above 120 °C, $BaTiO_3$ is paraelectric in the cubic phase (point group *m*3*m*), thus centrosymmetrical. The structure of $BaTiO_3$ in the cubic phase is a simple one with Ba^{2+} ions at the cube corners, Ti^{4+} ions at the body centers, and O^{2-} ions at the face centers. Below the Curie temperature $BaTiO_3$ undergoes a phase transition to the tetragonal phase with the point group 4*mm*, and the material becomes ferroelectric. In the tetragonal phase $BaTiO_3$ is noncentrosymmetric. The most important change in the structure is a shift of Ti^{4+} ion from the center of the octahedron toward an oxygen ion at one of the face centers of the cubic unit cell. The fourfold rotation axis is thus the polar direction, identified with the tetragonal *c* axis. The *c* axis lies parallel to the direction of one of the original cubic <100> directions. The tetragonal unit cell results from the distorted cubic cell, with one edge elongated to become the tetragonal *c* axis and the other two cube edges compressed to become the tetragonal *a* axes. Because there are six equivalent <100> axes in the cubic phase, the polar axis can be parallel to any of these six equivalent directions, giving rise to complicated domain patterns. The tetragonal phase is stable from 120 °C to about 5 °C. When the temperature is below 5 °C, the $BaTiO_3$ phase is orthorhombic with symmetry belonging to the point group *mm*. This phase is still ferroelectric, but the direction of spontaneous polarization is parallel to one of the original cubic <110> directions. The orthorhombic phase of $BaTiO_3$ is stable from 5 °C to about −90 °C. At −90 °C another phase transition occurs and the symmetry changes to rhombohedral. The point group of this phase is 3*m* and the polar axis lies along one of the original cubic <111> directions. The angle between the rhombohedral axes differs from 90° by 12′. The tetragonal phase of $BaTiO_3$ is of interest for applications because it is stable at room temperature. At 20 °C, the lattice parameters are $c = 4.0361$ Å. $BaTiO_3$ in its tetragonal phase, at room temperature, is a negative uniaxial crystal with relatively high refractive indices n_o, n_e, if compared with most other solids. The ordinary and extraordinary refractive indices of undoped $BaTiO_3$ at 20 °C are $n_o = 2.412$ and $n_e = 2.360$. The birefringence of the bulk $BaTiO_3$ crystal is then $n = n_e - n_o = -0.052$. The temperature derivative of the extraordinary refractive index is $dn_e/dt = 140 \times 10^{-6}\ K^{-1}$ at 514.5 nm wavelength. For a $BaTiO_3$ crystal in its tetragonal phase (point group 4*mm*), there are three independent components of the electro-optic tensor: r_{13}, r_{33}, and $r_{42} = r_{51}$ (in contracted notation) (Petraru 2003).

For instance, the $BaTiO_3$ films with a thickness of about 1 μm have been grown on MgO substrates forming planar optical waveguides. Pulsed laser deposition is used, which allows the growth of the epitaxial $BaTiO_3$ films on MgO with two different orientations of the films with respect to the substrate (*c*-axis and *a*-axis oriented films) by changing the deposition parameters (laser energy, substrate temperature, and oxygen pressure), because the kinetic energy of the ablated particles depends strongly on laser energy and oxygen pressure. This modifies the growth conditions and the orientation of the ablated films. The *a*-axis films have a smaller and positive birefringence $n = n_{TM} - n_{TE} = 0.012$ compared to the *c*-axis films ($n = -0.027$). The typical value of the loss coefficient is 3 dB/cm for the c-axis films and about 4 dB/cm for the a-axis films (Petraru 2003).

Mach-Zehnder waveguide modulators have been fabricated using standard lithography and ion beam etching. The waveguides are of ridge type and ensure single-mode propagation in the wavelength range 633–1,550 nm, having a ridge width of 2 μm and a ridge height of 70 nm. In the case of Mach-Zehnder modulators fabricated on c-axis films, the external electric filed was applied perpendicular to the optical axis of the film. The Mach-Zehnder modulators fabricated on a-axis films have a half-wave voltage of $V = 9.5$ V at the fiber-optic communication wavelength (1.5 μm) and a high extinction ratio of 20 dB (Petraru 2003).

8.6 Electro-Optic Polymer Materials and Formed Waveguides

8.6.1 Electro-Optic Polymer Materials

Electro-optic (EO) polymers are a classification of second-order NLO EO materials whose optical properties (e.g., the refractive index) can be modulated by an external electrical field. Such materials are members of the more general class of organic NLO materials wherein the NLO properties arise from the ability of the molecules and atoms to change, in a nonlinear way, their polarization under an external electrical field. Since the late 1960s, organic molecules have attracted an increasing amount of interest because of their potential applications in NLO devices and, in particular, in second-order nonlinear optics. This interest is motivated not only by the large, fast, and electronic-in-origin NLO response but also by the versatility, ease of processing, and possibility of tailoring the physicochemical properties by the molecular engineering approach. In organic NLO materials, the power series expansion of polarization with an electric field ($\vec{E}$) can be expressed either in terms of molecular polarization ($\vec{p}$) or macroscopic polarization ($\vec{p}$) (Zhou 2004):

$$p_i = p_{i0} + \alpha_{ij}E_j + \beta_{ijk}E_jE_k + \ldots \tag{8.15}$$

$$P_I = P_{I0} + \chi_{IJ}^{(1)}E_J + \chi_{IJK}^{(2)}E_JE_K + \ldots \tag{8.16}$$

where β is the molecular first hyperpolarizability, and $\chi^{(2)}$ is the macroscopic second-order NLO susceptibility. In order to have second-order nonlinear properties at the molecular level ($\beta \neq 0$), the molecules should present noncentrosymmetric symmetry. Although second-order NLO activity can arise from higher order octupolar symmetry, such materials have yet to be used in the fabrication of devices. Thus, for practical purposes, organic second-order NLO materials can be considered to be made up of dipolar molecules organized into noncentrosymmetric chromophore lattices. This requirement of ordered chromophore lattices, which is indispensable for having macroscopic second-order nonlinear properties ($\chi^{(2)} \neq 0$), is the most daunting materials requirement to be satisfied in the development of electro-optic material (Dalton 2002).

Dipolar chromophores with larger and larger dipole moments have been successfully synthesized using a (electron donor)-(π-electron connective segment)-(electron acceptor) structure. Several ways have been explored by which noncentrosymmetric lattices have been achieved including Molecular Self-Assembly, Sequential Synthesis Methods, and the Electric Field Poling method. Among these methods, the most popular and successful method has been electric field poling of dipolar NLO chromophores containing polymers. In this method, the NLO chromophores are first dispersed into a host polymer, which served as the polymer matrix, either dissolved as guests (guest-host type) or by being chemically connected to the polymer molecules to form side-chain/main-chain polymer (linear type), cross-linked polymer (cross-linked type), or hyperbranched/dendritic polymer (dendrimer type). Then the chromophore-containing polymer is cast into a thin film, and an external electrical field is applied to orient the chromophores employing the interaction strength of dipole moment with an applied external electric field. Several such orientation techniques have been developed including (a) static field poling, (b) photoassisted poling, and (c) all optical poling. Among them, the static field poling technique is the most successful. In this technique, the chromophore-containing polymers are heated up to around the glass transition temperature, T_g, of the polymer matrix: the chromophore molecules then become mobile and are aligned in the same direction by an applied static electric field. After the polymer is cooled down under a high electric field, the alignment of the electric dipoles is "frozen," and macroscopic optical nonlinearity is achieved. Many static-field poling techniques have been developed, such as electrode (contact) poling (electrodes are deposited for applying the poling electric field), corona (corona discharge is used to create the poling electric field), photothermal poling (a laser beam with wavelength lying in the material absorption band is used to heat the thin film), and electron beam poling (the poling field is generated by bulk charges provided by a monoenergetic electron beam with an energy of 2–40 keV). Contact and corona poling are the commonly used approach. After poling, the initially isotropic EO polymer becomes uniaxial, with the extraordinary optical axis parallel to the direction of the poling field. This direction is also chosen as the z-axis in the principal coordinate system. Because of the cylindrical polar symmetry induced by the electrical filed poling, and the fact that only one component of the molecular hyperpolarizability tensor, β_{zzz}, is important for dipolar chromophores, the corresponding EO tensors have only two independent nonzero components. They take the following form (Li and Yu 2003; Zhou 2004):

$$\begin{pmatrix} 0 & 0 & \gamma_{13} \\ 0 & 0 & \gamma_{13} \\ 0 & 0 & \gamma_{33} \\ 0 & \gamma_{13} & 0 \\ \gamma_{13} & 0 & 0 \\ 0 & 0 & 0 \end{pmatrix} \tag{8.17}$$

where

$$r_{33} = \left|2Nf(\omega)\beta_{zzz} < \cos^3\theta > /n_e^4\right| \tag{8.18a}$$

$$r_{13} = \left|2Nf(\omega)\beta_{zzz} < \cos^3\theta > /n_o^4\right| \tag{8.18b}$$

and N is the chromophore number density; $f(\omega)$ is a product of local field factors, taking into account that the applied fields are attenuated by the local environment of the chromophore; and n_e are the extraordinary and n_O ordinary indexes of refraction respectively; and θ is the angle between the chromophores' dipole direction and the z-axis in the principal coordinate system. For electrically poled polymer materials, it is common to assume that $\gamma_{33} = 3\left(\frac{n_O}{n_e}\right)^4\gamma_{13}$. However, this is not always the case and is seldom the case with the use of laser-assisted (photochemical) poling (Zhou 2004).

The index ellipsoid for EO polymers in the presence of an electric field is represented by (Zhou 2004):

$$x^2\left(\frac{1}{n_o^2}+r_{13}E_z\right)x^2 + y^2\left(\frac{1}{n_o^2}+r_{13}E_z\right) + z^2\left(\frac{1}{n_e^2}+r_{33}E_z\right)$$
$$+ 2yzr_{13}E_y + 2xzr_{13}E_x = 1 \tag{8.19}$$

where E_x, E_y, and E_z are the three components of the applied external electrical field. Equation (8.19) shows that the most efficient direction to apply the modulation electrical field is the poling (z-axis) direction. For optical polarization along the z-axis (TM mode) and perpendicular to the z-axis (TE mode), the optical refractive indexes become (Zhou 2004)

$$n'_{TM} = \left(\frac{1}{n_e^2}+\gamma_{33}E_z\right)^{-\frac{1}{2}} \approx n_e - \frac{1}{2}n_e^3\gamma_{33}E_z \tag{8.20}$$

and

$$n'_{TM} = \left(\frac{1}{n_o^2}+\gamma_{13}E_z\right)^{-\frac{1}{2}} \approx n_o - \frac{1}{2}n_o^3\gamma_{13}E_z \tag{8.21}$$

respectively. Equations (8.20) and (8.21) show that EO polymer-based devices are generally polarization-dependent. For modulation application, TM optical modes are desirable since they are much more efficiently modulated than are the TE optical modes because of the relatively larger value of γ_{33} over γ_{13}.

Electro-optic (EO) polymers are unique materials having many advantages over inorganic materials for a wide range of applications—from optical network components and optical interconnects to millimeter and microwave photonics. They have very large EO coefficients, low optical loss (in the 1.3 and 1.55 μm telecommunication windows), low dispersion of refractive index between optical frequencies and millimeter waves, as well as high bandwidth. They are easy to process and have relatively low cost. Moreover, the environmental stress test for EO polymer modulators guided by the industry standard Telcordia GR-468 environmental stress specifications with particular attention paid to test regarding thermal and/or photochemical stability demonstrated excellent device performance. Specifically, the devices were thermally stable for over 5,000 h at 85 °C. Furthermore, high optical power (500 mW @ 1,550 nm) tests showed the photochemical stability of the EO polymer material in hermetically sealed packages. There is a broad range of EO devices that can be fabricated based on these polymers having a waveguide structure as the main building block. Different waveguide design configurations have different advantages and limitations. For example, to create a miniature EO light modulator for the highest on-chip integration, size and power efficiency would be critical. This requires integration with silicon photonics and use of either slotted or photonic crystal waveguide (PCW) covered with EO polymer. For an optical network (Internet) the size of the light modulator is less significant, however operating voltage and bandwidth are more important. Therefore, all-polymer and hybrid polymer/sol–gel configurations might be the better choice. Different parameters should be optimized, corresponding waveguide configuration that can be used (Pyayt 2011).

8.6.2 *Electro-Optic Polymer Waveguides*

Many applications need optical waveguides that confine light in two directions and allow propagation in the third direction. Such waveguides consist of several thin polymer films. In vertical direction light confinement is done by choosing appropriate refractive indices of the core and the claddings. To create refractive index contrast in horizontal direction at least one of the polymer films has to be patterned. There are several possible fabrication techniques that can be used-photoresist-based patterning, direct photolithographic or e-beam patterning, micromolding in capillaries and microcontact printing.

1. Photolithography-patterned waveguides
   An example of the photoresist-based patterned EO polymer waveguide is shown in Fig. 8.2a with 3.2 μm polyurethane lower cladding shown in blue. This layer was patterned using photolithography and then 1.4-mm-high CLD-1/poly-(methylmethacrylate) (PMMA) guiding layer (shown in red), and a 2.9-mm-high ultraviolet light–curable upper cladding layer (white) were spin-coated on top. Figure 8.2b, c illustrate the side view of the device and the typical mode propagating in such waveguide. This device demonstrates the half-wave voltage V_π (the voltage needed to switch the device from on- to off-configuration) less than 1 V at 1.31 μm wavelength. This is one of the critical requirements for ultra-high-speed light modulators for fiber-optic communication link, because the link

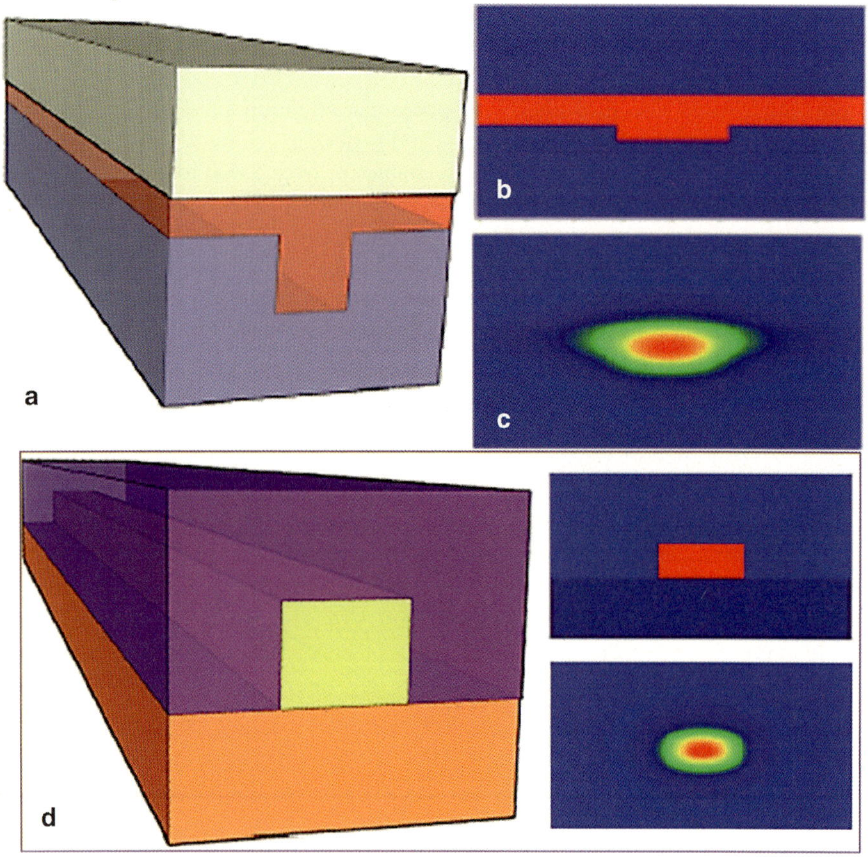

Fig. 8.2 Three-layer EO polymer waveguides (Pyayt 2011): (**a**) patterned with photolithography, (**b**) the side view, and (**c**) the guided mode; as well as the one patterned with e-beam lithography, the side view and the guided mode (**d**)

gain is inversely proportional to V_π^2 and the noise figure is directly proportional to V_π^2 in the low-gain limit. The link usually includes the optical source, modulator, fiber and optical receiver and has RF input and output ports. The gain is defined as a measure of efficiency in that it is the ratio of the output RF signal power to the input RF signal power. Noise figure expresses the extent to which the signal-to-noise ratio degrades between the input and the output of the link (Pyayt 2011).

2. Direct e-beam written waveguides

   Photolithography gives great results for the simple straight waveguides, but micro-ring resonators and other high Q structures require creation of the waveguides with ultra-smooth walls for very low scattering losses. Direct e-beam patterning of EO polymer waveguides have been developed in PMMA polymer matrix doped with organic nonlinear chromophores. The polymer in the exposed regions is removed with the standard electron beam resist developer.

This technique allows creation of high quality smooth waveguide walls and prevents the chromophores from being damaged by the e-beam. The fabricated micro-ring resonators with extinction ratios more than 16 dB and quality factors can range from $10^3 \sim 10^4$. The device structure, side view, and the typical mode profile are illustrated in Fig. 8.2d (Pyayt 2011).

3. Hybrid sol–gel polymer waveguides

   There is another important design constraint that always should be taken into consideration while fabricating EO polymer devices. The material has to be poled to be able to change refractive index under applied voltage. This means that the chromophore molecules have to be aligned in the same direction. It requires heating the EO polymer and applying high voltage for initial alignment. Then the polymer is cooled down and the voltage is turned off, so that the orientation of the molecules is fixed and the material is ready for operation. There are two steps that require presence of the electrodes in close proximity to electro-optic polymers during the poling and during the device operation. These electrodes are not necessary the same. For example, a layer of the electro-optic polymer can be corona-poled or contact-poled, and only several fabrication steps later the second electrode can be added to the device (Pyayt et al. 2010).

   Traditionally, when the EO properties of the thin polymer film are measured, the film is sandwiched between two electrodes and poled, and then the top electrode is removed and the EO activity is optically tested. Fabrication of high quality devices requires separation of the electrodes from the EO polymer layers by two optical claddings. This might affect poling, since the voltage applied between the electrodes would produce an electric field over thicker polymer films. Since it is critical to have most of the voltage drop on the EO layer, the optical claddings should have much higher conductivity than EO core. Due to much higher conductivity of sol–gel in comparison with EO polymer, this device configuration allowed achieving ideal 100 % poling efficiency in comparison with the best EO coefficient values measured for the single EO polymer films. The in-device EO coefficients are around 140–170 pm/V. The second important property of that device is that the coupling loss is only about 0.6 dB per facet due to excellent mode match to a standard optical fiber. This requires an advanced waveguide design. As can be seen from Fig. 8.2, the typical waveguide mode for the EO polymer waveguide is not symmetric. It can be around 5–6 μm wide and only 2–3 μm high. In comparison with the modes supported by the single-mode fibers with the circular cores 10 μm in diameter this would produce significant losses due to the mode mismatch. The design made based on making hybrid polymer sol–gel waveguides is shown in Fig. 8.3a. The waveguide interfacing the input and output fibers are made with sol–gel core and cladding with refractive indices of 1.5 and 1.487, respectively. The mode is well matched with the fibers. After coupling to the sol–gel core and initial propagation in the sol–gel waveguide, the light is adiabatically coupled into the EO polymer (refractive index 1.6–1.7) in the active region of the waveguide, and then back to the sol–gel (Pyayt 2011).

4. Photobleached and e-beam bleached waveguides
   When it is needed to avoid wet chemicals, such as different solvents and developers, alternative approaches to waveguide patterning have to be used. One such approach is based on bleaching of the EO polymers. Decomposition of the chromophore molecules under direct UV or electron beam irradiation reduces the refractive index of chromophore-containing polymers. High energy photons or electrons break the bonds, chromophore molecules decompose, and the material color bleaches out. Created refractive index contrast can be ~0.06–0.07, which is enough for efficient light guiding. Both photo- and electron-bleaching approaches are used to fabricate high quality micro-ring resonators. An interesting aspect of photobleaching is that it's not only changes the refractive index but also the thickness of the polymer film. For the waveguide fabricated using photobleaching with a photolithographic mask. The waveguides can be defined as opaque lines on the mask thus being protected from UV radiation. The waveguide surface is directly defined by the diffraction pattern on the photolithographic mask. The height variation is about 150–200 nm (Pyayt 2011).
5. Field-induced waveguides
   Local change of refractive index of EO polymer under narrow stripe electrode can be used to create waveguides dynamically, as shown in Fig. 8.3b. Field-Induced Guiding (FIG) was initially demonstrated in $LiNbO_3$ and GaAs-based heterostructures. The advantages of FIG-based design are temperature and wavelength independence, improved linearity and extremely high bandwidth. Wide variety of the devices (light modulators, splitters, scanners, etc.) based on this approach have been developed. The waveguide could be completely turned on by applying 7 V and switched off by -1 V. If the EO polymers with 300–500 pm/V are used, the voltage would have decreased proportionally. FIG waveguides can also be used as building blocks for ultra-flexible reconfigurable systems with dynamic waveguide structure. Figure 8.3b shows schematic structure of the FIG—optical layers are not patterned, waveguide is dynamically created by the local filed between two electrodes shown in yellow (Pyayt 2011).
6. Slotted waveguides
   In addition to their excellent EO and photonic properties, EO polymers have great mechanical properties. They can be easily spin-coated on different smooth and patterned surfaces and fill in even very narrow nano-scale slots and nano-holes in photonic crystals. This allows creating waveguides with the width of the guiding mode in order of 100 nm in contrast to several microns typical for all-polymer devices. Furthermore, this greatly improves on-chip integration with silicon photonics and allows creating fast miniature components for on-chip optical interconnects. The schematic view of a silicon slotted waveguide covered with EO polymer is shown in Fig. 8.4a. The typical mode supported by this waveguide has its maximum in the slot filled with the polymer. Figure 8.4b demonstrates modification of the slotted waveguides fabricated in photonic crystals. It was experimentally demonstrated that slots as narrow as 75 nm can be infiltrated with EO polymer. However, fabrication of these nano-scale

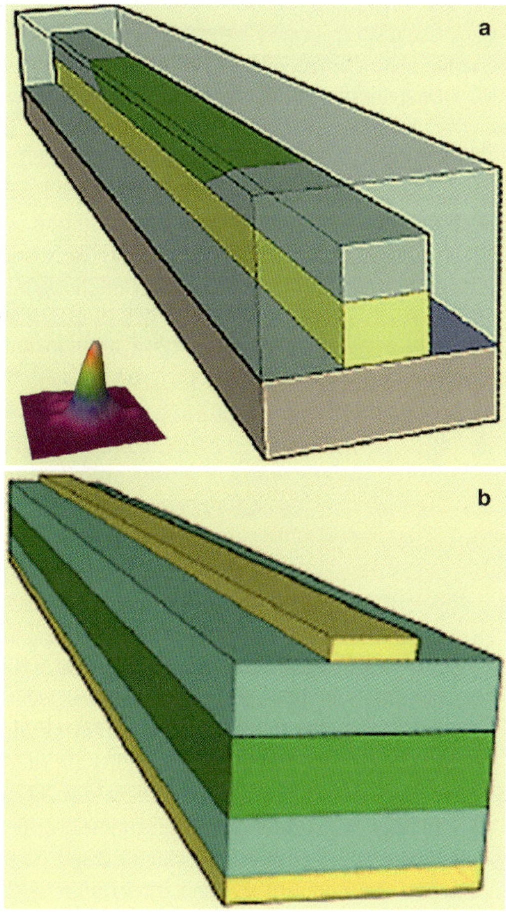

Fig. 8.3 Examples of EO polymer waveguides (Pyayt 2011): (**a**) hybrid sol–gel waveguide (the core of the sol–gel waveguide is shown in *yellow*; the electro-optic polymer layer in *green*; and *gray* portion is low refractive index sol–gel cladding); and (**b**) structure of the field-induced guiding—optical layers are not patterned, waveguide is dynamically created by the local field between two electrodes shown in *yellow*

features requires challenging e-beam lithography and reactive ion etching steps. Another important concern is that due to the high leakage current the poling efficiency is decreased and the overall r_{33} value is only 10–30 % compared to the value for EO polymer in thin film. Further improvement can be achieved by decreasing the leakage current. This can be organized by either using EO polymers with lower conductivity or by increasing width of the slot. The latter would decrease separation of electrodes and increase the operating voltage. The optimum slot width is 300–350 nm. For this width 30 % of the light still propagates through EO polymer, and the poling can be done efficiently. Optimization of the slot-waveguide-based optical modulators for high-speed

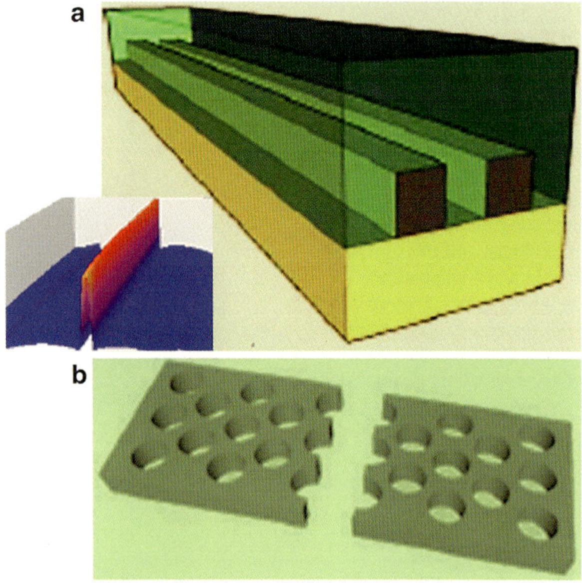

Fig. 8.4 Schematic view of a slot waveguide (oxide cladding shown in *yellow*, silicon slot in *gray*, and EO polymer cladding in *green*) with supported mode profile (**a**); and schematic of a photonic crystal slot waveguide (**b**) (Pyayt 2011)

performance requires additional considerations. The modulator fabricated from a silicon strip-loaded slot waveguide and nonlinear polymer cladding demonstrated the bandwidth 3 dB-bandwidth of 3 GHz and a $V_\pi L$ figure of merit of 0.8 V · cm. The electrodes are the forming parts of the waveguide, and the driving voltage is applied across a 200 nm slot (Pyayt 2011).

There are a number of fabrication approaches for differently designed EO polymer waveguides that can be used depending on the available equipment, material compatibility, and integration with silicon photonics. Furthermore, by picking the right configuration, it is possible to create miniature, ultra-fast, and power efficient devices. On the way towards commercialization the main goal is optimization of multiple parameters simultaneously and fabrication of devices demonstrating all these exceptional properties at the same time.

8.7 Liquid Crystal Electro-Optic Waveguides

Liquid crystal electro-optic waveguide platform has exhibited unprecedented electro-optical phase delays (>1 mm), with very low loss (<0.5 dB/cm) and rapid response time (sub millisecond). This technology is based upon a unique liquid-crystal waveguide geometry, which exploits the tremendous electro-optic response of liquid crystals while circumventing historic limitations of liquid crystals. The exceedingly large optical phase delays accessible with this technology enable the design and construction of a new class of previously unrealizable photonic devices, such as 1-D nonmechanical, analog beamsteerer with an 80° field of regard, a chip-scale widely tunable laser, a chip-scale Fourier transform spectrometer (FTS) (<5 nm resolution demonstrated), widely tunable micro-ring resonators, tunable lenses, ultra-low power (<5 μW) optical switches, true optical time delay (up to 10 ns), and many more. All of these devices may benefit from established manufacturing technologies and ultimately may be as inexpensive as a calculator display. Furthermore, this integrated photonic architecture has applications in a wide array of commercial and defense markets including: remote sensing, micro-LADAR, OCT, laser illumination, phased array radar, and optical communications, etc. (Davis et al. 2008).

Numerous applications require active control over light, such as robotic-vision, optical computing, telecommunications, holographic data storage, remote sensing, cold-atom optics, and industrial process analysis. In response to this need a diverse array of technologies, such as micro-electro mechanical systems (MEMs), photonic crystals, thermo-optics, and electro-optic materials such as inorganic crystals and organic poled-polymers. While tremendous progress has been made, there are still numerous applications, including beamsteering and large optical phase delay, where bulky and power-consumptive macroscopic opto-mechanical techniques are still the best. This is, at least in part, because typical electro-optic approaches do not realize sufficient control over light (*R* matrix values of typically <100 pm/V) to replace traditional opto-mechanics. MEMs techniques are still inherently mechanical, which imposes vibration and inertia design challenges, and developing MEMs architectures that provide macroscopic (>1 mm) control over optical phase has been challenging. Liquid-crystal (LC) waveguide architecture provides unprecedented voltage control over optical phase (>1 mm), orders of magnitude more than any other technology, e.g., liquid-crystal optical-phased arrays 10 or MEMs. This previously unrealizable level of control makes possible new devices with remarkable performance attributes. LCs have the largest electro-optic response ($\Delta n > 0.2$ over 5 V for a typical LC, which corresponds to 10^5–10^6 pm/V, i.e., several orders of magnitude larger than any other approach), are environmentally stable, amenable to high-volume manufacturing, and inexpensive. A typical "display-like" LC-optic is shown in Fig. 8.5a. The light traverses a thin (<20 μm) LC layer that is sandwiched between glass sheets. Transparent electrodes are used to apply an electric field, which, in combination with polarizers, may be used to either block or transmit the light. While undeniably potent for information displays, this traditional LC-optic has two significant limitations. First, the light must transmit through transparent electrodes, which in turn limits the total optical power that may be controlled. Second, and arguably more significant, the LC layer must be extremely thin. The LC material is rendered a single-domain crystal via thin alignment layers. The LC-molecules adjacent these alignment layers are highly ordered, which means low scattering loss, and fast response to changes in voltage. If one were to make the LC cell thicker, the bulk LC material would become prohibitively slow and opaque. Therefore, even though the LC material has a tremendous electro-optic effect, the necessarily short interaction length

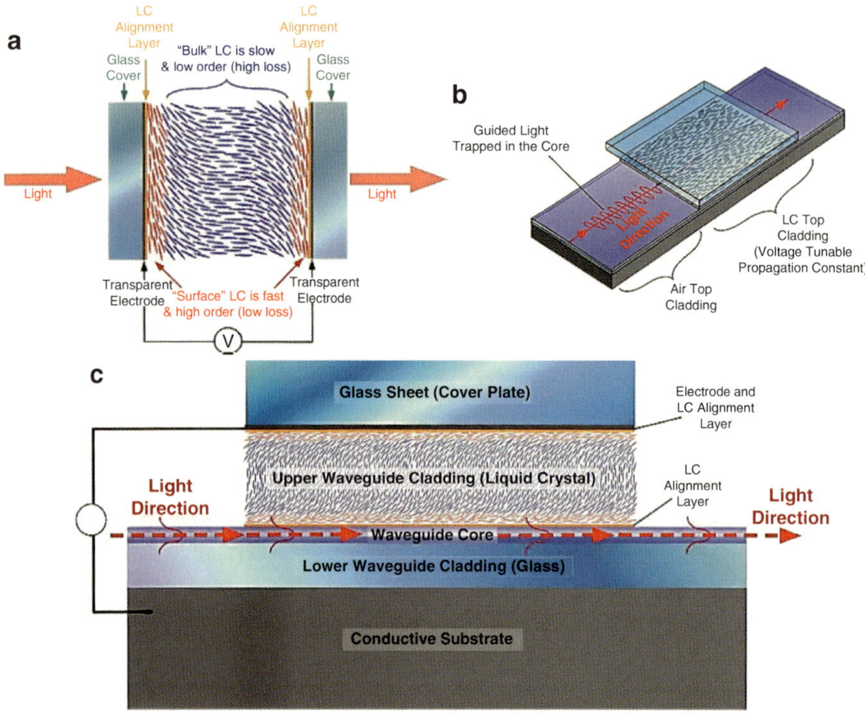

Fig. 8.5 Schematic illustrations of liquid crystal (LC) waveguides (Davis et al. 2008): (**a**) typical LC-optic used in the ubiquitous LC-display; (**b**) the basic geometry of an LC waveguide, the light is confined to a core and the LC is an electro-optic upper cladding. As the index of refraction of the upper cladding is tuned the effective index of the guided mode is also tuned; and (**c**) a side view of a liquid crystal waveguide. In a slab waveguide the light is guided in the x dimension, but is free to propagate as Gaussian beams, sheets, or even 1D images in the plane

mitigates this effect. In order to circumvent these limitations the LC-clad waveguide architecture has been developed (Davis et al. 2008).

Rather than transmit through an LC cell, which by design must be thin (typically <20 μm), the LC can be utilized as an active cladding layer in waveguide architecture, i.e., the light skims along the surface of an LC layer, as shown in Fig. 8.5b. This electro-evanescent architecture circumvents limitations of traditional LC-optics: (a) the light never crosses a transparent electrode; (b) the light only interacts with the well-behaved LC-surface layer via the evanescent field, and (c) the interaction length is now decoupled from the LC-layer thickness. For a given liquid crystal and waveguide structure, the mode model includes: LC-surface energy, pretilt, elastic coefficients of the LC, electrical properties of the LC (dielectric constants), optical properties of the LC (birefringence), electrode spacing, and electrical properties of the waveguide materials. The index modulation is the magnitude of the difference between the effective index at zero volts and the effective index and a higher voltage. Furthermore, the model shows that $\Delta n_{\text{eff}} = 0.05$ is possible by using

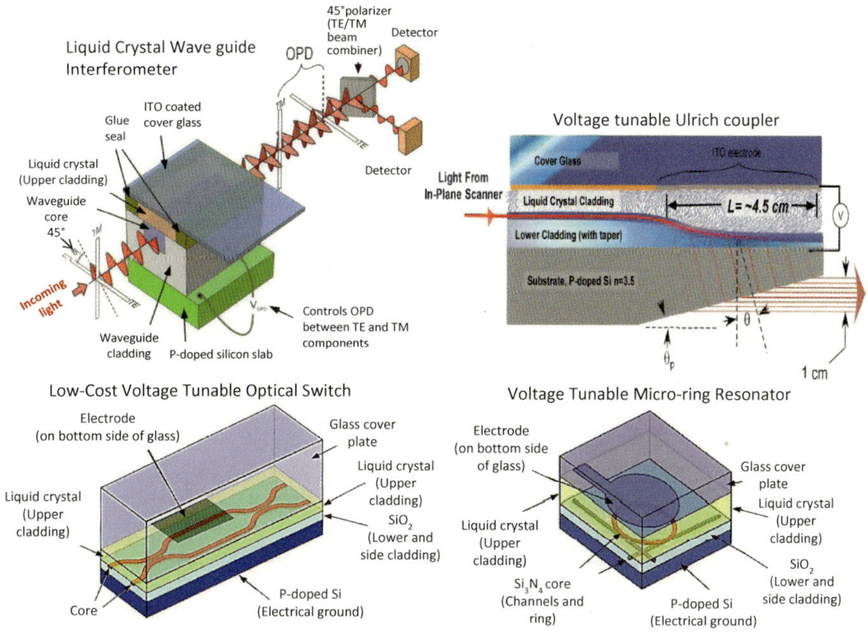

Fig. 8.6 Application examples of liquid crystal waveguides (Davis et al. 2010)

highly birefringent liquid crystals and by keeping the ratio of core thickness to wavelength less than one. While this represents a fourfold decrease in birefringence from the raw liquid crystal, this is more than offset by a possible 10,000-fold increase in the interaction length. Figure 8.6 illustrates possible photonic devices based on this technology (Davis et al. 2008; Davis et al. 2010).

In addition to these LC waveguides may be used for optical coherence tomography, true optical time delays, tunable lenses, optical switches, and many more. A device that demonstrates the utility of giant electro-optic control is a completely nonmechanical FTS. When the LC-waveguide is placed between polarizers the transmitted intensity undergoes extrema as the voltage across the LC layer is tuned. Different wavelengths experience different numbers of extrema and this can be used to analyze the wavelength content of the light. Moreover, a diverse array of technical approaches has been directed toward nonmechanical beam control: (a) planar electro-optic prisms constructed from KTP, Lithium Niobate, ferroelectric domain $LiTaO_3$, and electro-optic polymers; (b) thermo-optic planar prisms; (c) diffractive liquid-crystal phased arrays, and (d) diffractive acousto-optic techniques. Each of these approaches has advantages and drawbacks. Electro-optic crystals are very fast but have extremely small electro-optic coefficients, which mean very small steering angles and kilovolts to operate. Furthermore, EO crystals are quite expensive. Liquid-crystal optical-phased array beamsteerers tend to be slow, provide noncontinuous diffractive steering, and have

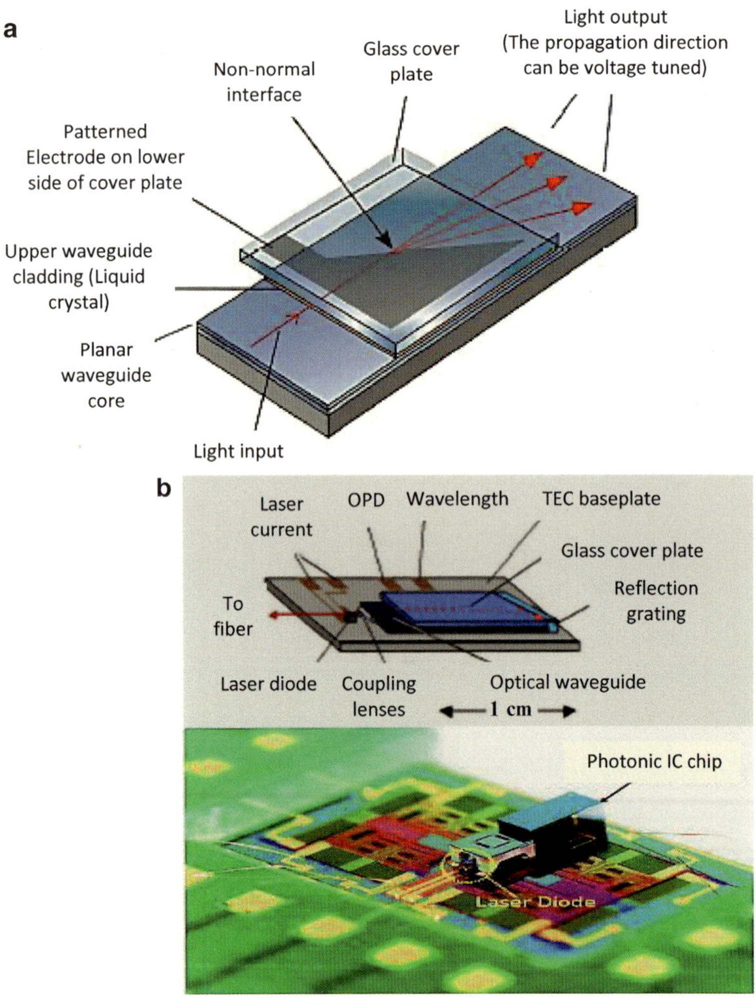

Fig. 8.7 (**a**) A 1-D LC-waveguide beamsteerer. A single control voltage is applied to a prism-shaped electrode having a non-normal interface to the beam propagation direction. As voltage is applied, the index under the patterned electrode is changed relative to the surrounding area and the beam is steered via Snell's law refraction; and (**b**) an LC-waveguide tunable diode laser. The *upper picture* shows that a typical cavity laser architecture is integrated onto an LC waveguide. The *lower picture* shows a micro-assembly of a laser (Davis et al. 2008)

a very limited steering range because thick LC layers are problematic. Acousto-optic beamsteerers have a larger steering range but are also diffractive, require very large power supplies and expensive crystals. LC waveguides provide an entirely new approach to electro-optic laser-beamscanning. The basic concept is shown in Fig. 8.7a wherein a lithographically patterned electrode provides a 1-D prism whose refractive index can be tuned relative to the surrounding area, providing truly analog steering, i.e., it is not diffractive. This lithographically patterned electrode is similar to, albeit much simpler than, the patterned electrodes required for traditional liquid crystal displays.

In principle, the pointing precision is only limited by the noise on the control voltage. In addition, LC waveguides may be used to construct tunable filters by placing beamsteering elements upstream from edge-bonded reflection gratings, or by placing LC over waveguides with Bragg gratings or ring resonators. An application of a tunable filter is for a tunable diode laser. The upper picture in Fig. 8.7b shows the design of a tunable diode laser based on a tunable Littrow grating filter. The lower of Fig. 8.7b shows an example of a prototype LC-waveguide laser. A typical external cavity laser architecture is integrated onto an LC waveguide. LC waveguide may also be used to provide a low power alternative to thermo-optic waveguide devices. A low-cost voltage tunable optical switch is shown in Fig. 8.6. In this example light from a single-frequency laser is coupled into one of the entrance ports. While voltage to the LC waveguide is ramped the intensity at an exit port is monitored. Finally, another type of thermo-optic device that may be rendered electro-optic with the LC-waveguide architecture is the tunable microring resonator. As shown in Fig. 8.6, the channel structures are constructed using the triplex waveguide design. The input and output channels were 3 μm wide and the channel in the ring was 15 μm wide and a lateral coupling scheme was used. The device can tune over nine resonances (>3 nm), with extremely minimal power consumption (<50 μW). All of these devices may ultimately be in small LCD-like packages that can ultimately be as low cost as a calculator display (Davis et al. 2008; Davis et al. 2010).

8.8 Strained Silicon as an Electro-Optic Material

The slow progress within silicon optoelectronics, where electronic and optical functionalities can be integrated into monolithic components based on the versatile silicon platform, is due to the limited active optical properties of silicon. However, a significant linear electro-optic effect can be induced in silicon by breaking the crystal symmetry. The symmetry is broken by depositing a straining layer on top of a silicon waveguide, and the induced nonlinear coefficient, $\chi^{(2)} \approx 15$ pm V^{-1}, makes it possible to realize a silicon electro-optic modulator. The strain-induced linear electro-optic effect may be used to remove a bottleneck in modern computers by replacing the electronic bus with a much faster optical alternative. The inversion symmetry of a non-strained silicon crystal prohibits the existence of a linear electro-optic effect. However, the symmetry can be broken by applying an asymmetric strain to the crystal by depositing a straining layer on top of it (Fig. 8.8a), hence creating a linear electro-optic effect in silicon. In other words, when silicon is properly strained, its bulk refractive index (n) varies linearly as a function of external applied electric field (E). For instance, when silicon nitride glass (Si_3N_4)

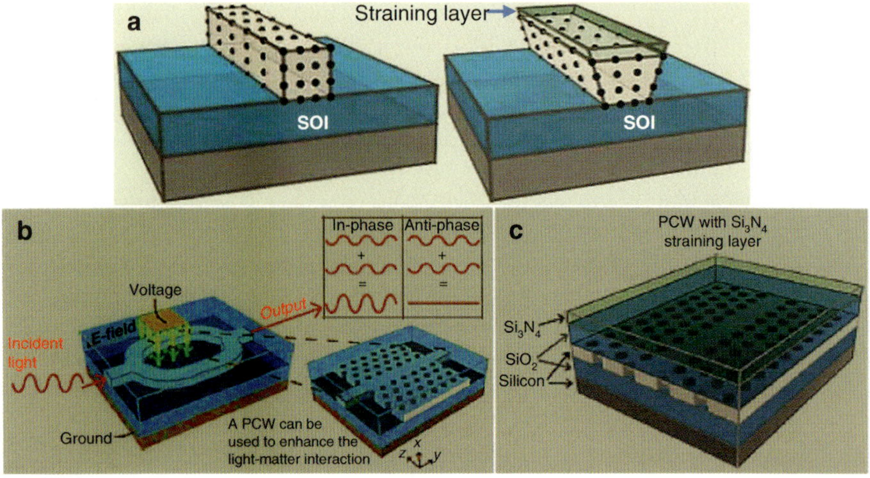

Fig. 8.8 Applying strain to crystalline silicon (Jacobsen 2006): (**a**) waveguides fabricated in the top layer of an SOI wafer without (*left*) and with (*right*) a straining layer deposited on top. The straining layer breaks the inversion symmetry and induces a linear electro-optic effect. (**b**) Diagram of a Mach-Zehnder modulator. Incident light is split into two waveguides. The output amplitude depends on the phase difference at recombination. In-phase recombination gives a “1” bit output while anti-phase recombination gives a “0” bit output. If the straight waveguide in the Mach-Zehnder modulator is replaced with a PCW, the effect of a material nonlinearity can be enhanced. (**c**) Schematic of PCW with Si_3N_4 straining layer

is used as a straining layer, the amorphous Si_3N_4 layer is compressively strained, that is, it tries to expand the structure underneath in both horizontal directions. The ability to induce a change in the refractive index can be applied to make an amplitude modulator using a Mach-Zehnder interferometer (Fig. 8.8b), where an applied electric field determines whether or not incident light is transmitted through the modulator. Such an electro-optic modulator is typically used when transmitting data, where transmitted light corresponds to a “1” bit and no light to a “0” bit. The effect of the material nonlinearity $\chi^{(2)}$ is enhanced if the waveguide possesses a permanently enlarged group index, n_g. It is theoretically predicted that the enhancement scales linearly with the group index. Here, the enhanced nonlinear coefficient $\chi^{(2)}_{\mathrm{enh}}$ is therefore defined as (Jacobsen 2006):

$$\chi^{(2)}_{\mathrm{enh}} = \frac{n_g}{n}\chi^{(2)} \tag{8.22}$$

An enlarged n_g can for instance be obtained by guiding the light in a PCW, which is a waveguide surrounded by a periodically microstructured material (Fig. 8.8b).

Here, the microstructure consists of silicon with holes that are partially filled with glass and air, giving rise to large material index variations. The period of the structure is on the same scale as the wavelength of the incident light. The result is a waveguide with extremely high group indices.

As the electric modulation field is applied in the *x*-direction (the coordinate system is shown in Fig. 8.8c) and the light is polarized in the *z*-direction (TE-polarized). In the structure as shown in Fig. 8.8c, a spacing layer of silica glass (SiO_2) is sandwiched between the silicon waveguide and the straining Si_3N_4 glass. By this arrangement, the guided light is confined well below the Si_3N_4 straining layer; that is, the optical field does not extend to this layer. This arrangement ensures that the only influence of the Si_3N_4 layer on the silicon waveguide is a physical strain, exerted through the SiO_2 layer. The difference in the induced nonlinear coefficient before and after depositing the Si_3N_4 layer is not only caused by the different magnitudes of strain but also by the different signs of the strain. The upper SiO_2 layer tries to contract, which is only possible if the whole structure bends like a deep plate. This deformation is only minor, as the thick supporting wafer is extremely difficult to bend. After depositing the Si_3N_4 layer, the sign of the strain is changed, and the top layers now try to expand in the horizontal directions. This expansion can be partially accommodated by a slight wafer bending and partially by introducing strong local deformations near each hole. As a result, the compressive strain is more effective in deforming the waveguiding silicon layer. A Si_3N_4 top layer is of course not required, as strain can also be obtained by altering the deposition conditions of the SiO_2 layer. In fact, another structure can be fabricated, in which the SiO_2 layer deposited directly on top of the PCW was grown with a surplus of silicon radicals. The magnitude of $\chi^{(2)}_{\mathrm{enh}}$ is governed both by the material's nonlinear coefficient and by the group index (8.22), where the group index is determined by the design of the waveguide. The advantage of silicon in comparison with $LiNbO_3$ is the possibility of applying cheap standard fabrication techniques used for fabrication of CMOS circuits. The silicon modulator can also be integrated monolithically with an electronic driver circuit and hybridized with existing semiconductor lasers. The strain-induced nonlinearity is not the only possibility for realizing an optical modulator in silicon. A modulator based on changing the carrier concentration has been demonstrated, and a significant increase of the modulation speed up to 10 Gbits has been obtained. However, changing the carrier concentration requires a large alternative current of 0.2 Å (r.m.s.) through the silicon structure. In contrast, in the strained silicon presented here, there is no current through the waveguide structure. Moreover, the speed of the strained silicon modulator is not limited by charge mobility or charge recombination times. Thus, the discovered linear electro-optic effect may provide a decisive step towards utilization of active silicon-based photonics (Jacobsen 2006).

8.9 Summary

Electro-optic waveguides are designed based on changing refractive index of the core layer with an external voltage. Materials used for electro-optic waveguides can be $LiNbO_3$, $LiTaO_3$, $BaTiO_3$, electro-optic polymers, liquid crystal, and strained silicon. The lithium niobate is a versatile material and used for various applications in guided wave optics, electro-optics, acousto-optics, and nonlinear optics. Its piezoelectric and photoelastic properties had been used for various technical developments. Lithium niobate has been used to fabricate integrated waveguides because of its high electro-optic coefficient, optical damage resistance, and low losses. Their strain-optic effects are used to implement wavelength-tunable polarization converters. The $LiNbO_3$ modulators have been used with bulky diode pumped YAG lasers and cavity RF filters to construct OEO's with ultra-low phase noise capability. The lithium niobate can be used to make QPM structure by reversing the spontaneous polarization under the influence of a sufficiently large electric field. The lithium niobate material has been extensively used to design switches with low loss, but their polarization dependency is the major concern. The choice of LN-based switches is proved as best for high and low speed systems. In addition, $LiNbO_3$ devices have successfully addressed the modulation requirements in digital fiber-optic TDM and WDM systems.

A promising material for applications at UV wavelengths is stoichiometric lithium tantalate. It is transparent down to 260 nm where most of the standard electro-optical materials, e.g., $LiNbO_3$ or $KNbO_3$, show large light absorption. There is an increasing interest in the use of $LiTaO_3$ for photorefractive and other applications at UV wavelengths. Besides lower absorption edge, stoichiometric $LiTaO_3$ as compared to $LiNbO_3$ shows higher Curie temperature, higher dark resistivity, smaller optical damage in the visible, smaller coercive fields, one order of magnitude longer dark-storage time of holograms and smaller birefringence, which can even be adjusted to zero by controlling the stoichiometry and/or the temperature. The near-stoichiometric lithium tantalite has shown some potential applications in the UV, such as tunable optical filters, nonvolatile holographical data storage, high-speed optical processing with high resolution, light-induced waveguides or frequency doubling into the UV using periodically poled crystals.

The optical ferroelectrics in the form of highly transparent thin films are promising materials for communication. Many ferroelectric thin film materials, such as $BaTiO_3$, $KNbO_3$, $LiNbO_3$, (Pb, La)TiO_3, (Na, K)NbO_3, Pb(Zr, Ti)O_3, and (Pb, La) (Zr, Ti)O_3 have been investigated for their optical and waveguiding properties. $BaTiO_3$ (BTO) is particularly an attractive candidate for thin film electro-optic modulators due to its large electro-optical coefficients, its high optical transparency, and its favorable growth characteristics.

Electro-optic (EO) polymers are unique materials having many advantages over inorganic materials for the wide range of applications from optical network components and optical interconnects to millimeter and microwave photonics. They have very large EO coefficients, low optical loss (in the 1.3 and 1.55 μm telecommunication windows), low dispersion of refractive index between optical frequencies and millimeter waves, as well as high bandwidth. They are easy to process and have relatively low cost. There is a broad range of EO devices that can be fabricated based on EO polymers having a waveguide structure as the main building block. Different waveguide design configurations have different advantages and limitations. For example, to create a miniature EO light modulator for the highest on-chip integration, size and power efficiency would be critical. This requires integration with silicon photonics and use of either slotted or PCW covered with EO polymer. For an optical network (Internet) the size of the light modulator is less significant, however operating voltage and bandwidth are more important. Therefore, all-polymer and hybrid polymer/sol–gel configurations might be the better choice. Different parameters should be optimized, corresponding waveguide configuration that can be used.

Liquid crystal electro-optic waveguide platform has exhibited unprecedented electro-optical phase delays (> 1 mm), with very low loss (<0.5 dB/cm) and rapid response time (sub millisecond). This technology is based upon a unique liquid-crystal waveguide geometry, which exploits the tremendous electro-optic response of liquid crystals while circumventing historic limitations of liquid crystals. The exceedingly large optical phase delays accessible with this technology enable the design and construction of a new class of previously unrealizable photonic devices, such as 1-D nonmechanical, analog beamsteerer, a chip-scale widely tunable laser, a chip-scale FTS (<5 nm resolution demonstrated), widely tunable micro-ring resonators, tunable lenses, ultra-low power (<5 μW) optical switches, and true optical time delay (up to 10 ns). All of these devices may benefit from established manufacturing technologies and ultimately may be as inexpensive as a calculator or other screen display. Furthermore, this integrated photonic architecture has applications in a wide array of commercial and defense markets including: remote sensing, micro-LADAR, OCT, laser illumination, phased array radar, optical communications, etc.

Strained silicon can be another electro-optic material. A significant linear electro-optic effect can be induced in silicon by breaking the crystal symmetry. The symmetry is broken by depositing a straining layer on top of a silicon waveguide, and the induced nonlinear coefficient, $\chi^{(2)} \approx 15$ pm V^{-1}, makes it possible to realize a silicon electro-optic modulator. The strain-induced linear electro-optic effect may be used to remove a bottleneck in modern computers by replacing the electronic bus with a much faster optical alternative.

References

Bhandare S (2003) Application of lithium niobate-based integrated optical circuits to optical communication. PhD diddertation, University of Paderborn, Paderborn, Germany

Castaldini D (2006) Study, fabrication and characterization of segmented waveguides for advanced photonics components on Lithium Niobate. PhD dissertation, Universita di Bologna

Dalton LR (2002) Nonlinear optical polymetic materials: from chromophore design to commercial applications. Adv Polym Sci 158:1–86

Das B (2003) Integrated optical distributed Bragg reflector and distributed feedback lasers in Er: $LiNbO_3$ waveguides with photorefractive gratings. Ph.D dissertation, University of Paderborn, Germany

Davis SR et al (2008) A new electro-optic waveguide architecture and the unprecedented devices it enables. Proc SPIE 6975:1–12

Davis SR et al (2010) Liquid crystal waveguides: new devices enabled by > 1000 waves of optical phase control. In: Liang-Chy Chien (ed) Liquid crystal technologies V. Proc of SPIE, vol 7618, 76180E, pp 1–14
Davis AA (2010) Design and analysis of a poled-polymer electro-optic modulator with a strip-loaded waveguide structure. PhD dissertation, University of Dayton, Dayton, OH
Dittrich P et al (2004) Deep-ultraviolet interband photorefraction in lithium tantalite. J Opt Soc Am B 21(3):632–639
Fellers TJ, Davidson MW (2012) Acousto-optic tunable filters (AOTFs). http://www.olympusmicro.com/primer/techniques/confocal/aotfintro.html. Accessed 03/07/2013
Fujii Y, Otsuka Y, Ikeda A (2007) Lithium niobate as an optical waveguide and its application to integrated optics. IEICE Trans Electron E90-C(5):1081–1089
Iodice M (2003) Transient analysis of a high-speed thermo-optic modulator integrated in an all-silicon waveguide. Opt Eng 42(1):169–175
Irvine SE, Elezzabi AY (2003) Wideband magneto-optic modulation in a bismuth-substituted yttrium iron garnet waveguide. Opt Commun 220(4):325–329
Jacobsen RS (2006) Strained silicon as a new electro-optic material. Nature 441:199–202
Juvalta F (2008) Light-induced waveguides, waveguide arrays and switches in photorefractive LiTO3 and Sn2P2S6. PhD dissertation, ETH Zurich, Switzerland
Li GL, Yu PKL (2003) Optical intensity modulators for digital and analog applications. J Lightwave Technol 21(9):2010–2030
Nguimdo RM (2008) Optical communications using chaotic carriers generated by electro-optical feedback devices. Master thesis, Universitat de les Illes Balears, Palma
Petraru AI (2003) Optical and electro-optical properties of $BaTiO_3$ thin films and Mach-Zehnder waveguide modulators. PhD dissertation, Der Universitat Koln
Pyayt AL (2011) Guiding light in electro-optic polymers. Polymers 3:1591–1599
Pyayt A, Luo J, Jen AK, Dalton L, Chen A (2010) Field-induced guiding optical devices made from electro-optic polymers. Appl Opt 49:892–896
Reed GT, Knights AP (2004) Silicon photonics: an introduction. Wiley, England
Selah BEA, Teich MC (2007) Fundamentals of photonics, 2nd edn. Wiley, New York
Singh G, Yadav RP, Janyani V (2010) Ti indiffused lithium niobate ($Ti:LiNbO_3$) Mach-Zehnder interferometer all optical switches: a review. In: Aleksandar Lazinica (ed) New advanced technologies, INTECH Publication, Austria, pp 311–322
Sohler W et al (2005) Erbium-doped lithium niobate waveguide lasers. IEICE Trans. Electron. E88-C (5):990–997
Zhou Q (2004) Electro-optic modulators based on polymeric Y-fed directional couplers. PhD dissertation, The University of Texas at Austin

Chapter 9
Polymer-Based Optical Waveguides

Abstract Polymer optical waveguides would play a key role in broadband communications, such as optical networking, metropolitan access communications, and computing systems, due mainly to their easier processibility and integration over inorganic counterparts. The combined advantages also make them an ideal integration platform where foreign material systems, such as yttrium iron garnet and lithium niobate, as well as semiconductor devices such as lasers, detectors, amplifiers, and logic circuits can be inserted into an etched groove in a planar lightwave circuit to enable full amplifier modules or optical add/drop multiplexers on a single substrate. Moreover, the combination of flexibility and toughness in optical polymers makes it suitable for vertical integration to realize 3D and even all-polymer integrated optics. This chapter would provide a brief review about polymer-based optical waveguides, including suitable polymer waveguide systems, their processing and fabrication techniques, and the integrated optical waveguide components and circuits derived from these materials.

9.1 Rationale of Polymers Used for Optical Waveguides

As discussed in previous chapters, depending on the phenómenon of total internal reflection guided wave optics can confine light in the optical waveguide, a material surrounded by other materials with lower refractive indices. Optical waveguides may be thin-film deposits used in integrated optical circuits or a filament of dielectric material usually circular in the cross section used in fiber optics. Depending on various possible patterns or propagating or standing electromagnetic fields, there are single-mode and multi-mode optical waveguides. Each mode is characterized by its frequency, effective refractive index, polarization, power distribution, electric field strength, and magnetic field strength.

With the rapid advance of integrated optics, the importance of optical waveguides, which are the fundamental elements of optical integrated circuits (ICs), has been widely recognized. For example, the further success of broadband

X.C. Tong, *Advanced Materials for Integrated Optical Waveguides*, Springer Series in Advanced Microelectronics 46, DOI 10.1007/978-3-319-01550-7_9,

communications in optical networking, metro/access communications, and computing will rely on the advancement of optical interconnects, optical components, such as splitters, combiners, multiplexers, and demultiplexers, optical switches/modulators, tunable filters, variable optical attenuators (VOAs), amplifiers, and integrated optical circuits that are based on optical waveguides. While the basic technologies for the design and production of many integrated optical waveguide devices are in place, a great many materials have also been developed. Today, glass optical fibers are routinely used for high-speed data transfer. Although these fibers provide a convenient means for carrying optical information over long distances, they are inconvenient for complex high-density circuitry. In addition to being fragile and vibration-sensitive, glass fiber devices are difficult to fabricate especially when they have a high port count, and as a result are quite expensive. InP-based quaternary semiconductor materials are widely employed for waveguide devices, due mainly to their potential for integration with active devices such as lasers and photodetectors operating at around 1,550 nm. However, semiconductor processing remains complex and expensive, especially at the large scale volumes presently experienced for optical components. By using manufacturing techniques closely related to those employed for silicon-integrated circuits, excellent optical components have already been made using silica-based planar lightwave circuits (PLCs). Currently, optical switches and arrayed waveguide gratings (AWGs) for multiplexing and demultiplexing multiple wavelengths in a dense wavelength division multiplexing (DWDM) system are the predominant applications of silica-based PLC technology. The cost issue, the high switching power needed in silica-based switching devices, the low wavelength tuning range, and the temperature dependence of the central wavelength of silica-based AWGs, however, are major problems with this technology. Among the candidate material systems, high expectations have been placed on polymers as the materials choice for highly integrated optical components and circuits. State-of-the-art optical polymers are particularly attractive in integrated optical waveguide devices because they offer rapid processibility, cost-effectiveness, high yields, high performance, such as lower optical loss and smaller birefringence compared to silica, power-efficient thermal actuation due to a larger thermo-optic coefficient than in silica, and compactness owing to a large refractive index contrast. Furthermore, polymers provide an ideal platform for the incorporation of more complex material functionalities through selective doping or reaction, thereby enabling amplification and electro-optic effects to be achieved, as the passive optical polymer technology is established (Ma et al. 2002).

On another hand, glass optical fibers are routinely used for high-speed data transfer. Although these fibers provide a convenient means for carrying optical information over long distances, they are inconvenient for complex high-density circuitry. In addition to being fragile, glass fiber devices are difficult to fabricate, especially when they have a high port count, and in most cases require active fabrication, resulting in high cost. Integrated optics is becoming the choice approach for advanced photonic components. Several material systems are being pursued as integrated optic platforms, such as silica, silicon, silicon oxynitride, sol–gels, polymers, lithium niobate, indium phosphide, and gallium arsenide. The most widely used integrated optic platform is silica on silicon. Whereas the choice of this platform in the early days of integrated optics was quite natural and justifiable, because of its kinship to silica fiber, the cons of this technology may outweigh the pros. Silica fiber has a propagation loss of 0.15 dB/km at 1,550 nm wavelength, a remarkably low number that is achieved due to the fact that the fiber has no stress and no roughness. When silica is grown on a silicon wafer, typically at high temperature by chemical vapor deposition (CVD), a high level of stress is obtained in the wafer when it cools down to room temperature due to the coefficient of thermal expansion (CTE) mismatch between silica and silicon, which stress results in polarization-dependent behavior, stress-induced scattering, and high coupling loss when pigtailing to fiber blocks due to warpage in the substrate. The reactive ion etching (RIE) patterning of waveguides in silica creates rough edge walls that create scattering loss and further polarization dependence. Another limitation is that the highest contrast achieved to date in this technology is only 1.5 %. And yields in this technology have historically been low, especially in large interferometric devices such as AWGs, where numbers below 10 % are the norm. Other platforms for integrated optics have been pursued, including silicon on insulator, silicon oxynitride, lithium niobate, indium phosphide, gallium arsenide, sol–gels, and polymers. Polymer is one of attractive materials of choice for integrated optics because, when synthesized and processed properly, it offers high performance (the loss in state-of-the-art polymers is slightly lower than the state-of-the-art loss achieved in silica; the birefringence is smaller than that of silica by two orders of magnitude), tenability (thermo-optic coefficient 30–40 times larger than in silica), environmental stability, high yields, and low cost (Eldada 2002).

Polymer is a long, repeating organic chain, formed through the linkage of many identical smaller molecules called monomers. A key feature that distinguishes polymers from other large molecules is the repetition of units of atoms (monomers) in their chains. This occurs during polymerization, in which many identical monomer molecules link to each other. One of the major advantages of these macromolecular systems is their ability to produce thin and stable films as well as coatings for large areas which may be deposited on a variety of substrates using sufficiently simple methods. Polymeric materials permit the mass production of low-cost high port-count photonic circuits in parallel on a planar substrate. The attractiveness of the process for polymer waveguide manufacture increases if it is compared with the processes required to produce silica waveguides. In conventional silica, the process takes several days, involves many fabrication steps, and has a low yield and deleterious stress-induced birefringence. In addition to use of low-cost environmentally friendly fabrication processes, polymer waveguides are quick and simple to fabricate. Polymer can be deposited by using spin coating or polymerization technique. Furthermore, the unique mechanical properties of polymers allow them to be processed by unconventional forming techniques such as molding, casting, stamping, and embossing, therefore permitting rapid and low-cost shaping for waveguide formation (Eldada 2003).

9.2 Polymeric Waveguide Materials

9.2.1 Current Perspectives

Classes of polymers used in integrated optics include acrylates, polyimides, polycarbonates, and olefins (e.g., cyclobutene). Some polymers, such as most polyimides and polycarbonates, are not photosensitive, and are typically processed using photoresist patterning and RIE. These polymers have most of the problems of the silica on silicon technology in terms of roughness-induced and stress-induced scattering loss and polarization dependence. Other polymers are photosensitive and as such are directly photopatternable, resulting in a full cycle time of about 30 min per multilayer optical circuit on a wafer. These materials have an obvious advantage in throughput, producing wafers between 10 and 1,000 times faster than other planar technologies. Furthermore, this technology uses low-cost materials and low-cost processing equipment, for instance, spin-coater and UV lamp vs. CVD growth system (Eldada and Shacklette 2000; Eldada 2002).

Optical polymers can be highly transparent, with absorption loss values below 0.1 dB/cm at all key communication wavelengths, such as 840, 1,310, and 1,550 nm. The scattering loss can be minimized in polymer waveguides by using direct photopatterning as opposed to surface-roughness-inducing RIE. The effect of the little roughness that is obtained can be further minimized by the use of a graded index, a natural process in direct polymer lithography where interlayer diffusion is easily achieved. The graded index profile results in weak confinement of the optical mode, and therefore causes its tails to penetrate well into the cladding, averaging out the effect of variations. The scattering loss can also be reduced by ensuring the homogeneity of the medium (i.e., no abrupt refractive index variations caused by phase separation, particles, etc.), and by minimizing intrinsic stresses. As opposed to planar glass technologies, polymer technologies can be designed to form stress-free layers regardless of the substrate composition (which can be silicon, glass, quartz, plastic, glass-filled epoxy printed-circuit board (PCB) substrate, etc.), and these films can be essentially free of stress-induced scattering loss and stress-induced polarization effects. These favorable characteristics can be observed when operating above the glass-transition temperature (T_g) in crosslinked polymer systems. The radiation loss can be reduced by using standard integrated optic design rules such as large radii of curvature and adiabatic modal transitions. The fiber pigtail loss can be minimized by matching the mode of the planar waveguide to that of the fiber (which is achieved by tuning the index contrast, the index profile, and the core dimensions), by optimizing the alignment of the waveguides to the fiber in six degrees of freedom (x, y, z, roll, tilt, yaw), and by minimizing the Fresnel reflections (by using appropriate index-matching adhesives at the interfaces). The total insertion loss achieved in planar polymer components can closely approach the value of the absorption loss in the waveguides when fabrication techniques are optimized. The polarization-dependent loss ($PDL = TE_{loss} - TM_{loss}$) varies with processing conditions. The TE loss measured in planar waveguides can be higher than the TM loss when the vertical walls of the core have a higher degree of roughness than the horizontal boundaries, and it can be lower when the vertical evanescent tails overlap with an absorptive substrate or superstrate. Waveguides that are well optimized by having minimal edge roughness and a well-confining material stack can have PDL values that are immeasurably small. The birefringence ($n_{TE} - n_{TM}$) can be extremely low in polymers that undergo little molecular orientation during processing, as is common in three-dimensionally crosslinked polymers (Eldada 2002).

The environmental stability of optical polymers (i.e., the stability of their optical and mechanical characteristics with temperature and humidity) is an important issue because most polymers do not have properties that are appropriate for operation in communication environments. A key characteristic for practical applications is the thermal stability of the optical properties since organic materials may be subject to yellowing upon thermal aging due to oxidation. The presence of hydrogen in a polymer allows the formation of H-halogen elimination products, which result in carbon double bonds, which are subject to oxidation. Fortunately, the absorbing species from thermal decomposition are centered near the blue region of the spectrum, and the thermal stability can be high at the datacom wavelength of 840 nm and even greater at the telecom wavelengths of 1,300 and 1,550 nm. On the humidity front, the resistance of polymers to water incursion is critical since optical absorption results from the overtone bands of the OH-stretch of water. However, polymers that stand up to 85 °C/85 %RH (relative humidity) conditions have been developed, and some polymers passed the Bellcore 1209 and 1221 environmental tests. Extensive materials research has yielded polymers that are highly transparent and reliable, to the extent that they are no longer the limiting factor in components lifetime. Another important feature of polymers is the tunability of the refractive index contrast, which can have values up to 35 %, enabling high-density compact waveguiding structures with small radii of curvature. Polymers also allow simple high-speed fabrication of 3D circuits with vertical couplers, which are needed with high-index-contrast waveguides where 2D circuits would require dimensional control, resolution, and aspect ratios that are beyond the levels achievable with today's technologies. Furthermore, the unique mechanical properties of polymers allow them to be processed by unconventional forming techniques such as molding, casting, stamping, and embossing; permitting rapid, low-cost shaping for both waveguide formation and material removal for grafting of other materials such as thin-film active layers, Faraday rotators, or half-wave plates (Eldada 2002).

The production of commercially viable polymeric optical components is a complex task because optical polymers need to meet simultaneously many key properties. The main properties that a polymer needs to meet in order to be a viable thermo-optic polymer with potential for commercial deployment (Eldada 2002): low absorption loss, low PDL, low chromatic dispersion, low polarization mode dispersion (PMD), low birefringence, low stress, low pigtail loss, variable refractive index difference (Δn), closeness of refractive index to silica fiber refractive index, ability to vary refractive index profile, low refractive index dispersion ($dn/d\lambda$), thermal stability, stability with humidity, stability with optical power, adhesion (to substrates, self, electrodes), full curability, low shrinkage, contrast in patterning, process latitude in patterning, patternability with low scattering loss, machinability (cleaving, dicing, polishing), compatibility with electrode patterning, large T/O coefficient (d*n*/d*T*), optimal thermal conductivity, and manufacturability with repeatable properties.

9.2.2 Materials Characterization and Performance Requirement

1. Refractive Index
   Based on the device design and waveguide geometry, the polymer used as the core material must have a refractive index higher than that of the cladding polymer. The appropriate index difference between core and cladding for either a single-mode or a multi-mode waveguide is strongly dependent on the dimension of the waveguide and the wavelength of the light source.

   (a) Polarizability, Packing Density, and Wavelength Dependence
   The refractive index of a material is related to the free volume-packing density, polarizability of the material, and the difference between the used optical wavelength and the maximum absorption wavelength of the material. Usually, dense packing or large polarizability results in an increase in refractive index. Three types of polarizability including electric, atomic, and dipole orientation contribute to the total polarization of a material. Electronic polarization is the slight skewing of the equilibrium electron distribution relative to the positive nuclei to which it is associated. Since only the movement of electrons is involved, this process can occur very rapidly and typically has a time constant of 10^{-15} s. The atomic polarization results from rearrangement of nuclei in response to an electric field. The positive nuclei are attracted to the negative pole of the applied field. However, the movement of heavy nuclei is more difficult to initiate opposite to that of electrons. As a result, it cannot follow an oscillating field as high a frequency as the electron response. Dipole orientation polarization results from the redistribution of charge when a group of atoms with a net permanent dipole moment reorient themselves in space in response to an electric field. Since large group masses must reorient, this process is usually slower than either electronic or atomic polarization and even in the gas phase will have time constants on the order of 10^{-9} s due to the larger inertia that must be overcome to reverse the direction of movement in each cycle of electric field oscillation. In the liquid or solid phases, large intermolecular forces must be overcome, which slows the process further and decreases the polarization possible under all but static conditions. The dipole orientation polarization is often the dominant mode of polarization contributing to the refractive index in polar liquids and gases. In solids, dipole movement is usually restricted to

the point where the dipole orientation polarization becomes less significant than the electronic mode. At optical frequencies, only electronic polarization is occurring and the refractive index can be fine-tuned through structure modification, physical aging, and guest doping of polymers. In general, aromatic polymers possess higher refractive indices than aliphatic ones due to better packing and electronic polarizability. Similarly, the incorporation of highly π-conjugated dyes raises a polymer's refractive index. The high-temperature densification of aromatic polymers leads to an increase in refractive index due to the decrease of free volume. The incorporation of fluorine atoms in the polymers can affect the refractive index in three ways simultaneously. First, the increase in free volume, which often accompanies fluorine substitution, may decrease the refractive index attributed to the greater steric volume of fluorine relative to hydrogen, which may interfere with efficient chain packing. Second, the electronic polarizability is always lowered with fluorine substitution because of the smaller electronic polarization of the C–F bond relative to C–H bond. And last, the larger difference between the measuring optical wavelengths and the blue-shifted absorption wavelength λ_{max} (due to fluorine substitution) also contributes to the lowering of the refractive index. As a result, an important feature of polymers is the controllability of the refractive index contrast, which can have values of up to 35 %, enabling high-density compact waveguiding structures with small radii of curvature. Compact structures are critical for the achievement of large scale photonic integration (Hougham et al. 1996; Ma et al. 2002).

(b) Birefringence

The birefringence ($n_{TE} - n_{TM}$) indicates the optical anisotropy of a material. In an isotropic material, birefringence is also related to the stress build-up within the material due to processing or thermal treatment. Conventional techniques used for measuring the TE and TM indices of a film with a prism coupling apparatus can only detect large birefringence in some materials. A very sensitive technique of evaluating the birefringence consists of measuring the polarization splitting in the reflection spectrum of a Bragg grating made of the material. In the case when the two reflections overlap too much to be readily resolved, the grating can be heated or cooled, which causes the reflections of the two polarizations to shift at different rates, and broadening/narrowing occurs. By evaluating the difference between the minimum reflection peak width and the width at the temperature where the birefringence can be determined. Unlike inorganic crystals or glasses, polymers can be molecularly engineered to achieve low birefringence. Some aromatic polymers, such as polyimides, exhibit a very large birefringence (up to 0.24) that is attributed to the strong preference of aromatic moieties to align with their planes oriented along the film surfaces. However, the birefringence can be extremely low (10^{-5}–10^{-16}, the limit of measurement) in polymers that undergo little molecular orientation during processing, as is common in three-dimensionally crosslinked polymers (Kihara et al. 1996b; Ma et al. 2002).

(c) Temperature Dependence

A fundamental difference between polymeric materials and more conventional optical materials, such as glass, is that their refractive index varies more rapidly with temperature (large thermo-optic (T-O) effect: dn/dT). The refractive index of polymers decreases in temperature at a rate of 10^{-4}/°C, which is one order of magnitude larger than that of inorganic glasses. This large T-O coefficient as well as poor thermal conductivity makes it possible to realize T-O switches with low power consumption and digital T-O switches based on adiabatic waveguide transitions, as opposed to the interferometric devices that are necessary in silica on silicon, for example. Furthermore, the T-O coefficient is negative, while that of inorganic glass is positive. The characteristic is used in the construction of temperature-independent waveguides. In these waveguides, the small, positive T-O effect in inorganic glass waveguides, used as the core, is cancelled out by using the negative T-O coefficient of polymers, used as cladding. In measuring dn/dT, an approach has been used to measure the return loss from the interface between the material to be measured and an optical fiber (Kihara et al. 1996a). Such samples can be easily prepared by curing the material around a well-cleaved optical fiber, the other end of which is connected to a return loss meter. If the fiber optical properties are well known, then by simply measuring the change in return loss upon heating, for example, the dn/dT of the material can be determined (Ma et al. 2002).

(d) Humidity Dependence

The return loss technique can also be used to look at the effects of humidity on the refractive index. The humidity-induced change of the refractive index would affect single-mode waveguide performance if core and cladding changes were different; they would also affect the return loss of the device if index matching were used as the method to reduce return loss. Even in the case where the core and cladding change is the same, if the humidity changes the effective index of the waveguide, it can then affect the performance of devices such as Bragg gratings and AWGs. The humidity dependence of the refractive index in the polymers was ascribed to the counterbalance between moisture absorption and swelling due to the existence of hydrophilic groups. The refractive index of d-poly(methylmethacrylate) (d-PMMA, d = deuteron) increases as the humidity increases at room temperature, while it decreases as the humidity increases at temperature higher than 60 °C. Although the humidity dependence of d-PMMA is as large as 10^{-5}/%RH, some hydrophobic polymers, such as silicone resin, fluorinated epoxy resin, and perfluorinated ether polymer, CYTOP, were affected by humidity to a much lesser degree than the acrylic polymer (Watanabe et al. 1998; Ma et al. 2002).

(e) Wavelength Dependence

Many optical systems rely on having no wavelength-dependent optical effects other than those geometrically designed into the system. Therefore, material dispersion (dn/dλ) is generally to be avoided. While the values for the polymers on the order of 10^{-6}/nm are comparable to those for SiO_2, they are much lower than those for semiconductors or doped glasses (Ma et al. 2002).

2. Optical Loss of Polymer Waveguides
   In general, all-optical waveguide devices need to have low optical loss, especially at the major telecom wavelengths (1,310 and 1,550 nm) and datacom wavelength (840 nm). With the utilization of optic-fiber telecommunications in the S-band (1,450–1,510 nm), C-band (1,525–1,560 nm), and L-band (1,570–1,620 nm), all-optical waveguide devices should possess low loss in these bands. There are several sources of optical loss including absorption, scattering, polarization dependence, reflections, radiation, and fiber coupling. Optical polymers have promising characteristics to reduce all of these kinds of loss (Norwood et al. 2001).

   (a) Absorption Loss
   For polymer materials, both electronic and vibrational absorptions are likely to contribute to optical loss. Polymeric media generally have large absorptions in the ultraviolet owing to fundamental excitations of their electrons. These absorptions tend to be in the deep ultraviolet (less than 200 nm) for polymers with predominantly aliphatic hydrogen atoms, and in the near-UV (200–400 nm) for polymers with significant numbers of aromatic hydrogen. Partially or fully fluorinated polymers tend to have their UV absorptions at higher energies. For energies well below electronic energy levels of the polymer, where the polymer is basically transparent, weak absorption can result from a number of sources including singlet–triplet absorption and vibration-mediated absorption inharmonic interactions. In general, electronic absorptions in polymers, with the viable exception of highly colored electro-optic polymers, are very unlikely to contribute significantly to optical losses in the major telecommunication windows near 1,300 and 1,550 nm. In the 1,300–1,600 nm range, absorptions coming from the overtones of fundamental molecular vibrations are dominant. Since the strength of the absorption tends to decrease by approximately an order of magnitude between each harmonic order, higher harmonics are generally weak enough to be of no concern (at least for waveguide applications). Clearly, the highest energy vibrations will be those that have high spring constants (stiff bonds, such as double bonds and/or small reduced masses). The smallest reduced mass occurs when one of the atoms is hydrogen, and the C–H aliphatic bond is typically used as the benchmark for infrared absorptions; its absorption is located at 3,390 nm. Both C–H and O–H overtones exhibit to be highly absorptive in the telecommunication windows, whereas C–F overtones, for example, show extremely low absorption throughout the range of interest, owing to their higher harmonic order. As hydrogen is removed through partial fluorination, the absorption of optical polymers reduces significantly. A good empirical quantity with which to compare different polymers is the molecular unit, which in the case of PMMA is approximately 12. For highly fluorinated optical polymers, this ratio can be well over 100 and obviously goes to infinity for perfluorinated materials. In general, it is difficult to directly determine the absorption of an optical material, since scattering contributions to the overall attenuation are

often indistinguishable from those coming from absorption. The one signature that distinguishes absorption from scattering is inevitable generation of heat in the absorption process. This observation has served as the basis for several unique techniques, the most relevant of which is photothermal deflection spectroscopy (PDS). PDS has been applied to the measurement of absorption in a number of important optical polymer systems. Optical systems can be highly transparent with absorption loss values below 0.1 dB/cm at all the key communication wavelengths (Jackson et al. 1981; Ma et al. 2002).

(b) Scattering Loss

There are numerous extrinsic contributions to scattering loss in optical materials. Chief among these are large inclusions such as particles, voids, cracks, and bubbles. Generally, an inclusion is considered large if it is greater than 1 μm in diameter, in which case the scattering intensity is largely wavelength-independent. Extrinsic scattering in polymers can result from unfiltered particles, dust, dissolved bubbles, and unreacted monomer. For extrinsic scattering to be eliminated it is necessary to follow rigorously clean procedures in the preparation of polymer formulations, and to perform all coating operations in a clean room. Intrinsic scattering results from density fluctuations and compositional inhomogeneities, both of which occur on very short length scales (0.1 μm or less). Polymer waveguides are typically formed by spin-coating process in which polymers are deposited from solution and subsequently dried by heating in an oven or on a hot plate. The resulting films are generally uniform, but can have a roughness that will contribute to scattering losses. For a slab waveguide, a loss of 0.03 dB/cm can be achieved by reducing RMS roughness of the cladding/core interfaces to 0.04 μm. Single-mode channel waveguides are typically made by a photolithographic process in which wet or dry etching defines the waveguide. Therefore, these etch processes must be performed well enough to keep the surface roughness well below 40 nm or so, to obtain waveguides with propagation losses on the order of 0.03 dB/cm or less. When thin films are deposited on substrates, an important aspect of the process is managing the stress that can develop. For planar glass waveguides, depositions are performed at high temperatures and there are subsequent annealing operations that also occur at high temperatures (on the order of 1,000 °C). Polymer waveguides are ordinarily deposited at low temperatures, but solvent bakes and annealing often occur at moderately high temperatures (several hundred degrees centigrade). These temperature excursions, when coupled with the CTE mismatch that exist between the film and the substrate, result in stress-induced scattering. As scattering often comes from a number of sources, experimental scattering data is often fit with an empirical law of the form (Ma et al. 2002):

$$\alpha_{\text{scatter}} = A + B/\lambda^2 + D/\lambda^4 \tag{9.1}$$

where A is the contribution from large particle scattering (i.e., $\gg\lambda$), B the contribution from inhomogeneities on the order of λ in size (Mie scattering), and D the contribution from small inhomogeneities ($\ll\lambda$, Rayleigh-like). This expression can be used to help deduce the source of the scattering loss if the loss can be measured at several well-separated wavelengths. The scattering loss can be minimized in polymer waveguides by using direct photopatterning as opposed to surface-roughness-inducing RIE. It also can be reduced by ensuring the homogeneity of the medium (for example, no abrupt refractive index variations caused by phase separation or particles) and by minimizing intrinsic stresses. As opposed to planar silica technologies, polymer technologies can be designed to form stress-free layers regardless of the substrate composition, which can be silicon, glass, quartz, plastic, or glass-filled epoxy PCB substrate. Moreover, these films can essentially be free of stress-induced scattering loss when operating above the glass-transition temperature (T_g) in crosslinked polymer systems. For channel waveguides, propagation loss measurements are almost exclusively performed using the cutback technique. However, losses in channel waveguides come from a number of sources, including the patterning process for the channels (both roughness and stress-induced losses by the patterning process). Therefore, it is useful to have some measure of the slab waveguide loss, as this will isolate the sources of loss to absorption, extrinsic scattering, film roughness, and stress-induced scattering resulting from the film deposition process. There are two primary methods, the scanning technique and the liquid prism. The major advantages of the scanning fiber technique are that it can be performed easily on any slab waveguide sample and it does not require extensive handling of the sample (just the coupling to the input prism). The biggest drawback to this method is that if the sample is not clean or has a minute amount of large particles, these defects will scatter significant amount of large particles, these defects will scatter significant amounts of light, causing the baseline for the analysis to vary greatly. However, for clean samples without defects this method can achieve accuracies of less than0.5 dB/cm. A common alternative approach to slab waveguide loss measurements is the moving prism technique, where a fixed in-coupling prism is used in conjunction with a movable out-coupling prism to achieve a measurement of transmission vs. propagation length. However, this method and its variations usually suffer from uncertainty with regard to the coupling loss as the out-coupling prism is moved from position to position. This uncertainty limits the accuracy of the technique. An approach to improving the accuracy is to adopt a liquid out-coupling prism. With this technique, accuracies well below 0.1 dB/cm are routinely achieved (Teng 1993; Ma et al. 2002).

(c) Polarization-Dependent Loss

PDL is the maximum difference of attenuation between any of the two polarization states. Frequently, for straight planar waveguides, PDL is the difference between the TE and TM loss. However, for devices, PDL is commonly defined as the maximum difference of loss between the elliptical polarization states, and it can also vary among the various types of optical devices. The TE loss measured in planar waveguides can be higher than the TM loss when the vertical walls of the core have a higher degree of roughness than the horizontal boundaries, and it can be lower when the vertical evanescent tails overlap with an absorptive substrate or superstrate. An additional source of PDL is the stress mentioned above, which can directly or indirectly lead to increased losses along either the TE or TM direction depending on the nature of the stress. By having a minimal edge roughness, a well-confined material stack, and low-stress processing a well-optimized waveguide can have PDL values that are very small, provided the waveguide material itself has a low birefringence (Ma et al. 2002).

(d) Insertion Loss
Insertion loss is defined as $-10\ \log_{10}(P_{out}/P_{in})$ with P_{in} the input and P_{out} the output power from particular ports of an optical device. Acceptable insertion loss is determined by the overall system loss budget, which will be a function of transmitter power, optical amplifier gain, and receiver sensitivity among other factors. The total insertion loss achieved in planar polymer components can approach the value of the material absorption loss when fabrication techniques are optimized (Ma et al. 2002).

(e) Return Loss
Return loss is defined as $-10\ \log_{10}(P_{refl}/P_{in})$ with P_{refl} the reflected and P_{in} the input power to the device. Return loss should be minimized both to improve the insertion loss and to prevent reflected light from traveling back up the transmission system or network, where it can have destabilizing and damaging effects on lasers, especially when passing through bi-directional amplifiers (Ma et al. 2002).

(f) Radiation Loss and Fiber Pigtail Loss
The radiation loss can be reduced by using standard integrated-optic design rules such as large radii of curvature and adiabatic modal transitions. The fiber pigtail loss can be minimized by matching the mode of the planar waveguide to that of the fiber, which can be achieved by tuning the index contrast, the index profile, and the core dimensions. It can also be minimized by optimizing the alignment of the waveguides to the fiber, and by minimizing the Fresnel reflections with appropriate index-matching materials at the interfaces (Ma et al. 2002).

3. Processibility and Mechanical Properties
Optical polymers provide for the flexible, large-area, and low-cost fabrication of waveguide devices through simple techniques (such as spin coating, doctor blading, extrusion, and lamination), low-temperature processing, and compatibility with semiconductor electronics. A wide range of rigid or flexible substrates can be used, including glass, quartz, oxidized silicon, glass-filled epoxy PCB substrate, and flexible polyimide film. By controlling the polymer/solvent ratio and the spin speed of film coating, film thickness can be obtained in the range of 0.1–100 μm.

Unlike other optical material systems, polymers are designed and synthesized by chemical modification of constituent molecules to have the desired characteristics, such as melt or solution processibility in the form of monomers or prepolymers, photo- or thermo-crosslinking-enhanced mechanical properties, and matched refractive index between core and cladding layers. Furthermore, these properties are adjustable through formulation variations. There are several polymer waveguide formulation techniques such as direct lithographic patterning, soft lithography, embossing, molding, and casting in addition to the conventional photoresist patterning. This permits the rapid, low-cost shaping for both waveguide formation and material removal for grafting of elements such as active films, Faraday rotators, or half-wave plates. This flexibility also makes polymers an ideal hybrid integration platform, where foreign material systems, such as YIG (yttrium iron garnet) as well as lithium niobate, and semiconductor devices, such as lasers, detectors, and logic circuits, can be inserted into an etched groove in a PLC to enable full amplifier modules or optical add/drop multiplexers (OADMs) on a single substrate. Moreover, the combination of flexibility and toughness in optical polymers makes them suitable for vertical integration to realize 3D and even all-polymer integrated optics (Ma et al. 2002).

4. Environmental Performance
   (a) Thermal Stability
   An important characteristic for practical applications is the thermal stability of optical properties because polymeric materials are subject to yellowing upon thermal aging. Typically, such aging results from the formation of partially conjugated molecular groups characterized by broad ultraviolet absorption bands, which tail off in intensity through the visible region. This yellowing is strongly influenced by the chemical structure of the original polymer. The chemical structures of various backbone segments in the optical polymers can vary substantially from simple aliphatic to aromatic, and the linkages can vary from ether, to ester or urethane. The choice of these linkages and the monomers or oligomers ultimately determines to a significantly degree the characteristics of the resulting polymer, including surface energy, hardness, toughness, modulus, water uptake, and stability to ward aging. In fully halogenated materials, yellowing becomes almost negligible at any wavelength because the absence of hydrogen prohibits the formation of H-halogen products that will result in carbon double bonds. These unsaturated double bonds are the major cause of yellowing when they are slowly oxidized under long-term thermal aging (Ma et al. 2002).
   (b) Reliability Assurance
   Of particular concern for polymers are extreme temperatures and humidity, as well as the broad range of temperatures that may bracket either the T_g or the sub-T_g relaxations of the polymer. However, polymers that stand up to 85 °C and 85 %RH conditions have been demonstrated, and some polymers have already passed the Telecordia GR1209 and GR1221

Table 9.1 Properties of conventional optical polymers (Ma et al. 2002)

Material	Refractive index (n)	T_g (°C)	Loss (dB/cm)
PMMA	1.49	105	0.2 at 850 nm
PS	1.59	100	–
PC	1.58	145	–
PU	1.56	–	0.8 at 633 and 1,064 nm
Epoxy resin	1.58	–	0.3 at 633 nm; 0.8 at 1,064 nm

environmental tests. Extensive materials research has yielded polymers that are highly reliable, to the extent that they are no longer the limiting factors in component lifetime (Ma et al. 2002).

9.2.3 Conventional Optical Polymers

The preparation of light-guiding films with polymers started in the 1970s. Typical conventional optical polymers are shown in Table 9.1 (Ma et al. 2002), such as PMMA, polystyrene (PS), polycarbonate (PC), polyurethane (PU), and epoxy resin. These polymers possess very different structures; for example, PMMA with an aliphatic backbone and ester side-chain, PC with an ester backbone, and polyurethane with an amino-ester backbone. As a result, their properties vary significantly in the aspects of refractive index, optical loss, and thermal stability.

9.2.4 Advanced Optical Polymers

Advanced optical polymers have been developed since early of 1980s. Table 9.2 shows some commercially available optical polymers (Ma et al. 2002). These polymers can be grouped into four major classes: deuterated and halogenated polyacrylates, fluorinated polyimides, perfluorocyclobutyl (PFCB) aryl ether polymers, and nonlinear optical polymers. Other important but less explored optical polymers include benzocyclobutene (BCB), perfluorovinyl ether cyclopolymer (CYTOP), tetrafluoroethylene and perfluorovinyl ether copolymer (Teflon AF), silicone, fluorinated poly (arylene ether sulfide), poly(pentafluorostyrene), fluorinated dendrimers, and fluorinated hyperbranched polymers. The current optical polymers are highly transparent with absorption loss values below 0.1 dB/cm at all major communication wavelengths (840, 1,310, and 1,550 nm).

9.2.4.1 Deuterated and Halogenated Polyacrylates

Deuterated or halogenated polyacrylates are a kind of photo-crosslinkable, optically transparent polymers based on the combinations of multifunctional halogenated acrylate monomers and oligomers in addition to various additives. These polymers

Table 9.2 Characteristics of advanced optical polymers (Ma et al. 2002)

Polymer type	Patterning techniques	Propagation loss, single-mode waveguide (dB/cm) (wavelength, nm)	Other properties (wavelength, nm)
Acrylate (Polyguide)	Diffusion	0.18 (800) 0.2 (1,300) 0.6 (1,550)	Laminated sheets Excimer-laser machinable
Acrylate	Photoexposure/ wet etch, RIE, laser ablation	0.02 (840) 0.3 (1,300) 0.8 (1,550)	Birefringence: 0.0002 (1,550) Crosslinked, T_g: 25 °C Environmentally stable
Halogenated acrylate	Photoexposure/ wet etch, RIE, laser ablation	0.01 (840) 0.06 (1,300) 0.2 (1,550)	Birefringence: 0.000001 (1,550) Crosslinked, T_g: −50 °C Environmentally stable
Halogenated acrylate	RIE	0.02 (830) 0.07 (1,310) 1.7 (1,550)	Birefringence: 0.000006 (1,310) T_g: 110 °C
Deuterated polysiloxane	RIE	0.17 (1,310) 0.43 (1,550)	Environmentally stable
Fluorinated polyimide	RIE	TE: 0.3, TM: 0.7 (1,310)	PDL: 0.4 dB/cm (1,310) Environmentally stable
Fluorinated polyimide (Ultradel)	Photoexposure/ wet etch	0.4 (1,300) 1.0 (1,550)	Birefringence: 0.025, crosslinked, thermally stable
Polyetherimide (Ultem)	RIE, laser ablation	0.24 (830)	Thermally stable
Fluorinated polyimide	Photoexposure/ wet etch	TE: 0.5, TM: 0.6 (1,300)	Birefringence: 0.009 (1,300), PDL: 0.1 dB/cm (1,300), T_g: 310 °C, thermally stable
Perfluorocyclobutane (XU 35,121)	Photoexposure/ wet etch	0.25 (1,300) 0.25 (1,550)	T_g: 400 °C
Benzocyclobutene (Cyclotene)	RIE	0.8 (1,300) 1.5 (1,550)	T_g: >350 °C
Perfluorovinyl ether cyclopolymer (CYTOP)			$n = 1.34$ T_g: 108 °C
Tetrafluoroethylene and perfluorovinyl ether copolymer (Teflon AF)			$n = 1.31$ (AF 1,600) $n = 1.29$ (AF 2,400)
Polycarbonate (BeamBox)	RIE	0.6 (1,550)	Thermally stable
(OASIC)	Photoexposure/ wet etch, RIE, laser ablation	<0.01 (840) 0.03 (1,300) 0.1 (1,550)	Environmentally stable

(continued)

Table 9.2 (continued)

Polymer type	Patterning techniques	Propagation loss, single-mode waveguide (dB/cm) (wavelength, nm)	Other properties (wavelength, nm)
(Gemfire)	Photoexposure/ wet etch	1.0 (1,550)	Birefringence: 0.0002 (1,550) Crosslinked
Fluorinated poly (arylene ether sulfide) (FPAESI)	RIE	TE: 0.42, TM: 0.4 (1,550)	Birefringence: 0.0003 (1,550), PDL: 0.02 dB/cm (1,550), crosslinked, thermally stable
Inorganic polymer glass (IPG)	RIE		Environmentally stable
PMMA copolymer (P2ANS)	Photobleaching	1.0 (1,300)	NLO polymer
Polycarbonate with CLD-1 chromophore (PC-CLD-1)	RIE	1.8 (1,550)	NLO polymer, $r_{33} = 70$ pm/V (1,310), pigtail loss = 1.5 dB/facet
Polyurethane with FTC chromophore (PU-FTC)	RIE	2.0 (1,330)	NLO polymer, $r_{33} = 25$ pm/V (1,310), pigtail loss = 5 dB/facet
Poly (methylmethacrylate) with CLD-1 chromophore (PMMA-CLD-1)	RIE	5.0 (1,300)	NLO polymer, $r_{33} = 60$ pm/V (1,300), pigtail loss = 3.5 dB/facet

are particularly suitable for practical low loss optical devices because acrylates have intrinsically very low stress-optic coefficients. Moreover, since these polymers can be photochemically processed directly from their neat monomers, they provide the resulting polymers with fairly low internal stress and dimensional shrinkage. This combination of material properties allows the creation of waveguides with low scattering losses and low PDLs. Upon exposure to actinic radiation (e.g., UV or e-beam), these monomer systems form highly crosslinked networks, which exhibit low intrinsic absorption in the wavelength range extending from 400 to 1,600 nm. By blending and copolymerizing with selected miscible monomers, this approach allows precise tailoring of the refractive index over a very broad range from 1.3 to 1.6, although it can be difficult to simultaneously achieve a given refractive index and low intrinsic absorption at the same time. This control of refractive index allows fabricating step-index or graded index optical waveguide structures with well-defined and reproducible refractive index differences (Δn) to within 0.0001. At the same time, this synthetic approach allows other physical properties of the material such as flexibility, toughness, surface energy, and adhesion to be tailed to meet the needs of specific applications. These materials have already passed most of the Telecordia protocols GR1209 and 1221. Deuterated polyfluoromethacrylate has been developed with high transparency, low birefringence, and good processibility. The material has been used to fabricate low optical loss single-mode and multi-mode optical waveguides. The propagation loss and waveguide birefringence of the single-mode waveguides can be as low as 0.10 dB/cm and -5.5×10^{-6} at 1.3 μm, respectively. The propagation loss of the multi-mode waveguides can be less than 0.02 dB/cm at both 0.68 and 0.83 μm, and 0.07 dB/cm at 1.31 μm. Polyguide has shown excellent layer quality and thickness control. Buried waveguide circuitry can be formed with relatively low optical losses using photolithographic patterning and lamination techniques. In addition, polyguide can be machined using excimer lasers to form mechanical structures with a high degree of accuracy allowing for connection to a mechanically transferable (MT-type) ferrule packaged with push/pull housing connected to a multichannel standard ribbon fiber. Polyguide possesses high refractive indices (1.48–1.51) and high optical losses (0.18 dB/cm at 0.8 μm, 0.2 dB/cm at 1.3 μm, and 0.6 dB/cm at 1.55 μm) due to its non-halogenated acrylate structures (Yoshimura et al. 1998; Ma et al. 2002).

9.2.4.2 Fluorinated Polyimides

Although acrylate-derived polymers have demonstrated many attractive properties, they do not possess the needed thermal stability (as high as 300 °C) for direct on-chip interconnect applications. In this regard, polyimides are a proven class of polymers in microelectronics industry due to their high thermal stability (>300 °C) and outstanding dielectric and mechanical properties. However, conventional polyimides used in the semiconductor industry have poor characteristics that make them unsuitable for use as optical materials. Several types of fluorinated polyimides that feature low optical losses in the near-infrared region, a broad range of refractive index control, and excellent heat resistance have been developed. For polyimides, there are two main causes for the observed waveguide losses: Firstly, ordering processes can lead to refractive index fluctuations producing scattering centers (domain formation/phase separation). Furthermore, ordering can favor the formation of charge-transfer complexes, thus leading to increased absorption. Secondly, the evaporation of complex (through H-bonding) or trapped casting solvents as well as water produced during imidization can cause voids or pinholes in the optical waveguide. Optical loss produced by the latter mechanism can be reduced by optimizing the curing procedure. However, it is also observed that annealing allows ordering to occur and thus introduces charge complication and refractive index fluctuations to exhibit high loss values, and rapid heating leads to increased optical losses accompanied by a small change in the absorption spectrum. As a result, it is assumed that the losses introduced by rapid curing are due to voids and/or pinholes whereas the losses introduced by annealing above the glass-transition temperature of the polyimide might originate from residual chain–chain interactions. Nevertheless, most of the ordering processes so typical for conventional polyimides seem to be suppressed in fluorinated polyimides. Therefore, these materials allow the production of thermally stable waveguides with low optical losses. One intrinsic problem that is still difficult to avoid, however, is the large birefringence and PDL that results from aromatic ordering in the polyimides (Kowalczyk et al. 1994; Ma et al. 2002).

9.2.4.3 Perfluorocyclobutyl Aryl Ether Polymers

For most spin-coating applications, the solubility of optical polymers or oligomers needs to exceed 50 wt% in common solvents if a reasonable film thickness and planarity are to be obtained. Such basic requirements severely limit the use of many fluoropolymers based on chain addition polymerization as well as polyimide condensation polymers. However, poly-(aryl ether) polymers based on PFCB repeating units possess both high-performance properties and processibility. These polymers are synthesized through the thermal cyclopolymerization of trifunctional and bifunctional aryl trifluorovinyl ether monomers to PFCB copolymers. PFCB polymers and copolymers possess a unique combination of properties well suited for optical applications such as high temperature stability, precisely controlled refractive index, low moisture absorption, excellent melt and solution processibility, a high thermo-optic coefficient, and low absorption at 1,300 and 1,550 nm. Copolymerization reactions also offer tailored thermal and optical properties in these polymers by simply choosing suitable comonomers. PFCB polymers can be solution- or melt-microfabricated via standard methods or be processed by soft-lithographic techniques. In addition, PFCBs also exhibit excellent thermal properties (T_g = 120–350 °C), low optical loss (<0.25 dB/cm at 1,550 nm), refractive index tenability (1.449–1.508 at 1,550 nm), low birefringence, and high optical stability with optimized polymerization and processing parameters (Ma et al. 2002).

9.2.4.4 Fluorinated Dendrimers and Hyperbranched Polymers

Dendrimers have a well-defined structure and branching at each repeat unit, which imparts high solubility and low viscosity for the materials, suggesting behavior that resembles that of molecular ball bearings. By taking advantage of the behavior found for fluoropolymers, such as hydrophobicity, lipophobicity, chemical resistance, and low cohesive forces, fluorocarbon units have been incorporated in the periphery of dendrimers to obtain amphiphilic fluorinated dendrimers with surfactant properties.

Crosslinkable fluorinated dendrimers are potentially very useful for photonic applications due to their good processibility, owing to high solubility and low viscosity to allow high solids content in formulating the solution samples, and low optical loss compared to linear polymers. Some crosslinkable fluorinated dendrimers have been developed exhibiting high thermal stability and low optical loss. Optical loss of <0.36 dB/cm at 1,300 nm, and <0.52 dB/cm at 1,550 nm can be obtained for a fluorinated dendrimer after it had been cured at 225 °C. In addition, crosslinkable fluorinated hyperbranched polymers have been developed, which may offer an easier way to tailor the required properties by simply modifying the functional groups at the branch ends. For example, incorporation of chlorinated functional groups can allow precise refractive index tuning while maintaining the low loss benefits of halogenations (Pitois et al. 2001; Ma et al. 2002).

9.2.4.5 Miscellaneous Highly Fluorinated Polymers

Perfluorinated ether-containing polymer, CYTOP, and copolymer between Teflon and fluorinated ketal are two interesting perfluorinated polymers that possess excellent optical properties for low loss polymer optical fibers at both visible and infrared wavelengths. However, due to their high hydrophobicity and low refractive index, there are some limitations for using these polymers for optical waveguide applications, such as (1) difficulty in integrating them in multilayer waveguide structures due to poor adhesion, (2) maintenance of overall dimensional or thermo-mechanical stability, and (3) difficulty in waveguide processing using established fabrication techniques. Although these problems are typically viewed as secondary issues after the primary performance, the latter limitation of processibility can be most critical when real production is considered (Zhao et al. 2000; Ma et al. 2002).

9.2.4.6 Electro-optic Polymers

Due to the rapid development of photonic technologies and the ever-increasing demand of faster switching speed and higher bandwidth in telecommunications, as discussed in Chap. 8, significant effort has been exerted in trying to commercialize optical waveguide devices such as E-O modulators and switches based on E-O polymers. Properties of electro-optic polymers include optical, structural, and mechanical characterization. The optical benefits are large non-resonant nonlinearity, low switching energy, sub-picosecond response times, broadband transparency, high optical damage threshold, and low absorption while structural and mechanical properties enable the benefits of low-cost materials, engineering at the molecular level, integrated optics, room temperature operation, chemical stability, and ease of processing and synthesis modifications along with resistance to radiation, heat, and shock (Mahanta 2005).

The second-order nonlinear optic (NLO) properties originate from non-centrosymmetric alignment of NLO chromophores, either doped as a guest or covalently bonded as side-chains in poled polymers. To obtain device-quality materials, three stringent issues must be addressed (Ma et al. 2002): (a) design and synthesis of high $\mu\beta$ (μ is the chromophore dipole moment and β is the molecular first hyperpolarizability) chromophores and realization of large macroscopic E-O activity in the chromophore-incorporated polymers; (b) maintenance of long-term temporal stability in the E-O response of the poled materials in addition to their high intrinsic stability toward the environment such as heat, light, oxygen, moisture, and chemicals; (c) minimization of optical loss from design and processing of materials to fabrication and integration of devices. In the context of addressing the first issue, the major task encountered is to efficiently translate the high molecular hyperpolarizability (β) values of organic chromophores into large macroscopic E-O activity (r_{33}) in polymers and also maintain their high thermal stability and low optical loss. Electric field poling of a polymer containing highly nonlinear chromophores is often hindered by the shape-dependent chromophore–chromophore electrostatic interactions through the large dipole moment of the molecules (Shi et al. 2000). Several very promising E-O dendritic material systems have been developed, ranging from the 3D shape dendritic chromophore, to the multiple-armed chromophore-containing dendrimers with NLO chromophores connected to the center core with crosslinkable trifluorovinylether as the crosslinkable periphery, to the side-chain dendrimerized NLO polymers. These dendritic material systems have shown great promise in achieving much high E-O activities (>100 pm/V, which is a factor of 3 larger than the commercial inorganic system, lithium niobate). This approach can provide the new paradigm for the design of highly efficient E-O materials for ultra-low drive-voltage E-O modulator (Ma et al. 2002).

In addressing the second issue, two approaches have been adopted to improve the long-term stability of molecular orientation in poled polymers. First, very high-T_g polymers such as polyimide and polyquinoline have been used as host matrices to incorporate NLO chromophores. After poling, these guest/host polymers have shown promising long-term stability at elevated temperatures. Moreover, the motion of chromophores in polymer matrices can be further restricted when the chromophores are covalently attached to polyimides or polyquinolines either in the form of a main chain or as a side-chain. This approach has led to polymers with exceptional thermal stability. Another alternative approach is to develop a lower-T_g material, which contains reactive functional groups on the chromophore or on the polymer backbone, that can be processed into optical-quality thin films, and can then be poled and crosslinked simultaneously as higher temperatures to lock the chromophores and polymers in place to achieve better control over the poling-induced polar order. The crosslinkable epoxy resins, polyurethanes, polyimides, and poly (perfluorocyclobutylaryl ether)s (PFCBs) have been used to enhance orientational stability of the poled polymers (Ma et al. 2002).

The third critical issue influencing the utilization of polymeric electro-optic materials is that of optical loss at the communication wavelengths of 1.3 and 1.55 μm. Both absorption and scattering of light from materials will contribute to optical loss. However, the scattering loss in E-O polymers can be reduced by controlling the homogeneity of chromophore dispersion in the E-O polymer films before and after poling to avoid phase separation, and the uniformity of the guiding and cladding films in the NLO polymeric waveguide. Minimization of optical loss associated with C–H vibrational absorptions is an approach to develop low loss passive polymer optical waveguides. Active polymers are unlikely to yield optical losses as low as passive polymers due to the requirement of chromophore incorporation and lattice hardening to lock poling-induced chromophore alignment in acentric order. Partial fluorination of NLO chromophores and the incorporation of chromophores into fluorinated polymers such as PFCB polymers have proved to be useful approach to minimize the absorption loss of the resulting materials, since the thermosetting reaction involved in PFCB formation does not generate any O–H, N–H, or aliphatic C–H bonds (Ma et al. 2002).

Electro-optic polymers continue to hold the promise of low-cost, high-performance devices, although in the reliability front polymer optic devices are likely to have to struggle for acceptance. Research is very much warranted in light of the fact that despite their questionability in high performance and high reliability, in some niche applications they may be the only option. Polymers could outperform rival technologies in cases such as optical waveguide switches and modulators, which require a fusion of lower cost and higher performance. Two effects, important for the functioning of communication devices, namely thermo-optic and electro-optic effects, decide the use of the material. While larger change in the refractive index with temperature (dn/dT) is a disadvantage for temperature-sensitive polymer devices such as directional couplers and interferometers, the large thermo-optic coefficient enables switches to operate at low switching power and to be insensitive to ambient temperature fluctuations so as to maintain a stable operating condition. This is helped further by the low thermal conductivity of polymers.

Electro-optic effects are important for achieving faster data transmission which means packing more wavelengths per fiber (as in wavelength division multiplexing) as well as running each wavelength at a higher bit rate. E-O polymer modulators have exhibited better characteristics for faster data transmission due to the low dielectric constants of polymers which allows velocity matching (equalization of the speeds of optical and microwave signals in a traveling microwave design), which is not available with any other electro-optic material. Polymer modulators have lower driving voltage and power requirements (Mahanta 2005).

9.3 Fabrication Process of Polymer Waveguides

Polymer waveguides constitute one means of providing interconnections between elements of optoelectronic integrated circuits. The techniques and relevant issues for fabrication of polymer optical waveguides are quite similar to those for polymers in microelectronic packaging (Hagerhorst-Trewhella et al. 1989). Common techniques for casting polymer films include spin coating, doctor blading, extrusion, and lamination. The major concerns when fabricating a film are that the film is of uniform thickness, is free of bubbles and striations, will adhere well to the substrate of interest, and that the film can be made to the desirable thickness. Each casting technique has its own advantages and disadvantages. For example, spin coating allows suitable control of thickness and uniformity; however, eliminating striations in films can be difficult. The techniques that have been used for patterning optical waveguides in polymer films include photoresist-based patterning, direct lithographic patterning, and soft lithography. Figure 9.1 compares photoresist-based patterning and direct lithographic patterning to generate polymer optical waveguides. Fabrication processes such as micro-molding in capillaries and micro-contact printing to achieve polymer optical waveguides are shown in Fig. 9.2 (Lindquist et al. 2012).

Some other methods for fabricating polymer waveguides include RIE, chemical etching, injection molding, electron beam irradiation, UV writing, pressure

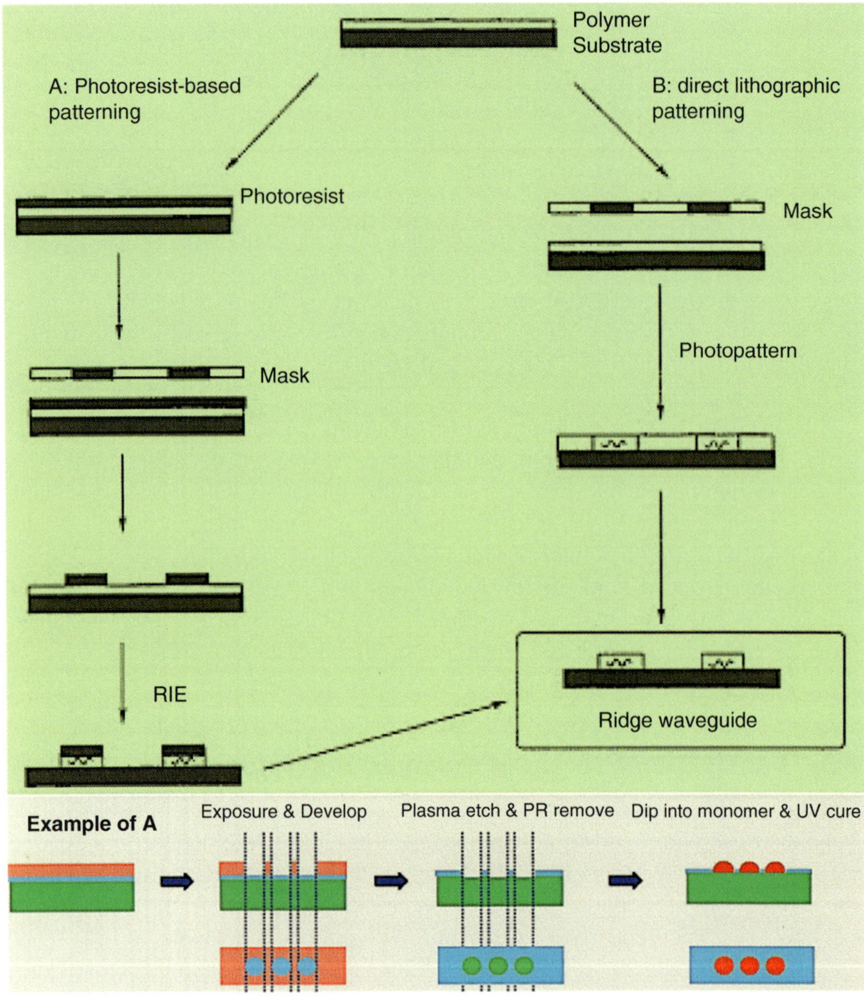

Fig. 9.1 Comparison of photoresist-based patterning and direct lithographic patterning to generate polymer optical waveguides (Ma et al. 2002)

dispensing, and doping. Many of these methods are expensive and time consuming and would limit the rapid development and testing of new materials and devices. Also, fabrication methods are needed that result in single-mode waveguides for active device applications. Polymers, in addition to the ease of shaping, offer a great deal of control in the structure, size, and composition for these devices. For example, the substrate for these devices could even be adapted to give flexible waveguides to further minimize the space taken up by an active device. A highly polished thin flexible ribbon of polymer for the substrate would take advantage of the flexibility of the core and cladding to create a waveguide that could be wound around it to take up less space (Kalajian 2003).

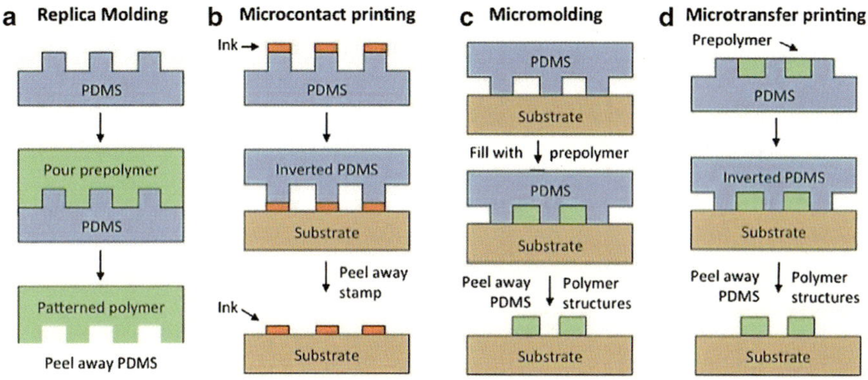

Fig. 9.2 Micro-molding and printing processes to generate polymer optical waveguides (Lindquist et al. 2012)

9.3.1 Photoresist-Based Patterning

In general, for the fabrication of polymer optical waveguides, standard semiconductor processing technology is adopted; for example, spin coating, baking, and RIE processes. However, very significant efforts have been made to optimize the waveguide fabrication process, particularly in the areas of deep, smooth, and vertical sidewall etching to minimize optical losses contributed from sidewall scattering. In conventional RIE, only one radiofrequency power source is applied to regulate both plasma density and ion energy (and direct current bias) and it is impossible to control them separately. Although the etch rate increases with radiofrequency power, the increased radiofrequency power can cause severe sputtering by energetic ions at the sample surface, leading to increased damage on the sample and decreased etch selectivity. In this regard, high-density plasma etching using an inductively coupled plasma (ICP) source offers an attractive alternative over the conventional RIE. ICP etching uses a secondary radiofrequency power source (ICP power) to generate a high-density plasma (typically two to three orders of magnitude higher than that generated by RIE) without correlating to the ion energy. Thus, ICP etching produces low surface damage while achieving high etching rates (Ma et al. 2002).

9.3.2 Direct Lithographic Patterning

Laser direct-writing can be used to fabricate optical waveguides from photosensitive polymers. The technique of laser direct-writing has the advantage of being maskless, allowing rapid and inexpensive prototyping in contrast to conventional mask-based photolithographic approaches in which a mask must first be designed and fabricated before waveguides can be produced. This technique is capable of patterning features with long and linear dimensions over comparatively large planar areas. This capability is matched to the requirements of producing an array of waveguides and switches, typical of optical switching fabrics, arranged over areas that are large, e.g., ~10–1,000 cm^2, in comparison to those characteristic of integrated circuit (IC) dice. However, it can be difficult to form important structures such as splitter Y-junctions and directional couplers with laser direct-writing. Laser-writing can play a role in the fabrication of large dimension parts where masks cannot be produced; for instance, several-meter-long polymer waveguides can easily be laser-written on large substrates such as rolls of flexible plastics (Ma et al. 2002).

Photobleaching is another kind of direct lithographic patterning and is useful for dye-containing optical polymer systems such as E-O polymers. An advanced fabrication technique is electron beam direct-writing, which has advantages over the conventional UV-photobleaching technique (Ma et al. 2002): (a) since the irradiation is operated in vacuum, the effect of dust can be neglected; (b) this technique can demonstrate the fabrication of sub-micrometer or even a nanometer pattern with high resolution; and (c) a computer-assisted-design system supports the writing of complex patterns for advanced optical devices such as E-O modulators.

9.3.3 Soft Lithography

Microfabrication using non-photolithographic techniques (soft lithography) has been used to fabricate optical waveguides. These techniques involve preparation of poly(dimethylsiloxane) (PDMS) molds/stamps by casting PDMS on a photolithographically generated silicon master with the desired micropattern. The PDMS mold, thus obtained, can be used in a variety of ways to replicate the micropatterns of the polymer on a given substrate. Two such examples, namely micro-contact printing and micro-molding in capillaries, are shown in Fig. 9.2. Both of the techniques were found to work very well with PFCB polymers (Ma et al. 2002).

9.3.4 Electron Beam Bombardment

Electron beam bombardment has been used to make polymer waveguides. For example, a highly transparent polymer, such as poly(diethylene glycol bis (allylcarbonate)), a PMMA polymer, is used as the substrate. The polymer is coated with a 20 nm thick Au–Pd layer to keep the electron beam from charging the polymer. A 10 mm × 10 mm × 1 mm section of the polymer substrate for instance is then irradiated using a scanning electron microscope. Energies in the range of 10–30 keV are typically used, allowing the electron beam to penetrate up to about 10 μm. The section of polymer substrate that is irradiated has an increased index of refraction compared to the unexposed area. This produces single-mode or multi-mode graded index waveguides based on the thickness of the waveguide. However, the scanning electron microscope needed to produce these waveguides makes this technique costly. Moreover, the waveguides are also relatively large for use in photonic integrated circuits (Kalajian 2003).

9.3.5 Injection Molding

Single-mode polymer waveguides with low transmission losses can be fabricated using injection-molding technology (Neyer et al. 1993). The first step is conventional UV-light lithographic structuring of photoresist layers on silicon, with the thickness of the resist determining the thickness of the final waveguides. Development of the photoresist results in the waveguide structure being transferred to the resist in the form of rectangular ridges. Electroplating the developed resist layers with nickel creates a metal mold that replicates the negative of the resist profile and will define the outline of the grooves in the device structure. This negative is then used as an insert in the injection-molding machine for the substrate. A polymer like polymethyl-methacrylate, or PMMA, of high optical quality is then used as the injection-molded material and creates the molded substrate polymer. The grooves left in the mold by the nickel profile are then filled with a waveguide core polymer like ethyleneglycol-dimethacrylate or EGDMA. This core polymer is selected because it has a higher index of refraction than the substrate as well as not react with the substrate and dissolve it. Another layer of the substrate polymer is pressed on top to remove the excess liquid core polymer and becomes the cover of the waveguide. The waveguide is then cured using UV light. The losses associated with this technique are 0.3 dB/cm at 1,300 nm. This method has the advantage of being easily mass-produced once the fabrication of the mold is finished. The core polymer could also be changed to achieve single-mode waveguides at different wavelengths. However, there are disadvantages to this system. The process is not only time consuming, but also expensive during the research and development process (Kalajian 2003).

9.3.6 UV Writing

UV-written waveguides in polymers on silicon or silica are another frequent method of fabrication (Koo et al. 2003). Polymers developed for waveguide applications such as the PMMA are often used in this fabrication process. Synthesis of these polymers involves thermally initiated free radical polymerization at an elevated temperature and the resulting polymers are precipitated into ether. The esterification step then results in a pure white powder substance. This polymer is then dissolved in a solvent and filtered before it is used to spin-coat the substrate. The writing of the waveguides is done using a frequency-doubled argon ion laser with a 244 nm wavelength. The beam of the laser is focused in the polymer core layer, and the substrate and the core layer are moved on computer-controlled translational stages to crosslink the polymer. The index of refraction is greater in the regions that the polymers have been crosslinked as compared to the regions not exposed to the UV beam. This method has the option of spin coating another layer on top of the UV-written layer to place many layers of waveguides on top of each other. The ability to change the polymers used in the substrates within the selection conditions makes this method more versatile. Although the layering of waveguides has many advantages in the development of a final extensive product, the development stage of any one project would be hindered by the expense of the frequency-doubled argon ion laser. The many steps involved also make this fabrication method time consuming. The losses are also higher for this waveguide fabrication method, being 0.4–0.9 dB/cm (Kalajian 2003).

9.3.7 Dispensed Polymer Waveguides

Direct dispensed polymer waveguides on glass substrates have been fabricated with optical adhesives that have a higher index of refraction than glass. They are dispensed onto the substrate directly and cured immediately with no further processing required (Keyworth 1995). Norland optical adhesive (such as NOA 88) can be used for its high index of refraction and the small amount of volume lost during curing under a UV light. The polymer is dispensed using a 160-μm inner-diameter syringe to form the core, and an inexpensive mercury light source is used for curing since there is no requirement for a small focused spot. As pressure dispenses polymer from the syringe a translation stage is moved beneath the tip to create the waveguide. These waveguides use air as the top cladding, although it is suggested that another polymer of lower index of refraction than the core could be placed on top of the waveguides. Compared to other fabrication methods, this technique requires much less expensive equipment for fabrication; however, these waveguides were multi-mode, and therefore less desirable for telecommunication uses. This technique also had the lowest losses of all the fabrication techniques. At 633 nm losses of approximately 0.2 dB/cm are determined for the V-groove waveguides and 0.45 dB/cm is found at 633 nm for the surface waveguides. These lower losses make this method a reasonable choice to improve upon (Kalajian 2003).

9.3.8 Doping of Polymers to Create Optical Waveguide Devices

Polymer waveguides doped with rare-earth ions as the active element can be used for optical amplification (Slooff et al. 2002). The trivalent ions such as erbium (Er), neodymium (Nd), europium (Eu), and terbium (Tb) are commonly used in optical applications. Er and Nd in particular are used as they have transitions at 1.53 and 1.34 μm, respectively. Optical gain is achieved with these rare-earth ions by first exciting the ion from its ground state into a higher short-lived state. The ion quickly decays through non-radiating means to a longer-lived excited state. When it decays back to the ground state, a photon is released. The Er salts are inorganic and cannot be directly dispersed into an organic polymer. First an organic liquid must encapsulate the Er^{3+} ions, and then the complex can be dispersed into the polymer film. A channel waveguide is fabricated with a high refractive index optical polymer doped with the Er^{3+} complexes and embedded in a low index polymer. When excited by a wavelength of 488 nm a gain of 1.7 dB/cm was observed. This could be applied to create a no loss splitter with a waveguide only a few centimeters in length. The process to create the polymers with the inclusion of the dopants with this technique is very time consuming and difficult. However, once the polymer is doped conventional waveguide fabrication techniques may be employed (Kalajian 2003).

9.4 Polymer-Based Optical Components and Integrated Optics

One of the earlier drivers for work with polymer waveguides was the concept of fiber-to-the-home, which raised the possible future need for low-cost passive optical components such as splitters. This perception of the need for low-cost devices has driven much of the early effort on the development of low-cost replication technologies. When this opportunity did not appear as soon as expected, other higher value applications requiring greater functionality were pursued. The early commercialization of polymer waveguide technology came with the introduction of thermo-optic switches. Integrated polymer components produced to date for the optical communication industry include couplers, switches, attenuators, filters, polarization controllers, modulators, lasers, amplifiers, and detectors. An important property of polymers is that they have a large negative thermo-optic coefficient ($dn/dT = -1$ to 4×10^{-4}) that is 10–40 times larger (in absolute value) than that of more conventional optical materials such as glass, resulting in low power consumption thermally actuated optical elements. With optical polymers, a variety of thermally actuated polymeric components have been made, including switches, VOAs, and tunable filters. The electro-optic components are also made such as modulators, and the use of birefringent polymers for polarization control. Developments in active polymeric components are underway, including lasers, amplifiers, and detectors (Eldada 2002).

9.4.1 *Switches*

All-optical networks will require complex optical switches and routers to direct signals to their proper destination. Compared to opto-mechanical switches, waveguide switches, including thermo-optic (T-O) and electro-optic (E-O) switches, offer entirely solid-state operation. T-O switches are slower than E-O witches, but they are easier to operate. Their millisecond response times are adequate for most types of optical switching, so they have been more practically applied over times. Another advantage of T-O switches is that they can generally be made to be polarization-independent. Meanwhile, electro-optic switches will find their niche in high-speed applications, especially in the military area (Ma et al. 2002).

The operating principle of polymer T-O switch is based on temperature of the refractive index of the waveguide material. The driving power and the response time of the T-O device depend mainly on the thermal properties (i.e., the thermal conductivity, the temperature coefficient of the refractive indices and the specific heat of the buffer, guiding, and cladding layers and the substrate material, and on the dimensions of the waveguide and heater). Thermo-optic $N \times N$ switches can be interferometric switches based on directional couplers or Mach–Zehnder interferometers (MZIs), or they can be digital optical switches (DOSs) based on X junctions or Y junctions. The most widely used switch design is the Y-junction-based DOS (Y-DOS), because of its simplicity and its digital behavior. The building-block element in a Y-DOS is a small-angle (typically about 0.1°) 1×2 splitter with heaters on its arms. A schematic diagram of such a device is shown in Fig. 9.3a. These splitters can be connected with bends and crossings to form $M \times N$ switching matrices. Each 1×2 splitter relies on adiabatic evolution of the mode profile in its two waveguides into the mode of the ON guide (the guide with the higher effective refractive index) when the OFF guide is heated to reduce its index, as shown in the computer simulation of Fig. 9.3b. The device is considered to have switched once it reaches the desired isolation value, which occurs at some level of electrical power dissipation in the electrodes, beyond which power level the device maintains the isolation, resulting in its digital behavior. DOSs in a 1×2^N configuration can be fabricated with $2^N - 1$ 1×2s, and one electrode at each stage needs to be heated to perform the switching. For instance a 1×4 switch can be built with three 1×2s, as shown in Fig. 9.3c where the upper electrode in the first stage and the lower electrode in the second stage are powered to switch the light from port 1 to port 3′. As another example, a 2×2 (or crossbar) DOS can be fabricated with four 1×2s arranged as shown in Fig. 9.3d. This switch is operated in the bar state by powering the four inner electrodes, while powering the four outer electrodes results in the cross state. Any size $N \times N$ switch can be built out of 1×2 switches. The number of 1×2 switches needed in a strictly non-blocking $N \times N(2^n \times 2^n)$ switch is $2N(2^{n-1} + 2^{n-2} + \cdots +1)$, which simplifies to $2N(N - 1)$ (Eldada 2003).

9.4.2 *Variable Optical Attenuators and Tunable Filters*

With the increasing complexity of WDM optical networks comes an increasing need for reliable, low-cost VOAs that adjust the power level of optical signals with high accuracy and repeatability. VOAs can be based on any switching principle including interferometry, modal transition, or mode confinement.

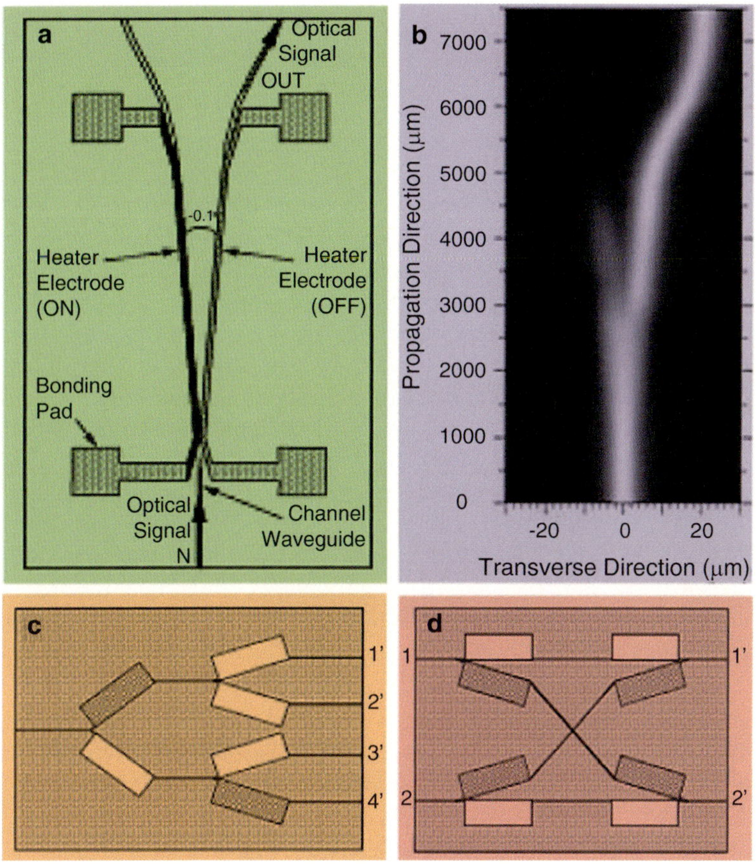

Fig. 9.3 Schematic diagram of (**a**) a 1 × 2 Y-DOS and (**b**) computer simulation of such a device where the left arm is heated, allowing the light to exit the right arm. Schematic layouts of (**c**) a 1 × 4 DOS and (**d**) a 2 × 2 DOS (Eldada 2003)

An interferometric approach involves using an MZI where heat can be applied to at least one of the arms to induce a phase shift between the two arms before they recombine, thereby controlling the level of optical power exiting the output guide. One performance specification that is typically difficult to achieve in VOAs is low PDL. The PDL achieved in polymeric VOAs is under 0.25 dB across the entire attenuation range, a value that is lower than that achieved in any other material system (Eldada 2003).

A wavelength tunable filter is an important device for WDM systems because of its multichannel selectivity. The requirements for a practical tunable filter are high wavelength resolution, low polarization dependence, low levels of cross talk (CT), and low insertion loss. A polymeric AWG-based thermo-optic tunable filter has been made of crosslinked silicone and operated at around 1.55 μm. This filter exhibited a TE–TM polarization shift of <0.03 nm, a cross talk of <−35 dB, insertion loss of <3 dB, and a tuning range of 8.8 nm in the 25–75 °C temperature region. In addition, two other types of polymeric AWG-based thermo-optic tunable filters have also been developed, each with a pair of triangular phase shifters. Both filters operated with a low cross talk of <−25 dB, one with a tuning range of >30 nm for a large scale system and the other with a tuning range of ~10 nm for a small-scale system. Although these results are quite impressive, the performance of AWG-based thermo-optic tunable filters is not as efficient as that of completing technologies such as the fiber Bragg gratings (FBGs) or thin film interference filters. However, by combining the advantages of planar devices, such as lithographic manufacturing with a filter performance approaching that of FBGs, photochemically produced Bragg gratings in polymer waveguides have been fabricated. The resulting polymer Bragg gratings (PBGs) have the potential for achieving excellent filter characteristics, relatively low insertion loss through proper integrated optic design, and modularity, a highly valued quality of FBGs. One PBG has been tuned by actuating a heater for a polymer with a $\mathrm{d}n/\mathrm{d}T$ of -3.1×10^{-4}/°C (about 30 times larger than in glass), resulting in a tuning rate of −0.36 nm/°C, therefore permitting tuning across the entire erbium C-band (1,528–1,565 nm) with a temperature range of about 100 °C (Ma et al. 2002).

Tunable filters can be based on Bragg gratings, diffraction gratings, AWGs, microring resonators, or photonic crystals (PCs). By alternating in a waveguide the refractive index periodically about an average effective refractive index of n, an in-line series of weakly reflecting mirrors (a Bragg grating) of spacing Λ is created. The cumulative effect of the mirrors is to maximally reflect wavelengths λ, equal to $1/N$ multiples of $2n\Lambda$, where $N \geq 1$ is an integer indicating the order of the grating period. Gratings in planar polymers can be produced by a variety of techniques such as casting, molding, embossing, e-beam writing, and photochemical processes. The first three techniques produce surface relief gratings while the last two can produce either relief gratings or bulk index gratings across the waveguide core. Photochemical fabrication processes for bulk index gratings utilize two-beam interference to induce an index modulation. This effect can be achieved either through the use of a phase mask (where two beams corresponding to the +1st and −1st diffraction orders are allowed to interfere) or through the use of direct interference of split laser beams. For a waveguide refractive index of $n = 1.5$, the period of the grating corrugation (the Bragg period Λ) is approximately 0.52 μm. Tuning of the grating is achieved by powering a heater that is fabricated in the proximity of the grating. Broad thermal tuning of polymeric gratings can be achieved because of the large $\mathrm{d}n/\mathrm{d}T$ (Eldada 2003).

9.4.3 Polarization Controllers and Modulators

Birefringent elements, typically made from birefringent crystals, can be produced out of birefringent polymers (e.g., polyimide). They are used in beam displacers, prism polarizers, isolators, circulators, switches, PMD compensators, and other precise optical components where polarization control is needed. Birefringent materials used for polarization splitting are typically crystals such as calcite, rutile, yttrium orthovanadate, and barium borate. Materials used for polarization rotation, such as in half-wave plates, include polyimide and $LiNbO_3$. The commercial availability of polyimide half-wave plates makes them commonly utilized to obtain polarization independence, which effect is achieved by inserting them in exact positions in the optical path of interferometric optical components. However, polyimide half-wave plates are hygroscopic, which makes the thin-film $LiNbO_3$ half-wave plates a more attractive option (Eldada 2002).

Some polymer formulations have been designed to have a large electro-optic coefficient (as large as 200 pm/V; the largest value achieved in any material system). These formulations are typically composed of standard polymers (e.g., polycarbonate) impregnated with specialty chromophores (e.g., CLD-1). They exhibit a large electro-optic effect once subjected to poling, a process where large electric fields (~200 V/μm) are applied to the material in order to orient the molecules. However, the result of the poling process is not stable with time or with environmental conditions, limiting the applications where polymer electro-optic modulators can be used (Eldada 2002).

For polymeric E-O materials to be adopted commercially, they must exhibit properties that are competitive with commercial materials such as lithium niobate E-O and gallium arsenide electro-absorptive materials. A well-recognized advantage of polymeric materials is that of bandwidth. The low and relative frequency-independent dielectric constants and refractive indices of organic materials naturally lead to intrinsic material bandwidth on the order of 350 GHz (measured by pulsed techniques for a 1 cm device). Devices fabricated from polymeric materials have been demonstrated to 113 GHz. Another putative advantage of polymeric E-O materials is that of processibility and ease of integration with disparate materials. Polymeric E-O devices have been successively integrated with VLSI semiconductor electronic circuitry and passive fiber optical circuitry. By using a modified optical push–pull MZI architecture in modular design, a half-wave voltage of 0.8 V and a half-wave voltage-interaction length product of 2.2 V cm have been achieved from a guest/host E-O PMMA using a high β polynee-type chromophore. The push–pull architecture improves the devices' modulation efficiency by 6 dB because it can reduce $V\pi$ by a factor of 2 when compared with single-arm modulation at the same interaction length. However, materials and devices reliability and stability are more challenging since the E-O effect in polymers derives from chromophores that are electrically aligned (poled) in the polymer matrix. By using a constant DC bias to keep the chromophores from orientational relaxation, an r_{33} as much as 3 times of the partially relaxed residual r_{33} after poling has been observed. Thus, it also reduces the half-wave voltage due to the utilization of the full potential of E-O polymers. The device based on a constant DC bias was operated as a birefringent modulator. The temperature of the modulators was maintained by a closed-loop system consisting of a temperature controller, a thermoelectric cooler (TEC), and a thermistor. The lower electrode was grounded, and a DC bias voltage was applied to the top electrode to create a constant electric field that actively aligns the dipole with the chromophores. The AC modulating voltage was coupled to the top electrode through a bias tee circuit consisting of a capacitor and a resistor. Light of 1.31 μm wavelength was coupled into the optical waveguide from the single-mode fiber pigtail of a 1.4 mW semiconductor laser. However, thus far their polarization-dependent response precludes any actual use in optical telecommunication switching systems. The traditional uniform poling procedure, a prerequisite processing step for E-O polymer-based devices whereby and electric field is applied at a temperature in the vicinity of the glass-transition temperature, results in uniform statistical polar orientation of the chromophores, and is, therefore, responsible for sandwich electrode configurations, any desired prerequisite modulation axis can, however, be imprinted onto the material, thus resulting in the possibility of balanced TE/TM polarization efficiencies. Monolithic integration of both electrode configurations with a Mach–Zehnder modulator is shown to result in an original polarization-insensitive E-O polymer amplitude modulator for integrated optics. A digital optical modulator based on an asymmetric Y-branch waveguide is fabricated by using an E-O polymer, which shows the potential for polarization- and wavelength-insensitive operation (Ma et al. 2002).

9.4.4 Lasers and Amplifiers

Rare-earth-doping is widely used to produce lasers and all-optical amplifiers that are simple, reliable, low cost, and have a wide gain bandwidth. Although rare-earth doping has been typically used in silica (mostly in fiber, and to a lesser degree in planar waveguides), there has been ongoing research work to develop viable and stable rare-earth-doped polymer lasers and amplifiers. The main rare-earth ions used are erbium and thulium. Erbium amplifiers provide gain in the C-band between 1,530 and 1,570 nm, thulium amplifiers provide gain in the S-band between 1,450 and 1,480, and gain-shifted thulium amplifiers provide gain in the S-band between 1,480 and 1,510 nm. The gain achieved with these technologies is not uniform across the gain bandwidth, requiring gain-flattening filters, typically achieved with an array of attenuators between a demultiplexer and a multiplexer. Since the gain shape of the amplifier is not stable with time (for example, due to fluctuations in temperature), tunable gain-flattening filters (TGFFs) are needed, where the static attenuators are replaced with VOAs. Laser dyes (e.g., Rhodamine B) are highly efficient gain media that can be used in liquids or in solids to form either laser sources with narrow pulse width and wide tunable range, or optical amplifiers with high gain, high power conversion, and broad spectral bandwidth. Laser dyes captured in a solid matrix are easier and safer to handle than their counterpart in liquid form. Dye-doped polymers are found to have better efficiency, beam quality, and optical homogeneity than dye-doped sol–gels. In optical fiber form (silica or polymer), the pump power can be used in an efficient way because it is well confined in the core area, it propagates diffraction-free, and it has a long interaction length. The reduced pump power is significant in optimizing the lifetime of solid-state gain media. The photostability is one of the main concerns in solid-state gain media and the higher pump intensity can cause a quicker degradation of the dye molecule (Eldada 2002).

9.4.5 Detectors

Photoconductive polymeric material systems have been developed, enabling the monolithic integration of photodetectors in polymeric optical circuits. As an example, a photoconductive complex was formed in the system poly-(*N*-vinyl carbazole) (PVK)–trinitrofluorenone (TNF). This complex is sensitive in the visible spectral region. The compound was deposited from a solution of chlorobenzene/tetrahydrofuran between thin gold electrodes. The photocurrent was measured and was observed to peak at a TNF:PVK molecular ratio of 1.9:1. The functionality of this system was demonstrated by placing electrodes covered with the photoconductor on top of a polymer waveguide film in order to collect scattered light, and the TE and TM mode spectra of the waveguide were measured by monitoring the change of photocurrent as a function of in-coupling angle. The TE and TM modes were well resolved on the collected spectra (Eldada 2002).

9.4.6 Optical Interconnects for Computing Systems

The speed and complexity of integrated circuits are increasing rapidly from very large scale integrated (VLSI) circuits to ultra-large scale integrated (ULSI) circuits. As the number of components per chip, the number of chips per board, the modulation speed, and the degree of integration continue to increase, electrical interconnects are facing fundamental bottle-necks, such as speed, packaging, fan-out, and power dissipation. Multichip module (MCM) technology is employed to provide higher clock speeds and circuit densities. However, the state-of-the-art technologies based on electrical interconnects fail to provide the required multi-Gbit/s clock speed and communication distance in intra-MCM and inter-MCM hierarchies. As a result, optical interconnects for MCMs, chips, boards, and backplanes have been fabricated. Figure 9.4 shows typical schematic of optical interconnects in computing systems (Flores and Wang 2010). Freestanding polymeric optical waveguide films with 45° mirrors and MT connectors (mechanically transferable spacing connectors) could be fabricated by spin coating, photolithography, and RIE. The 45° mirror was formed in the waveguide by using a 90° V-shaped diamond blade, and the mirror loss was 0.3–0.7 dB/cm depending on the measurement condition. Freestanding polymeric waveguide films were mounted in an MT-compatible connector made of Pyrex glass by using UV curable resin for instance. The insertion loss can be about 1 dB/cm or lower, which is sufficiently low for practical use (Ma et al. 2002).

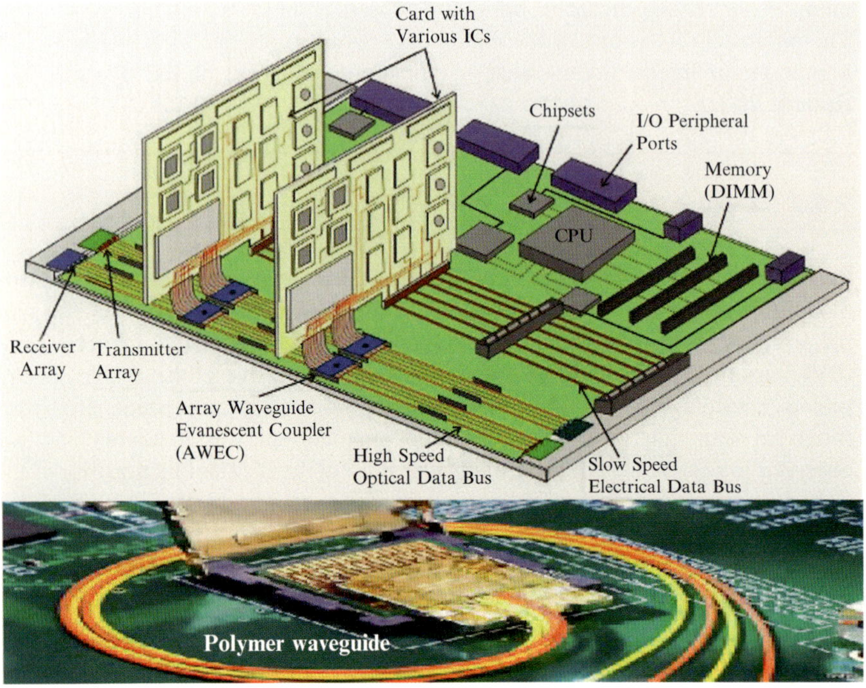

Fig. 9.4 Schematic illustration of optical interconnect for card-to-backplane motherboard in computing systems (Flores and Wang 2010)

9.4.7 Planar Optical Connects for Wavelength Division Multiplexing Telecommunication Systems

The simplest passive optical component is the splitter, a device that distributes optical power incident on input port into specified power fractions at the output ports. A systemmetrical $1 \times N$ splitter divides the power into N ports with a fraction $1/N$ of the incident power each (assuming no other sources of loss). Thus an ideal 1×8 splitter, for example, has outputs with power reduced by 9 dB compared to the input power level. In practice, ideal splitters have been demonstrated for some time using fused biconic taper (FBT) fiber technology. However, the drawbacks are the relatively narrow band performance, vibration sensitivity, and difficulty in scaling to very high port counts. Planar splitters, as realized in silica on silicon or polymer waveguides, benefit from ease of fabrication (especially at high port counts) and reduced wavelength dependence, since splits can be based on Y junctions as opposed to extended directional couplers (as in FBT). However, planar splitters generally suffer from increased insertion loss compared to FBT splitters, at least at low port counts, owing to fiber-waveguide coupling loss. Therefore, propagation loss within the planar splitter must be kept to an absolute minimum to remain competitive. As there are no significant performance advantages of polymer technology over planar glass technology for this application, the waveguide losses must be equal or less than those of planar glass, or on the order of 0.02 dB/cm at 1,550 nm. A directional coupler is also one of the most important elements in optical integrated circuits. It is used for constructing many components, such as power splitters, mux/demuxes, and optical switches (Ma et al. 2002).

Another passive optical component of increasing interest is the AWG, a cleverly designed device that uses a concept employed by phased array radar to obtain highly efficient wavelength multiplexing and demultiplexing. Planar glass AWGs are already highly successful products used in large volume in optical network, so it would be expected that polymer-based AWGs would have to attain the same performance as silica-based AWGs to achieve acceptance, once again driving interest toward very low loss waveguide materials. An additional requirement for AWGs is that the polarization dependence of the waveguides be very low, so that PDLs and polarization-dependent shifts in the filter performance are minimized. One potential advantage of polymer waveguide technology for AWG is the ability to make completely polymer-based AWGs that are temperature-independent, thus potentially requiring no costly and cumbersome temperature control technology, as generally needed for the planar glass devices. By matching the positive CTE of the polymer substrate and the negative thermo-optic (T-O) coefficient of the waveguide material, the implementation of a polarization-independent and temperature-independent AWG has been demonstrated (Ma et al. 2002).

Modulators and optical switches are two of the most important active components that operate on optical signals passing through fiber optic systems. They are active in two senses: they require external electrical input power and alter the signals in a variable way—functions that are becoming increasingly important as fiber optic systems grow more sophisticated (Hecht 2001).

The VOA has attracted extensive attention due to its promising applications in DWDM systems. It is used to control the gain of optical amplifiers, thus equalizing the channel powers at add/drop nodes of DWDM systems. It can also dynamically regulate the channel powers regardless of fluctuations in the input light power and polarization. It is expected that an array structure containing multiple attenuators on a single chip is preferable for future DWDM system applications. One of the most critical issues for the implementation of the array is the electrical power consumption of the unit device involved. In this regard, polymer-based T-O VOAs are quite promising. VOAs can be based on any switching principle, including interferometry, mode transition, or mode confinement. Low power consumption T-O VOA that incorporates a membrane-type asymmetric branch waveguide in polymers has been reported. The measured electrical consumption of the attenuator is as small as 25 mW at 1,550 nm, which is reduced by about 50 % compared to that of conventional polymer devices. In addition, an eight-channel polymeric VOA array based on MZI exhibits an ultra-low power consumption of about 1.5 mW for 30 dB attenuation.

Also, a VOA based on the Gemfires leaky mode design has been demonstrated with very promising results (Ma et al. 2002).

The rapid expansion of optical telecommunication technology also increases the need for planar optical amplifiers that can be used to compensate losses in splitters, multiplexers, switches, and other devices. These losses could be due to imperfections in the waveguide, inefficient coupling and decoupling, mismatch between the laser mode and the waveguide mode, or decrease in the intensity due to splitting of the beam. Planar optical amplifiers have been widely studied and they often use the rare-earth ions erbium or neodymium as the active element because these ions exhibit intra-4*f* transitions around 1,550 and 1,340 nm, respectively. Amplification is obtained by optical pumping of the rare-earth ions in order to create population inversion. Stimulated emission induced by the signal light then results in optical amplification. An important aspect of polymer-based waveguide amplifiers is the potential for the incorporation of various rare-earth species at higher concentrations than can be achieved in inorganic glass. Since the rare earth is molecularly incorporated, the local environment is better enrolled than in the inorganic glass case, further providing for selective co-doping of energy transfer agents (such as Yb^{3+} for Er^{3+}). There are two kinds of approaches, butt-end coupling and distributed coupling, to realize the pumping of an optical waveguide amplifier. Butt-end coupling is done by coupling the pump beam into the waveguide at the input facet of the waveguide. The pump light is absorbed by the rare-earth ions as it travels through the waveguide, resulting in decrease in pump power over the entire length of the waveguide; relatively high pump powers are coupled into the input section of the waveguide. This pumping scheme can be successfully used for materials in which high pump powers do not affect the pumping efficiency. However, in several materials systems, an optimum pump power for amplification exists. Such systems include Er-doped waveguides in which cooperative up-conversion and excited-state absorption take place, or systems in which the rare-earth ions are excited via an energy transfer from a sensitizer. In these systems, butt-end coupling is not efficient. The excess pump power at the beginning of the waveguide will result in pump power absorption, which does not contribute to the optical gain (Slooff 2000).

Distributed coupling is based on the coupling between two adjacent waveguides, where pump light is gradually coupled from a non-absorbing pump waveguide into the amplifier waveguide. The coupling between the waveguides in such a configuration is calculated using an improved coupled mode theory (CMT). The proposed distributed coupling scheme can enhance the optical gain in systems that exhibit reduced pump efficiency at high pump power. An example of this enhancement has been shown for a sensitized neodymium-doped polymer waveguide amplifier, in which the optical gain increases from 0.005 to 1.6 dB by changing from the conventional butt-coupling to the distributed coupling (Slooff 2000).

9.4.8 Planar Optical Waveguides for Sensors

Optical waveguide formats offer a flexible option to fabricate a compact, disposable, and integrated sensor with an appropriate waveguide structure and wide tolerance in applications. Planar optical waveguides are very sensitive to changes in parameters such as refractive index, absorption and emission spectra due to temperature, humidity, pressure, and chemical species. Parameter changes cause modulation of light traveling within the optical waveguides, which is useful for optical sensing in industrial, environmental, medical, civilian, and military applications. A planar geometry utilizing waveguide-coupled surface plasmon resonance (SPR) may have a theoretical sensitivity of $\Delta n = 2 \times 10^{-5}$. Grating structures in planar waveguides, coated with thin, highly specific, and responsive chemical coatings that can respond to sub-mole layer film formation by adsorption or binding of antibody molecules within seconds with a theoretical minimum index change of $\Delta n = 2 \times 10^{-6}$, have been developed for biochemical sensors. An MZI has been used to detect antibodies absorbed onto waveguide surfaces at a concentration of 5×10^{-11} M HCG (human chorionic gonadotropin) at 2.5 ng/mL in 20 min. Polymer-based optical waveguide sensors have been explored because in inorganic waveguide systems it is difficult to control stress, refractive index, and uniformity. By using polymer-based waveguide systems, optical sensors have been achieved with small residual stress, easy fabrication, and low cost. For an opto-mechanical pressure sensor using polymer multi-mode interference (MMI) couplers, a high sensitivity of 8.2 ppm/Pa has been obtained. Sensitivity of several parts per million humidity concentrations has also been achieved for a polymer waveguide sensor with a symmetric multilayer configuration (Ma et al. 2002).

9.4.9 Integrated Planar Lightwave Circuits

To fully realize the function of polymer optical waveguide devices, the integration of light sources, optical transmission waveguides, optical components, and optical sensors/detectors is necessary. On the other hand, the performance of optical components can also be enhanced by their integration. The fabrication of Bragg grating-based multichannel OADMs in polymers has been designed by cascading multiple stages of single-channel mux/demuxes. Each stage consists of an MZI with a Bragg grating across its arms. The 3 dB couplers in each MZI are MMI couplers, which offer excellent tolerance to polarization and wavelength variation, and relaxed fabrication requirements compared to alternatives such as directional couplers, adiabatic X- or Y-junctions, and diffractive star couplers. The gratings were printed photochemically, achieving index differences as high as 10^{-3} and reflective as large as 33 dB. The 3D bandwidth is about 0.2 nm. Anodization is used to reduce side lobes in the reflection spectrum. The Bragg gratings in the four MZIs have slightly different periods, the grating closest to the port reflects at wavelength λ_1, so light at that wavelength launched at the input port exits the drop port (λ_1) while light at that wavelength launched into the add port (λ_1) exits the pass port. The resulting four-channel OADMs exhibit the proper 400 GHz channel separation and International Telecommunications Union (ITU) grid wavelength alignment, as well as excellent channel isolation and uniform channel response. Another way to form an OADM is to place a grating between optical circulators. Tunable OADMs can be achieved by combining thermally tunable planar PBGs with optical circulators. The grating exhibits better than 45 dB reflection with no detectable out-of-band reflection. These OADMs have a high-bandwidth utilization factor of 0.92, with a minimum channel spacing of 75 GHz. They are tunable at a rate of −0.256 nm/°C (Ma et al. 2002).

Photonic crystals (PCs) have a periodic dielectric structure with high-index contrast, designed to control photons in the same way as conventional crystals in solids control electrons. They have very large dispersion characteristics (500 times that of a regular prism and more than 50 times stronger than that of a conventional grating), which makes them very attractive for WDM applications. Using polymer waveguide technology and block copolymer photonic bandgap materials, it is possible to fabricate a waveguide superprism within polymeric planar waveguide circuits. New optical devices and the miniaturization of current integrated optical devices may result. A simple integrated WDM demultiplexer circuit can be fabricated from a quadrilateral 2D photonic crystal element. This integrated optical demultiplexer reduces component size by hundreds of times compared to conventional devices. Significantly, there is no restriction on the free-special range caused by the high-order diffraction in the demultiplexer based on a conventional grating (Ma et al. 2002).

DOSs, which have been realized using different materials, show a cross talk (CT) of about −20 dB, which is still insufficient for network applications. An approved CT value was achieved with a W-shaped DOS or a cascade of DOSs and attenuator. Moreover, a polymer DOS with an integrated attenuator showed a cross talk value of −46 dB, but the higher power consumption and the additional insertion losses need to be reduced. Based on Y-branched wide-angle DOSs, a polymer 1 × 2 DOS integrated with a VOA showed low switching power without adversely affecting the cross talk. This switch exhibited a low cross talk value of about −40 dB, and the switching power was ~170 mW at 1.55 μm (Ma et al. 2002).

Organic light-emitting diodes (OLEDs) utilizing fluorescent dyes or conjugated polymers have attracted great interest because they are capable of emitting light that can cover almost the whole visible spectral range and highly efficient, requiring only a quite low driving voltage (2–10 V). OLEDs have also been demonstrated to have a long lifetime (>20,000 h) and excellent durability for flat-panel display applications. An additional advantage is that they are simple to fabricate on various kinds of substrates, including polymer and glass substrates. The combination of polymer waveguides and organic optical devices will provide huge advantages for fabricating optical integrated circuits. Since polymer waveguides have a low transmission loss in the near-infrared and red light region, use of light-emitting devices at these wavelengths will provide an efficient tool for optical integration (Ma et al. 2002).

Conventional two-dimensional (2D) guided wave integrated optics provides a reliable and structurally stable method for interconnecting optical devices in photonic circuits and for the optical interconnection of high-density integrated electronics. However, in two dimensions, the complexity of the photonic circuits is limited by the size of the substrate and by the difficulty of connecting a large number of input and output fibers or electrical connections. The possibility of adding the third dimension to integrated optics would greatly increase the integration density and would overcome some of the input–output problems while preserving the reliability and structural stability of guided wave optics. Three-dimensional (3D) integrated optics consists of vertically stacked layers of horizontal 2D integrated optics with properly placed vertical waveguide interconnects between the layers. For 3D optical integration, optical polymers have some significant advantages over other materials such as semiconductors or silica. The vertical features of optical polymers can be etched by using shadow mask etching or a photoresist as the etch mask. In addition, optical polymers can have excellent adhesion to a number of surfaces, and by varying the doping level or by combining multiple polymer systems, precise control of the index of refraction is possible. Furthermore, a series of vertical waveguide bends, waveguide power splitters, waveguide polarization splitters, and coupler switches have been fabricated based on optical polymers. In addition, 3D integrated Mach–Zehnder modulators, directional couplers, optical fan-out devices, and E-O devices have also been realized in polymer material systems (Ma et al. 2002).

The commercial production of integrated-optic-based components requires efforts in a wide range of disciplines, including material synthesis, material formulation, material analysis, optical design, optical device fabrication, electronic design, PCB fabrication and assembly, firmware, digital signal processing, fiber pigtailing, packaging, functional testing, tooling, automation, quality, reliability, and qualification. From materials to processing to devices, polymer optical waveguides have shown great promise for integrated optical circuits due to their processibility and integration advantages. These advantages also make them an ideal integration platform where foreign material systems, such as lithium niobate, and semiconductor devices, such as lasers, detectors, amplifiers, and logic circuits, can be inserted into an etched groove in a PLC to enable full amplifier modules or OADMs on a single substrate. Moreover, the combination of flexibility and toughness in optical polymers makes them suitable for vertical integration to realize 3D and even all-polymer integrated optics. Due to the capability of being produced in large volumes at low costs, polymer optical waveguide devices will play a major role in broadband communications, in areas such as optical networking based on DWDM metro/access communications, and on computing systems. In addition, highly integrated polymer optical waveguide components and circuits also need to demonstrate their full advantages such as size reduction, low cost, performance improvement, higher yields, and shorter time to market.

As a result, the optimization of polymer optical components, the integration of polymer passive and active optical waveguide devices, and the integration of high sources, optical transmission waveguides, optical components, and optical circuits will grow. Besides, the introduction of photonic bandgap materials and nanolithographic processing to polymer passive or active optical waveguide devices will present great opportunities for enhancing the performance of polymer-based integrated optics (Ma et al. 2002).

9.5 Summary

Polymer optical waveguides would play a key role in broadband communications, such as optical networking, metropolitan access communications, and computing systems, due mainly to their easier processibility and integration over inorganic counterparts. The combined advantages also make them an ideal integration platform where foreign material systems such as YIG and lithium niobate, as well as semiconductor devices such as lasers, detectors, amplifiers, and logic circuits can be inserted into an etched groove in a PLC to enable full amplifier modules or OADMs on a single substrate. Moreover, the combination of flexibility and toughness in optical polymers makes it suitable for vertical integration to realize 3D and even all-polymer integrated optics.

Classes of polymers used in integrated optics include acrylates, polyimides, polycarbonates, and olefins (e.g., cyclobutene). Some polymers, such as most polyimides and polycarbonates, are not photosensitive, and are typically processed using photoresist patterning and RIE. These polymers have most of the problems of the silica on silicon technology in terms of roughness-induced and stress-induced scattering loss and polarization dependence. Other polymers are photosensitive and as such are directly photopatternable, resulting in a full cycle time of about 30 min per multilayer optical circuit on a wafer. These materials have an obvious advantage in throughput, producing wafers between 10 and 1,000 times faster than other planar technologies. Furthermore, this technology uses low-cost materials and low-cost processing equipment (e.g., spin-coater and UV lamp vs. CVD growth system). Typical conventional optical polymers include PMMA, polystyrene (PS), polycarbonate (PC), polyurethane (PU), and epoxy resin. These polymers possess very different structures; for example, PMMA with an aliphatic backbone and ester side-chain, PC with an ester backbone, and polyurethane with an amino-ester backbone. As a result, their properties vary significantly in the aspects of refractive index, optical loss, and thermal stability. Advanced optical polymers can be grouped into four major classes: deuterated and halogenated polyacrylates, fluorinated polyimides, PFCB aryl ether polymers, and nonlinear optical polymers. Other important but less explored optical polymers include BCB, perfluorovinyl ether cyclopolymer (CYTOP), tetrafluoroethylene and perfluorovinyl ether copolymer (Teflon AF), silicone, fluorinated poly(arylene ether sulfide), poly(pentafluorostyrene), fluorinated dendrimers, and fluorinated hyperbranched polymers. The current optical polymers are highly transparent with absorption loss values below 0.1 dB/cm at all major communication wavelengths (840, 1,310, and 1,550 nm).

Polymer waveguides constitute one means of providing interconnections between elements of optoelectronic integrated circuits. Common techniques for casting polymer films are spin coating, doctor blading, extrusion, and lamination. Some other methods for fabricating polymer waveguides include RIE, chemical etching, injection molding, electron beam irradiation, UV writing, pressure dispensing, and doping. Many of these methods are expensive and time consuming and would limit the rapid development and testing of new materials and devices. Also, fabrication methods are needed that result in single-mode waveguides for active device applications. Polymers, in addition to the ease of shaping, offer a great deal of control in the structure, size, and composition for these devices. For example, the substrate for these devices could even be adapted to give flexible waveguides to further minimize the space taken up by an active device. A highly polished thin flexible ribbon of polymer for the substrate would take advantage of the flexibility of the core and cladding to create a waveguide that could be wound around it to take up less space.

The commercial production of integrated-optic-based components requires efforts in a wide range of disciplines, including material synthesis, material formulation, material analysis, optical design, optical device fabrication, electronic design, PCB fabrication and assembly, firmware, digital signal processing, fiber pigtailing, packaging, functional testing, tooling, automation, quality, reliability, and qualification. From materials to processing to devices, polymer optical waveguides have shown great promise for integrated optical circuits due to their processibility and integration advantages. These advantages also make them an ideal integration platform where foreign material systems, such as lithium niobate, and semiconductor devices, such as lasers, detectors, amplifiers, and logic circuits, can be inserted into an etched groove in a PLC to enable full amplifier modules or OADMs on a single substrate. Moreover, the combination of flexibility and toughness in optical polymers makes them suitable for vertical integration to realize 3D and even all-polymer integrated optics. Due to the capability of being produced in large volumes at low costs, polymer optical waveguide devices will play a major role in broadband communications, in areas such as optical networking based on DWDM metro/access communications, and on computing systems. Therefore, the optimization of polymer optical components, the integration of polymer passive and active optical waveguide devices, and the integration of high sources, optical transmission waveguides, optical components and optical circuits will grow. In addition, the introduction of photonic bandgap materials and nanolithographic processing to polymer passive or active optical waveguide devices will present great opportunities for enhancing the performance of polymer-based integrated optics.

References

Eldada L (2002) Polymer integrated optics: promise vs. practicality. In: Kippelen B, Bradley DD (eds) Organic photonic materials and devices IV. Proceedings of SPIE, San Jose, CA. SPIE—The International Society for Optical Engineering, vol 4642, pp 11–22. Accessed 18 January 2002

Eldada L (2003) Hybrid integrated photonic components based on a polymer platform. In: Heyler RA, Robbins DJ, Jabbour GE (eds) Photonics packaging and integration III. Proceedings of SPIE, San Jose, CA. SPIE—The International Society for Optical Engineering, vol 4997, pp 88–102. Accessed 25 January 2003

Eldada L, Shacklette LW (2000) Advances in polymer integrated optics. IEEE J Sel Top Quant Electron 6(1):54–68

Flores A, Wang MR (2010) Soft lithographic fabrication of micro optic and guided wave devices. In: Wang MR (ed) Lithography. InTech, New York

Hagerhorst-Trewhella JM et al (1989) Polymeric optical waveguides. Proc SPIE—Int Soc Opt Eng 1177:379–386

Hecht J (2001) Modulators and switches are key to all-optical networks. Laser Focus Word 35 (6):85–88

Hougham G, Tesoro G, Viehbeck A (1996) Influence of free volume change on the relative permittivity and refractive index in fluoropolyimides. Macromolecules 29:3453–3456

Jackson WB et al (1981) Photothermal deflection spectroscopy and detection. Appl Opt 20:1333–1344

Kalajian JM (2003) Towards a single-mode dispensed polymer optical waveguide. Master thesis, University of South Florida

Keyworth BP (1995) Computer-controlled pressure-dispensed multimode polymer waveguides. IEEE Trans Components Packaging Manuf Technol B 18:572–577

Kihara M et al (1996a) Return loss characteristics of optical fiber connectors. J Lightwave Technol 14:1986–1991

Kihara M et al (1996b) Characteristics of thermally expanded core fiber. J Lightwave Technol 14 (10):2209–2214

Koo J et al (2003) UV written waveguides using crosslinkable PMMA-based copolymers. Opt Mater 23:583–592

Kowalczyk TC et al (1994) Loss mechanisms in polyimide waveguides. J Appl Phys 76:2505–2508

Lindquist NC et al (2012) Engineering metallic nanostructures for plasmonics and nanophotonics. Rep Prog Phys 75:036501. doi:10.1088/0034-4885/75/3/036501

Ma H et al (2002) Polymer-based optical waveguides: materials, processing, and devices. Adv Mater 14(19):1339–1365

Mahanta K (2005) Polymer waveguides for quantum information. AFRL-IF-RS-TR-2005-4. Air Force Research Laboratory, Rome, New York. http://www.dtic.mil/dtic/tr/fulltext/u2/a430134.pdf. Accessed 3 March 2013

Neyer A et al (1993) Fabrication of low loss polymer waveguides using injection molding technology. Electron Lett 29:399–401

Norwood RA et al (2001) Sources of loss in single-mode polymer optical waveguides. In: Northwood RA (ed) Design, manufacturing, and testing of planar optical waveguide devices, vol 4439. SPIE, Bellingham, pp 19–28

Pitois C et al (2001) Functionalized fluorinated hyperbranched polymers for optical waveguide applications. Adv Mater 13:1483–1487

Shi YQ et al (2000) Low (sub-1-volt) halfwave voltage polymeric electro-optic modulators achieved by controlling chromophore shape. Science 288:119–122

Slooff LH (2000) Rare-earth doped polymer waveguides and light emitting diodes. PhD dissertation, Ultrecht University, Netherlands

Slooff LH et al (2002) Rare-earth doped polymers for planar optical amplifiers. J Appl Phys 91:3955–3980

Teng CC (1993) Precision measurements of the optical attenuation profile along the propagation path in thin-film waveguides. Appl Opt 32:1051–1054

Watanabe T et al (1998) Influence of humidity on refractive index of polymers for optical waveguide and its temperature dependence. Appl Phys Lett 72(13):1533–1535

Yoshimura R et al (1998) Low-loss polymeric optical waveguides fabricated with deuterated polyfluoromethacrylate. J Lightwave Technol 16:1030–1037

Zhao YG et al (2000) Polymer waveguides useful over a very wide wavelength range from the ultraviolet to infrared. Appl Phys Lett 77:2961–2963

Chapter 10
Hollow Waveguides

Abstract Hollow-core waveguides (HCWs) are comprised of a central hole surrounded by a highly reflective inner wall. The core can be filled with air, inert gas, liquid, or vacuum, allowing these waveguides to transmit a broad range of wavelengths with low attenuation. HCWs are of particular interest for the transmission of infrared (IR) to THz radiation, where it is otherwise difficult to find materials that have the optical, thermal, and mechanical properties required for use in solid-core optical fibers. Therefore, IR-transmitting hollow waveguides can be an attractive alternative to solid-core IR fibers. Hollow waveguides can be made from plastic, metal, or glass tubes that have highly reflective coatings deposited on the inside surface. These waveguides have losses as low as 0.1 dB/m at 10.6 mm and may be bent to radii less than 5 cm. For use in high-power laser delivery applications, the waveguides have shown to be capable of transmitting up to 3 kW of CO_2 laser power. They are also finding uses in both temperature and chemical fiber sensor applications. This chapter will give a brief review about the progress in hollow waveguide technology with emphasis on the available hollow waveguides that have been developed.

10.1 State of Art and Perspectives

The concept of using hollow pipes to guide electromagnetic waves was first described by Rayleigh in 1897 (Sanghera and Aggarwal 1998). Further understanding of hollow waveguides was delayed until the 1930s when microwave-generating equipment was first developed and hollow waveguides for these wavelengths were constructed. The success of these waveguides inspired researchers to develop hollow waveguides to the IR region. Initially these waveguides were developed for medical uses especially high-power laser delivery. Later on they have been used to transmit incoherent light for broadband spectroscopic and radiometric applications (Merberg 1993).

X.C. Tong, *Advanced Materials for Integrated Optical Waveguides*, Springer Series in Advanced Microelectronics 46, DOI 10.1007/978-3-319-01550-7_10,

Infrared (IR) optical fibers are defined as fiber optics transmitting wavelengths greater than approximately 2 mm. The first IR fibers were fabricated in the mid-1960s from chalcogenide glasses such as arsenic trisulfide with losses in excess of 10 dB/m. During the mid-1970s, the interest in developing an efficient and reliable IR fiber for short-haul applications increased partly in response to the need for a fiber to link broadband, long wavelength radiation to remote photodetectors in military sensor applications. In addition, there was an ever-increasing need for a flexible fiber delivery system for transmitting CO_2 laser radiation in surgical applications. Since around 1975, a variety of IR materials and fibers have been developed to meet these needs. These included the heavy metal fluoride glass (HMFG) and polycrystalline fibers as well as hollow-core waveguides (HCWs) (Harrington 2000).

Hollow waveguides present several advantages over solid-core optical fibers with key features including their ability to transmit wavelengths well beyond 20 μm, their inherent advantage of having an air core for high-power laser delivery, and their relatively simple structure and potentially low cost. In general, hollow waveguides exhibit advantages of high laser power damage thresholds; low nonlinear effects; low insertion loss (no Fresnel reflections from the end face of a HCW when radiation is coupled from free space into the air core); no end reflection, ruggedness, and small beam divergence; as well as low materials absorption because most of the radiation they transmit is confined to the hollow core. These advantages have led to the use of HCWs in applications requiring the transmission of radiation at wavelengths where it is otherwise difficult to find materials that have the optical, thermal, and mechanical properties required for use in solid-core fibers. Potential disadvantages, however, include an additional loss on bending and a small numerical aperture (NA). Nevertheless, they are one of the best alternatives for both chemical and temperature sensing as well as for power delivery in IR laser surgery or in industrial laser delivery systems with losses as low as 0.1 dB/m and transmitted CW laser powers as high as 2.7 kW (Harrington 2000; Bowden 2007).

HCWs can be classified into two categories: attenuated total internal reflection (ATIR) waveguides and leaky-type HCWs. ATIR waveguides consist of a hollow core surrounded by wall material with a refractive index, n, that is less than one for the transmitted wavelength. ATIR waveguides guide light by total internal reflection in the same manner as the silica core/clad optical fibers used in the telecommunications industry. The spectral range where a material has $n < 1$ is called an anomalous dispersion region. In this region the extinction coefficient, k, is also high. Evanescent waves propagating in the wall material are attenuated due to the wall's high k, limiting the performance of ATIR guides. One example of an ATIR waveguide is the hollow sapphire fiber that transmits CO_2 laser radiation with a loss of approximately 0.5 dB/m. The other type of HCW utilizes a highly reflective wall to confine light to the hollow core. Leaky modes propagate in these HCWs since the wall is not a perfect reflector; thus they are referred to as leaky-type HCWs. Increasing the wall's reflectivity increases the degree to which light is confined to the waveguide's hollow core, thus decreasing the attenuation of radiation propagating along its axis. Leaky-type HCWs can be divided into different

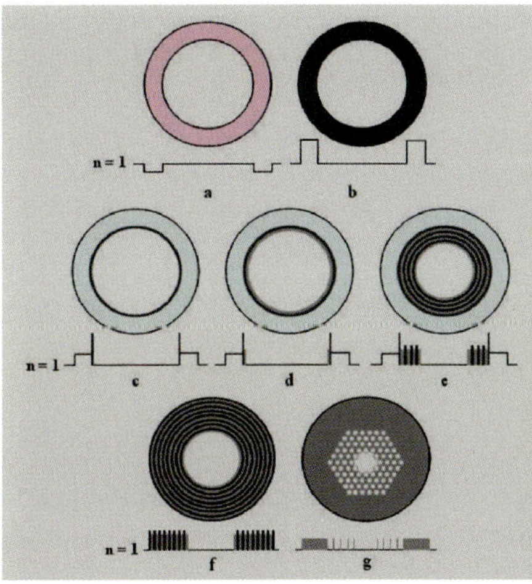

The basis of hollow waveguides is a tube, which can be made of plastic, metal, special or ordinary glass or crystal.

Base tube		Film/layers	Fabrication/method
Metal	Silver tube	AgBr	Wet chemistry
	Stainless	$AgPbF_2$	Wet chemistry
	Steel + metal strip inside nickel tube	Ag ZnS	Sputtering of metallic films → etching → electroplating
	Phosphor bronze strip	ZnS PbTe/PbF_2	Evaporation → soldering strips
Plastic	Polyethylene	Ag AgI	Wet chemistry
	Polycarbonate	Ag AgI	Wet chemistry
	Plastic sleeve-Polyimide	Ag AgI	Etching wall electroless
	Polyimide Teflon	Ag AgI	Wet chemistry
Hollow glass	Silica glass	Ag AgI	Wet chemistry
	Silica glass	Ag FCP	Wet chemistry
	Silica glass	Ag COP	Wet chemistry
	Silica glass	Ag COP	Wet chemistry

Fig. 10.1 Typical types and schematic cross sections/refractive index profiles of hollow-core waveguides (Bowden 2007): ATIR waveguide (**a**), metal tube (**b**), metal-coated hollow glass waveguide (HGW) or hollow polymer waveguide (HPW) (**c**), metal/dielectric HGW or HPW (**d**), multilayer metal/dielectric HGW or HPW (**e**), hollow Bragg fiber (HBF) (**f**), and photonic crystal fiber (**g**)

categories based upon how the wall of the waveguide is designed to achieve high reflectivity. The most basic design is to use a smooth metal surface to form the inner wall of the waveguide. This can be accomplished using a metal tube or by depositing a metal film on the smooth inner surface of a glass or plastic tube. A dielectric layer is then added over the metal to provide enhanced reflectivity within specific wavelength ranges due to an interference effect. These waveguides are called metal/dielectric HCWs. Multilayer metal/dielectric HCWs are fabricated by adding multiple dielectric layers over the metal to form an alternating high/low refractive index structure. In the absence of materials absorption, roughness, and structural imperfections, the transmission loss of these waveguides decreases exponentially as the number of layer pairs increases. For an infinite number of layer pairs, radiation is completely confined to the hollow core and the metal is no longer necessary to achieve high reflectivity. A waveguide that consists of a hollow core surrounded by an alternating high/low refractive index dielectric structure with no metal layer is called a hollow Bragg fiber (HBF). Photonic crystal fibers (PCFs) utilize a two-dimensional periodic dielectric structure to confine light to the air core. Also called holey fibers, these waveguides generally consist of a series of air channels stretching along the length of the waveguide that are arranged in a hexagonal lattice. Several air holes are removed from the center of the waveguide to create a defect in which light can propagate with low loss. The typical types of HCWs and their refractive index profiles are shown in Fig. 10.1 (Bowden 2007).

The theory of hollow waveguide transmission has been described from the viewpoint of both wave and ray optics (Harrington 2000). Wave optics has been used to predict attenuation coefficient α for either metallic or dielectric waveguides:

$\alpha \sim 1/a^3$, where a is the bore radius. Bending the hollow waveguides increases the total loss. The additional bending loss varies as $1/R$, where R is the bending radius. Therefore, in contrast to the solid-core fibers, the loss of a hollow waveguide depends strongly on its diameter and bending radius. For the thin film waveguide structures with dielectric coatings deposited over a metallic layer, the attenuation coefficient is given by (Miyagi and Kawakami 1984)

$$\alpha_{\infty} = \frac{\lambda^2}{a^3}\left(\frac{U_0}{2\pi}\right)^2\left(\frac{n}{n^2+k^2}\right)_{\text{metal}} F_{\text{film}} \tag{10.1}$$

where $\propto$ is the loss for a straight guide, U_0 is a mode-dependent parameter which for the lowest-order HE11 mode equals 2.405, n and k refer to the optical constants of metal film, and F_{film} is a term which accounts for the loss due to the dielectric film(s).

One of the challenges that remain in the development of IR waveguides is the fabrication of ultralow-loss waveguides for the delivery of high-power IR laser radiation. Reducing the transmission loss of the HCWs could be a driving force for their application in high-power IR laser delivery systems. The development of IR and THz waveguide technologies is at vastly different stages of maturity. THz radiation covers the spectral range from 100 to 3,000 mm (3–0.1 THz). Interest in THz radiation has grown considerably with the emergence of stronger transmitters and more sensitive receivers. The unique interaction of THz radiation with matter, in particular gases and materials that contain water, has led researchers to investigate applications in imaging and spectroscopy. Molecules with similar chemical bonds can be difficult to distinguish using IR spectroscopy. The rotational transitions and associated THz absorption spectrum of molecules can be more readily distinguished. THz time domain spectroscopy (TDS) allows the rapid collection of an analyst's spectral response over the range of 0.1–3 THz. THz TDS spectroscopy is therefore a powerful tool in the chemical analysis of gases and liquids.

It may be possible to achieve added functionality in systems utilizing THz radiation by incorporating low-loss optical waveguides. For instance, a flexible waveguide could be coiled to create a small footprint, long path length cell for gas sensing using TDS. Many approaches have been taken to achieve low-loss THz waveguides, such as PCFs, hollow polymer fibers, metal tubes, Cu-coated hollow polycarbonate waveguides (HPWs), metal/dielectric HCW, and metal wires (Bowden 2007).

10.2 Hollow Waveguide Design and Materials Selection

The advantages of HCWs over solid waveguides have led their uses in applications requiring the transmission of radiation at wavelengths where it is difficult to find materials that have the optical, thermal, and mechanical properties required for use in solid-core fibers. HCWs have also been used to transmit the infrared, ultraviolet, and soft X-ray radiation. The design target for a hollow waveguide is to fabricate a low-loss, hollow, flexible waveguide for the propagation of the required wavelength, such as terahertz radiation. Generally electromagnetic radiation is guided by dielectric or metal waveguides. Metals are highly reflective but not flexible. Dielectrics are flexible but not lossless in terahertz regime. For instance, depositing a layer of highly reflective metal inside a flexible tube will make an efficient waveguide for the transmission of terahertz radiation. A dielectric layer is then added over the metal to provide enhanced reflectivity within specific wavelength ranges due to an interference effect. These waveguides are so-called metal/dielectric HCWs (Doradla et al. 2012).

10.2.1 Design Principle

For example, hollow cylindrical leaky-type waveguides propagate light by reflection just like a normal mirror. To produce a waveguide that is able to efficiently transmit at one or more specific wavelengths, reflection at that wavelength is maximized by depositing one or more dielectric layers with the correct optical thickness for enhanced reflection. These layers create interfaces from which multiple reflections occur. The phase shift of the waves is 180° when going from lower index of refraction to a higher refractive index layer. The calculated thickness of the topmost layer (adjacent to the air core) also takes into account the phase shift. Thus, these waves constructively interfere to produce greater reflection, at regularly spaced wavelengths, than is possible from the silver surface alone. Thus, a thin film interference pattern is created. The parameters most commonly used to characterize the performance of hollow waveguides are straight transmission loss and loss on bending. The total loss, for either single or multilayer waveguides made with any metallic and dielectric materials, has three contributions: straight loss, α_∞; bending loss, α_{bend}; and surface roughness, α_{surface}. The loss on bending is an additional contribution that is added to the loss measured when the waveguide is straight. The straight loss is directly dependent on the square of the wavelength of operation and the mode(s) being propagated. It is inversely dependent on the cube of the bore radius. Surface roughness and film thickness variation are processing-related parameters which contribute to the measured loss. The dependence of the total loss on bending, core size, wavelength of operation, optical characteristics of the reflective metal layer, the individual dielectric layers, surface roughness, and index contrast between the dielectrics is given by (Johnson 2007)

$$\alpha_{\text{total}} = \alpha_\infty + \alpha_{\text{bend}} \tag{10.2}$$

where

$$\alpha_\infty = \left(\frac{U_m}{2\pi}\right)^2 \left(\frac{\lambda^2}{a^3}\right) \left(\frac{n_{\text{Ag}}}{n_{\text{Ag}}^2 + \kappa_{\text{Ag}}^2}\right) * F_{\text{film}} \tag{10.2a}$$

$$\alpha_{\text{bend}} \propto \frac{a^3}{R} \tag{10.2b}$$

$$\alpha_{\text{bend}} \propto \left(\frac{\sigma}{\lambda}\right)^2 \tag{10.2c}$$

and U_{nm} is a parameter that is determined by the mode being propagated, λ is the wavelength of interest [μm], a is the inner radius [μm], n_{Ag} is the real part of the index of refraction for silver layer at λ, κ_{Ag} is the extinction coefficient for silver layer at λ, and F_{film} is the contribution to the loss from the dielectric layer, which are usually made using only the real part of the index of refraction and do not include the absorption. F_{film} term is generally dependent on propagating mode, dielectric film material, and the number of dielectric thin films. HE11 mode is lowest loss mode in dielectric-coated hollow glass waveguides (HGWs).

For a single dielectric layer, F_{film} may be expressed as (Johnson 2007)

$$F_{\text{film}} = \frac{1}{2}\left(1 + \frac{n_{\text{low}}^2}{\left(n_{\text{low}}^2 - 1\right)^{1/2}}\right)^2 \tag{10.3}$$

and for a multilayer dielectric stack, F_{film} may be estimated by (Johnson 2007)

$$F_{\text{film}} = \alpha_{\text{bend}}\left[1 + \frac{n_{\text{low}}^2}{\left(n_{\text{low}}^2 - 1\right)^{1/2}}\left(\frac{n_{\text{low}}}{n_{\text{high}}}\right)^{2m_p} C^{-m_p}\right] \tag{10.4}$$

where n_{low} is the index of refraction of the low-index material and n_{high} is the index of refraction of the high-index material. $C = (n_{\text{low}}^2 - 1)/(n_{\text{high}}^2 - 1)$, where m_p is the number of pairs of low- and high-index dielectric layers. For the bending loss, α_{bend}, the approximation is appropriate when the bending radius, R, is much greater than the inner core radius. The loss due to surface roughness of the film denoted by α_{surface}, will strongly increase with surface roughness, σ. The roughness from all layers contributes to the final roughness.

Dielectric materials are chosen based on optical properties and ease of deposition forming process. Pairs of dielectrics are chosen with high index contrast. In addition to the dependence of loss on the size of the core, the shape of the core and the symmetry of the mode of propagation directly determine the loss in hollow waveguides. The explicit loss dependence on the bore size and cross-sectional geometry (circular or rectangular) is due to the fact that the core is assumed to be essentially lossless as the loss is due mainly to the thin films. Specifically, it is the materials absorption, the deviation from the optimized thickness, and surface roughness of each layer that increase the loss. The spatial distribution of the electric field energy within the circular core varies for the different modes of radiation. The ideal distribution would have a zero energy density at the waveguide wall so that there would be a minimal amount absorption by the wall material. The HE11 mode has a distribution in which most of the energy is concentrated in the center of the core with less at the waveguide wall. Among the energy distributions for the first five modes, HE11 is the lowest loss mode. Surface roughness increases with increased deposition time (thicker layers) and with an increasing number of layers. Thus, for multilayer coatings, surface roughness becomes a more significant consideration because each additional layer follows the existing surface topography and adds its own roughness. Greater surface roughness of the layer in contact with the hollow core also means increased scattering which can be significant at shorter wavelengths. Nonuniform film thickness also increases the loss. The appropriate dielectric layer thickness for a given wavelength, λ_{opt}, in a single dielectric layer waveguide is calculated using the equation (Rabii et al. 1999):

$$d = \frac{\lambda_{opt}}{2\pi\sqrt{\left(n_d^2 - 1\right)}} \tan^{-1}\left[\frac{n_d}{\left(n_d^2 - 1\right)^{1/4}}\right] \tag{10.5}$$

where d is the physical thickness of the layer being considered [μm], λ_{opt} is the wavelength [μm] for which the waveguide design is optimized, and n_d is the real part of the index of refraction of the dielectric layer being considered at λ_{opt}.

10.2.2 Materials Selection and Structure Design

Metal/dielectric HGW, for instance, consists of a glass capillary tube with an inner metallic coating and thin dielectric film. Metals that are commonly employed in HGWs include silver, copper, gold, and aluminum. The dielectric layer is formed either by a subtractive process in which the metal is reacted with other chemicals to form a dielectric, as is the case with Ag/AgI and Cu/CuI HGWs, or the dielectric is added directly over the metal surface as in Ag/CdS and Ag/COP HGWs. Metal/Dielectric HGWs are most commonly used for the delivery of infrared laser radiation. For application to the transmission of THz radiation, it is necessary to optimize the waveguide structure, such as the layer thickness of the dielectric needs only be scaled proportionally to achieve low loss at the desired wavelength so long as the metal is thicker than its skin depth for the radiation that is to be transmitted in the waveguide. Since wavelengths in the THz spectrum are one to two orders of magnitude longer than those in the infrared, a corresponding increase in film thicknesses is necessary. Stresses induced during deposition prevent films of such large thickness from being deposited by the subtractive iodization process or the additive process used to create metal sulfide films. The dynamic liquid-phase deposition process used to create polymer films may be the best candidate for scaling to the thicknesses required for THz radiation (Johnson 2007).

1. Selection of Base Material
   In general, silica glass tubing is the base material for terahertz waveguides. Oxide glasses, such as fused silica, are the basis for long-haul telecommunications fibers. HMFGs like ZBLAN received much attention in the 1980s due to their low theoretical loss in the near-infrared, but interest has waned due

to difficulties in fabricating fibers that are free of inclusions. Chalcogenide glasses, which include arsenic trisulfide, have been used in infrared transmitting fibers. Chalcohalide glasses, for instance, TeX, are also of interest for infrared transmission and have applications in rare earth-doped fiber amplifiers. Optical fibers made from thermoplastic optical polymers like PMMA and CYTOP are being implemented in low bandwidth telecommunications applications as a low-cost alternative to fused silica fiber. The only materials that are drawable and have low transmission losses for CO_2 laser radiation are chalcogenide and chalcohalide glasses. The similarities of chalcogenide and chalcohalide glasses does allow them to be co-drawn, so it may be possible to fabricate an HBF with alternating layers of each glass type. The range of refractive indices from which to choose the n_L and n_H materials for an HBF operating at 10.6 mm may thus be taken as 2.0–3.4. Infrared-transmitting crystalline chalcogenides, such as PbSe, $n = 4.74$ at 10.6 mm, can have much higher refractive indices than those demonstrated in glass-forming chalcogenides (Johnson 2007).

Since glass is not flexible at large bore diameters, polycarbonate (PC) is usually chosen to be the base material for flexible terahertz waveguides, due to its flexibility even at large bore diameters (up to 6.3 mm). PC has an advantage over polyethylene and Teflon as the extrusion process for glass and polycarbonate capillaries creates a smooth inner surface with surface roughness (order of nm) (Doradla et al. 2012).

2. Selection of Metal

The complex refractive index of metal is defined as $N = n - ik$, where n is the refractive index and k is the extinction coefficient. In the THz regime, metals have high reflectivity because of their high refractive index and extinction coefficients. Commonly used metals for high reflection coatings are silver, gold, aluminum, and copper. The figure of merit, $F = n/(n^2 + k^2)$, is the quantity used to characterize the performance of the waveguide, which is a measure of metal's reflectivity. For metals, the figure of merit is inversely proportional to its reflectivity. For example, minimum loss at a wavelength of 215 μm can be achieved by choosing either silver (Ag) with $F = 0.684 \times 10^{-3}$ or gold (Au) with $F = 0.878 \times 10^{-3}$ due to their lower figure of merit values (Doradla 2012). The plasma frequency ω_p and the damping frequency ω_τ are related to the metal's optical constants n and k by the equations (Johnson 2007):

$$\omega_\tau = \frac{2\omega nk}{1 - n^2 + k^2} \tag{10.6}$$

and

$$\omega_p^2 = \left(1 - n^2 + k^2\right)\left(\omega^2 + \omega_\tau^2\right) \tag{10.7}$$

3. Selection of Dielectric

Hollow-core metal-coated waveguides utilize a highly reflective wall to confine light to the air core. By increasing the wall's reflectivity, the degree to which light is confined to the waveguide's air core increases, thus decreasing the attenuation of radiation propagating along its axis. Since the wall is not a perfect reflector, leaky modes tend to propagate in these hollow silver- or gold-coated waveguides. A dielectric layer is then added over the metal to improve reflectivity within specific wavelength ranges due to the interference effect.

4. Optimized HCW Structures
As the refractive index of each layer is complex, the reflectivity design of a hollow waveguide must consider two effects that reduce the reflectivity coefficients: transmission of energy though the system, T, and absorption of energy by the system, A. A simple definition of the reflectivity of the system is thus (Johnson 2007)

$$R = 1 - T - A \tag{10.8}$$

Metallic reflectors have high R due to the high refractive index and extinction coefficients of the metals commonly used as reflectors such as, Ag, Au, Cu, and Al. In the case of a metallic reflector, $T = 0$ as long as the metal is thicker than its skin depth. A is nonzero because any radiation that is not reflected is absorbed by the metal. Greater R values can be achieved using multilayer dielectric reflectors. The layer thicknesses of a multilayer reflector are designed to achieve very low T for specific wavelengths. A is also low when dielectrics with low extinction coefficients are chosen. Hybrid metallic/dielectric mirrors have one or more dielectric layers deposited over a metallic substrate. High R can be obtained using a hybrid reflector design with fewer layers than an all-dielectric design. Typical examples include the single dielectric thin film on a metal substrate hybrid reflector and the all-dielectric multilayer Bragg reflector.

(a) All-Dielectric HBFs
HBFs are based upon the design of the multilayer Bragg reflector which is shown in Fig. 10.2a. The Bragg reflector design consists of alternating high refractive index, n_{H}, and low refractive index, n_{L}, layers. The Bragg reflector is also characterized by the number of layer pairs, N, of in the dielectric stack. Structures with n_{L} layers nearest the incident medium and substrate, thus having a total of $2 \times N + 1$ layers, have higher reflectance at near grazing incidence (a characteristic of the low-loss modes) than those with an even number of layers or those with n_{H} layers at the incident medium and substrate interfaces. When the arbitrary film layer j is odd layers, which has the refractive index n_{L} and the j is even layers have the refractive index n_{H}. Under the assumption that the layers are nonabsorbing ($k = 0$), the optimum thickness of each layer, with the exception the layer nearest the incident medium ($j = 0$), is given by the quarter wave condition at glancing incidence (Miyagi and Kawakami 1984):

$$d_{\mathrm{H}} = \frac{\lambda_0}{4\sqrt{(n_{\mathrm{H}}^2 - 1)}} \tag{10.9}$$

where d_{H} is the thickness of the n_{H} layers ($j = 2 \ldots J - 1$) and

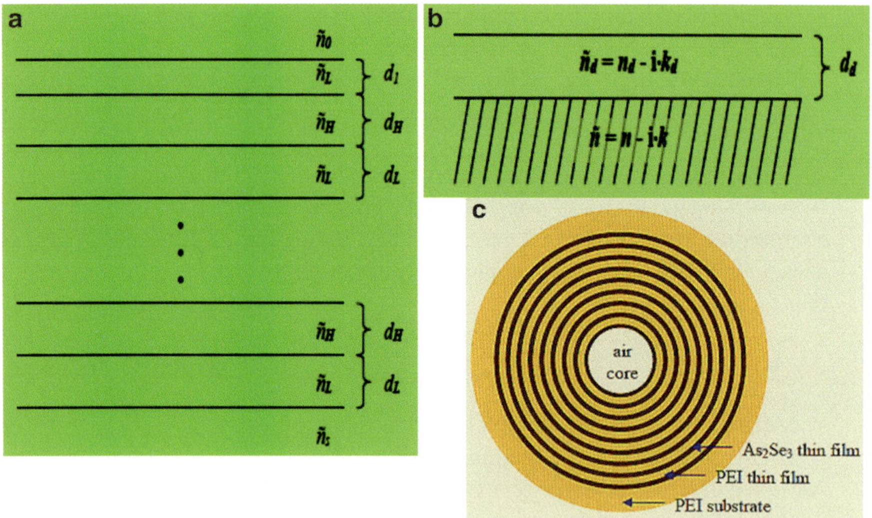

Fig. 10.2 Examples of some hollow waveguide design (Bowden 2007): (**a**) a schematic diagram of the multilayer alternating nL/nH all-dielectric Bragg reflector design, (**b**) single dielectric layer on a metal substrate, and (**c**) OmniGuide HBF cross section

$$d_{\mathrm{L}} = \frac{\lambda_0}{4\sqrt{(n_{\mathrm{L}}^2 - 1)}} \tag{10.10}$$

where d_{L} is the thickness of the $j \neq 1$, n_{L} layers ($j = 3 \ldots J$). The thickness of the $j = 1$ layer is given by (Miyagi and Kawakami 1984)

$$d_1 = \frac{\lambda_{\mathrm{opt}}}{2\pi\sqrt{(n_{\mathrm{L}}^2 - 1)}} \tan^{-1}\left[\frac{n_{\mathrm{L}}}{(n_{\mathrm{L}}^2 - 1)^{1/4}} \left(\frac{n_{\mathrm{L}}}{n_{\mathrm{H}}}\right)^N \left(\frac{n_{\mathrm{L}}^2 - 1}{n_{\mathrm{H}}^2 - 1}\right)^{-N/2} \left(\frac{n_s}{n_s^2 - 1}\right)^{-1/2}\right] \tag{10.11}$$

It is important to consider the magnitude of the reflectance and its spectral variation when designing coatings for a HCW for practical use. If the waveguide is to be used for laser power delivery, it is critical to design coatings to achieve maximum reflectivity, providing the lowest loss, at the laser's operating wavelength. The small spectral width of laser sources dictates that the HCW needs only to have low loss over a narrow spectral region centered on the operating wavelength. The HBF suits this criterion well. A wall design that provides low loss over a broad range of wavelengths is favorable for waveguides that are intended for spectroscopic applications. Metal/dielectric HGWs and HPWs have the flexibility to provide low loss at a single design wavelength and relatively low loss over spectral range that extends from the design wavelength to longer wavelengths (Johnson 2007).

(b) Metal/Dielectric HGWs

The design of the single dielectric layer on metal reflector, as shown in Fig. 10.2b, is very simple and is the basis for the design of metal/dielectric HGWs. The thickness of the metal, d_{metal}, must be thicker than its skin depth to prevent radiation leakage through the metal. For example, the skin depth of silver in IR is between 10 and 30 nm. A dielectric added over the metal greatly enhances its reflectivity. The optimum thickness, d_d, at glancing incidence of a nonabsorbing ($k = 0$) dielectric film on a metal substrate is given by (Miyagi and Kawakami 1984)

$$d_d = \frac{\lambda_0}{2\pi\sqrt{(n_d^2 - 1)}} \tan^{-1}\left[\frac{n_d}{(n_d^2 - 1)^{1/4}}\right] \tag{10.12}$$

The refractive index and extinction coefficient of the dielectrics layer(s) of both structures are critical to the fabrication of low-loss waveguides. Minimum loss is achieved for HBFs when the refractive index contrast is maximized and the extinction coefficients of the layers are minimized. Incorporating additional layer pairs results in an exponential decrease in loss to the extent that loss is limited by the extinction coefficients of the layers. For metal/dielectric HGWs, silver provides the lowest theoretical losses at 119 mm. The optimum values of the dielectric's optical constants are interdependent. The lowest theoretical losses are possible when the extinction coefficient is low and the refractive index of the dielectric layer is between 1.3 and 2.0. For instance, the dielectric polystyrene meets this criterion. In theory Ag/PS HGW with a 12.7-mm-thick PS has an HE11 mode loss of 0.25 dB/m, representing over an order of magnitude improvement with respect to an Ag-only HGW. Chalcogen HBFs have smaller low-loss spectral bands and are therefore more sensitive to the variation in layer thickness than HGWs (Johnson 2007).

10.3 OmniGuide Hollow Bragg Fibers

OmniGuide HBFs are based on radial periodic structure of high- and low-index concentric rings, which is provided by high refractive index chalcogenide glasses and low refractive index polymer. They have only one central air hole, where the majority of optical power is carried axially by the radial photonic Bragg condition. Similarly hollow IR-transmitting hollow fibers carry the optical fiber through the central air hole by reflective metal or metal-insulator coating on the inner surface of the holey fiber. Even though practical applications of these two types of fibers, OmniGuide fibers and IR-transmitting hollow fibers, have been limited to high-power IR laser delivery in a relatively short distance, these structures could provide high potential for ultimately low-loss transmission for long-haul optical communications (Oh et al. 2005).

OmniGuide fiber is the first practical demonstration of an HBF (Temelkuran et al. 2002), which consists of a hollow core surrounded by alternating layers of arsenic triselenide (As_2Se_3) glass, $\tilde{n} = 2.74 - i1.3 \times 10^{-6}$, and polymer polyetherimide (PEI), $\tilde{n} = 1.69 - i0.046$, supported by a thick outer PEI cladding as shown in Fig. 10.2c. The OmniGuide HBF derives its name from the fact that over a range of wavelengths the bandgap of its Bragg reflector wall is omnidirectional, meaning that it extends over all angles of incidence and polarizations (Fink et al. 1998).

The fabrication of the fiber begins with the thermal evaporation of As_2Se_3 glass on either side of a PEI film. The coated film is then rolled into a spiral and covered with a thick PEI cladding to form a drawable preform that closely approximates the ideal HBF structure. An interesting feature of the OmniGuide fabrication process is its wavelength scalability. It is obvious that the ratio d_H/d_L does not change with λ_0 based on (10.9) and (10.10). During the draw process, the OmniGuide HBF preform is heated and pulled reducing the inner and outer diameters and layer thicknesses in proportion. Since the ratio of the As_2Se_3 layer thickness to the PEI layer thickness does not change during the draw process, the OmniGuide fiber's low-loss wavelength can be adjusted simply by changing the diameter of the final hollow waveguide. The draw speed can be altered during the draw process allowing fibers that operate at many different wavelengths to be drawn from a single preform (Bowden 2007).

The ray optics has been used to calculate the theoretical loss of the HE11 mode at 10.6 mm as a function of N for the OmniGuide fiber. A minimum theoretical loss of 0.18 dB/m is achieved for $N > 10$. In practice, OmniGuide HBFs with $a = 350$ mm tuned for transmission at 10.6 mm have losses as low as 0.5 dB/m (Anastassiou et al. 2004). The loss of the real fiber is expected to be higher than the calculated HE11 mode loss due to the presence of some higher-order modes and structural imperfections from the fabrication process. The low loss of the OmniGuide HBF at 10.6 mm is remarkable in light of the fact that the extinction coefficient of PEI at 10.6 mm corresponds to materials absorption of over 200,000 dB/m. To put this in perspective, CO_2 laser radiation focused on the outside of an OmniGuide HBF can ignite its PEI cladding. The low loss of the OmniGuide HBF, despite its high materials absorption, is possible because of its high refractive index contrast ($n_H/n_L = 2.74/1.69 = 1.62$) (Bowden 2007).

The significance of omnidirectionality is its potential to reduce bend losses in HCWs. The coupling of energy to higher-order (hence higher loss) modes is responsible for increased loss on bending and is fundamental to all HCWs. There are two contributions to bend loss: (a) a higher mode parameter fundamentally indicates higher loss, and (b) the value of average reflection loss R_{avg} decreases as the principal angle of incidence θ_z decreases for most HCW reflective wall designs. Bend losses are highest in metal-coated HGWs because the reflectivity of the metal decreases dramatically as θ_z approaches the metals principal angle of incidence. For instance, a dielectric AgI layer added over an Ag coated HGW increases the wall's reflectivity at 10.6 mm for incidence angles near silver's principal angle of incidence and thus reduces the HGWs bend loss. If materials absorption is ignored, R_{avg} of the OmniGuide HBF is near unity for all angles of incidence which would provide reduced bend losses with respect to Ag- and Ag/AgI-coated HGWs. The omnidirectionality of the OmniGuide's wall can be diminished due to materials absorption. The bend loss of the OmniGuide at 10.6 mm has been found to increase linearly with the inverse of the bend radius at a rate of 0.015 dB m for a 2.5-m-long 700-mm-diameter fiber subjected to 90° bends initiated 15 cm from the output. The fabrication techniques for the OmniGuide HBF have been refined to the point that fibers with near theoretical losses can be produced, and they are gaining acceptance in the medical community for the delivery of CO_2 laser radiation for surgical applications. The OmniGuide HBF may not be able to meet the low-loss requirements of high-power CO_2 laser delivery applications like laser cutting and welding due to limitations imposed by its materials absorption and the low-power threshold of the polymer layers. Application of the OmniGuide manufacturing techniques to materials with high index contrast and low absorption is a logical step towards the demonstration of even lower loss HBFs (Bowden 2007; Doradla et al. 2012).

10.4 Metal/Dielectric-Coated Hollow Waveguides

Metal/dielectric hollow waveguides have been fabricated using a variety of techniques, such as physical vapor deposition (PVD) of silver and dielectric layers on metallic substrates; sputtering of metallic, dielectric, and semiconductor films on a leachable mandrel followed by electroplating; and liquid-phase formation of coatings inside plastic tubing and glass tubing. Most often the cross section of the waveguides is circular but early rectangular waveguides continues to be of interest. The advantage of the circular cross section is the ease of bending and the small overall size compared to rectangular or square cross section guides. The earliest circular cross section hollow waveguides were formed using metallic and plastic tubing as the structural members. Tubing made from stainless steel or nickel was used in many of the early waveguides.

Some metal/dielectric-coated hollow waveguides have been developed and their typical structures are shown in Fig. 10.3 (Harrington 2000). Figure 10.3a illustrates fabricated nickel tubes with inner Ge, ZnS, or PbF_2 dielectric coatings by sputtering the dielectric layer onto an aluminum substrate, electroplating a thick nickel layer over the dielectric, and then dissolving the aluminum tube. This process specifically includes three steps (Harrington 2000): (a) a pipe made typically of aluminum is placed in a sputtering chamber and a dielectric layer followed by a metallic film is deposited on the pipe, (b) the coated pipe is put into an electroplating tank where a thick nickel layer is deposited on top of the

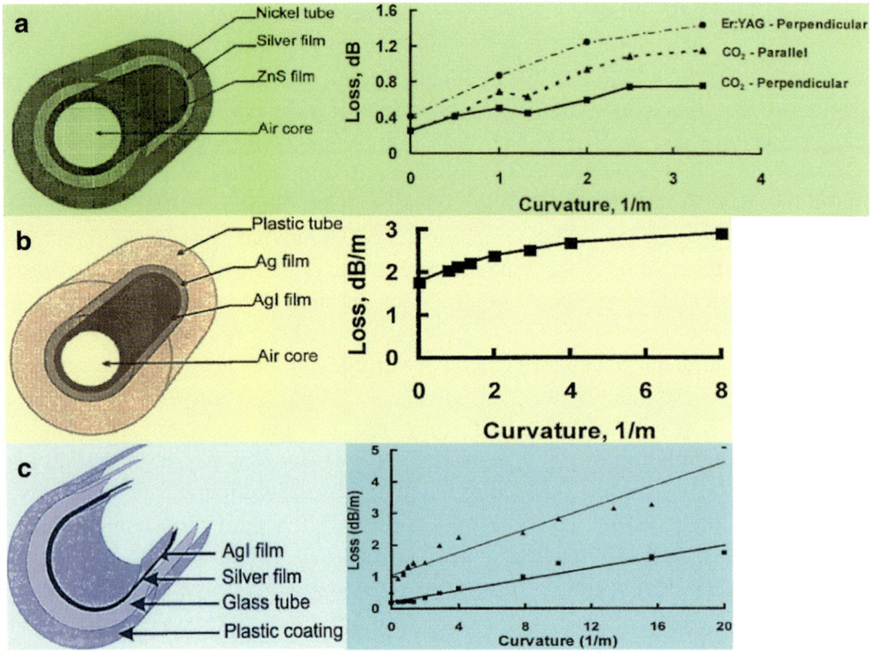

Fig. 10.3 Typical structures of hollow waveguides (Harrington 2000): (**a**) metal/dielectric; (**b**) plastic/dielectric; and (**c**) glass/dielectric hollow waveguides

sputtered layers, and (c) the pipe is etched away leaving the final structure. In Fig. 10.3b, dielectric coatings (AgI or ZnS) over silver are deposited on the inner surface of plastic tube, such as Teflon or polyethylene. This leads to a very flexible structure that is inexpensive to fabricate and durable enough that a reasonable laser power (safe limit <25 W) can be transmitted through the waveguides (Harrington 2000). Ag or PbF_2 layers can also deposited on a thin metal strip, which is then rolled into a tube and inserted into a stainless steel sleeve. Instead of depositing a metallic layer inside a hollow mandrel, a silver tube is extruded and then deposited with a silver halide film on the inside of the glass tube as shown in Fig. 10.3c. Moreover, a more direct technique in which the inner surface of a silver tube is partially converted to dielectric AgI, AgBr, or AgCl using a gas- or liquid-phase reaction. The surface of the silver tube was chemically polished prior to the addition of the dielectric film. A smoother silver surface, and therefore improved performance, can be achieved by depositing a silver coating on the inside of a smooth glass or plastic tube, such as polyethylene and Teflon tubing using an electroless liquid-phase chemical deposition (ELPCD) process. Since the films are added over the silver layer, multiple layers, such as CdS and PbS, can be deposited to form a periodic alternating high/low refractive index structure. In theory, a multilayer dielectric waveguide should have significantly lower loss than a waveguide consisting of only a single layer of either dielectric material. In practice, the potential loss reduction by depositing multilayer dielectric films is compromised by the increase in roughness as the overall thickness of the dielectric stack increases (Bowden 2007).

A significant challenge in the application of metal/dielectric HCW technology to THz waveguides is that much greater dielectric layer thicknesses are required for low loss. The optimum dielectric layer thickness for metal/dielectric HCWs is proportional to the design wavelength. The wavelength of THz frequency radiation is 10–100 times longer than IR radiation; hence the dielectric film thickness of a metal/dielectric HCW designed to transmit THz frequency radiation must be 10–100 times greater than those designed for IR radiation (Harrington 2000). The increase in roughness upon depositing 10–100 times thicker films may therefore be prohibitive in the application of Ag/AgI or Ag/metal sulfide technology to HCWs designed for THz frequency radiation. However the increase in scattering resulting from rougher films will be at least partially offset by the decrease in scattering losses at these longer wavelengths. In practice it is difficult to form thick dielectric coatings limiting the choices for fabricating a viable THz HCW. The refractive indices of AgI, CdS, and PbS at IR wavelengths are 2.2, 2.3, and 4, respectively. Assuming the refractive indices of these materials are similar at THz frequencies, the dielectric film thicknesses for single-layer metal/dielectric HCWs optimized for 119 mm are 9.7, 9.3, and 5.4 mm for AgI, CdS, and PbS, respectively. The growth rate of the silver films deposited by the ELPCD process is about 1 mm/h, so it will take at least 10 h to fabricate a single Ag/AgI HGW optimized for 119-mm radiation. The growth rates of CdS and PbS films are 0.14 and 0.4 mm/h, respectively. To fabricate a single-layer HGW, the deposition times for CdS and PbS are therefore about 66 and 14 h, respectively. Application of the fabrication processes used to make Ag/AgI and Ag/metal sulfide IR-transmitting HCWs to THz radiation would result in increased roughness and long deposition times and would also increase the hazardous waste generated. Therefore, metal/polymer-coated HGWs have been developed as an alternative to HCWs with inorganic dielectric thin films. Polymers have refractive indices that are close to 1.4, which provides the lowest theoretical loss for metal/dielectric HCWs assuming that the polymer has no absorption. Polymer films are deposited by the liquid flow coating (LFC) process. The LFC process is a promising alternative to the dynamic liquid-phase chemical deposition process for fabricating HCWs optimized for the transmission of THz radiation (Bowden 2007).

In addition to decreased roughness, on another hand, the LFC process addresses the other major issues associated with extending the ELPCD process to HCWs designed for THz radiation. The volume of solution required to deposit a polymer film using the LFC process is less than 10 mL so very little waste is generated. Also, the deposition time is independent of the film thickness because the value of solution concentration can be varied to achieve any film thickness at a given coating rate. Assuming a typical coating rate of 4 cm/min, it takes only 25 min to deposit a polymer film inside a 1 m long HCW. Clearly the LFC process has significant advantages over the ELPCD process for the fabrication of HCWs designed to transmit THz radiation (Bowden 2007).

Moreover, metal/polymer HGW, such as the polyimide (PI)/silver HCW, has been fabricated with the combination of ELPCD and LFC processes (Iwai et al. 2006): A silver layer is formed inside a fused silica tube by the ELPCD process. The PI film is then added over the silver by a flow of 3.9 wt. % solution through the waveguide at a rate of 12.5 cm/min. A thin liquid film remains on the surface of the silver coating after the liquid has passed through the tube, a phenomenon called the Jamin effect. The waveguide is heated to evaporate the solvent leaving a thin uniform film of PI. The waveguide thus fabricated has a loss of about 1 dB/m at the Er:YAG laser wavelength of 2.94 mm. Cyclic olefin polymer (COP) has also used because it has low absorption at the Er:YAG and CO_2 laser wavelengths.

10.5 Hollow Glass Waveguides

The HGW is one of the most popular hollow waveguides. This hollow glass structure has the advantage over other hollow structures because it is simple in design and extremely flexible and, most importantly, has a very smooth inner surface. As discussed earlier, one of HGWs has a metallic layer of Ag on the inside of silica glass tubing, and then a dielectric layer of AgI or a polymer-like cyclic olefin polymer is deposited over the metal film identical to that used to make the hollow plastic guides. Figure 10.4 illustrates typical structure of the HGWs showing the metallic and dielectric films deposited inside silica glass tubing (Bowden 2007; Harington 2009). The fabrication of Ag/AgI HGWs, for instance, generally begins with silica tubing which has a polymer (UV acrylate or polyimide) coating on the outside surface. A wet-chemistry technique is employed to first deposit a silver film using standard Ag plating technology. Next, a very uniform dielectric layer of AgI is formed through an iodization process in which some of the Ag is converted to AgI. Using these methods, HGWs with bore sizes ranging from 250 to 1,000 mm and lengths as long as 13 m have been made. The spectral response for HGWs depends critically on the thickness of the dielectric film. Generally, for the AgI films, the film thickness ranges from 0.2 to 0.8 mm (Harrington 2000).

In general, the spectrum losses are less than 0.5 dB/m at 10 mm for bore sizes larger than ~400 mm. For broadband applications and shorter wavelength applications, a thinner AgI coating would be used to shift the interference peaks to shorter wavelengths. For such HGWs the optical response will be nearly flat without interference bands in the far IR fiber region of the spectrum. The bending loss depends on many factors such as the quality of the films, the bore size, and the uniformity of the silica tubing. While there is an additional loss on bending for any hollow guide, it does not necessarily mean that this restricts their use in power delivery or sensor applications. Normally most fiber delivery systems have rather large bend radii and, therefore, a minimal amount of the guide is under tight bending conditions and the bending loss is low. An additional important feature of hollow waveguides is that they are nearly single mode. This is a result of the

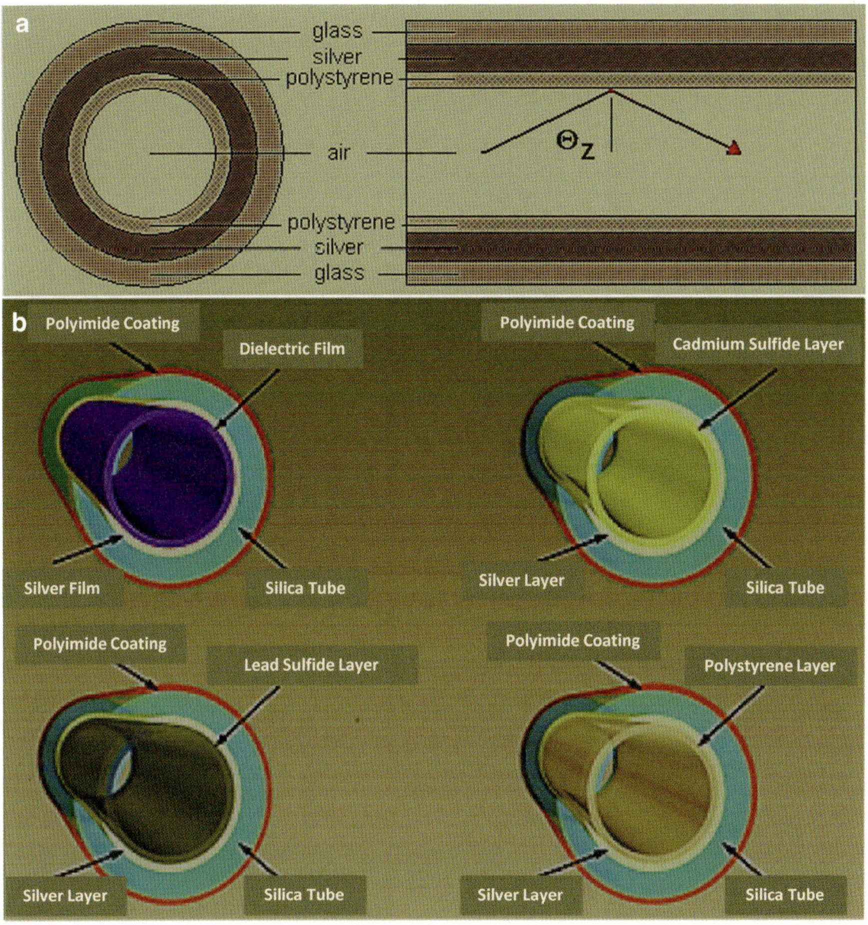

Fig. 10.4 Structure of the HGWs showing the metallic and dielectric films deposited inside silica glass tubing (Bowden 2007; Harington 2009)

strong dependence of loss on the fiber mode parameter. That is, the loss of high-order modes increases as the square of the mode parameter, so even though the guides are very multimode, in practice only the lowest-order modes propagate. This is particularly true for the small bore (<300 mm) guides in which virtually only the lowest-order HE11 mode is propagated (Harington 2009).

HGWs have been used quite successfully in IR laser power delivery and in some sensor applications. Modest CO_2 and Er:YAG laser powers below about 80 W can be delivered without difficulty. At higher powers, water-cooling jackets have been placed around the guides to prevent laser damage. The highest CO_2 laser power delivered through a water-cooled, hollow metallic waveguide with a bore of 1,800 mm was 2,700 W and the highest power through a water-cooled, 700-mm bore HGW was 1,040 W. Sensor applications include gas and temperature measurements. A coiled HGW filled with gas can be used in place of a more complex and costly white cell to provide an effective means for gas analysis. Unlike evanescent wave spectroscopy in which light is coupled out of a solid-core-only fiber into media in contact with the core, all of the light is passing through the gas in the hollow guide cell making this a sensitive, quick-response fiber sensor. Temperature measurements may be aided by using a HGW to transmit blackbody radiation from a remote site to an IR detector. Such an arrangement has been used to measure jet engine temperatures (Harington 2009).

10.6 Chalcogenide Glass Hollow Bragg Fibers

The technology used to fabricate the OmniGuide HBFs, as shown in Fig. 10.5 (Jacobs et al. 2007), has produced fibers with near-theoretical losses that are much lower than the loss of the materials from which the fiber is made. The high refractive index contrast of the OmniGuide HBF allows it to transmit CO_2 laser radiation with low loss, because the fields are largely confined to fiber's air core. However, the OmniGuide HBF's performance as a waveguide for CO_2 laser radiation is limited by the high extinction coefficient of PEI. To achieve a fiber with ultralow losses for CO_2 laser radiation, the fabrication techniques used to make OmniGuide HBFs must be applied to a pair of materials with similar refractive

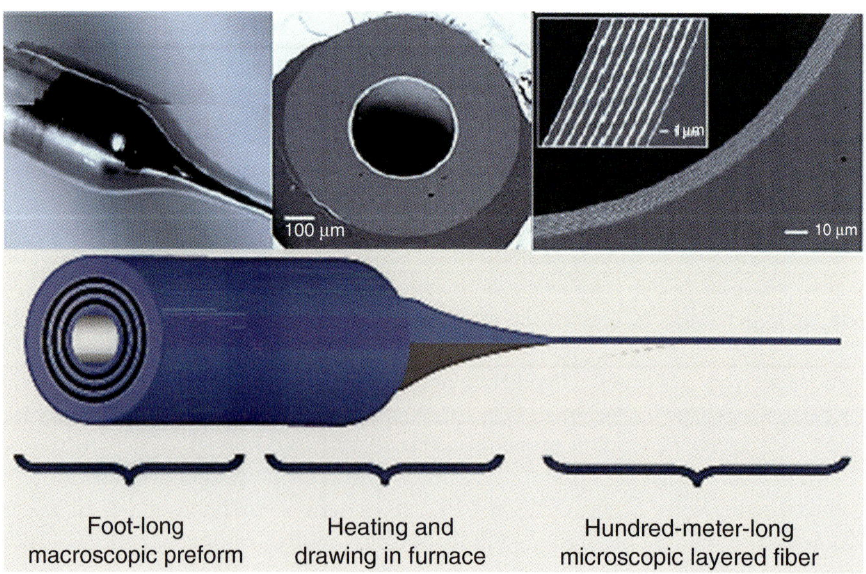

Fig. 10.5 Schematic illustration of the manufacturing process of OmniGuide fibers: the magnified detail of the middle panel of the OmniGuide fiber clearly shows the one-dimensional photonic crystal (Jacobs et al. 2007)

contrast but much lower extinction coefficients at 10.6 mm. Chalcogenide and chalcohalide glasses are the only glassy materials that have the requisite thermal and optical properties for this application (Bowden 2007).

Chalcogenide glasses have non-oxide compositions that contain at least one of the elements S, Se, and Te. These chalcogen elements (group 16) are commonly alloyed with elements from groups 14 and 15, such as Ge and As, to form thermally stable glasses. The chalcogen elements are generally two coordinated and have the ability to form homopolar bonds creating chain-like structures linking the higher coordinated atoms. This bonding behavior is responsible for the wide range of compositions that form chalcogenide glasses. In contrast, oxide glasses have stricter stoichiometry. Chalcohalide glasses are formed by adding halogen (group 17) elements to chalcogenide glass compositions. The halides act as chain-ending units, decreasing the connectivity of glass, and hence have the effect of reducing the glass transition temperature. Chalcogenide glasses are formed by melting together the pure elemental constituents in an inert atmosphere or more commonly in an evacuated ampoule. These special atmospheric conditions are needed to prevent contamination of the glass with oxide impurities, which have strong absorption bands in the infrared. The elimination of oxide impurities is critical when fabricating chalcogenide glasses for use in the transmission of infrared radiation. The raw materials for chalcogenide glass fabrication can be subjected to special purification measures like distillation or surface oxide removal to achieve low levels of oxide impurities. Chalcogenide glasses have unique optical, thermal, and electronic properties that have led to their use in numerous applications. Chalcogenide glasses have good IR transparency so they have been applied in transmission windows, optical fibers, and chemical sensors operating in the 2–12-mm spectral range. The photo-induced amorphous to crystalline-phase transition behavior of some chalcogenide glasses has led to applications in optical memory devices. Rare earth-doped chalcogenide fibers are of interest for infrared fiber amplifiers and lasers (Bowden 2007).

It is challenging to identify glass compositions that have a high index contrast and transparency at a common wavelength because their transmission windows, the range of wavelengths between their electronic band edge and the multiphonon edge, tend to correlate strongly with refractive index. For instance, oxide glasses have low refractive indices and transmit in the visible to near-infrared, while telluride glasses have very high refractive indices and transmit only in the mid to far infrared. Among chalcogenide glass-forming systems, only those based on Se and Te have low absorption at 10.6 mm. The refractive index of selenide glasses is generally greater than 2.6, while tellurides have refractive indices greater than 2.8. Chalcohalide glasses have significantly lower refractive indices yet still maintain good IR transparency. From an optical standpoint chalcohalide glass compositions are ideal for the low-index material of an HBF operating at 10.6 mm, but the high vapor pressure of the halogen atoms is prohibitive to the deposition of stoichiometric chalcohalide thin films. It is thus necessary to identify a low refractive index chalcogenide glass with a transmission window that extends beyond 10.6 mm, effectively limiting the search to selenium-based chalcogenide glasses. The criteria for selecting the chalcogenide glass pair that can be used in an HBF designed for low loss at 10.6 mm are as follows (Bowden 2007): (a) The compositions must have high refractive index contrast for 10.6-mm radiation; (b) both compositions must have low absorption (low k) at 10.6 mm; and (c) the compositions must be co-drawable, which implies that they have similar glass transition temperatures, T_g, and high crystallization temperatures, T_x.

10.6.1 Germanium Selenide Glass

Germanium selenide (Ge_{1-x} Se_x) glasses have unique thermal and optical properties that make them an excellent candidate for the low refractive index composition in a chalcogenide glass HBF. Optimum glass formation is realized when the number of constraints placed on each atom by the surrounding glass network is equal to its spatial degrees of freedom (3 for a 3-dimensional network). This condition is satisfied when the average coordination number, $<r>$, of each atom is 2.4. The composition $Ge_{20}Se_{80}$ corresponds to $<r> = 2.4$ and is in the range of Ge-Se compositions that form glasses most readily. The T_g of Ge-Se glasses increases continuously with increasing germanium content and takes on values from 125 to 223 °C over the range of $0.75 < \times < 0.85$. Ge-Se glasses could be used in a similar fabrication method as the OmniGuide fiber based upon their thermal stability and the similarity of their T_g to As_2Se_3, which has a $T_g = 173$ °C. The transmission window of selenide-based glasses that are free of oxygen and hydrogen impurities typically extends beyond 10.6 mm, but their refractive indices are generally too high to be considered for the low refractive index material in an HBF. Ge-Se glasses have lower refractive indices than typical selenide glasses. Ge-Se bonds have lower susceptibilities than either Se-Se or Ge-Ge, which explains the refractive index decrease upon the addition of Se to pure Ge, or Ge to pure Se. The fraction of Ge-Se bonds is maximized for the composition $GeSe_2$, which has a refractive index of 2.4 corresponding to the minimum value for Ge-Se glasses. The composition $Ge_{20}Se_{80}$ has a slightly higher refractive index of 2.46, but it is a much better glass former. $Ge_{20}Se_{80}$ has a T_g of 152 °C and does not crystallize below 422 °C. The chalcogenide glass composition $Ge_{20}Se_{80}$ is an excellent glass former, has a relatively low refractive index among selenide-based glasses, and has low transmission loss at 10.6 mm. Therefore, $Ge_{20}Se_{80}$ has been used as the low refractive index composition for a chalcogenide glass HBF (Bowden 2007).

10.6.2 High Refractive Index Chalcogenide Glasses

If n_L is taken as 2.45, n_H must be greater than 3.9 to equal the refractive index contrast of the OmniGuide fiber. A chalcogenide glass HBF with low materials absorption requires less refractive index contrast to achieve similar or even lower losses than the OmniGuide fiber. It is therefore not necessary to achieve an equal refractive index contrast to the OmniGuide fiber, but there are benefits to identifying a high refractive index composition to pair with $Ge_{20}Se_{80}$. One important benefit is a reduced number of layers pairs needed to achieve low loss. The highest refractive index chalcogenide glasses are those containing the heavy elements like Sb, Te, Pb, and Tl, which are more polarizable than the lighter elements from the same groups. Glasses containing large mol fractions of these elements, however, tend to exhibit poor glass-forming ability, not to mention high toxicity, thus limiting their usefulness for HBFs (Bowden 2007).

10.6.3 Silver-Arsenic-Selenide Glasses

The Ag-As-Se system has high density and stable glass-forming compositions with similar T_g to $Ge_{20}Se_{80}$; therefore it has been examined for use in chalcogenide HBFs. The smaller of the glass-forming regions contains compositions enriched with As and Ag that exhibit more stable glass formation than the compositions in the Se-enriched glass-forming region. The difference in T_g and T_x, denoted DT, is a good indication of the suitability of a glass composition for the fiber drawing process. The glasses with high arsenic and silver content have high DT and similar T_g to $Ge_{20}Se_{80}$ indicating their potential to be co-drawn. The density of these glasses is close to 6 g/cm^3 indicating that their refractive index is likely to be greater than three based upon the correlation between the density and refractive index of chalcogenide glasses. Of particular interest is the composition $Ag_{25}As_{41.7}Se_{33.3}$ that has DT = 159 °C, the highest of the glasses investigated, and a density of 5.93 g/cm^3. A chalcogenide glass HBF consisting of at least seven $Ge_{20}Se_{80}$/$Ag_{25}As_{40}Se_{35}$ layer pairs could have improved performance over the OmniGuide HBF for the transmission of CO_2 laser radiation, with a fabrication method for chalcogenide glass HBFs that is capable of achieving similar microstructure precision to that which is possible using the OmniGuide HBF fabrication process (Bowden 2007).

10.6.4 Chalcogenide Glass HBF Preform Fabrication and Drawing

Chalcogenide glasses have been investigated for application to HBFs both for their optical and thermal properties. Fiber drawing is a key aspect of the OmniGuide HBF fabrication process because it allows precise control of HBF microstructure, long lengths of fiber can be made continuously, and fibers designed for many different wavelengths can be drawn from a single preform. A freestanding chalcogenide glass thin film is too brittle to be rolled into a preform, as is done with an As_2Se_3-coated PEI thin film to make the OmniGuide preform. To make a drawable chalcogenide glass HBF preform, chalcogenide glass thin films have been obtained by PVD techniques, including thermal evaporation, sputtering, and pulsed laser deposition. Film stoichiometry is important because small changes in the composition of the film can change the optical and thermal properties significantly. A high deposition rate is needed to deposit films with thicknesses that are in proportion to the bore diameter of the preform taking into account the desired bore diameter of the HBF. For instance, the film thickness of a $Ge_{20}Se_{80}$ layer of an HBF designed for low loss at 10.6 mm is 1.18 mm; thus the $Ge_{20}Se_{80}$ layers of a 12-mm bore diameter preform that will be drawn into a 700-mm bore diameter fiber must be $1.18 \times 12/0.7 = 20.2$ mm thick. The slow deposition rates of sputtering and PLD may be prohibitive to their application in fabricating chalcogenide glass HBFs. These methods also require sophisticated equipment, whereas a rudimentary thermal evaporation system can be fabricated using a simple vacuum chamber and a heat source.

Figure 10.6a shows a thermal evaporation system constructed to deposit chalcogenide glass films onto a water-cooled aluminum substrate tube. To deposit a chalcogenide glass layer, the glass is first crushed and dispersed along the bottom of the 10″ deposition zone in the fused silica evaporation chamber. A ¼″ diameter aluminum substrate tube is suspended above the chalcogenide glass and sealed within the chamber using differentially pumped rotational vacuum fittings. During the deposition process the tube is rotated at 5 RPM, while cooling water (~0 °C) flows along its axis at a rate of 1 L/min. A turbo pump draws vacuum on the chamber as it is heated at a rate of 5 °C/min to 150 °C. This temperature is held for 1 h to eliminate physically adsorbed water molecules from the chamber walls. The furnace is then heated at 10 °C/min to the final deposition temperature of 400–500 °C. The furnace is held at the final deposition temperature for 4 h to ensure that all of the material has been deposited on the aluminum substrate. This process is repeated to build up alternating layers of different chalcogenide glass compositions. The thickness of the layers is determined by the mass of the glass dispersed in the bottom of the fused silica chamber. The fabrication of a mechanically stable all-chalcogenide glass HBF preform would require the deposition of an impractical number of layers. Instead, the chalcogenide glass films and aluminum substrate tube are covered with a thick PEI coating to provide mechanical support for the chalcogenide films. A 12″ × 36″ × 0.003″ thick PEI film is rolled onto the substrate forming a 1/16 inch thick coating that covers a 12 inch long section of the tube. The last layer of the polymer coating is tacked in place using a soldering iron. The coating is consolidated under vacuum by placing it in an evacuated fused silica tube and then passing a small hot zone tube furnace set at ~260 °C along the length of the coated substrate at a rate of 1 cm/min. This consolidation furnace is illustrated in Fig. 10.6b. The portions of the aluminum substrate tube that extend beyond the consolidated PEI are cut off, and the remaining aluminum is dissolved by submersing it in a 3 M hydrochloric acid solution. The chalcogenide glass film remains bonded to the PEI film when the dissolution of aluminum is complete. The chalcogenide-coated PEI tube is drawn at a temperature ~305 °C at a draw rate of about 1 m/min, producing tens of meters of hollow fiber from each preform. A draw furnace with a small ~1-cm hot

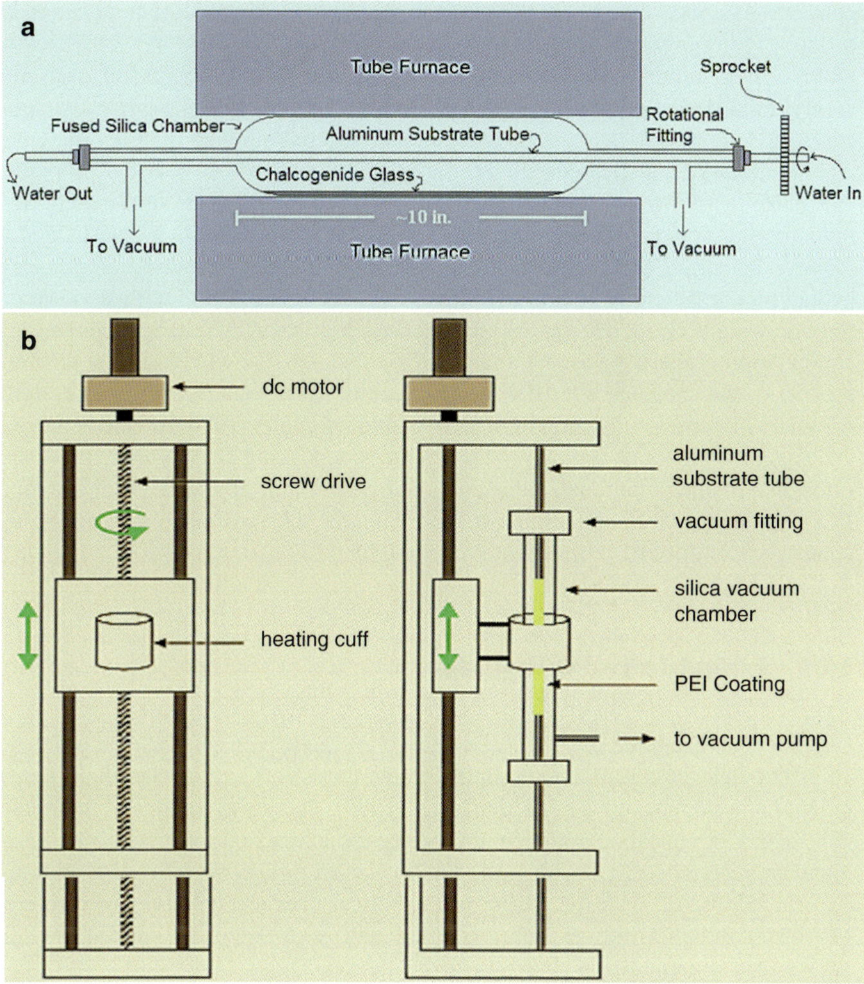

Fig. 10.6 Schematic diagram of chalcogenide glass processing (Bowden 2007): (**a**) the thermal evaporation setup used to deposit chalcogenide glass films on the outside of a rotating aluminum tube and (**b**) the front (*left*) and side (*right*) of the polymer cladding setup used to consolidate PEI films to form a protective coating for chalcogenide glass films deposited on the outside of and aluminum mandrel

zone is used to minimize collapse of the air core. Chalcogenide glass layers, such as $As_{40}Se_{60}$, $Ge_{20}Se_{80}$, and $As_{40}Se_{60}$, can be deposited. The thicknesses of the films are uniform around the circumference of substrate and their interfaces can be well defined. $Ag_{25}As_{40}Se_{35}$ glass cannot be deposited using this thermal evaporation system because silver has a much lower vapor pressure than As or Se. The molten glass in the evaporation chamber becomes increasingly concentrated in Ag as As and Se evaporate, which leaves a crystalline deposit that has a high silver content. $As_{40}Se_{60}$ can be used in place of $Ag_{25}As_{40}Se_{35}$ to that layers with different compositions deposited on top of one another. A more sophisticated thermal evaporation system incorporating separately heated boats for each element would be required to deposit $Ag_{25}As_{40}Se_{35}$ layers with the correct stoichiometry (Bowden 2007).

In summary, chalcogenide glasses have unique optical and thermal properties that make them a good candidate for application to the fabrication of a CO_2 laser radiation transmitting HBF. Ge-Se and Ag-As-Se glasses have been investigated to identify a pair of glass compositions with high refractive index contrast, low extinction coefficient at 10.6 mm, and appropriate T_g and T_x for application to the fiber drawing process. The pair of glasses $Ge_{20}Se_{80}$ and $Ag_{25}As_{40}Se_{35}$ has the best combination of these properties among Ge-Se and Ag-As-Se glasses. The theoretical loss of the chalcogenide HBF is significantly lower than the OmniGuide HBF for microstructures consisting of at least eight layer pairs. The basic elements of a chalcogenide HBF fabrication process includes depositing chalcogenide glass thin films on a rotating aluminum substrate tube by thermal evaporation, adding a thick PEI cladding for mechanical support, dissolving the substrate tube, and drawing the chalcogenide film/PEI preform into a hollow fiber.

10.7 Liquid-Core Waveguides

Liquid-core waveguide (LCW) is an optical waveguide with a liquid medium as the core. The light propagation was studied in the nineteenth century by total internal reflection in streams of water (Altkorn et al. 1997). These waveguides were initially considered for optical communications before the arrival of the optical fibers. Small diameter glass tubes filled with high refractive index liquids were developed for this purpose. However, with the advent of the optical fiber technology, which offered low-loss solid-core optical fibers coupled with easy handling capabilities, the LCWs for communication purposes died a natural death.

The unavailability of suitable cladding materials put a huge constraint on the choice of the liquid to be considered as a core. Glass capillaries were considered for the waveguides which had a refractive index of $n \geq 1.46$. The waveguiding in the case of glass capillaries filled with liquid happened at the glass/air interface. Although these waveguides were reported to have a relatively low loss lying between 18 and 8 dB/m, there were many disadvantages associated with using them (Tsunoda et al. 1989). As the guiding was done at the air/glass interface, the glass surface had to be kept optically clean to avoid any scattering-based losses from the surface of the capillary tube. The dimensions of the capillary wall had to be significantly lower than the core diameter so as to confine the dominant modes of propagation in the liquid core making these capillaries extremely fragile to handle. Thus the choice of liquids was limited to those having a refractive index greater than that of water (1.33). Waveguides which are based on highly reflective surfaces have also been considered for the LCW design. Glass capillaries coated with metallic coatings to act as reflecting surfaces or glass capillaries externally coated with metal surface were studied. As these waveguides do not depend on the refractive indices, these designs could accept light over a very large solid angle. Such waveguides have been used successfully in the microwave and infrared regions for visible absorbance spectroscopy (Wang et al. 1991). However, these waveguides have been reported to have large losses and were rendered unsuitable for fluorescence collection and Raman applications (Tsunoda et al. 1988). Long path length waveguides could not be constructed with this arrangement owing to the large attenuation in a very short distance. Losses as high as 250 dB/m were reported for capillaries externally coated with silver filled with water (Tsunoda et al. 1989). In addition to these, the possibility of tarnishing of the metallic coatings by the liquids of interest is always present which would eventually ruin the waveguiding properties of the system. For externally metallized glass/quartz tubes, this aspect of metal tarnishing is reduced; however the attenuation is even higher (Korampally 2007).

Although LCWs were never put to use for optical communications, these waveguides show a great promise in various optical spectroscopic methods. For example, such waveguides may be used as fluorescence-based biosensors. With the development of a new class of polymeric materials of the class tetrafluoroethane (PTFE)-type Teflon and amorphous fluoropolymer (Teflon AF 2400) with lower refractive indices than water, interest in aqueous core waveguides has once again picked up. Teflon AF 2400 has been reported to have a refractive index of $n = 1.29$ which was the lowest reported refractive index for any known polymer. The use of LCWs based on these polymeric materials for Raman spectroscopy has already been developed. Teflon tubes or glass capillaries internally/externally coated with Teflon were used for various spectroscopic applications. With a surge of interest in the MEMS-based lab-on-a-chip devices, many attempts in the microfabrication of Teflon AF-based LCWs have been made (Korampally 2007). A microfabricated Teflon AF-based LCW system has been formed by etching channels in silicon followed by coating and bonding them with Teflon AF 2400-coated glass substrates (Datta et al. 2003). A similar microfabrication design is carried out with channels formed in soda lime glass (Manor 2003).

Although Teflon has established itself as a versatile material for LCW design, it is not the ideal material, especially in applications in fluorescence-based biosensor design and Raman spectroscopy. Teflon offers little chance for chemical functionalization desired in most biosensing applications. The refractive index contrast between Teflon and water is at most 0.04 which translates to an acceptance angle of just 18°. For applications in biosensors and Raman-based chemical sensors, this poses a big limitation. As the fluorescence is generated in all directions within the waveguide, it is advantageous for the waveguide to have a large acceptance cone thus guiding most of the generated fluorescence through the length of the waveguide. Thus for a more efficient LCW bio/chemical sensor, new materials with ultralow refractive indices have to be engineered. Nanoporous silica is one such material which would be an ideal choice as the cladding material. With pores in the order of 2–4-nm diameters, these materials have an extremely low refractive index. Refractive index, as low as 1.15, has been reported, which corresponds to an acceptance angle of 42° (Korampally 2007). These dielectric materials have excellent adhesion to the common substrates and easy functionalization of these surfaces with established methods makes them all the more attractive for biosensor applications. The feasibility of using nanoporous dielectric as low-index cladding material for aqueous core planar waveguides has already been demonstrated (Risk et al. 2004). Two fused silica substrates coated with nanoporous dielectric (0.8 mm thick) were held facing each other, separated by 50-mm-thick polyester spacing. The space between the two substrates was filled with an aqueous solution containing fluorescent beads. Waveguide loss characterization for the setup was done by transversely illuminating the aqueous core with the excitation signal. Fluorescence was collected at the edge of the waveguide by coupling an optical fiber connected to a spectrometer. A waveguide loss of 0.55–0.56 dB/mm has been achieved. The attractive properties of the nanoporous dielectric combined with the massive multiplexing capability provided by the MEMS-based technologies would result in novel and cost-effective sensors. Therefore, fabrication of such a sensor with extreme sensitivity has been investigated, which is suitable for mass production (Korampally 2007).

LCWs have also been used to maximize path length while minimizing volume in many systems. Combining the technology of the LCW and microchannels opens doors for microscale fluorescence spectroscopy. Important aspects of the LCW fabrication include the photolithography process, etching parameters, Teflon AF adhesion and coating, as well as assembly bonding. The SU-8 mask must be processed correctly during the photolithography step or premature lifting may occur, causing ripples, or additional roughness along the channel wall. Not every glass is appropriate for etching microchannels. Extreme roughness causes increased signal loss and can be avoided by using the right glass and etching solution. After etching the proper thickness, spinning and baking parameters of Teflon AF can increase the chance of having quality channels. The microchannels may not be as successful if the coating is too thick, thin, or not uniform. Microfabricating LCWs show promise as components in high sensitivity detection and analysis systems for micro total analysis systems (Manor 2003; Manor et al. 2003).

Moreover, LCW systems can be constructed using Teflon® AF coatings allowing measurement of analytes in aqueous solutions. The amount of light collected in a transverse illumination scheme is significantly enhanced through the low-index coating. Correlation has been done between loss in the LCW microchannels and the Teflon coating used to coat the channel. Uncoated channels proved to have the highest loss of 0.79 dB/mm. This result was expected since the refractive index of the glass ($n = 1.47$) is much greater than that of the liquid core ($n = 1.33$). These refractive indices do not obey the laws for TIR and therefore would have a higher loss than guides that have a cladding of the lower refractive index than their core, such as with the loss results of Teflon AF 1600 and 1601 which have loss measurements of 0.61 and 0.53 for TAF 1600 and 0.31 for TAF 1601. Teflon AF 1601 rendered a thicker coating of Teflon allowing the roughness of the microchannel to be filled in. With proper design, LCW microchannels can be integrated into a microanalysis system (Manor 2003; Manor et al. 2003).

On another hand, the major highlights in new fields like optofluidics are the ability to combine optics and fluidics in same platform, opening new frontiers in detection of biological molecules in aqueous solutions. For a more compact and robust optofluidic lab-on-a-chip, integration of liquid and optical elements in the same system is needed. LCWs in optofluidics offer fully planar optofluidic lab-on-a-chip devices. The major challenge in using LCW, however, is to use a solid cladding material with refractive index below 1.33. Most extended technologies in this field are Teflon AF-cladded LCWs and ARROW waveguides (Manor et al. 2003).

Besides, LCWs have been used to increase the length of the optical path, i.e., the sensitivity of various spectroscopic measurements, such as spectrophotometry, fluorometry, and Raman spectroscopy. Therefore, various types of liquid-core waveguides are developed. If the refractive index of the capillary material is lower than that of the liquid core, excitation or Raman light that hits the capillary wall under an angle equal to, or larger than the critical angle to the normal, will be captured inside the capillary. Based on this consideration, four types of LCW can be distinguished: (a) internally or externally metal-coated tubing, (b) uncoated glass tubes, (c) internally or externally plastic-coated tubes, and (d) entirely plastic tubes. Although metal-coated LCWs can increase the optical path length and have the advantage of guiding light through virtually any transparent liquid, regardless of its refractive index, the approach suffers from high attenuation. As regards the uncoated glass tube LCWs, here light is reflected at the outer glass/air interface rather than at the liquid/glass interface. As a consequence, the light guiding is independent the use of the refractive index of the liquid. However, such LCWs are very fragile, and the loss characteristics are largely dependent on the outer glass surface. Therefore, much care has to be taken to create and maintain a clean and smooth surface. LCWs internally or externally coated with a plastic material have the advantage that they are more robust than the uncoated glass tubes. However, most plastics have a refractive index larger than that of water. The use of Teflon AF (2,2-bistrifluoromethyl-4,5-difluoro-1,3-dioxole/tetrafluoroethylene), whose refractive indices ($n_D = 1.29$ (AF-2400) and $n_D = 1.31$ (AF-1600)) are lower than that of water ($n_D = 1.33$), has widened the applicability of the LCW technique to a great extent, since the AF tubing can make it possible to form the LCW with water. Two types of LCWs using Teflon AF have been developed. One is Teflon AF capillary tubing where a solvent-core and an AF-clad waveguides are formed (Type 1), and the other is glass capillary tubing whose outer surface is coated with Teflon AF (Type 2). In Type 2 LCW, total internal reflection occurs at the interface between the outer glass capillary surface and the Teflon AF coating. Type 2 LCW has been applied to the spectrophotometric determination of nitrite with a FIA system and achieved the detection limit of 2 nmol dm^{-3} nitrite with 4 m of the LCW. Although Type 2 LCW has various advantages, such as less adsorption of analytes to the cell surface, Type 1 LCWs have so far mainly been used for analytical purposes because of their easier optical configuration (Takiguchi et al. 2006).

10.8 Applications of Hollow Waveguides

10.8.1 Hollow Waveguides for Optical PCB Technology

Hollow waveguide-based optical circuit board (PCB) technology has the potential to significantly improve the performance of optical circuits. It can also make them more rugged, more compact, and much cheaper to manufacture. This technology has further developed to demonstrate a miniaturized laser spectrometer system for the measurement of atmospheric carbon dioxide, and fabricate devices for microsensors, and for a very wide range of laser- and optics-based space instruments both for earth science and planetary exploration (Jenkins et al. 2007).

As sown in Fig. 10.7, the hollow waveguide PCB represents the optical equivalent of the electronic printed circuit board. Hollow waveguides formed in the surface of a dielectric substrate are used to guide light through a circuit of optical components, each of which is mounted in a precision alignment slot. Computer-controlled milling techniques are used to create square-section channels which are the basis of the hollow waveguide circuits and the component alignment slots. A lid which caps the substrate forms the upper wall of all the waveguides. The use of deep reactive ion etching in conjunction with silicon substrates also facilitates the formation of much smaller cross-sectional waveguide circuits and alignment slots compatible with micro-optical components. In this manner, optical PCBs with the size of a SIM card can be realized. In conjunction with automatic pick-and-place equipment, the approach could lead to high-volume low-cost mass production of a very wide range of integrated optic sensors. The hollow waveguide circuit board has a number of advantages compared with conventional solid-core integrated circuits. The hollow core has very broad waveband, high-power, transmission characteristics. The fact that there is no refractive index step to bridge means discrete components can be easily integrated. The hollow core can also provide the basis of flow cells for the analysis of gases and liquids using absorption, spectroscopic, and interferometric techniques (Jenkins et al. 2007).

More specifically, current optical circuit manufacturing techniques lead to instruments which are bulky and sensitive to misalignment and are expensive to manufacture. The advantages of the hollow waveguide PCB technology are that they can lead to optical and laser systems which are more compact, more rugged, and much cheaper to manufacture. Hollow waveguide optical PCBs also provide an integration platform for a very wide range of discrete optical and micro-optical components of both an active and passive nature. These include beam splitters, lenses, wave plates, polarizers, absorption cells, lasers, amplifiers, detectors, and modulators. These will enable the realization of integrated systems for a diverse range of applications in the fields of space, environment, health, security, and defense, including (Jenkins et al. 2007) (a) spectrometers, interferometers, and radiometers for earth science and planetary exploration; (b) laser radars, including lidar-on-a-chip based on hollow silicon technology for space and defense applications; (c) biochemical sensors, including lab-on-a-chip for DNA analysis, for health, defense, and pharmaceutical development; (d) gas sensors for

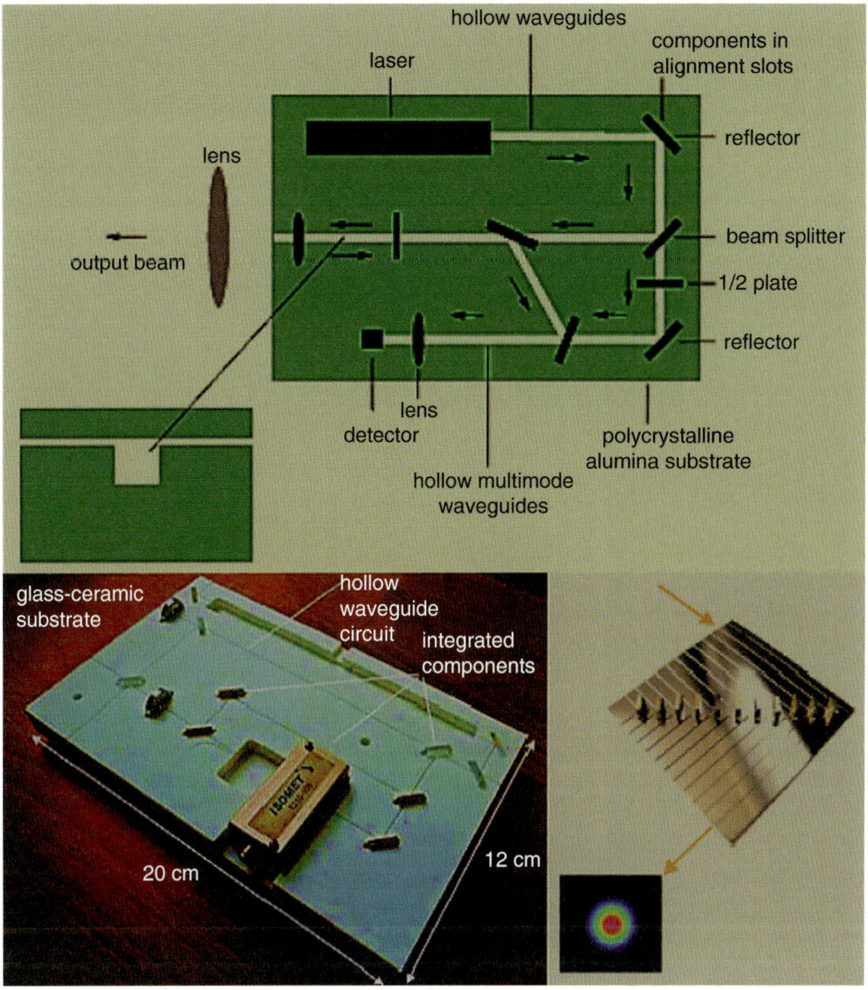

Fig. 10.7 Schematic illustration of hollow waveguide optical circuit board (Jenkins et al. 2007)

environmental and industrial monitoring and health care; (e) fluidic sensors for pollution monitoring and health care; (f) telecommunications modules based on micro-optical components integrated on fiber pig-tailed hollow waveguide PCBs; and (g) generic optical test and measurement instrumentation.

10.8.2 Hollow Waveguides for Medical Applications

Optical waveguides are used in medicine for a diverse range of applications, from sensing, diagnostics, to therapeutics. The main advantage of hollow waveguides in medicine is their ability to transmit wavelengths for which traditional solid-core waveguides are not transparent. For example, traditional silica-based waveguides cannot transmit wavelengths above approximately 2.1 mm because the materials absorption of silica becomes too large. These longer wavelengths, such as the Er: YAG laser wavelength at 2.94 mm and the CO_2 laser wavelength at 10.6 mm, have significant clinical advantages over the shorter wavelengths accessible to silica-based waveguide. The interaction of light and the different components of biological tissue encountered in medical applications is determined mainly by the absorption of the light by these components. For wavelengths above 2 mm, the absorption of water, the main constituent of biological tissue, rises sharply. This strong absorption of laser light by water can be used for very efficient cutting and ablation of soft tissue. The operating wavelengths of CO_2 lasers (10.6 mm) and Er:YAG lasers (2.94 mm) offer particularly strong water absorption and are, therefore, especially well suited for precise cutting and ablation; the strong absorption prevents the laser energy from penetrating tissue much beyond the point of application. In contrast, wavelengths less than 2 mm penetrate as much as 1 cm beyond the intended target. CO_2 lasers, in particular, are reliable and commercially available and have been used in medicine for a long time. In addition, the CO_2 laser wavelength offers excellent coagulation capability to stop bleeding during laser-based surgery, a capability not shared by the Er:YAG laser wavelength. This unique combination of precise cutting, limited penetration beyond the point of application, and coagulation is found only with the CO_2 laser wavelength, making it the optimal wavelength for many surgical procedures. In spite of the advantages associated with CO_2 lasers, their use in medicine has been relatively limited because of the lack of a flexible medium to transmit the laser power to a target within the human body. Without a flexible delivery medium, the use of the CO_2 laser was confined to procedures in which a direct line of sight could be established between the laser and the target, primarily in dermatology and ear, nose, and throat (ENT) surgery with a rigid laryngoscope. Thus, since 1976, various alternative waveguides and fibers have been developed, which would allow the use of CO_2 lasers in a much wider variety of medical applications. For the most part, these attempts have been met with only limited success. Products were developed that used either complicated and bulky articulated arms or restricted line-of-sight micromanipulators. Some commercial hollow metallic waveguides were developed, mainly as an articulated arm substitute. Some solid-core fibers were also developed, based on materials that are relatively transparent in the mid-IR, but these fibers proved either too lossy or too brittle (Jacobs et al. 2007).

Hollow-core waveguides and fibers using an omnidirectionally reflective 1D photonic bandgap to confine light have been developed and used successfully in a number of minimally invasive procedures with CO_2 lasers. These OmniGuide fibers are flexible enough to be introduced into the body through flexible endoscopes, delivering CO_2 laser radiation to regions not accessible by such lasers before. In one particularly interesting example, an OmniGuide fiber was used through the working channel of a flexible endoscope to successfully remove a large malignant tumor from the bronchus intermedius of a patient with lung cancer (Bueno et al. 2005; Jacobs et al. 2007).

10.8.3 Prospective Telecommunication Applications

Various oxide glasses have materials absorption losses below 100 dB/km and indices of refraction near 1.6 at the typical telecom wavelengths. When combined with the high-index glass formed from As_2Se_3, a very low-loss OmniGuide fiber can be produced. Indeed, with approximately 20 bilayers and the proper choice of the hollow-core radius, the theoretical loss for the TE01 mode is below 0.01 dB/km. Losses this low would reduce the number of relatively expensive amplifier modules needed in long-haul systems, changing the fundamental economics of this market. In addition, nonlinear effects in the hollow core would be greatly reduced, allowing for higher launch power and dense wavelength division multiplexing without the usual concerns raised by four-wave mixing (Johnson et al. 2001; Jacobs et al. 2007).

A relatively large hollow-core radius was employed in OmniGuide fiber for telecom applications to achieve the benefits of extremely low loss and nonlinearity (Johnson et al. 2001). Of course, an OmniGuide fiber with this core size is multimoded but, because of large differential mode attenuation, behaves as essentially single moded, with the lowest loss mode serving as the operating mode. The lowest loss mode for OmniGuide fiber is the TE01 mode; the use of a large core radius, therefore, necessitates the use of this mode as the operating mode. The TE01 mode, unlike the doubly degenerate HE11 mode, has the advantage of being free from PMD but does not couple at all to the TEM00 field output from conventional telecom lasers. Thus, its use in telecom systems will require the coincident development of efficient HE11, TE01 mode converters. In addition, this kind of telecom fiber would need to have very low surface roughness to prevent scattering of power from the operating mode to higher-order lossy modes (Jacobs et al. 2007).

Two-dimensional, hollow-core PCF also offers the possibility of extremely low loss and nonlinearity. In addition, because the cladding matrix of this fiber is low-loss silica glass, there is no need for the use of a large core to achieve the desired loss level; true single-mode behavior can be obtained, with the doubly degenerate HE11-like operating mode coupling efficiently to the output from typical telecom lasers. On the other hand, this fiber supports lossy surface modes, which couple easily to the operating mode and provide an additional loss mechanism. In addition, the double degeneracy of the operating mode will produce PMD in the presence of perturbations that split the degeneracy. The complexity of the 2D photonic crystal mirror, with attendant difficulty in maintaining the desired structure during manufacture, enhances the likelihood of the existence of perturbations that both induce coupling between the operating and surface modes and split the degeneracy of the operating mode, giving rise to PMD (West et al. 2004).

The wealth of parameters that can be modified to fine-tune the confinement mechanism in OmniGuide fiber, combined with the virtual absence of materials dispersion, suggest that OmniGuide fiber can be tailored to have certain desirable dispersion properties. For instance, modification to the multilayer cladding structure of an OmniGuide fiber can be made to obtain a dispersion-compensating fiber (DCF) with a theoretical dispersion/loss figure of merit up to 5 times larger than conventional DCFs (Engeness et al. 2003). Similar design freedom in 2D PCF has also been exploited to provide theoretically high-performance DCF (Birks et al. 1999). Hollow-core 2D PCF for dispersion compensation is also commercially available (Jacobs et al. 2007).

10.8.4 Hollow Waveguides as Gas Cells

Hollow waveguides can also be used as gas cells. Interaction of light with gases leads to a variety of complex and important phenomena of both academic and commercial interest in fields such as nonlinear optics, chemical sensing, quantum optics, and frequency measurement. Because of the low density of gaseous media, maximization of interactions with light is often a challenge. Typically, the signature strength of the phenomena being explored increases with the intensity of the probing beam and with the interaction length. To enhance intensity in a conventional gas cell, one can focus the laser beam more tightly. Unfortunately, that leads to a smaller diffraction length, which in turn implies shorter interaction lengths. The use of hollow waveguides as gas cells eliminates these restrictions on intensities and interaction lengths. If the laser beam is confined as a guided mode of a hollow waveguide, diffraction is prevented, and one can explore very small beam profiles and, therefore, very high intensities. Moreover, simply for practical considerations, lengths of conventional gas cells are typically limited to less than approximately 1 m. In contrast, the lengths of gas cells implemented via hollow waveguides are much less limited in that respect; the only fundamental limitation is the transmission loss of the waveguide. For example, losses of approximately 0.3 dB/m allow for propagation lengths of about 10 m. In addition, hollow waveguides can often be bent into loops of fairly small radii, thereby enabling very compact gas cells that nevertheless have very long interaction lengths (Jacobs et al. 2007).

In chemical sensing applications, hollow waveguides offer significant benefits over conventional gas cells because they require a drastically smaller volume of the sample gas. Moreover, the response time is faster since such smaller volumes fill up faster. For example, a photonic bandgap hollow waveguide is used to detect ethyl chloride at concentration levels of 30 ppb with a sample volume of 1.5 mL, and a response time of 8 s, representing an increase in sensitivity by almost three orders of magnitude over conventional gas cells (Charlton et al. 2005).

Hollow waveguides have also enabled many impressive accomplishments in nonlinear optics. For example, stimulated Raman scattering has been observed in hollow-core PCFs filled with hydrogen gas, with up to 92 % quantum efficiency and threshold pulse energies six orders of magnitude lower than what was previously reported (Benabid et al. 2004). In addition, slow-light effects have been observed in acetylene-filled hollow-core PCFs (Ghosh et al. 2005); similar fibers filled with acetylene have also been used to implement frequency locking of diode lasers (Benabid et al. 2005; Jacobs et al. 2007).

10.8.5 Applications of Hollow Waveguides for Remote Sensing

Hollow waveguides can alternatively be used as passive elements to deliver radiation from a source being sensed or studied, to the remote sensor, which then analyzes the radiation. Applications of interest include temperature sensing (analysis of blackbody radiation) and chemical sensing. This kind of remote sensing can be useful in hazardous environments (e.g., battlefields) or when the point of sensing is difficult to reach (e.g., inside a machine). It is also useful when many points within an environment need to be analyzed often, but not continuously (bottoms of oceans, industrial plants); in that case, the number of sensors that needs to be deployed is significantly reduced compared to the situation in which one sensor is placed at each point that needs monitoring. The motivation to use hollow waveguides is that they can often transmit broader bandwidths and access wavelength regimes that would be difficult to explore using any other solid-core waveguide (e.g., for chemical sensing, the 3- to 20-mm regime is of particular interest). Applications of interest include environmental sensing (pollution, etc.), homeland security, process monitoring, and biomedical sensing (e.g., breath analysis for detection of asthma) (Worrell et al. 1992; Jacobs et al. 2007).

10.8.6 Industrial Applications

CO_2 lasers are used in many industrial applications such as cutting, welding, and marking. It is a challenge, however, to bring the laser beam from the source to the working area, as the path is usually obstructed by other equipment. The use of articulated arms is the only method for beam delivery, even though these bulky systems require a large working space and use mirrors that require constant maintenance and alignment. Early efforts at using solid-core waveguides to deliver high-power CO_2 beams have suffered from thermal damage, particularly at the air-fiber interfaces at the input or the output end of the waveguide. Some of the promising solid-core fibers also suffered from short lifetimes because of degradation of the guiding material when carrying high-power laser beams. To overcome these problems, high-power CO_2 laser delivery systems have been investigated based on hollow-core fibers. Initial attempts based on rectangular and circular metal-coated hollow-core waveguides succeeded in transmitting up to 3 kW of laser power. However, these power levels were achieved with waveguides having large core radii, which exhibited poor output mode quality when subjected to movements and bends. Most of the industrial applications require well-defined, low-order output modes so that the cut or mark is clean and sharp. The only way to achieve this type of modal quality with hollow waveguides is to have a smaller core size that will filter the higher-order modes and transmit only the fundamental mode. Of course, as the core size is reduced, the losses of the fiber increase and the power capacity decreases. These problems limited the utilization of hollow waveguides to lower-power industrial applications, like marking or cutting paper or plastic products, applications that require powers in the range of 1–100 W, which can be delivered with the small-core waveguides. The flow of gas through the waveguide core and the use of a water-circulating jacket are often employed to improve the power handling capacity of these waveguides, but they still find little use in industrial processing. An interesting solution to the problem of waveguide-motion-induced mode mixing in a marking application was using eight computer-controlled lasers coupled to eight stationary waveguides to form the desired mark (Harrington 2009; Jacobs et al. 2007).

10.9 Summary

Hollow waveguides are becoming an attractive fiber optic for the delivery of high-power laser radiation as well as for PCB integration, medical, telecom, gas cell, sensor, and industrial applications. In general, these guides enjoy losses of a few tenths of a dB/m and are quite flexible. Because the energy is carried in the hollow core, there is no core material that might be easily damaged by high peak or average laser powers as there would be in a comparable solid-core IR fiber. In addition, there is no Fresnel loss; this is especially important in comparison to the chalcogenide glass fibers where Fresnel losses can be as high as 25 %/surface. The optical principles are different between a HCW and a solid-core fiber. The most important distinctions are that the hollow waveguides have a loss that varies as the reciprocal of the bore radius cubed and that there is an additional loss on bending which varies as $1/R$. These properties are not shared by conventional solid-core fibers. However, most applications of IR fibers do not require tight bending radii so that the additional loss on bending is not prohibitive. Two other advantages of hollow waveguides are important: these guides are nearly single mode and there seems to be no loss due to aging.

Hollow waveguides may be grouped into two categories: (a) those whose inner core materials have refractive indices greater than one (leaky guides) and (b) those whose inner wall material has a refractive index less than one (attenuated total reflectance, i.e., ATR, guides). Leaky or $n > 1$ guides have metallic and dielectric films deposited on the inside of metallic, plastic, or glass tubing. ATR guides are made from dielectric materials with refractive indices less than one in the wavelength region of interest. However, hollow waveguides have two apparent drawbacks: high attenuation when bend and high sensitivity to the laser beam's angle of incidence. Moreover, each waveguide is tuned for delivering only a specific wavelength. One of the ways to overcome these drawbacks is to introduce the notion of photonic crystals.

Hollow waveguides seem to be the best choice as power delivery fiber for CO_2 lasers. As an example, OmniGuide HBFs are based on radial periodic structure of high- and low-index concentric rings, which is provided by high refractive index chalcogenide glasses and low refractive index polymer. They have only one central air hole, where the majority of optical power is carried axially by the radial photonic Bragg condition. Similarly hollow IR-transmitting hollow fibers carry the optical fiber through the central air hole by reflective metal or metal-insulator coating on the inner surface of the holey fiber. Even though practical applications of these two types of fibers, OmniGuide fibers and IR-transmitting hollow fibers, have been limited to high-power IR laser delivery in a relatively short distance, these structures could provide high potential for ultimately low-loss transmission for long-haul optical communications.

LCWs in optofluidics offer fully planar optofluidic lab-on-a-chip devices and have also been used to increase the length of the optical path, i.e., the sensitivity of various spectroscopic measurements, such as spectrophotometry, fluorometry, and Raman spectroscopy. The use of Teflon AF, whose refractive indices are lower than that of water, has widened the applicability of the LCW technique to a great extent, since the AF tubing can make it possible to form the LCW with water.

References

Altkorn R et al (1997) Low-loss liquid-core optical fiber for low refractive-index liquids: fabrication, characterization, and application in Raman spectroscopy. Appl Opt 36(34):8992–8998

Anastassiou C et al (2004) Fibers deliver CO_2 laser beams for medical applications. Photon Spectra 38(3):108–109

Benabid F et al (2004) Ultrahigh efficiency laser wavelength conversion in a gas-filled hollow core photonic crystal fiber by pure stimulated rotational raman scattering in molecular hydrogen. Phys Rev Lett 93:123903

Benabid F et al (2005) Compact, stable, and efficient all-fiber gas cells using hollow-core photonic crystal fibers. Nature 434:488–491

Birks TA et al (1999) Dispersion compensation using single-material fibers. IEEE Photon Technol Lett 11(6):674–676

Bowden BF (2007) Design theory, materials selection, and fabrication of hollow core waveguides for infrared to THZ radiation. PhD dissertation, The State University of New Jersey, Rutgers

Bueno R et al (2005) Flexible delivery of carbon dioxide lasers through the omniguide photonic bandgap fiber for treatment of airway obstruction: safety and feasibility study. Chest Meet Abstr 128:497S

Charlton C et al (2005) Mid-infrared sensors meet nanotechnology: trace gas sensing with quantum cascade lasers inside photonic band-gap hollow waveguides. Appl Phys Lett 86:194102–194104

Datta A et al (2003) Microfabrication and characterization of Teflon AF-coated liquid core waveguide channels in silicon. IEEE Sens J 3:788–795

Doradla P et al (2012) Propagation loss optimization in metal/dielectric coated hollow flexible terahertz waveguides. In: Sadwick LP, O'Sullivan CM (eds) Terahertz technology and applications V. Proceedings of SPIE, San Francisco, CA, 21 Jan 2012. SPIE—The International Society for Optical Engineering, vol 8261, pp 1–10

Engeness T et al (2003) Dispersion tailoring and compensation by modal interactions in OmniGuide fibers. Opt Express 11:1175–1198

Fink Y et al (1998) A dielectric omnidirectional reflector. Science 282:1679–1682

Ghosh S et al (2005) Resonant optical interactions with molecules confined in photonic band gap fibers. Phys Rev Lett 94:093902

Harington JA (2009) Infrared fibers. In: Michael B (ed) Handbook of optics (volume V) atmospheric optics, modulators, fiber optics, X-ray and neutron optics. McGraw-Hill Professional, New York, pp 12.1–12.13

Harrington JA (2000) A review of IR transmitting, hollow waveguides. Fiber Integr Opt 19:211–217

Iwai K, Miyagi M, Shi YW, Matsuura Y (2006) Uniform polymer film formation in hollow fiber by closed loop coating method. Proc SPIE 6083:60830J1–60830J8

Jacobs SA et al (2007) Hollow-core fibers. In: Mendes A, Morse TF (eds) Specialty optical fibers handbook. Elsevier, Boston, pp 315–360

Jenkins RM et al (2007) Hollow waveguide optical PCB technology for micro-optical sensors. In: Sixth ESA round table on micro & nano technologies for space applications, Toulouse, France

Johnson VS (2007) Hollow glass waveguides with multilayer polystyrene and metal sulfide thin film coatings for improved infrared transmission. PhD dissertation, The State University of New Jersey, Rutgers

Johnson SG et al (2001) Low-loss asymptotically single-mode propagation in large-core OmniGuide fibers. Opt Express 9:748–779

Korampally V (2007) A novel ultra-low refractive index nanoporous dielectric based aqueous core light waveguide system. PhD Dissertation, University of Missouri-Columbia

Manor RM (2003) Microfabricated liquid core waveguides. Master degree thesis, Texas Tech University

Manor R et al (2003) Microfabrication and characterization of liquid core waveguide glass channels coated with Teflon AF. IEEE Sens J 3:687–692

Merberg GN (1993) Current status of infrared fiber optics for medical laser power delivery. Lasers Surg Med 13(5):572–576

Miyagi M, Kawakami S (1984) Design theory of dielectric-coated circular metallic waveguides for infrared transmission. J Lightwave Technol LT-2:116–126

Oh K et al (2005) Novel hollow optical fibers and their applications in photonic devices for optical communications. J Lightwave Technol 23(2):524–532

Rabii C et al (1999) Processing and characterization of silver films used to fabricate hollow glass waveguides. Appl Opt 38(21):4486–4493

Risk WP et al (2004) Optical waveguides with an aqueous core and a low-Index nanoporous cladding. Opt Express 12(26):6446–6455

Sanghera JS, Aggarwal ID (1998) Infrared fibers optics. CRC Press, Boca Raton

Takiguchi H et al (2006) Liquid core waveguide spectrophotometry for the sensitive determination of nitrite in river water samples. Anal Sci 22:1017–1019

Temelkuran B et al (2002) Wavelength scalable hollow optical fibres with large photonic band gaps for CO_2 laser transmission. Nature 420:650–653

Tsunoda K et al (1988) Long capillary cell with the use of successive total reflection at outer cell surface for liquid absorption spectrometry. Anal Sci 4:321–323

Tsunoda K et al (1989) The possibility of signal enhancement in liquid absorption spectrometry with a long capillary cell utilizing successive total reflection at the outer cell surface. Appl Spectrosc 43:49–55

Wang T et al (1991) Nanoliter-scale multireflection cell for absorption detection in capillary electrophoresis. Anal Chem 63:1372–1376

West J et al (2004) Surface modes in air-core photonic band-gap fibers. Opt Express 12 (8):1485–1496

Worrell CA et al (1992) Remote gas sensing with mid-infra-red hollow waveguide. Electron Lett 28(7):615–617

Chapter 11
Metamaterial Optical Waveguides

Abstract Optical metamaterials are man-made composite materials constructed with nanometer-sized periodic structures containing both dielectric and metal materials. These structures can produce materials with negative index of refraction—a unique material property that does not occur naturally. The theoretical breakthroughs made in this new class of electromagnetic materials are closely linked with progress in developing physics-driven design, novel fabrication, and characterization methods. For the optical waveguiding, a perfect control of the interaction between light and matter has been brought closer by the advances in fabrication of optical metamaterials. The unusual electromagnetic properties of metamaterials are expected to enable a new generation of miniaturized passive and active optical devices based on novel optical waveguides. In developing design strategies and new concepts for such devices, it is paramount that anisotropic properties of metamaterials are considered along with their other material features. Moreover, even the ways in which common devices operate require revisions when ordinary materials in their design are replaced by anisotropic metamaterials. Therefore, these metamaterials provide a route to creating potential devices through artificially engineered structures with negative average relative permittivity and permeability. The electromagnetic response of a metamaterial can be designed to produce desired waveguide properties. One particularly interesting metamaterial device is planar metamaterial waveguide structure that has potentially exciting applications. Properties of metamaterial waveguides when the limitations arise from fabrication techniques and physical principles have been taken into account. A considerable amount of theoretical effort has also been devoted to the analysis of optical propagation through different types of metamaterial structures, including uniaxial dielectrics and indefinite media, metal–dielectric heterostructures and superlattices, as well as strongly anisotropic waveguides. This chapter will give a brief review about perspective and prospective of the metamaterial optical waveguides.

X.C. Tong, *Advanced Materials for Integrated Optical Waveguides*, Springer Series in Advanced Microelectronics 46, DOI 10.1007/978-3-319-01550-7_11,

11.1 Historical Perspectives

Optical metamaterials have attracted considerable attention due to the exciting potential applications ranging from perfect lenses and cloaking devices to subwavelength optical waveguides and the enhancement of magnetic resonance imaging (MRI). Metamaterials are a class of composite materials artificially fabricated to achieve unusual electromagnetic properties not available in nature. Their unusual properties are not determined by their constituent materials but by man-made structures that are built from periodically arranged unit cells—resonators smaller than the electromagnetic wavelength involved. These unit cells act like artificial atoms and can be coupled to both the electric and magnetic field components of the electromagnetic waves, leading to novel optical properties, such as a negative index of refraction, as shown in Fig. 11.1 (Kim 2004; Reza 2008). Figure 11.1a illustrates positive and negative refractions. Angles α and β are the incidence and refraction angles, respectively. Referring to the positive refraction for a comparison, metamaterials are also called materials with a negative refractive index (NRI). Figure 11.1b shows reversed energy flows inside and outside of a metamaterial open waveguide. Metamaterials are also called backward (BW) wave materials or materials with a negative phase velocity (NPV). As shown in Fig. 11.1c, Lycurgus Cup is made of ruby glass, the first optical metamaterials. The cup has a greenish color when light reflected from the glass; however, it has a reddish color when light transmitted through the glass. Figure 11.1d gives an example of light (electromagnetic wave) manipulation by three-dimensional photonic crystals.

The origin of metamaterials can be dated back centuries ago; for example, medieval ruby glass might be considered as the earliest optical metamaterial. Ruby glass contains nanoscale gold droplets (5–60 nm) that give an unusual color to the glass. Actually, a resonance of the surface plasmon on the gold droplets causes the extraordinary color of the glass. Ruby glass appears neither golden nor transparent, but depending on the size of the droplets and the direction of the light beam, looks red or green, as shown in Fig. 11.1c. Furthermore, the work on the rotation of the plane of polarization by artificial twisted structures in 1898 or artificial dielectrics for microwave antenna lenses in 1945 could be other examples of metamaterials. However, the modern metamaterials caused attention when Dr. Smith and his team presented evidence of both a negative electric permittivity and a negative magnetic permeability at a specific microwave range of frequencies in 2000. The idea of metamaterials with negative permittivity and permeability was first proposed by Victor Veselago in 1968, who termed such media left-handed material (LHM) because the electric and magnetic fields and the wave vector form a left-handed set of vectors rather than the usual right-hand set. In his study, Veselago showed that the Poynting vector of an electromagnetic wave is antiparallel to the wave vector in an LHM; thus, light propagates in the opposite direction to the energy flux. This leads to a number of unusual properties in metamaterials such as negative-index refraction, the reversal of Cherenkov radiation, the reversal of the Doppler shift, and the reversal of Snell's law. These revolutionary properties give remarkable control over electromagnetic fields

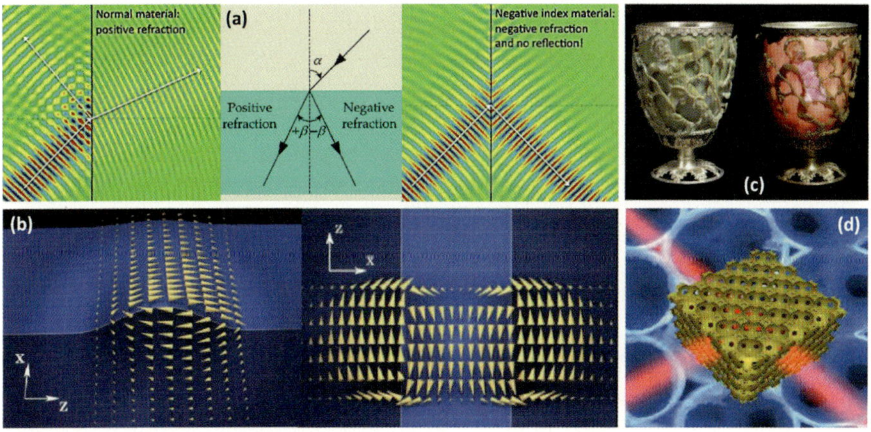

Fig. 11.1 Schematic illustration of metamaterial optical properties (Kim 2004; Reza 2008): (**a**) positive and negative refractions. Angles α and β are the incidence and refraction angles, respectively. Referring to the positive refraction for a comparison, metamaterials are also called materials with a negative refractive index (NRI). (**b**) Reversed energy flows inside and outside of a metamaterial open waveguide. Metamaterials are also called backward (BW) wave materials or materials with a negative phase velocity (NPV). (**c**) Lycurgus Cup is made of ruby glass, the first optical metamaterials. The cup has a greenish color when light reflected from the glass; however, it has a reddish color when light transmitted through the glass. And (**d**) example of light (electromagnetic wave) manipulation by three-dimensional photonic crystals

with incredible possible applications such as invisibility devices. As a further consequence of negative refraction, metamaterials produce unusual optical effects. For example, a bulb located in front of a metamaterial slab, would appear to be in front of the slab; a fish swimming in "water" made of metamaterial would appear to be moving above the water surface. Negative refraction could even lead to "negative space." For example, if a distance between two objects is properly filled with a material with negative refraction and a conventional material, the two objects would appear to be beside each other (Reza 2008).

Since metamaterials possess various extraordinary electromagnetic responses that cannot (or are difficult to) be observed in conventional media, their unrevealed or modified physical phenomena have been investigated, such as superluminality, a negative Goos-Hänchen shift, reversal of circular Bragg phenomena, reversal of circular dichroism, total external reflections (TER), and negative Hartman effect. In addition, metamaterials have also been applied to various electromagnetic wave devices and systems for the development of novel or enhanced engineering performances, including perfect lenses for future image processing, near-field imaging/focusing, and near-field lithography, radar absorbing materials, directive antennas, leaky wave antennas, subwavelength resonators, various filter applications, phase shifters, directional couplers, MRI, microstrip lines, optical storage systems, and many more. Although the practical fabrication of metamaterials is under development, a lot of methods have been explored, such as thin wire arrays plus SRR structures, transmission line approach, which involves the artificial additive loading of a series capacitance and shunt inductance to host transmission lines; and photonic crystals that are regarded as candidates for the fabrication of three-dimensional electromagnetic wave manipulating structures, including metamaterials optical waveguides (Kim 2004).

Given that the electromagnetic response of a metamaterial can, in principal, be engineered, there has been considerable interest in developing metamaterial optical waveguides. The modes supported by metamaterial waveguides exhibit a wide range of properties with applications to improving the design of optical and near ultraviolet systems. The properties of modes in both a slab and cylindrical waveguide with a lossy metamaterial cladding have been investigated. Energy loss is present in metamaterials due to the nature of the materials used to construct them and due to effects caused by the artificial structure built into them. These losses could prove to be a major obstacle in the implementation of metamaterial optical waveguides. However, a metamaterial–dielectric waveguide can support surface and ordinary TM modes as well as hybrid ordinary-surface TM modes. The metamaterial guide is able to support these hybrid modes over a much wider frequency region than was observed for a metal waveguide. Some modes have rapidly changing attenuation curves where the change depends on the structural parameters, thereby enabling a metamaterial–dielectric waveguide to act as a frequency filter. A metamaterial–dielectric waveguide can support surface modes with a lower attenuation than is possible for the same mode in a metal–dielectric waveguide, thereby opening up the possibility of low-loss metamaterial waveguide devices. Additionally, the low-loss surface mode is reminiscent of the near-zero-loss surface mode observed on a single-interface metamaterial–dielectric waveguide. This single-interface waveguide is used to confine the fields in order to enhance optical nonlinearity for all-optical switching on a photonic level (Moiseev et al. 2010). In contrast to a single-interface waveguide, the low-loss surface mode of a cylindrical waveguide confines the fields in the transverse direction, which could further enhance the required optical nonlinearity.

In addition, metamaterials often derive their extraordinary properties from surface plasmon waves which are collective oscillations of free-electrons on the surface of metallic nanostructures. These surface plasmon waves are characterized by their extremely short wavelength and thus provide a natural interface to couple light to much smaller nanoscale devices more effectively. The so-called plasmonic metamaterials operate simply by harnessing properties of resonant surface plasmons. Amplification of evanescent waves, achievement of NRI such as in fishnet metamaterial, extraordinary transmission enhancement, and enhanced Raman scattering are some of the surface plasmons driven phenomena that offer great opportunities for some applications. Optical imaging with subdiffraction resolution, nanolithography, and detection of chemical and biological species with single molecule sensitivity represent some of the possibilities. For example, a planar silver (Ag) film (termed as superlens) is one of the simplest forms of plasmonic metamaterials with an extraordinary ability to beat the diffraction limit through amplification of evanescent waves. Artificial plasmonic metamaterials also offer an opportunity to engineer surface modes over a wide range of frequency by simple surface patterning. Patterning also allows strong confinement and enhancement of resonant plasmon modes compared to flat metal films. These metallic patterns are often utilized in surface-enhanced Raman spectroscopy (SERS) as sensing substrates. The intricate structure of these novel metamaterials and devices is derived from physics-driven design for desired properties and applications. These designs require development of viable manufacturing and novel characterization techniques. These optical metamaterials have been explored for applications in optical regime, namely telecommunication including optical modulator and waveguides, optical imaging beyond diffraction limit, and chemical sensing (Chaturvedi 2009).

11.2 Fabrication Techniques of Optical Metamaterials

On the path to new prospects for manipulating light, fabrication techniques of artificially engineered metamaterials have been explored with a great deal of attention. When rationally designed, metamaterials demonstrate unprecedented electromagnetic properties that are unattainable with naturally occurring materials; by changing the design of the unit cell or "meta-atom," the optical properties of the metamaterial can be tailored, e.g., the values of permeability, μ, and permittivity, ε, can be controlled (Boltasseva and Shalaev 2008). Figure 11.2 shows history of metamaterial fabrication: from SRR to fishnet structure; and other types of unit cells/meta-atoms (Reza 2008; Bayatpur 2009).

As shown in Fig. 11.2, a first step towards the realization of a material with a negative index of refraction was done through periodically arranged split-ring resonators (SRRs) and parallel wires. Combining the negative-permeability structures with the negative-permittivity material (for example, simple metal wires arranged in a cubic lattice) can result in a negative-index material. Such a metamaterial was straightforward to accomplish in the microwave region. The approaches for moving to shorter wavelengths were initially based on concepts from the microwave regime (such as SRRs) with scaled down unit cell sizes. The main idea was that the magnetic resonance frequency of the SRR is inversely proportional to its size. Changing the double SRR to a single one and reducing the size of the structure, such magnetic structures were successfully realized from the microwave, to THz and near-infrared (IR) optical range, 100 and 200 THz (Reza 2008). Using single SRRs, this approach works up to about 200 THz. However, this scaling breaks down for higher frequencies for the single SRR case because, for wavelengths shorter than the 200 THz range, the metal starts to strongly deviate from an ideal conductor. For an ideal metal with infinite carrier density, hence infinite plasma frequency, the carrier velocity and the kinetic energy are zero, even for finite current in a metal coil. For a real metal, velocity and kinetic energy become finite. For a small SRR, nonideal metal behavior leads to a modified scaling law where the frequency approaches a constant and becomes independent of the SRR size. This scaling limit combined with the fabrication difficulties of making nanometer-scale SRRs along with metal wires led to the

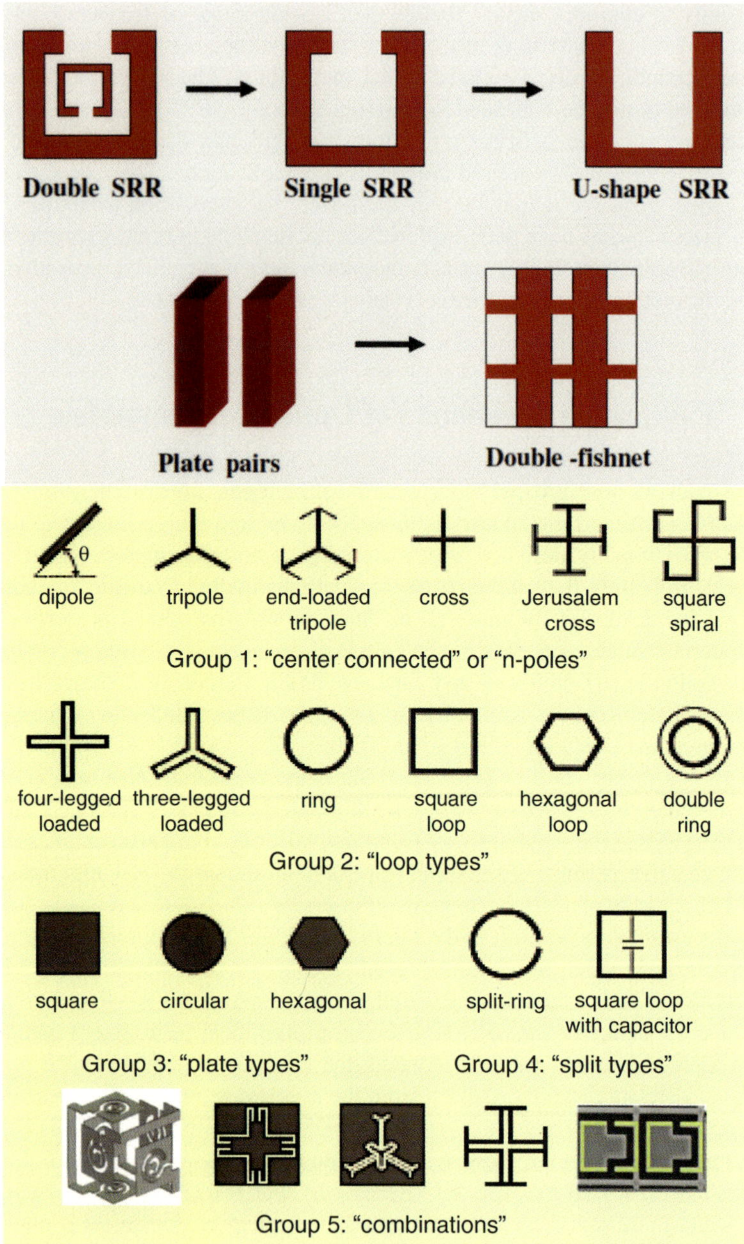

Fig. 11.2 History of metamaterial fabrication: from SRR to fishnet structure; and other types of unit cells/meta-atoms (Reza 2008; Bayatpur 2009)

development of alternative designs that are more suitable for the THz and optical regimes. One suitable design for optical NIMs is based on pairs of metal rods (also called "cut-wires") or metal strips, separated by a dielectric spacer. Such structures can provide a magnetic resonance $\mu < 0$ originating from antiparallel currents in the strips. An electric resonance with $\varepsilon < 0$ can result from the excitation of parallel currents in the same strips. However, normally it is difficult to get the $\varepsilon < 0$ and $\mu < 0$ regions to overlap so a different design was proposed, the so-called fishnet structure. This structure combines the magnetic coupled strips (providing $\mu < 0$) with continuous "electric" strips that provide $\varepsilon < 0$ in a broad spectral range. NIMs at optical wavelengths were successfully demonstrated by using coupled cut-wires and fishnet structures. To create a metamaterial at optical wavelengths, one should deal with small periodicities (about 300 nm and less) and tiny feature sizes (about 30 nm) to ensure effective-medium-like behavior. Thus, the fabrication of optical NIMs is challenging when aiming at high precision, high-throughput, and low-cost manufacturing processes. Feature sizes for metamaterials operating in the infrared or visible range can be smaller than the resolution of state-of-the-art photolithography (due to the diffraction limit), thus requiring nanofabrication processes with 100- or sub-100-nm resolution. Due to the limitations of current nanolithography tools, fabricated metamaterials do not really enter the real meta-regime where the unit cell size is orders of magnitude smaller than the wavelength. However, the feature size is typically small enough compared to the wavelength so that one can describe such material, at least approximately, as a medium with effective permeability, μ and permittivity, ε, which is in contrast to photonic crystals where the lattice period matches the wavelength (Boltasseva and Shalaev 2008).

In addition to subwavelength resolution, careful material selection is required for NIM fabrication. The selection of the metal for NIMs is crucial in the optical regime because the overall losses are dominated by losses in the metal components. Thus, their lowest losses at optical frequencies make silver and gold the first choice metals for NIMs. Since the refractive index is a complex number $n = n' + in''$, where the imaginary part n'' characterizes light extinction (losses), a convenient measure for optical performance of an NIM is the figure of merit (FOM), defined as the ratio of the real and imaginary parts of n: $F = |n'|/n''$. Aside from the proper metal choice, loss compensation by introducing optically amplifying materials can be considered for achieving low-loss NIMs. To make use of the optical properties of metamaterials, where meta-atoms are designed to efficiently interact with both the electric and magnetic components of light, one has to create a slab of such a material with an interaction length of many light wavelengths. For all exciting applications to be within reach, the next issue to address (in addition to losses) is the development of truly three-dimensional (3D) NIMs in the optical range. The challenges are to move from planar structures to a 3D slab of layered metamaterial and to develop new isotropic designs. With the development of low-loss large-scale optical NIMs, many device concepts will emerge empowered by optical magnetism and new regimes of linear and nonlinear light–matter interactions. Materials to manipulate an object's degree of visibility, subwavelength imaging systems for sensors and nanolithography tools, and ultracompact waveguides and resonators for nanophotonics are only a few examples of possible metamaterial applications. For example, among the most exciting new applications for 3D low-loss metamaterials are those based on transformation optics, including hyperlenses that enable subwavelength far-field resolution and designs for optical cloaking (Boltasseva and Shalaev 2008). A great many efforts have been done in fabrication of metal–dielectric nanostructured metamaterials at optical wavelengths, and development of alternative manufacturing techniques that can be adapted for future NIM fabrication.

11.2.1 2D Metamaterial Structures

1. Standard method: electron-beam lithography (EBL)
   Due to the fact that the required feature sizes for optical NIM fabrication are smaller than the resolution of state-of-the-art photolithography, 2D metamaterial layers are normally fabricated using EBL. In EBL, a beam of electrons is used to generate patterns on a surface. Beam widths can be on the order of nanometers, which gives rise to the high nanoscale resolution of the technique. EBL is a serial process wherein the electron beam must be scanned across the surface to be patterned. The EBL technique is quite versatile at the point of initial design and preliminary studies of optical properties of metamaterials since it offers subwavelength resolution and almost complete pattern flexibility. Different NIM structures have been successfully fabricated by EBL. For example, A refractive index in a fabricated fishnet structure (lattice constant 600 nm) based on a sandwich of Ag (45 nm)–MgF_2 (30 nm)–Ag (45 nm) was obtained with $n' = -2$ at $\lambda \approx 1.45$ μm. Moving to shorter wavelengths, nano-fishnets were made in either Ag–MgF_2–Ag or Ag–Al_2O_3–Ag multilayer structures with the smallest features below 100 nm. The fabricated structures had good large-scale homogeneity. However, due to sub-100-nm features required for visible-range NIMs (68 nm minimum in-plane feature size at 97 nm thickness of the Ag–MgF_2–Ag sandwich), the aspect ratio (height/width) for these NIM stacks can exceed unity. This gives rise to significant fabrication challenges connected to the lift-off procedure, and to increased sidewall roughness. Even though EBL is normally used to write relatively small areas, large-area all-dielectric planar chiral metamaterials can also be fabricated by EBL. In this approach, the use of a charge dispersion layer solves stitching errors during the serial writing process, which enables the fabrication of good-quality structures without the limitations normally encountered by stitching errors and field alignment. The fabrication of optical NIMs is quite challenging due to the fabrication requirements of small periodicities (near and below 300 nm) and tiny feature sizes (down to several tens of nm). Thus, EBL is still the method of choice for fabricating metamaterials despite low throughput of the serial point-by-point writing and high fabrication costs. Since only small areas (of the order of 100×100 μm) can normally be structured within reasonable time and at reasonable costs, EBL does not offer a solution for the large-scale NIM fabrication

required by applications, where many square centimeters have to be nanopatterned (Boltasseva and Shalaev 2008).

2. Rapid prototyping: focused-ion beam (FIB) milling
For rapid prototyping of metamaterials, FIB milling techniques can be used. In FIB, a focused beam of gallium ions is used to sputter atoms from the surface or to implant gallium atoms into the top few nanometers of the surface, making the surface amorphous. Because of the sputtering capability, the FIB is used as a micro-machining tool, to modify or machine materials at the micro- and nano-scale. This technique has been used for fabricating magnetic metamaterials based on SRRs. Scaling of the SRR structure requires sub-100-nm gap sizes (down to 35 nm) for a 1.5 μm resonance wavelength, thus requiring state-of-the-art nanofabrication tools. For such small features, EBL-based fabrication requires time-consuming tests and careful optimization of writing parameters and processing steps, leading to relatively long overall fabrication times. In contrast, the rapid prototyping of complete structures can be fabricated via FIB writing in times as short as 20 min. The process is based on FIB writing and corresponds to an inverse process where the FIB removes metal (20 nm of Au) deposited on a glass substrate. After FIB writing the structure is ready and no further post-processing steps are required. FIB nanofabrication might be of preference for fabricating specific (e.g., SRR-based) designs of metamaterials as in the above example. However, for creating an optical NIM, SRRs have to be combined with other metallic structures that can provide negative permittivity. This will give rise to additional processing steps and delicate considerations regarding the choice of process parameters and materials. Moreover, moving SRRs from the telecom to the visible range will bring the process to the limit of size scaling in SRRs. Thus, due to design and material limitations combined with low throughput, FIB has certain limitation in its use for fabricating optical NIMs. However, for some specific designs and proof-of-principle experiments, for example, including nonlinear materials, FIB is the first choice for rapid prototyping (Boltasseva and Shalaev 2008).
3. Large-scale fabrication: interference lithography
The only large-scale manufacturing technique used so far by the integrated circuit industry is optical lithography (OL). Nowadays, OL continues to be extended and offers new ways of increasing its resolution, for example, by using immersion techniques that meet the industry needs for 45-nm half-pitch nodes. One type of OL, namely interference lithography (IL), is a powerful technique for the fabrication of a wide array of samples for nanotechnology. This fabrication technique is based on the superposition of two or more coherent optical beams forming a standing wave pattern. Being a parallel process, IL provides a low-cost, large-area (up to ~cm^2) mass-production capability. Moreover, multiple exposures, multiple beams, and mix-and-match synthesis with other lithographic techniques can extend the range of IL applicability. IL offers high structural uniformity combined with considerable, but not total, pattern flexibility while its resolution is now approaching the 20-nm scale. Since NIM

fabrication requires a periodic or quasiperiodic pattern, IL is an excellent candidate for large-area metamaterial fabrication. For example, a large-scale NIM created by making elliptical voids in an Au (30 nm)-Al_2O_3(75 nm)-Au(30 nm) multilayer stack was found to exhibit $n' \approx -4$ at 1.8 m. IL technique is compact, robust, does not require expensive cleanroom equipment and can provide sample areas up to many square centimeters by up-scaling the apertures of the optics. Given the simplicity and robustness of making a high quality, single layer of a metamaterial using IL, one can envision further investigations aiming at piling 2D layers to create a 3D structure. Such a transition to 3D fabrication will turn parallel IL process into a step-by-step procedure that would require alignment of subsequent layers. Even though for compact versions of IL such a transition might result in a time-consuming fabrication process due to multiple alignments, proper technique development would make it possible to optimize and automate the alignment procedure. Thus, IL can be considered to be one approach for making 3D optical NIMs (Boltasseva and Shalaev 2008).

4. High-resolution large-scale fabrication: nanoimprint lithography (NIL)
Another promising direction for the fabrication of production-compatible, large-area, high-quality optical NIMs at low processing cost and time is offered by NIL. As a next generation lithography candidate, NIL accomplishes pattern transfer by the mechanical deformation of the resist via a stamp rather than a photo- or electro-induced reaction in the resist as in most lithographic methods. Thus the resolution of the technique is not limited by the wavelength of the light source, and the smallest attainable features are given solely by stamp fabrication. Moreover, NIL provides parallel processing with high throughput. Since NIM fabrication requires high patterning resolution, NIL is well suited for large-scale production of optical NIMs, providing wafer-scale processing using standard cleanroom procedures combined with simplicity and low cost. For example, two types of NIMs operating at near- and mid-IR frequencies, respectively, were fabricated via NIL. The first structure comprised ordered “fishnet” arrays of metal–dielectric–metal stacks that demonstrated negative permittivity and permeability in the same frequency region and hence exhibited an NRI of $n' \approx -1.6$ at a wavelength near 1.7 m. In the mid-IR range, the metamaterial was an ordered array of fourfold symmetric L-shaped resonator (with a minimum feature size of 45 nm) that were shown to exhibit negative permittivity and a magnetic resonance with negative permeability near wavelengths of 3.7 and 5.25 m, respectively (Wu et al. 2007). The smallest achieved feature sizes were about 100 and 45 nm for near- and mid-IR NIMs, respectively. A more simplified NIL method for creating metallic 2D structures was also used based on direct, hot embossing into metals (for example, Al) using hard templates like SiC. In this approach, metal nanostructures can be obtained by printing directly on metal substrates without any further processing step (when compared to standard lithography and etching/deposition steps), thus simplifying the processing steps and lowering the production cost. Using these hard molds, successful pattern transfer directly to Al substrates as well as Au films has been achieved by pressing at room temperature. However, this method has

some challenges in the technical development of this approach, mostly connected to the fabrication of the hard stamps. Moreover, for the existing NIM designs it cannot be directly applied to the fabrication of optical NIMs due to specific requirements on geometry and materials (Boltasseva and Shalaev 2008).

11.2.2 3D Metamaterials

1. Making multiple metamaterial layers
   Multilayer NIM design is initially based on an NIM slab consisting of multiple layers of perforated metal–dielectric stacks (for 100 and 200 layers) that exhibit a small imaginary part of the index over the wavelength range for negative refraction. This established an approach for thick, low-loss metamaterials at infrared and optical frequencies. The modified structures with up to three functional layers (seven actual layers): The silver-based samples were fabricated by standard EBL, metal and dielectric depositions and a lift-off procedure with processing steps similar to those for creating a single NIM layer. The measured performance of the fabricated NIMs was found to be close to theory, and the retrieved optical parameters ($n' = -1$ at 1.4 m) did not change much with the number of functional layers, as expected for an ideal metamaterial. While the realization of such a three-functional-layer NIM can be seen as a first step towards three-dimensional photonic metamaterials, the fabrication of thicker NIM slabs using this approach will be increasingly difficult. This issue is related to the fact that in a standard deposition, lift-off procedure, the total thickness of the deposited layers is limited by the thickness of the patterned e-beam resist. For a successful lift-off procedure, the total deposited thickness should normally be at least 15–20 % less than the thickness of the resist, as shown in Fig. 11.3a. This value is usually not more than a couple 100 nm (for e-beam writing structures with sub-100-nm features). As mentioned above (2D fabrication via EBL), an aspect ratio exceeding unity poses significant fabrication challenges. Moreover, this fabrication procedure results in non-rectangular sidewalls, typically with an angle of about 10° with respect to the substrate normal on all sides (Fig. 11.3a). Obviously, this effect becomes particularly large for thick, multilayer structures. To overcome problems connected to the lift-off procedure and non-rectangular sidewalls, and to create thicker NIM stacks, an alternative approach, a stack of 3D optical NIMs, was realized through a layer-by-layer technique similar to that developed for 3D photonic crystal fabrication (Subramania and Lin 2004). Even though the developed layer-by-layer technique can be seen as a general method for the manufacture 3D optical NIMs, such a step-by-step method requires additional process development (like planarization techniques) and careful lateral alignment of different layers that are crucial for successful stacking. Together with EBL fabrication of a single layer, alignment procedures lead to increased fabrication time. Thus, this

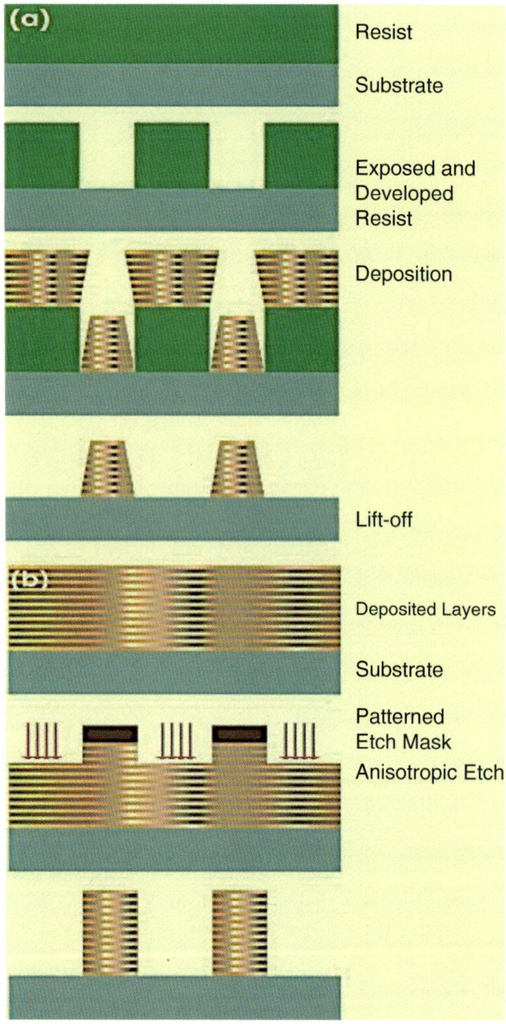

Fig. 11.3 Schematic of two possible ways of making multiple metal–dielectric layers (Boltasseva and Shalaev 2008): (**a**) the standard deposition—lift-off procedure that provides trapezoidal final structures and has a total deposited thickness limitation (it has to be at least 15–20 % less than the thickness of the resist for a successful lift-off process) together with (**b**) the etch-based procedure where a thick planar stack of metal–dielectric layers is deep etched to create a 3D metamaterial slab

approach is still too costly and has a low throughput for creating large-scale 3D metamaterial slabs for possible applications. Another way to truly create a 3D, multiple-layer NIM could be a process based on deep, anisotropic etching. In such an approach, one would first fabricate planar, alternating metal, and dielectric layers of any desired thickness, after which a deep etch would be performed using an etch mask pre-patterned to any required design by lithographic means (Fig. 11.3b). However, this approach requires both heavy material and process developments including the careful choice of etch-resistant mask materials and anisotropic etch optimization so that both the metal and dielectric layers can be etched (Boltasseva and Shalaev 2008).

2. Two-photon photopolymerization (TPP) technology
   The two-photon photopolymerization technique has been used extensively to realize 3D patterning. TPP involves the polymerization of a material via a nonlinear, multiphoton process that only occurs at the focal point of a tightly focused laser beam, thereby providing 3D control over the location of polymerization. TPP enables the fabrication of complex objects with diffraction resolutions since the absorption of light in the material occurs only at the focal region of the laser beam. This technique is a step forward from standard, single-beam laser writing into a polymer matrix, which is time-consuming and thus unlikely to be adopted for large-scale fabrication. The method, however, enabled the simultaneously writing of more than 700 polymer structures that were uniform in size. The metallization of the structures was then achieved through the deposition of thin films composed of small silver particles by means of electroless plating. A hydrophobic coating on the substrate prevented silver deposition in unwanted areas and allowed the formation of a large number of isolated and highly conducting objects. This of metallization is flexible in that it can produce either polymer structures covered with metal or numerous isolated, insulating polymer objects spread over a metallic film, depending on the resin properties and treatment procedures (Formanek et al. 2006). TPP-based laser writing is one of the most promising methods for manufacturing of large-area, true 3D metamaterials. Offering intrinsic 3D parallel processing capability with acceptable resolution (100 nm), TPP can be successfully combined with selective metal deposition by electroless plating. While TPP enables arbitrary sculpting below the micrometer scale, electroless plating can provide thin coatings of various metals. Going in this direction, process development and optimization is required in order to include metal–dielectric layers into a polymer template made by TPP. In addition to electroless deposition, multimaterial deposition (including metal like silver) can be achieved via chemical vapor deposition (CVD), which is a chemical process used to deposit high-purity materials. Both approaches need further advances and developments for creating thin, uniform, and smooth layers of different materials on a TPP-fabricated polymer matrix (Boltasseva and Shalaev 2008).
3. Fabrication of complex 3D structures
   Complex 3D metal–dielectric nanostructures have been fabricated by several techniques. Two typical methods include direct electron-beam writing (EBW) and focused-ion beam chemical vapor deposition (FIB-CVD). These methods offer 3D fabrication that is not possible using traditional layered optical and electron-beam lithographic techniques. The use of EBW was demonstrated for building structures of multiple layers with line width resolutions of 80–100 nm using an electron beam to cause direct sintering of 2–10 nm nanoparticles, while various free-space Ga- and W-containing nanowirings were successfully fabricated by FIB-CVD also called free-space-nanowiring fabrication technology. In the EBW process, an additive-layer build technique for multiple-material functionality was developed. The processing steps involved spin-coated or drawn down solutions of thiol-capped nanoparticles (either Ag or Au) onto silicon wafers, glass slides, or polyimide films. The resultant nanoparticle

films were patterned by electron-beam and showed conductivity for the patterned metals within one order of magnitude of their bulk material properties. Multiple-layer fabrication was successfully demonstrated by repeating the processing steps for subsequent layers spun over previously patterned features. Further material and process optimization (for example, to allow combinations of multiple materials) may lead to EBW adoption for fabrications of complex metamaterials. While these and similar nanofabrication techniques offer unique possibilities for making very complex 3D structures, they all suffer from severe material limitations in terms of what materials can be patterned/deposited. Moreover, such methods are complex and time-consuming. Thus they can only be used for making first prototypes or single structures for proof-of-principle studies.

Complex 3D metallic structures can also be fabricated using layer-by-layer repetition of standard planar processes. For example, 3D tungsten photonic crystals were fabricated using a modified planar silicon MEMS process. In this approach, a patterned silicon dioxide mold is filled with a 500-nm-thick tungsten film and planarized using a chemical mechanical polishing process.

This process is repeated several times, and at the end of the process the silicon dioxide is released from the substrate, leaving a freely standing thin patterned tungsten film. This method allows the patterning of multilayered (up to 60) samples of large area (~cm^2). Another process for the parallel fabrication of microcomponents is the LIGA microfabrication technique. LIGA is an acronym referring to the main steps of the process, i.e., deep X-ray lithography (DXRL), electroforming, and plastic molding. These three steps make it possible to mass-produce microcomponents at a low cost. For instance, the LIGA method was used to fabricate 3D photonic lattices. In this approach, the 3D lattice is patterned using DXRL to create a series of intersecting channels in a polymer. This technique makes it possible to create large area molds that can be filled with gold by electroplating techniques. Large-scale woodpile structures with gold as the matrix material were processed and demonstrated (Boltasseva and Shalaev 2008).

4. 3D structures by nanoimprint

As an ultrahigh resolution patterning technique simultaneously offering sub-50-nm resolution and sub-10-nm layer alignment capability, NIL can be used in multilayer processes for creating 3D structures. NIL has the potential for very large-scale and inexpensive manufacture of structures via multilayer lithography, alignment, and plating techniques. This rapid-prototyping contact lithography approach provides a platform for investigating new structures, materials, and multilayer alignment techniques that are critical for device designs at optical operational wavelengths. Three-dimensional polymer structures can also be realized by reverse-contact UV NIL (Kehagias et al. 2007), which is a combination of NIL and contact printing lithography. In this process, a lift-off resist and a UV cross-linkable polymer are spin-coated successively onto a patterned UV mask-mold. These thin polymer films are then transferred from the mold to the substrate by contact at a suitable temperature and pressure. The whole assembly is then exposed to UV light. After separation of the mold and the substrate, the unexposed polymer areas are dissolved in a developer solution leaving behind the negative features of the original stamp. This method delivers resist pattern transfer without a residual layer, thereby rending unnecessary the etching steps typically needed in the imprint lithography techniques for three-dimensional patterning. This method is reproducible over millimeter-scale surface areas and has already provided encouraging results for fabricating 3D woodpile-like structures for polymer photonic devices (Kehagias et al. 2007). Combined with multimaterial deposition similar to the TPP case, this approach might also be adapted for future NIM fabrication (Boltasseva and Shalaev 2008).

5. Self-assembly
Similar to the case of metamaterials, fabrication of 3D photonic crystals requires different techniques for creating periodic structures built of materials with alternating refractive indices. The fabrication of 3D photonic crystals tends towards synthesis methods based on self-assembly to realize such materials in the optical range. In the self-assembly approach, opals have received a strong backing from their ability to be used as scaffoldings for further templating of other materials (Galisteo et al. 2005). In the fabrication procedure, the formation of the templates (opals) is followed by the subsequent synthesis of guest materials such as semiconductors, metals and/or insulators and, if desired, additional 2D patterning for the design of new structures. Accurate amounts of silicon, germanium, silica, etc., can be grown in the interior the opal structures by CVD, allowing the fabrication of multilayer systems of different materials. The use of opals to pattern the growth of other materials (including metals) and subsequent vertical and lateral engineering of the fabricated structures have also been demonstrated (Galisteo et al. 2005). This method, though not directly applicable to metamaterial fabrication at the moment, comprises fabrication steps that are of vital importance for future 3D metamaterials, namely pre-patterned, controllable growth of different materials including metals. For the realization of 3D, metallic, periodic structures on a large scale, the self-organization of metal-coated colloid particles was successfully employed (Chen et al. 2004). However, similar to opal structures, this method is difficult to employ for metamaterial fabrication requiring the creation of specially designed shapes. Another self-assembly approach was suggested for the realization of 3D NIMs for the microwave and optical frequencies (Logeeswaran et al. 2006). By using a metal–dielectric, stress-actuated, self-assembly method, periodic arrays of metal flap SRRs and hinges were fabricated by lithographic patterning combined with metal deposition, lift-off and etching procedures. This approach offers large-scale fabrication and can in principle be scalable from microwave (100 GHz) to optical frequencies (300 THz). However, for creating optical NIMs, the careful choice of materials as well as the close control of geometrical parameters and release etching chemistry is required. Moreover, combined with the fabrication challenges, moving to the optical range will bring up the limit of size scaling in SRR functionality (Boltasseva and Shalaev 2008).

11.2.3 Thin Metal Film Deposition for Fabrication of Metamaterials

Aside from the requirements of nanometer-scale resolution and high throughput, highly controllable thin-film deposition methods are needed for the realization of good performance (low loss) NIMs. Whatever approach is chosen as a manufacturing method for optical NIMs, the possibility of creating thin metal and dielectric films with reduced surface roughness is vital. High roughness of the metal film is the main limit to obtaining lower-loss metamaterials since it leads to increased scattering losses in the system and can annihilate the negative-index effect. Depending on the chosen approach, different deposition techniques have to be studied and optimized. For example, in the standard lithography—deposition—lift-off procedure, the roughness of the e-beam-evaporated metal films can be reduced by the proper choice of the deposition conditions. The quality of the deposited metal film depends also on the quality of the dielectric spacer layer. Especially in the case of an unstable metal like silver, the roughness of the initial dielectric structure is very important since clusters and lumps on a dielectric surface can, for example, work as seeds for the induced silver restructuring. The conventional way to improve the quality of the metal film is to use a lower deposition rate and a dielectric material with stronger adhesion and better surface quality. For example, the decrease of the deposition rate from 2 to 0.5A°/s resulted in improved surface roughness and, hence, better optical performance of a magnetic metamaterial. However, the decrease in the deposition rate leads to heating of the structure during deposition that makes the following lift-off process more difficult. Thus, higher complexity, multistep deposition procedures were required. In the case of silver, which is the most-used metal for optical NIMs due to its optical properties, air exposure is also a problem since silver degrades under ambient conditions. The deposition of a dielectric layer on top of the structure can help to prevent silver deterioration. In order to improve the structure further, annealing can also be applied. When considering approaches that are feasible for future 3D NIM fabrication, specifically direct laser writing based on TPP or 3D interference lithography yielding complex polymer structures, deposition methods different from standard deposition and sputtering techniques must be developed. Here, metal deposition by electroless plating seems to be one of the possible directions. This approach has been applied to selectively deposit metal on a polymer structure. Electroless plating has also been shown to be a promising way of achieving controllable deposition of thin metal films with low roughness (average roughness below 2 nm). In light of future NIM manufacturing where controllable coating of complex 3D structures is required, alternative metal deposition techniques like CVD, polymer-assisted and nanoparticles-assisted deposition have to be explored as well (Boltasseva and Shalaev 2008).

Consequently, the fabrication of optical negative-refractive-index metamaterials is quite challenging, due to the requirements of 100- and sub-100-nm feature sizes of the meta-atoms and small periodicities on the order of 300 nm or less. Due to the high-resolution requirement, EBL is still the first choice for fabricating small-area metamaterials. Writing larger areas requires long e-beam writing times and hence, boosts the operation cost. Thus, this approach is only suitable for proof-of-principle studies. Similar to EBL, other serial processes, for example, the FIB milling technique, are not considered to be feasible for the large-scale metamaterial fabrication required by applications. One approach to manufacturing high-quality NIMs on a large scale is provided by interference lithography. To increase the resolution, IL can be combined with self-assembly techniques. Moreover, this technique could also be applied to the fabrication of future 3D metamaterials by piling 2D layers into a 3D structure. Another promising approach to create large-scale, high-quality metamaterials is NIL. NIL offers nanoscale resolution; it is a parallel process with high throughput and, hence, a good candidate for NIM fabrication. Since NIL requires a stamp made by other nanofabrication techniques (like EBL) it is ideal for parallel production of already optimized metamaterials, when the preliminary test structures were patterned via EBL. Thus, NIL can be seen as a large-scale, low-cost process of making EBL-written structures that offers solutions to the intrinsic EBL drawbacks. The first steps towards the realization of a 3D NIM were made by creating a multilayer structures (instead of a single functional layer) and by utilizing a layer-by-layer technique. Both approaches of making stacked metamaterials still have limitations (like challenging lift-off procedure in the first method and alignment requirements in the second). Another possible direction of making multiple-layer NIMs could be to consider deep anisotropic etching of a pre-fabricated multilayer stack. While complex 3D nanostructures can be fabricated by several techniques (for example, direct EBW and FIB-CVD), these methods are too complex and time-consuming to be adapted for large-scale NIM fabrication. A fabrication method that is now considered to be one of the most promising approaches for future manufacturing of large area, true 3D metamaterials is based on two-photon photopolymerization techniques. While offering sub-diffraction resolution (down to 100 nm) due to a nonlinear multiphoton process, this technique possesses intrinsic 3D processing capability. In addition to direct single-beam laser writing of complex structures into a polymer matrix, large-scale 3D polymer structures for future real-life applications can be realized via a 3D, multiple-beam TPP technique. When combined with selective metal deposition by electroless plating or via CVD, TPP can enable arbitrary 100-nm scale sculpturing of metal–dielectric structures. However, this approach needs further advances and developments for creating thin, uniform, and smooth layers of different materials on a TPP-fabricated polymer matrix. Three-dimensional, multilayered, polymer and metallic structures can also be realized by NIL. This method offers high reproducibility over large areas (millimeter scale) and has been used for fabricating 3D cubic arrays of gold cubes and 3D woodpile-like polymer structures for photonic crystal-based devices. Combined with multimaterial deposition, these approaches might also be adapted for future NIM fabrication. To reach real NIM applications, several tasks have to be fulfilled: loss reduction, large-scale 3D fabrication, and new isotropic designs. Careful material choice (for example, new crystalline metals with lower absorption instead of traditional silver and gold) and process optimization (reduced roughness and high uniformity of the materials) can help on the way to creating low-loss optical NIMs. Another possibility is to introduce a gain material into the NIM, thus compensating for losses. Even though it is still a long way to truly three dimensional, isotropic, negative-index metamaterials at optical frequencies, several fabrication approaches do seem to be feasible. With emerging techniques such as nanoimprint, contact lithography, direct laser writing, and possibly new types of self-assembly, it seems likely that truly 3D metamaterials with meta-atom sizes much smaller than the wavelength can be created. In the next generation of optical NIMs, for any chosen manufacturing approach, the careful choice of materials and process optimization will be required in order to obtain high-quality structures. Thus, the selection of a fabrication tool needs to be based on careful considerations of the structural quality able with the suggested method and the associated cost (Boltasseva and Shalaev 2008).

11.3 Metamaterial Waveguiding Principle

A plane wave propagating in a metamaterial, with electric and magnetic fields can be described as (Reza 2008)

$$\varepsilon(r,t) = R_{\mathrm{e}}\{\boldsymbol{E}\exp[i(\boldsymbol{k}\cdot\boldsymbol{r}-\omega t)]\} \tag{11.1a}$$

$$H(r,t) = R_{\mathrm{e}}\{\boldsymbol{H}\exp[i(\boldsymbol{k}\cdot\boldsymbol{r}-\omega t)]\} \tag{11.1b}$$

From the Maxwell curl equations,

$$\boldsymbol{k}\times\boldsymbol{E} = i\frac{\omega}{c}\mu\boldsymbol{H} \tag{11.2a}$$

$$\boldsymbol{k}\times\boldsymbol{H} = -i\frac{\omega}{c}\varepsilon\boldsymbol{H} \tag{11.2b}$$

where ε and μ are the relative electric permittivity and magnetic permeability, respectively, and $\boldsymbol{k}$ is the wave vector. It is defined by

$$k^2 = \frac{\omega^2\varepsilon\mu}{c^2} \tag{11.3}$$

As evidenced from (11.2a) and (11.2b), $\boldsymbol{E}$, $\boldsymbol{H}$, and $\boldsymbol{k}$ form a left-handed set of vectors when ε and μ are simultaneously negative, and form a right-handed set of vectors when ε and μ are positive. However, it is meaningless to define handedness of the system when only ε or μ is negative because the wave vector is imaginary in this case. Moreover the phase velocity of a wave (represented by the wave vector) is opposite to the movement of the energy flux of the wave in a metamaterial medium. In a metamaterial with both negative ε and μ; the phase velocity of a wave is always opposite to the movement of the energy flux of the wave. In addition, from (11.3), the index of refraction n can be expressed as (Reza 2008)

$$n^2 = \frac{k^2}{k_0^2} = \varepsilon\mu \tag{11.4}$$

where negative ε and μ leads to negative index of refraction. Both negative and positive index of refraction solutions satisfy Maxwell's equations and boundary conditions for refraction of light. However, the negative solution for the index of refraction is selected in metamaterials to satisfy the requirement that in the refracted beam the energy flows away from the interface.

In a lossy metamaterial, ε and μ are complex and expressed as (Reza 2008):

$$\varepsilon = \varepsilon' + i\varepsilon'' \tag{11.5a}$$

$$\mu = \mu' + i\mu'' \tag{11.5b}$$

where $\varepsilon' < 0$; $\mu' < 0$ and $\varepsilon'' > 0$; $\mu'' > 0$, assuming that a metamaterial medium is absorbing. Using (11.4), the index of refraction for a passive medium with small absorption is

$$n = \pm\sqrt{\varepsilon\mu} = \pm\sqrt{(\varepsilon'\mu' - \varepsilon''\mu'') + i(\varepsilon'\mu'' - \varepsilon''\mu')} \approx \pm\left[\varepsilon'\mu' + i\frac{\varepsilon'\mu'' + \varepsilon''\mu'}{2\varepsilon'\mu'}\right] \tag{11.6}$$

The $\varepsilon' < 0$; $\mu' < 0$ and $\varepsilon'' > 0$; $\mu'' > 0$ conditions lead to

$$\frac{\varepsilon'\mu'' + \varepsilon''\mu'}{2\varepsilon'\mu'} < 0 \tag{11.7}$$

Since the imaginary component of refraction index is positive in a lossy medium, (11.7) demands that the negative sign of the square root in (11.6) be chosen. Therefore, the real component of index of refraction must be negative if both ε and μ are negative.

For a slab waveguide that consists of three layers, the center layer is the core, and the width of the waveguide, w, is the thickness of the core. If choosing the width of the slab waveguide to be $w = 4\pi c/\omega_e$, the frequency at which w is equal to the wavelength of the field is near the middle of the frequency range in which the metamaterial is metallic-like. The core is surrounded on both sides by a cladding. These cladding layers are assumed to be sufficiently thick to substantially contain the evanescent fields, as the effective width of the waveguide (the width of the core plus the skin depth at the interface) is typically only slightly larger than w. The permittivity and permeability are constant in the core. Assuming

(a) The slab waveguide supports two distinct classes of modes, transverse magnetic (TM) and transverse electric (TE). The TM modes are those modes where the magnetic field components are non-zero only in the transverse direction (H_z, $x = 0$) whereas the TE modes have electric field components which are non-zero only in the transverse direction (E_z, $x = 0$).
(b) The existence of TM surface modes is partially dependent on the cladding having $\varepsilon' < 0$ and $\mu' > 0$, whereas the TE modes require $\varepsilon' > 0$ and $\mu' < 0$ in the cladding.

For the TM modes, the dispersion relation can be expressed as (Lavoie et al. 2012)

$$\frac{\varepsilon_1}{\gamma_1}\left(\frac{\varepsilon_2}{\gamma_2}+\frac{\varepsilon_3}{\gamma_3}\right) = -\left(\frac{\varepsilon_1^2}{\gamma_1^2}+\frac{\varepsilon_2\varepsilon_3}{\gamma_2\gamma_3}\right)\tanh\gamma_1\omega \tag{11.8}$$

which must be satisfied for the existence of a propagating mode. Here

$$\gamma_j = \sqrt{\left(k_z^2 - \omega^2\varepsilon_j\mu_j\right)} \tag{11.9}$$

are the complex wave numbers for the transverse components of the fields for $j = 1, 2, 3$ referring to the core and two cladding layers, respectively.

For a cylindrical waveguide that consists of a dielectric cylindrical core of circular cross-section with radius a. A cladding layer surrounds the core and is assumed to be arbitrarily thick. In general the modes of a cylindrical waveguide are not TE or TM but in-between modes that contain all of the field components and are called HE or EH. The names HE and EH allude to the fact that all components of both electric and magnetic fields are non-zero. The TM and TE modes of the cylindrical waveguide are special cases of the general solution. Using the definitions (Lavoie et al. 2012)

$$J'_m(ak_1) = \left.\frac{\partial J_m(rk_1)}{\partial r}\right|_{r=a} \tag{11.10}$$

and

$$K'_m(a\gamma_2) = \left.\frac{\partial K_m(r\gamma_2)}{\partial r}\right|_{r=a} \tag{11.11}$$

where J_m and K_m are the Bessel function and modified Bessel function, respectively, the dispersion relation for the cylindrical guide is (Yeh and Shimabukuro 2008)

$$\frac{k_z^2 m^2}{a^2\omega^2}\left(\frac{1}{k_1^2}+\frac{1}{\gamma_2^2}\right)^2 = \left(\frac{\mu_1}{k_1^2}\frac{J'_m(ak_1)}{J_m(ak_1)}+\frac{\mu_2}{\gamma_2^2}\frac{K'_m(a\gamma_2)}{J_m(a\gamma_2)}\right)\left(\frac{\varepsilon_1}{k_1^2}\frac{J'_m(ak_1)}{J_m(ak_1)}+\frac{\varepsilon_2}{\gamma_2^2}\frac{K'_m(a\gamma_2)}{J_m(a\gamma_2)}\right) \tag{11.12}$$

Here γ_2 is the complex wave number in the radial direction for the cladding and is given by (11.9), and $k_1 = i\gamma_1$ is that for the core, whereas m is an integer characterizing the azimuthal symmetry. With the dispersion relations for both geometries, the propagation constants can be determined for the allowed modes.

A planar metamaterial waveguide can support surface polariton modes, modes with an evanescent profile decay from the interface. These types of modes had been observed at metal surfaces and had been used for near-field imaging. However, these modes have richer properties in metamaterials; for example, the metal waveguide can guide only transverse magnetic, TM, waves while the metamaterial slab can guide both polarizations of light. Thus, the existence of surface polariton modes in metamaterials opened the door for further applications. Not only can the metamaterial waveguide support surface polariton modes but also there are new types of modes (complex surface or leaky modes) with immediate applications to leaky antennas. Moreover, the properties of the normal guided modes of the metamaterial waveguide are completely different than in a conventional waveguide. For example, no cut-off thickness for the first mode of metamaterials waveguide. Thus, this feature provides a solution to the problem of energy transmission with lateral cross-section below the diffraction limits. Moreover, the fundamental mode (the mode with no node in the modal profile) does not exist in a metamaterial waveguide. By using this feature, concept of surface wave suppression has been introduced which leads to enhance radiation efficiency for microstrip antennas or introduce a one dimensional periodic waveguide with a complete band gap. Usually, 3D photonic crystals (PhC) have complete band gap; however, 1D PhC containing matamaterials also exhibit this property.

A multilayer system containing metamaterials supports new types of transmission band gaps, the so-called zero-n gaps, with unique properties. The novel gap is completely different from the conventional Bragg reflection gap, as it is insensitive to the direction or polarization of the beam.

In the 1D PhC containing metamaterials, the spontaneous emission of an atom can be completely suppressed because metamaterials provide the possibility of vanishing optical path length between two points. The vortex-like energy flux structure is the most important property of a metamaterial waveguide structure (MWS). Almost all unusual properties of metamaterial waveguides such as field localization, super-waveguiding (a theoretical design which can transmit light with high-power density), and trapped-rainbows (an idealized structure which could stop light over broadband range of frequency) come from this feature. This phenomenon happens at the interfaces of the metamaterial waveguide because the energy in a metamaterial waveguide core flows in the opposite direction to that in the cladding. One can potentially slow down or even stop the light passing through an MWS due to the vortex-like energy structure. For example, a theoretical wedge-shaped MWS could stop light over a broadband range of frequencies at room temperature producing the so-called trapped-rainbow. By controlling the thickness of the waveguide, the optical path of the light could be set to zero because of the negative Goos-Hanchen shift in metamaterials. The trapped-rainbows have also been created in a hybrid metamaterial-photonic crystal structure, in a surface plasmon in a graded metallic structure, in graphene for an electronic system, and in an anisotropic MWS. In addition, not only can light be slowed down in an anisotropic metamaterial waveguide but the velocity of a particular slow-light mode is almost constant over a broadband range of frequencies. Moreover, the effect of material loss is an inherent feature of metamaterials, as all currently methods to create negative-index metamaterials are based on the use of metallic structures and all real metals introduce some losses particularly at optical frequencies. The other methods rather than lossy metallic structures might be used to create negative-index metamaterials. One could employ a photonic crystal in its negative-index regime or introduce chirality or optically amplifying materials (gain mechanism) to remove losses in negative-index metamaterials. However, loss with dispersion is the inherent features of negative-index materials and it is impossible to compensate loss completely over a broadband range of frequencies. To ensure a positive energy density and to satisfy causality, any negative-index material must have dispersion (Reza 2008).

11.4 Modes of Metamaterial Waveguide Structures

The metamaterial–dielectric slab waveguide can support a number of TM modes with three distinct mode behaviors: ordinary, surface, and hybrid modes. Each mode can display more than one behavior, with different behaviors supported for different frequency regions. The mode behavior is determined by the relative sizes of the wave number γ_1 perpendicular to the interface inside the core. Ordinary wave, surface wave, and hybrid wave are used when referring to the character, or behavior, of a mode and use the term mode or explicitly name the mode (TM_j) for a specific solution to the dispersion relation. Ordinary waves are caused by traveling electromagnetic fields inside the core reflecting from the core-cladding interfaces. The reflected fields overlap with the incident fields causing an interference effect, with constructive interference corresponding to the allowed modes. The condition for ordinary wave behavior is Im $(\gamma_1) \gg$ Re (γ_1), which ensures that, in the slab guide, the fields have an oscillating wave pattern in the transverse direction of the core (Lavoie et al. 2012).

Surface waves are the result of electrons at the core-cladding interface carrying energy along the waveguide. The energy from the fields is transferred to the electrons causing them to oscillate, providing a medium for energy transport. This type of coherent oscillation of electrons at a surface is called a surface polariton or surface plasmon polariton (SPPs). The electromagnetic field of the surface mode exponentially decays as a function of the distance from the interface. Mathematically the condition for surface waves is Re $(\gamma_1) \gg$ Im (γ_1), which causes the intensity to be concentrated around the interfaces between the core and cladding and a relatively small intensity through most of the core. In slab guides only two surface wave TM modes are supported, namely symmetric (TMs) and antisymmetric (TMa) according

Table 11.1 The different regions of the metamaterial, determined by the parameter choice, are identified by the sign of the real part of both the permittivity and the permeability (Lavoie et al. 2012)

Frequency range	Sign of ε', μ'	Property
$\omega_a < \omega$	$\varepsilon' > 0, \mu' > 0$	Dielectric-like
$0.3\omega_a \leq \omega < \omega_a$	$\varepsilon' < 0, \mu' > 0$	Metal-like
$0.2\omega_a \leq \omega \leq 0.3\omega_a$	$\varepsilon' < 0, \mu' < 0$	Negative-index media
$\omega \leq 0.2\omega_a$	$\varepsilon' < 0, \mu' > 0$	Metal-like

to the symmetry of the electric field component normal to the core-cladding interface (Lavoie et al. 2012).

In frequency regions where both Re (γ_1) and Im (γ_1) are comparable, the mode is a hybrid wave. Hybrid waves can be understood as a product of two features in the transverse direction, namely an evanescent feature and an oscillatory feature. To see this consider the components of the electric field in the core of the slab waveguide, which have the form (Lavoie et al. 2012)

$$E = E_0 e^{\gamma_1 x} e^{i(k_z z - \omega t)} = E_0 e^{\gamma'_1 x} e^{i\gamma''_1 x} e^{i(k_z z - \omega t)} \tag{11.13}$$

where E_0 is a constant, the complex wave number $\gamma_1 = {\gamma_1}' + i{\gamma_1}''$, with ${\gamma_1}'$ and ${\gamma_1}''$ both real, and the terms $e^{\gamma'_1 x}$ and $e^{i\gamma''_1 x}$ represent surface wave and ordinary wave behavior, respectively. The hybrid waves are then a combination of surface waves, i.e., fields being carried by electron oscillations, and ordinary waves due fields traveling in the core and reflecting off the interfaces thereby resulting in interference effects.

For hybrid waves, a substantial portion of the energy is transferred along the interfaces as well as in the core, so it is like a combination of ordinary waves and surface waves. Hybrid waves are not seen in dielectric–dielectric waveguides and only for a very narrow frequency range of the TM_a mode in metal–dielectric waveguides, where the mode changes from a surface wave to an ordinary wave. Table 11.1 shows that the different regions of the metamaterial, determined by the parameter choice, are identified by the sign of the real part of both the permittivity and the permeability. The metamaterial–dielectric cylindrical waveguide can support the same three mode types as the slab waveguide though they differ slightly from those of the slab waveguide, as the cylindrical waveguide has a circular symmetry. The fields of the ordinary wave TM modes in the cylindrical waveguide have the oscillating wave pattern of a vibrating circular membrane. The azimuthal symmetry is characterized by the integer m, which determines the number of oscillations in the fields over an angle of 2π. The fields in the cladding decay exponentially, as Re $(\gamma_2) \gg$ Im (γ_2). Due to the circular symmetry of the cylindrical waveguide it can support only one surface wave TM mode. However, the cylindrical waveguide can support a number of surface wave HE modes of different orders with the symmetry determined, as with the ordinary wave TM modes, by the azimuthal parameter m (Lavoie et al. 2012).

The metamaterial waveguide supports the surface polariton modes for both the TE and TM polarizations. In contrast, the conventional waveguide cannot support surface polariton modes. At the metallic interface, only the magnetic transverse, TM, modes can be supported; however, a metamaterial waveguide can support both TM and TE surface polariton modes; moreover, it supports complex modes. The complex modes are the complex solution of the dispersion equation above the cut-off frequencies. The propagation wave number is always complex in this type of mode, even in a lossless structure. However, a lossless MWS always supports proper complex modes. Moreover, it is possible for both TE and TM modes to have the same number of nodes since one of them propagates in the same direction of the group velocity, (forward propagating mode), and the other propagate in the opposite direction of the group velocity, (backward propagating mode); hence, both forward and backward modes are supported by the metamaterial slab. In addition, the first bound mode, TE_0, does not exist in MWSs. This property gives an opportunity to introduce one dimensional periodic waveguides with a complete band gap, introduce the concept of surface wave suppression which leads to enhance radiation efficiency for microstrip antennas, and introduce the slow-light modes under the single mode operation. However, in the realistic metamaterial waveguide, which should be dispersive and lossy, and also by considering the role of the complex modes, this property could not been seen anymore. Finally, not only does the metamaterial waveguide support evanescent and oscillatory field distributions, but also supports linear field distribution. This unique mode is a transitional mode between the first magnetic or electric transverse, TE_1 or TM_1; oscillatory and evanescent modes. The electric field distributions for the TE mode lead to constant longitudinal magnetic field for these modes. The constant longitudinal magnetic field gives more degrees of freedom for potential applications in a metamaterial waveguide (Reza 2008).

A rich variety of modes (bound, surface polariton, and complex modes) exist in the MWS. The bound modes are restricted between the vacuum light-line and metamaterial light-line. The TE_1 mode behaves like a bound mode whenever its propagation constant. As the propagation constant of TE_1 goes below the metamaterial light-line, it behaves like a surface polariton mode. The TE_2; and TE_3 modes behave like a bound mode when the propagation constant of the dispersion curve is purely real and behave like a complex leaky mode when the propagation constant is complex. Even for a lossless structure, the propagation constant has a complex value for complex modes. In the complex modes, the energy is not confined in the core and leaks out through the cladding. As the complex modes do not carry energy in the lossless MWS, they do not play an important role in the structure. Moreover, there are domains of frequencies where there are no bound modes; for instance, there is no bound mode in the range of 2 and 3 GHz. In contrast with the dielectric waveguide modes, the modes in the metamaterial waveguide are not restricted to lie between the vacuum light-line and the metamaterial light-line. For example, the complex modes go above the vacuum light-line and the TE_1 mode goes below the metamaterial light-line. The TE_1 mode appears at frequencies lower than the TE_0 mode. As the propagation constant increases, the TE_1 mode changes from a bound mode to a surface polariton mode. There is a transition point on the metamaterial light-line between the TE_1 bound mode and surface polariton mode. At this point, the electric field distribution is linear or the magnetic field is constant in the core of the waveguide structure. Thus, the TE_1 mode takes on different diverse types of modes, such as bound mode, complex mode, and surface polariton mode, at different propagation constants. Although the shape of TE_0 is almost insensitive to changes in the thickness of the waveguide's core, the TE_1 curve changes dramatically when the thickness is changed. Nevertheless, there is no limit on the reduction of the slab thickness: all the modes are preserved by changing the thickness of the slab. This feature, which is contrary to that in a conventional waveguide, leads to possibility to design ultrathin slab waveguides. When decreasing the slab thickness, the dispersion curves shift to lower frequencies and concentrate near resonance frequency region. However, in the dispersionless waveguide structure, modes shift to higher frequencies with decreasing slab thickness (Reza 2008).

In summary, many of the exciting properties of dispersionless and lossless MWS are preserved for the dispersive MWS although the dispersion changes the dispersion curves of the bound modes. The MWS supports a rich variety of modes with unusual properties. These properties provide more control over electromagnetic fields and open a door to exciting potential applications, such as miniaturizing devices and systems with slow-light modes. Dispersion and loss are inherent features in metamaterials. The properties of modes can change greatly when loss comes into play. Metamaterials can change the electromagnetic properties of many different physical structures. Even a slab waveguide containing metamaterial shows exotic features that may have important potential applications, such as a vortex-like energy flux structure, and stopped-light modes. A double vortex-like energy flux pattern forms at the two interfaces of a waveguide with the core made of metamaterial and claddings made of positive-index-of-refraction materials. The vortex-like energy structure forms because the propagating energy in a metamaterial waveguide core flows in the opposite direction to that in the cladding. This property plays a key role in controlling light propagation in a metamaterial waveguide. The vortex-like energy flux structure property leads to a slowing down or even a stopping of the light traveling through the MWS. Such an anomalous feature has aroused strong interest due to its potential applications in data storage, optical buffers, and optical sensing. Moreover, the vortex-like surface wave property may lead to an extremely high power density in a metamaterial waveguide, which may find potential applications in medical treatments, and industry (Reza 2008).

11.5 Metamaterial Modulators

Metamaterials have been demonstrated as promising candidates for optical modulation. These artificial materials consisting of discrete set of metal–dielectric composite structures are used to mimic the properties of bulk materials. These discrete elements can be designed to achieve a desired response in a frequency

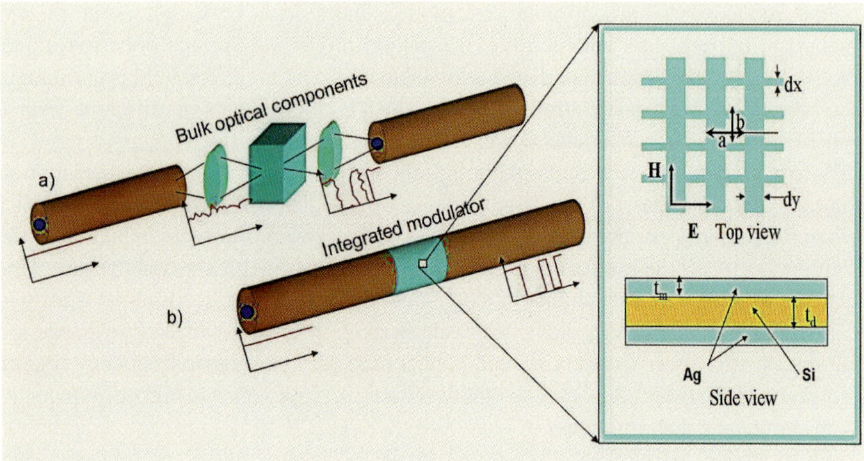

Fig. 11.4 Schematic illustration of metamaterial modulator integrated directly to an optical fiber (Chaturvedi 2009): (**a**) conventional fiber-optic communication systems require bulk optical components such as lenses, modulator, and amplifiers to generate and transport information; (**b**) an integrated modulator design eliminates the need for bulk components. Inset zoom: schematic of Ag–Si–Ag fishnet structure. Vector **H** and **E** denote the directions of magnetic and electric field, respectively

range not readily accessible with natural materials. In particular, the response can be tuned optically or electrically by including active elements in the structural unit. For example, excitation of charge carriers in a constituent semiconductor layer or the substrate can lead to modulation of optical properties such as effective refractive index and resonance frequency of the metamaterial. The inherent resonant nature of the metamaterial response enhances the effect of active elements. Moreover, since the structural unit of metamaterial can be very small compared to wavelength, the realization of compact photonic devices only several wavelengths in footprint is a distinct possibility. For example, using gold SRR on a thin semiconductor substrate a tunable optical response was initially achieved in the terahertz regime. These active metamaterial devices find tremendous potential in telecommunication and fiber-optic systems. While SRR design shows a strong promise as a terahertz modulator, however, fiber-optic systems require devices operating in near infrared (IR) wavelengths. Scaling SRR to optical frequencies and obtain photoconductive switching in IR is extremely challenging fabrication-wise, as it requires very small structural dimensions. Furthermore, linear scaling of resonant wavelength of SRR design with its structural dimensions breaks down and resonant response starts to saturate near optical frequencies. This happens because of deviation of metal from perfect electric conductor behavior at higher frequencies. One of the improved designs as fishnet metamaterial operating in near-IR regime is a metal–dielectric–metal sandwich structure (Fig. 11.4b inset). This sandwich structure arranged in the form of cross-wires can be designed to have simultaneous negative values of magnetic permeability (μ) and electric permittivity (ε);

a feature leading to NRI. Relative ease of fabrication, operation in near-IR frequency range, and metal–dielectric composite structure which allows switching by modulating the dielectric layer, makes this design a promising candidate for optical modulation. Although the metamaterial itself is small in size (~10–20λ, where λ is free-space wavelength), optical fiber communication systems require several bulk components (e.g., lenses, alignment optics) to couple light out of a fiber into the modulator and then back into the optical fiber, as shown in Fig. 11.4a. Correspondingly, the free-space propagation introduces additional losses and noise into the signal. Optical amplifiers are often required to compensate for these losses. However, a metamaterial modulator can be integrated directly to an optical fiber, thus eliminating the need for bulk optical components. The modulator design is based on silver (Ag)–silicon (Si)–silver fishnet structure (Fig. 11.4b) that allows modulation in near-IR frequency range with photoexcitation of carriers in silicon layer (Chaturvedi 2009).

11.5.1 Free-Space Fishnet Metamaterial Modulator

To design an integrated fiber-optic modulator based on fishnet metamaterial, the free-space modulation effect has been investigated with full scale three-dimensional numerical simulations by CST Microwave Studio to obtain the geometric parameters of Ag–Si–Ag fishnet metamaterial and its performance as an optical modulator. Silver is modeled as a dispersive lossy metal with permittivity governed by Drude model, whereas Si is modeled as a nondispersive lossless material with $\varepsilon = 11.9$. The structure is embedded in free-space and is excited by waveguide simulator, which allows modeling of the free-space problem as a bounded simulation (Weiland et al. 2001). The conduction current density (small compared to displacement current) in the two metal strips is oppositely directed, thus forming a current loop and generating a magnetic resonance. At magnetic resonance, the structure assumes simultaneous negative values of magnetic permeability and electric permittivity and hence the structure is better impedance matched to surroundings ($Z(\omega) = \mu(\omega)/\varepsilon(\omega)$, where Z represents impedance at frequency ω). This leads to a dip in reflection at 1,760 nm. However, away from resonance, magnetic permeability achieves positive values while electric permittivity is still negative. This causes a large impedance mismatch and therefore transmission achieves a minimum. The spectral location and quality factor of this resonance is strongly dependent on the complex refractive index of the sandwich layer and this is central to the function of fishnet metamaterial as an optical modulator. For example, the refractive index of Si layer can be modulated by utilizing a plasma-dispersion effect, thereby detuning the resonance wavelength and achieving optical modulation. For one fishnet layer with the dimensional parameters of the fishnet structure as illustrated in Fig. 11.4, $t_m = 28$ nm, $t_d = 80$ nm, $d_x = 108$ nm, $d_y = 250$ nm, $a = b = 550$ nm, by modulating the refractive index of Si by just 1.7 %, transmission is modulated by 6 dB (75 % change) and phase undergoes a shift of 35°.

In comparison, an equally thick Si layer by itself undergoes a transmission modulation of less than 0.2 dB (4 %). By stacking five layers of this sandwich structure, it is possible to build a Mach-Zehnder interferometer with a total interaction length of just 680 nm, an order of magnitude smaller than other nonlinear materials. Therefore, fishnet metamaterial design can be an efficient and compact stand-alone component. However, to build an optical modulator for fiber-optic communication systems would require several other bulk components which to a certain extent nullifies the advantages of having a compact central unit. By integrating the modulator directly onto the fiber, for instance, the bulk optical components can be eliminated, leading to significant cost reduction. This integrated design offers smaller footprint, low loss, high efficiency, self-alignment, and is less prone to electromagnetic interference (Chaturvedi 2009).

11.5.2 Integrated Fishnet Metamaterial Modulator

A fishnet optical modulator made on side-polished fiber is illustrated in Fig. 11.5a (Chaturvedi 2009). By fabricating this modulator onto a fiber flat (after removal of a section of the fiber's cladding) permits evanescent interaction of guided modes of the fiber with the device. An external optical modulation signal, thus allows the possibility of modulating transmitted output signals from the fiber. Fishnet metamaterial was fabricated using FIB milling (Fig. 11.5b) onto the flat side of a commercially available D-shaped fiber. This integrated design is based on mode coupling between fishnet and fiber-guided modes. The transmission and reflection characteristics of a fishnet designed to be resonant near 1,500 nm (reflection dip). Near the metamaterial resonance, light is coupled into the fiber through surface modes and is guided by it. Light guided by a fiber interacts with fishnet only in the form of evanescent modes. The evanescent light is coupled to the resonant mode of the metamaterial (Chaturvedi 2009).

Optical modulation of the effective properties of the fishnet metamaterial can be accomplished with Ag–Si–Ag heterostructure by photoexcitation of carriers in Si layer. The resonant nature of the structure allows modulation of the effective refractive index by as much as 40 % with small changes in Si refractive index (~1.7 %) as indicated by free-space simulations shown earlier. One such case of modulation of effective properties of fishnet can be achieved with pumping the Si intermediate layer with visible light. While on one hand the design offers extremely small footprint, it has certain drawbacks. First, the fiber-guided mode couples weakly to the resonance mode of the metamaterial compared to free-space coupling and second, relatively moderate modulation depths. Under fiber-guidance light is incident on the metamaterial at steep angles, and this reduces the resonance strength as some diffracted modes start to propagate through the metamaterial. Resonance strength is also low because of losses in the metamaterial which is also responsible for broadening of resonance. Modulation depth is rather limited because of the fact that the switching layer is buried underneath a metal film which requires relatively

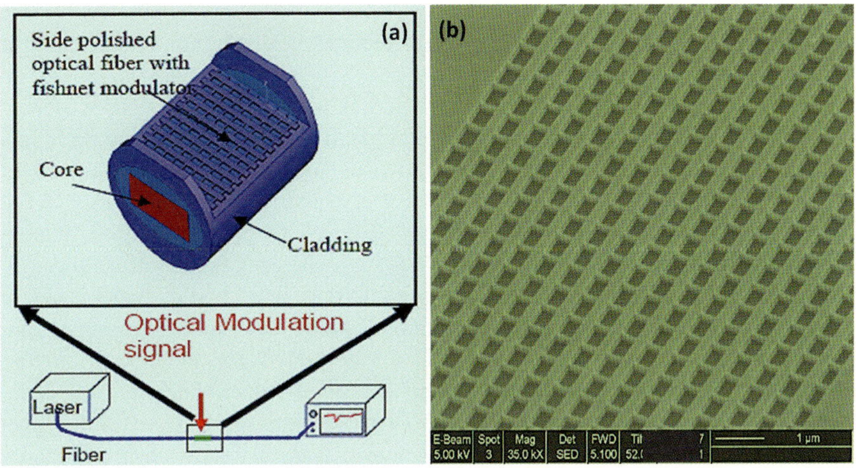

Fig. 11.5 Illustration of fishnet modulator integrated on polished optical fiber (Chaturvedi 2009): (**a**) assembly sketch; and (**b**) secondary electron image depicts fishnet metamaterial fabricated onto a single D-fiber

higher pump intensities to bring the modulation effect. One of the most fundamental challenges with the prevalent designs of metamaterials is the presence of losses. These losses originate from intrinsic absorption of constituent materials, specifically metals which are highly lossy at optical frequencies. Resonant nature of the metamaterial and topological effects such as surface roughness also contributes to the losses. These losses severely hinder the performance of metamaterials and restrict their range of practical applications. One of the approaches to compensate loss is inclusion of an optically pumped gain media. Gain media can be incorporated in close proximity of the metamaterial or can be an integral layer of the metamaterial itself. The former approach was investigated theoretically and experimentally where an $In_{0.786}Ga_{0.214}As_{0.465}$ $P_{0.535}/In_{0.53}Ga_{0.47}As$ quantum well structure was used as a substrate providing gain (gain coefficient, $g = 3{,}000\ \mathrm{cm}^{-1}$) to the fishnet metamaterial fabricated on top (Ponizovskaya and Bratkovsky 2009). It was observed that even at a nominal gap of 20 nm (in the form of a spacer layer) between the gain media and fishnet metamaterial, the gain had a very little effect (<2 %) on the properties of the metamaterial. This is because electromagnetic field is concentrated mainly inside the dielectric layer of the fishnet structure and it penetrates weakly into the quantum well structure, leading to poor coupling between the metamaterial and underneath gain media. An alternative to this approach is to replace the passive dielectric layer in fishnet metamaterial with active gain media. This can be accomplished by using dye molecules/quantum dots in a polymer matrix as the dielectric layer. Yet another approach to reduce losses in fishnet metamaterial is to cascade multiple fishnet layers in the direction of wave propagation. While it may seem counterintuitive, but increasing the number of lossy metal layers indeed leads to reduction in overall loss and increased FOM of the metamaterial. This is due to destructive interference of antisymmetric conduction currents in the metal films which effectively cancels out current flow in intermediate layers and reduces the ohmic losses. To reduce losses in the metamaterial at oblique incidences, the individual magnetic resonator unit of fishnet is flipped. As the wave propagates through the fiber, the individual flipped units experience different phase, which allows the metal films to have antisymmetric currents and hence low loss. This design also offers an enhanced modulation effect as the switching media is directly exposed to pump radiation, thereby improving the overall efficiency of integrated modulator (Chaturvedi 2009).

As a result, fishnet metamaterial has been used as an optical modulator with an on/off ratio of 1.8 dB for on-fiber communication and information processing applications. This design offers small footprint (~10λ) and integration on fiber eliminates the need for bulk optical components. To reduce the losses associated with fishnet metamaterial and improve coupling to fiber-guided modes, a flipped fishnet design is used, which has metal–dielectric–metal sandwich in a direction perpendicular to conventional fishnet. This design offers several advantages: Reduction in ohmic losses, as the antisymmetric currents in adjacent metal strips lead to destructive interference at oblique incidences; secondly, enhanced modulation effect, as the switching layer is directly exposed to pump radiation. With less than 12 % of incident pump power compared to conventional fishnet, flipped fishnet shows a modulation depth of 2.05 dB of fiber-guided modes. This small footprint, high efficiency metamaterial opens avenues for telecommunication applications (Chaturvedi 2009).

11.6 Superlens

One of the most striking properties of NIM materials is that a slab of NIM can be a perfect lens in which the evanescent waves, instead of decaying, are in fact enhanced through the slab. This offers the possibility of restoring or recovering the "lost treasures," theoretically making a perfect image without any deterioration. The light emitted or scattered from an object includes not only propagating waves but also evanescent waves, which carry the subwavelength detail of the object. The evanescent waves decay exponentially in any medium with a positive refractive index so that they cannot be collected at the image plane by a conventional lens, and this results in a diffraction-limited image. However, if a lens made of NIM is placed close to an object, the near-field evanescent waves can be strongly enhanced across the lens. After emerging from the NIM lens, the evanescent waves decay again until their amplitudes reach their original level at the image plane. On the other hand, the propagating waves pass through the NIM lens with both negative refraction and a reversed phase front, leading to zero phase change at the image plane. By completely recovering both propagating and evanescent waves in phase and amplitude, a perfect image is created, as shown in Fig. 11.6 (Pendry 2000; Zhang and Liu 2008).

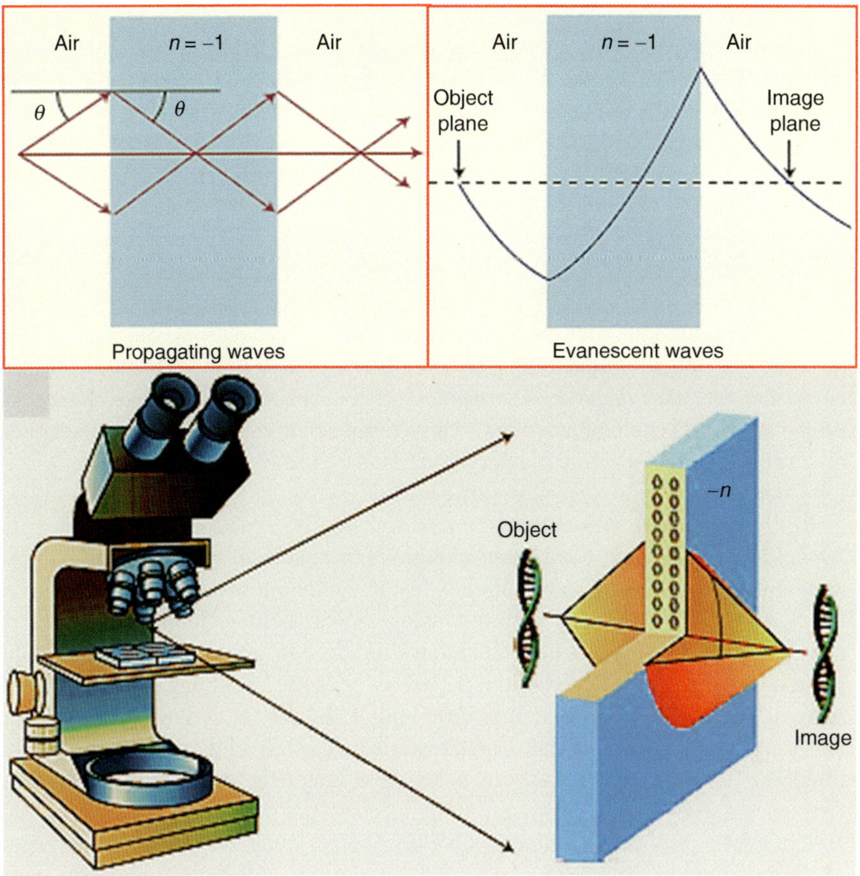

Fig. 11.6 A microscope based on an ideal NIM lens should focus both propagating and evanescent waves into an image with arbitrarily high resolution (Zhang and Liu 2008)

Such intriguing behavior relies on the fact that the NIM supports resonant surface waves, one example of which is SPPs. Evanescent waves can be efficiently coupled into surface modes and enhanced by their resonant nature when their wave vectors are matched. Because the amplitude of the coupled evanescent waves measures the stored energy in the material, the enhancement can be built up by a few cycles of electromagnetic oscillations in the surface resonance. Because the evanescent waves do not carry any net energy flux, the energy can never be amplified; only the distribution of the energy or field will be modified across the space. Resonance-based negative-refractive-index media constructed from realistic materials, however, are inherently associated with substantial energy dissipation or loss (that is, the imaginary part of ε and μ), which hinders the resolution of the "perfect image." The term "superlens" is used for lenses that take this practical limit into account. Unlike conventional lenses, the resolution of the superlens is not limited by diffraction, but rather determined by the enhancement and by how many evanescent modes can be restored. Another consequence of the material loss and imperfections is that the distance between the slab and both the object and its image, as well as the slab's thickness, must all be small compared with the wavelength if meaningful resonant enhancement of the evanescent waves can be obtained. When their scale is deeply subwavelength, the electrostatic limit can be applied. In this case, the electric and magnetic responses of a material decouple, and only one material property (ε or μ) needs to be negative to support resonant surface waves for one specific polarization (transverse magnetic, TM, or transverse electric, TE mode). In this special case, the superlens effect therefore remains valid in a single negative-index medium for one polarization of light. As a natural candidate having negative ε at optical frequencies, a slab of silver (because of low loss in metal) was suggested for achieving the superlens effect for TM waves in optics. Superlens properties have been characterized in terms of resolution, inherent loss, dielectric mismatch, and also geometrical optimizations (Smith et al. 2003; Zhang and Liu 2008).

More specifically, conventional optical imaging is capable of focusing only the far-field or propagating component of light. The near-field or evanescent component with subwavelength information is lost in a medium with a positive refractive index, giving rise to diffraction-limited images. Near-field scanning optical microscopy and cathodoluminescence techniques are able to image surface and optical properties with far better resolution. However, being scanning techniques, the images have to be acquired in a point-by-point fashion. In contrast, a thin planar lens made up of a negative-index metamaterial is capable of parallel subdiffraction imaging. It is not easy to ensure a negative magnetic permeability at optical frequencies. Fortunately, however, in the electrostatic near-field limit, the electric and magnetic responses of materials are decoupled. Thus, for transverse magnetic (TM) polarization, having only negative permittivity suffices to obtain the near-field "superlensing" effect. This makes metals with relatively lower losses such as silver (Ag), natural candidates for superlensing at optical frequencies. Exciting quasistatic surface plasmons of a thin silver film allows the recovery of evanescent waves, thus providing subdiffraction images in the near-field. The device termed as silver superlens demonstrated parallel subwavelength imaging capability for arbitrary nano-objects. While the planar silver superlens can resolve deep subwavelength features, the imaging is limited to near field. This is because planar superlens doesn't alter the evanescent decaying nature of subwavelength information. A far-field superlens (FSL) operating at near-infrared (IR) wavelengths that allows resolving subwavelength features in the far field. By utilizing evanescent enhancement provided by plasmonic materials such as silver nanorods and Moiré effect, subwavelength information of an object can be converted to far-field or propagating information which in turn, can be captured by conventional optical components. A simple image restoration algorithm can then be used to reconstruct the object with subwavelength resolution. This unique class of optical superlenses with potential molecular scale resolution capability will enable parallel imaging and nanofabrication in a single snapshot, a feat that is not yet available with other nanoscale imaging and fabrication techniques such as atomic force microscopy and EBL (Chaturvedi 2009).

11.6.1 Superlensing in the Near Field

Superlenses have been realized in both microwave and optical frequencies but with different designs. A microwave metamaterial was constructed from wires and SRRs forming a two-dimensional waveguide that can be treated as homogeneous because the 3-cm wavelength is much larger than the lattice constant (Houck et al. 2003). With an input antenna placed 2 cm away from a rectangular slab of metamaterial, a focused image was obtained by scanning the detector on the imaging side of the objective sample. By using a negative-index planar transmission-line structure, an image below the diffraction limit (at $\lambda/5$) was reported at ~1 GHz frequency. Although these metamaterials experienced significant loss and are very sensitive to frequency and their electromagnetic environment, the experimental results demonstrated the superlens effect in the microwave range (Zhang and Liu 2008).

The optical superlensing effect was observed in 2005 using a thin slab of silver, a single negative-ε material that could effectively image 60-nm features ($\lambda/6$), well below the diffraction limit. The sub-diffraction-limited image was recorded by optical lithography at 365 nm wavelength as shown in Fig. 11.7a–c. The presence of the silver superlens improved the image resolution remarkably in the near field by excitation of surface plasmons. Using a similar scheme, superlensing effect was confirmed in silver films. The concept of the basic single-layer silver superlens was also improved by lamination of the metal into many layers (Melville and Blaikie 2005; Zhang and Liu 2008).

Theoretically, silver superlens is capable of $\lambda/20$–$\lambda/30$ image resolution (where λ is incident wavelength). However, challenges remain to realize such a high resolution imaging system, such as minimizing the information loss due to evanescent decay, absorption, or scattering. The thickness of spacer layer (separating the object and the lens) and that of silver film are the two major governing factors that determine subwavelength information loss due to evanescent decay and material absorption. Particularly, the surface morphology of silver film plays a significant role in determining the image resolution capability. Below a critical thickness silver is known to form rough islandized films. Rougher films perturb the surface plasmon modes causing loss of subwavelength details and hence diminished resolution. Producing thin, uniform, and ultrasmooth silver films has been a holy-grail for plasmonics, molecular electronics, and nanophotonics. Smoothing thin silver films have resulted in novel approaches capable of producing ultrathin silver films with atomic-scale roughness. An intermediate ultrathin germanium (Ge) layer (~1 nm) is introduced before depositing Ag. Utilizing Ag–Ge surface interactions, smooth superlens down to 15 nm Ag thickness has been fabricated. Introducing the Ge layer drastically improves Ag surface morphology and helps minimize the island cluster formation. The RMS roughness of Ag (over 1 × 1 μm scan area) deposited on quartz substrate

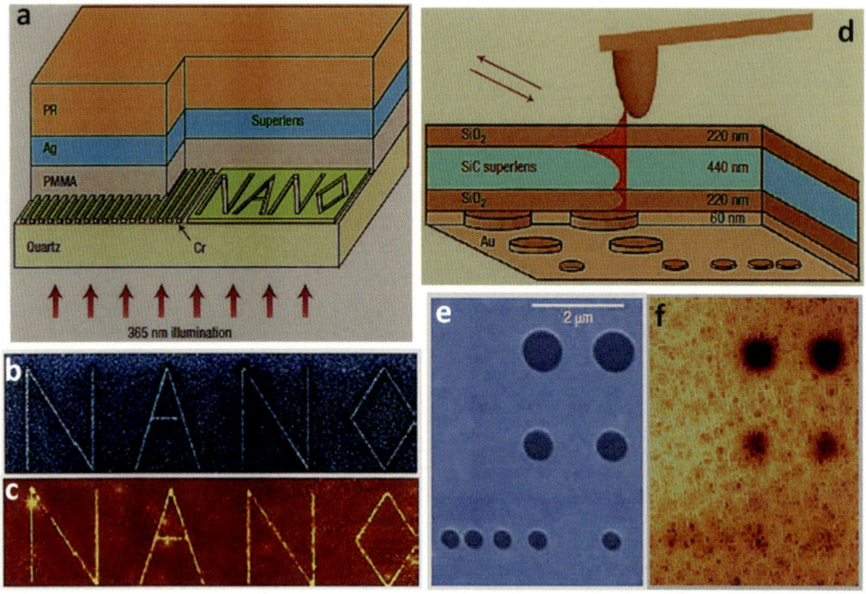

Fig. 11.7 Near-field optical superlenses (Zhang and Liu 2008): (**a**) schematic configuration of optical silver superlens. PR-photoresist layer; PMMA-polymethylmethacrylate. The superlens image can be revealed by means of photolithography at wavelength $\lambda = 365$ nm. (**b**) Focused ion-beam image of the object with a 40-nm linewidth. (**c**) Atomic force microscopy of the developed image on photoresist with a silver superlens. The presence of the superlens improved the resolution to 89 nm from an average line width of 321 ± 10 nm without the superlens. (**d**) Schematic configuration of the SiC superlens. The image transferred through the superlens can be recorded by near-field scanning microscopy. (**e**) Scanning electron microscope (SEM) image of the object. (**f**) Amplitude image through the SiC superlens measured by near-field scanning microscopy at $\lambda = 10.85$ μm

improves from 2.7 nm down to 0.8 nm by introducing Ge. 20 nm thick Ag is capable of transferring a broad range of strongly evanescent modes and can exceed $\lambda/11$ half-pitch resolution. Adding Ge is generally unfavorable at UV wavelengths, as it is absorptive. However with only 1 nm thick Ge in Ag–Ge superlens, the evanescent decay is only significant for feature sizes below $\lambda/12$. Computed PSF of such a superlens has full-width at half-maximum (FWHM) of 23 nm. An object grating constructed with FWHM of 30 nm at 60 nm pitch 1 μm when convoluted with this PSF gives an image grating with FWHM = 37 nm. Moreover, the intensity contrast appearing in the image ($r = I_{max}/I_{min} \sim 3$) is sufficient to resolve this object with most commercial photoresists (PR) using superlens photolithography. In contrast, a near-field lens without Ag layer (e.g., spacer 27 nm thick) gives a PSF with FWHM of 45 nm (Chaturvedi 2009).

An SiC superlens at mid-infrared frequency, using optical phonon resonance enhancement, showed even better feature resolution in terms of wavelength ($\lambda/20$) because of the low loss in the SiC material, as shown in Fig. 11.7d–f. Despite challenges in fabricating ultra-flat surfaces on either side of the superlens (surfaces roughness being detrimental to the surface resonances and enhancement), these experimental realizations of the optical superlens clearly showed the potential for imaging and lithography below the diffraction limit. Negative refraction and imaging in photonic crystals using low-loss dielectric materials, can be an approach to counter the loss that NIM must deal with. The key to the superlens effect in photonic crystals is to design appropriate dispersion so as to achieve negative refraction for all angles. Although the lensing effect is similar to the NIM case, it arises from a different mechanism: Bragg scattering in a photonic crystal rather than refraction in NIM. The ultimate image resolution therefore is limited by the period of the crystal, and the image contains some distortions as well, because the evanescent components cannot be uniformly enhanced by the surface mode for all wave vectors (Zhang and Liu 2008).

11.6.2 Superlenses Projecting Far-Field Images

When a beam of light hits an object, the object's information is transferred to the scattered light with various wave vectors which comprise both propagating and evanescent components. The propagating waves carry large feature information and can reach the far field, whereas evanescent waves carry fine details, but are confined to the near field. If a conventional glass lens is used to collect the scattered light, the evanescent waves are permanently lost before reaching the image plane. Thus, the resolution of the final image is always "diffraction limited." However, the evanescent fields can be enhanced and thus contribute to a sub-diffraction-limited image if a superlens with NRI is placed close to the object. If the superlens only possesses either single negative ε or single negative μ, the superlens effect remains valid only for transverse magnetic or transverse electric modes of light, respectively. By adding a coupling element to the superlens, the enhanced evanescent waves can be coupled into propagating waves, making far-field detection possible. High-resolution images can be reconstructed by collecting the far-field signals from such a lens. Taking another approach, the hyperlens uses a piece of artificial metamaterial to transfer the deep subwavelength information into the far field. The evanescent waves from the object can become propagating waves in such a strongly anisotropic metamaterial. With the help of the hyperlens geometry, the waves gradually reduce their wave vector values along the propagation direction in the metamaterial, and thus the waves can continue to propagate even after leaving the hyperlens (Zhang and Liu 2008).

FSL, that project a sub-diffraction-resolution image into the far field, has been made based on a silver superlens with additional nanoscale corrugations on its top surface as shown in Fig. 11.8a. This lens not only enhances the evanescent waves but also converts them into propagating waves. A proper transfer function of the FSL is crucial, because it ensures the "uniqueness" required for the reconstruction process to form sub-diffraction-limit images. The concept was first tested with a carefully designed silver FSL geometry working at 377 nm wavelength (Liu et al. 2007a, b).

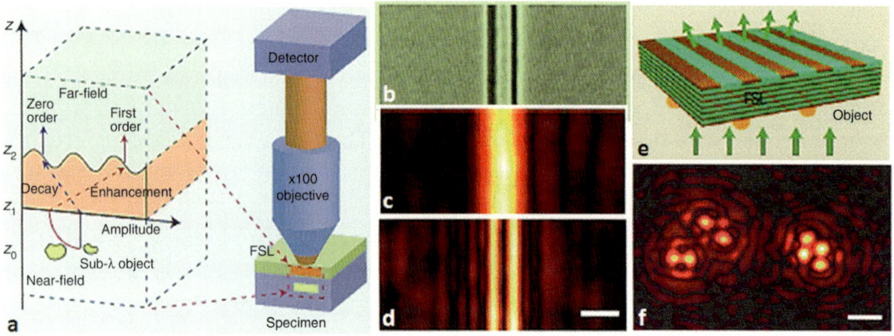

Fig. 11.8 Far-field optical superlens (Zhang and Liu 2008): (**a**) a far-field superlens is constructed by adding a subwavelength grating onto a silver slab superlens. It can selectively enhance the evanescent waves from the object and also convert them into propagating waves. An FSL microscope can be produced by inserting an FSL between the specimen and objective lens of a regular microscope. (**b**) SEM image of a nanowire pair object with 50-nm-wide slit in Cr film separated by a 70-nm gap. (**c**) Diffraction-limited image from a regular microscope cannot resolve the two nanowires (numerical aperture = 1.4, $\lambda = 377$ nm). (**d**) Reconstructed FSL image clearly resolves the sub-diffraction details at $\lambda = 377$ nm. Scale bar, 200 nm. (**e**) A new generation of FSL in which the original silver slab is replaced by a silver-dielectric multilayer structure so that the lens works at visible wavelengths. (**f**) A numerical demonstration of two-dimensional sub-diffraction-limited imaging at 405 nm wavelength using the multilayer FSL. The image was reconstructed by collecting signals in the far field from six different FSL grating orientations. The object consists of particles with a 40-nm radius and 100-nm smallest center-to-center distance. Scale bar, 200 nm

After inserting the FSL into a conventional optical microscope, two nanowires of 50 nm width separated by a 70-nm gap were clearly imaged (Fig. 11.8b, d). As a comparison, without the FSL, the same optical microscope cannot resolve the details of such nanostructures, as shown in Fig. 11.8c (Zhang and Liu 2008).

A tunable FSL is capable of working at any visible wavelength as shown in Fig. 11.8e. The silver slab in the original design was replaced by a slab of silver dielectric multilayers. The mechanism of evanescent wave enhancement for a broad range of wave vectors comes from surface plasmon mode splitting, so that a multilayer-based FSL can be designed to work at any frequency below the surface plasmon frequency (Xiong et al. 2007). Similar multilayer structures without coupling elements have been used for near-field superlenses, but the working wavelength has to be close to the surface plasmon frequency and the bandwidth is rather narrow. A two-dimensional arbitrary particle of 40 nm radius can be imaged with sub-diffraction-limited resolution by a multilayer FSL, as shown in see Fig. 11.8f (Zhang and Liu 2008).

Another approach is to utilize Moiré effect mediated by excitation of surface plasmons allowing recovery of subwavelength information in the far field. By carefully designing a subwavelength grating, it is possible to achieve a "frequency mixing" of evanescent fields from the object and grating. For example, a metamaterial substrate consisting of periodic array of silver nanorods has been designed, which have plasmonic resonance in IR regime and the near-field enhancement associated with this plasmonic resonance fulfills the key requirement for frequency mixing of evanescent fields from the nanorods and the object. This near-field frequency mixing leads to formation of Moiré features that are of propagating nature and can be recorded with a conventional microscope. A simple image reconstruction algorithm can then be utilized to recover subwavelength spatial details of the object from the acquired far-field image.

Object features corresponding to 2.5 μm period are recorded in far field with an incident wavelength of 6 μm, indicating a far-field imaging resolution capability of $\lambda/2.4$. This imaging scheme can be easily interfaced with FTIR microscopes and would enable real-time imaging with ultrahigh resolution (Chaturvedi 2009).

11.6.3 Hyperlens as an Optical Turbine

The FSL obtains a sub-diffraction-limited image in the far field using a two-stage process: evanescent wave enhancement via surface resonance, and subsequently, conversion into a propagation wave at the exit surface via a designed surface scatter. An alternative approach uses an anisotropic medium in curved multilayer stacks. The evanescent waves can tunnel through flat films to produce a high-resolution image. However, the image exists only at the near field of the final layer. An anisotropic metamaterial with hyperbolic dispersion (Fig. 11.9a) could provide magnification in a cylindrical geometry. When ordinary evanescent waves enter such anisotropic media, they immediately become propagating. What is interesting is that their large transverse wave vectors can be gradually compressed as they propagate outwards, until they are small enough to be truly propagating in air or surrounding dielectrics outside this cylindrical medium, projecting a magnified image into the far field. The cylindrical anisotropic lens has been termed a "hyperlens" because of the hyperbolic function of the dispersion in such a metamaterial. Practically, the hyperbolic curve in Fig. 11.9a needs to be designed to be nearly flat so that the waves with different transverse wave vectors will all propagate with the same phase velocity along the radial direction, which is important in forming an undistorted image in the far field. The hyperlens essentially works as an optical compressor or turbine that continuously compresses or transforms the large wave vectors of the original evanescent waves into smaller ones that propagate to the far field. The first optical hyperlens was successfully created by conformal film deposition of alternating Ag and Al_2O_3 thin films on a prefabricated quartz substrate molded with a cylindrical cavity (see cross-section scanning electron micrograph in Fig. 11.9b) (Liu et al. 2007a, b). On combining such a hyperlens with a conventional optical microscope, an image with 130-nm resolution, well below the diffraction limit of 260 nm, was directly observed in the far field (Fig. 11.9c, d). In addition to the magnification in one dimension shown in this work, a two-dimensional hyperlens is possible with a spherical geometry. A different approach to making a magnifying superlens used two dimensional surface

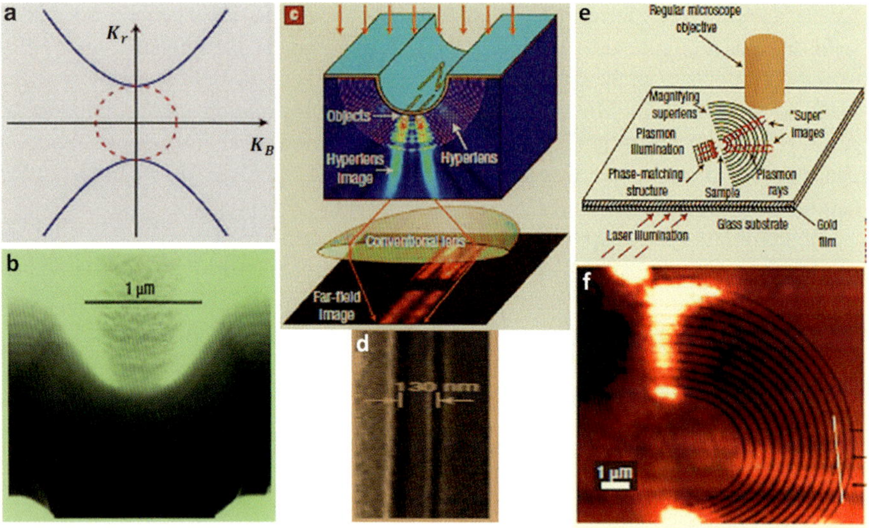

Fig. 11.9 Optical hyperlens and magnifying superlens (Zhang and Liu 2008): (**a**) for a material with $\varepsilon_r < 0, \varepsilon_\theta > 0$, the k_r and k_θ relationship shows a hyperbolic curve. The propagating waves in such medium can take on arbitrarily large values of k_θ. (**b**) SEM image of the cross-section of a hyperlens made by depositing 16 periodic silver (35 nm) and Al_2O_3 (35 nm) layers on a cylindrical cavity in quartz substrate. (**c**) Schematic of the experimental setup for the first hyperlens demonstration. (**d**) SEM image of the pair object (the two *dark lines*) with 50-nm linewidths separated by an 80-nm gap. (**e**) Schematic of the magnifying superlens for surface plasmons. The lateral distance between the images produced by the alternating layers of effective negative refractive index grows with distance along the radius. The magnified images are scattered by the surface roughness and captured by a regular microscope. (**f**) Composite of the atomic force microscope image of the sample and the corresponding image obtained by regular microscopy

plasmon waves confined by a concentric polymer grating placed on a metal surface (see Fig. 11.9e). The magnified images of the surface waves were scattered by the surface roughness and collected by a microscope as shown in Fig. 11.9f (Drezet et al. 2007). For a hyperlens, the resolution could be further improved at least several times by increasing the ratio of the outer to inner radii and by using thinner films, but at the expense of a smaller field of view and a reduced total transmission. On the other hand, although an FSL does not have the limitation of a small imaging area, its magnification is dependent on the coupling grating and the bandwidth for propagating waves. To obtain the very large bandwidth in Fourier space needed to reconstruct a deep sub-diffractional image, the information might have to be collected from multiple measurements (Zhang and Liu 2008).

Overall, the superlens, with its unique ability to overcome the diffraction barrier, has vast potential for future applications. The major limitation of present designs of the FSL or hyperlens is that the object must be in the near field of the superlens, although the image can be projected into the far field. This is because one has to make sure that the evanescent waves do not decay too much before reaching the superlens and being enhanced or converted into propagating waves. An entirely new approach to this issue may be needed, although practical applications are still possible with the object near the lens. Another fundamental challenge is the loss, as most superlens schemes involve resonances in metal, which limits both the practical resolution and transmission. Imperfect sample fabrication further aggravates the performances because small perturbations are often strong enough to distort the delicate transformations of large wave vectors. The quest for materials with truly negative index at optical frequencies presents the possibility of building an optical superlens for both polarizations. However, one may question whether this direction is better than single negative metamaterials in terms of losses, simplicity of design and fabrication. It has yet to be seen in bulk optical metamaterials, where a truly NRI can be defined, even though the ultimate resolution in metamaterials is still limited by the unit size of the meta-atom. An acoustic superlens could greatly improve underwater sonar sensing and ultrasonic imaging for medical diagnostics.

A possible futuristic optical nanoscope using superlenses may extend the optical microscope into the nanometer scale, with similar impact to the way in which optical microscopy transformed modern biology. Although there are currently quite a few imaging technologies, such as scanning electron microscopy, that offer nanometer resolution, an optical nanoscope is essential for real-time in situ observation of complex molecular machinery in living biological cells, which will accelerate discoveries in biology and medicine. Another possible application is the extension of optical lithography to ultrasmall scales, which is the key to scaling down integrated circuits for high-performance electronics. Optical and magnetic data storage, as well as biosensing may also benefit from the ability to write or read information within ultrasmall volume, thereby increasing storage densities or sensing resolution (Zhang and Liu 2008).

11.7 Metamaterial Sensors

Metamaterials exhibit a strong localization and enhancement of fields so that they can be used to actually improve the sensor selectivity of detecting nonlinear substances and to enable detection of extremely small amounts of analytes. Based on this property, metamaterials have been used for sensing applications, such as surface plasmon resonance (SPR) sensors, high frequency sensors, 2D subwavelength resonator, dielectric sensing with ε near-zero narrow waveguide channels, and microwave nondestructive evaluation sensor with metamaterial lens to detect material defects even as small as a wavelength. The design of sensing devices needs to follow (Chen et al. 2012): (a) the sensors must have an operating frequency low enough to avoid the background and substrate absorption; (b) the sensors must produce a strong and measurable readout signal with a resonant behavior sharp enough to accurately track the shift in transmission spectra; (c) the linearity of sensing should be pertained, which is related to the quality factor of sensors; and (d) the sensor sensitivity should be high enough. If there are a limited number of data points in one frequency scan of the network analyzer, it is easier to resolve smaller shifts in the transmission spectra in response to the externally applied load when the sensitivity is higher. Metamaterial-based sensors possess much higher sensitivity than traditional sensors. Moreover, metamaterial-based sensors have potential applications in environmental sensing, homeland security, and biosensing.

11.7.1 Biosensors

Biosensors are essential in many areas, such as disease diagnostics, environmental monitoring, and food safety, and they are also vital tools in the investigation of biological phenomena. Biosensing technologies based on metamaterials have attracted significant attentions from the microwave to optical frequency because of their cost-efficient and label-free biomolecule detection. According to operating frequency of sensing biomolecule and component, the metamaterial-based sensors are classified into three types (Chen et al. 2012): microwave biosensor, terahertz biosensor, and plasmonic biosensor.

1. Microwave biosensor

   Among metamaterial components, an SRR can be used to produce a negative magnetic permeability material (NMPM) in a time-varying H-field component of perpendicularly polarized wave incident on its surface. In particular, the key characteristic of SRRs used to synthesize an effective NMPM is a small electrical size, which can be made smaller than the signal wavelength at resonance (Martin et al. 2003). Hence, SRRs can be considered as quasilumped elements suitable for the miniaturization of planar microwave devices, such as filters and antennas. SRRs have also been used for the fabrication of sensors and devices. For example, an SRR-based biosensor with a small electrical size to detect the occurrence of biomolecular binding consists of two pairs of SRRs and a planar microwave transmission line, as shown in Fig. 11.10a. The planar microstrip transmission line produced a time-varying H-field component in a direction perpendicular to the surface of the SRRs. Moreover, the line was an open conduit for wave transmission and the electromagnetic field was not entirely confined to it. In addition, there existed a small E-field component along the axis of the line. Hence, the propagating mode of microwaves through the line was not a pure transverse electromagnetic (TEM) mode but a quasi-TEM mode, as shown in Fig. 11.10b. According to Faraday's law, when a time-varying H-field component was perpendicular incident on the surface of SRR, the SRR will generate resonance. This was because the SRR can be considered to be a simple LC resonant circuit with a resonant frequency of $f_0 = 1/[2\pi(\mathrm{LC})]^{1/2}$. Based on the formula, the changes in resonant frequency depended on the changes in the inductance and/or capacitance (Chen et al. 2012).

   Generally, a biosensor consists of the sensitive biological element, transducer or the detector element, and associated electronics or signal processors. The SRR-based biosensor can be directly used for label-free biomolecular detection.

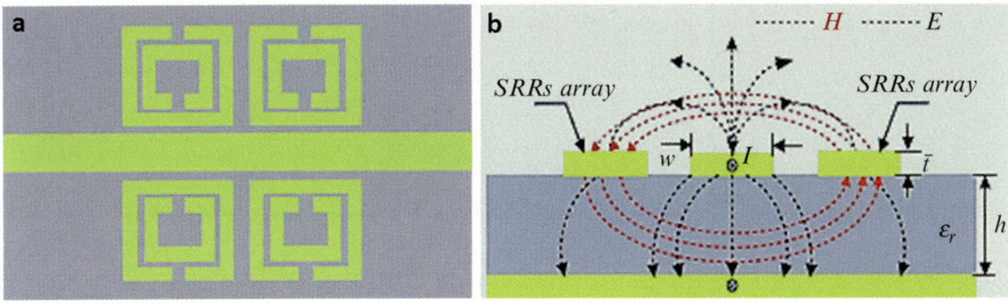

Fig. 11.10 The structure of biosensing based on SRR array (Chen et al. 2012): (**a**) top view of microstrip transmission line and (**b**) cross-section of microstrip transmission line with a pair of SRRs along electromagnetic field distribution

Even though biosensors based on the resonator array have exhibited a remarkable resonant frequency shift, the occupied areas of the devices were relatively large and their biomolecular concentration was relatively high (~1 μg/mL). The mechanism of such biosensor mainly depends on the change in the resonant frequency due to the binding of two different biomolecules onto LC resonators, when the resonator is excited by time-varying magnetic fields of microstrip transmission line (Chen et al. 2012).

2. Terahertz biosensor

Sensing the complex dielectric properties of a sample in the terahertz frequency range can directly identify the chemical or biochemical molecular composition either by detecting the resonant absorption of molecular or phonon resonances for small molecular compounds. For instance, a label-free sensing method was based on the change of the transmittance of terahertz radiation through a thin metallic mesh, when a sample substance was loaded on the mesh openings. The transmittance of the thin metallic mesh does not change due to the absorption. But, dominantly because of the variation of the refractive index of the sample substances near the openings, a distinct shift of the transmission dip frequency was observed for 500 pgmm^2 of horseradish peroxidase printed on the metallic mesh. To achieve a higher sensitivity, the sensor needs a sharp edge in its frequency response and a point of high concentration of electric field to enable the detection of small changes in the dielectric environment. Therefore, a terahertz frequency-selective surface (FSS) made from asymmetric double split-ring resonators (aDSR) of metamaterial was proposed to sense small amounts of chemical and biochemical material, as shown in Fig. 11.11a (Christian et al. 2008). The high sensitivity terahertz sensors were typically based on resonant structures whose frequency response was shifted by dielectric loading. Such a frequency shift can depend very sensitively on the dielectric properties of material placed in the environment of such a structure. A terahertz paper-based metamaterial (MM) device can be potentially utilized for quantitative analysis in biochemical sensing applications, as shown in Fig. 11.11b (Tao et al. 2011). Planar metallic resonators with minimum features of less than 5 μm have been fabricated on paper, using a photoresist-free shadow mask deposition technique. The fabricated paper MM devices show unique electromagnetic resonant

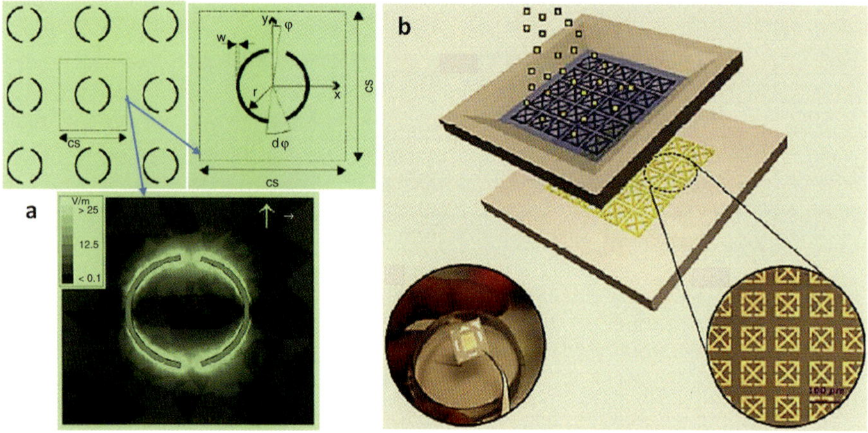

Fig. 11.11 Illustration of terahertz biosensor (Chen et al. 2012): (**a**) aDSR-based FSS adopting a square lattice of unit cell with radius $r = 50\,\mu m$, width $w = 5\,\mu m$, cell size cs $= 220\,\mu m$, asymmetry angle $\varphi = 4°$, and gap angle $d\varphi = 20°$. The E-field in the resonator plane shows a strong concentration (bright) at the ends of the arcs. $f = 875$ GHz, amplitude of excitation 1 V/m. (**b**) Micrometer-sized metamaterial resonators sprayed on paper substrates with a predefined microstencil; photograph of a paper-based terahertz metamaterial sample (holding with hand); and optical microscopy image (enlarged) of one portion of an as-fabricated paper metamaterial sample

responses at predefined frequencies, which depend on the size of resonator and could be utilized as a signature for biochemical sensing applications. In this sensing structure, the paper acted as the dielectric substrate providing both support and a material to sample and embed analytes, while the patterning metamaterial on paper substrates would offer a platform where the resonance shifts, mainly because of alterations in the SRR capacitance induced by the added analytes, can be utilized for quantitative biochemical sensing applications (Chen et al. 2012).

3. Plasmonic Biosensor

   The surface plasmons are known to be extremely sensitive to the refractive index of the dielectric medium within the penetration depth of the evanescent field. This remarkable property has been used for the development of label-free plasmonic biosensors, which emerged as a leading modern technology for detection and investigation of binding events between the target analyte and its corresponding receptor on a metal surface. Most of the SPR biosensors have used SPPs. However, the SPPs can only provide an extremely small detection limit exceeding 10^{-5} refractive-index units (RIU), due to the resonant photon-SPP coupling conditions. Moreover, an SPR system demands optical couplers (e.g., prisms and gratings), displays narrow operation ranges and performs over short detection distances, thus impeding its integration with low-cost, real-time, and high-throughput biochips for rapid bio-analytical measurements of quantity-limited samples. Therefore, the SPP-based approach needs an improvement in sensitivity for the detection of small analytes and satisfies modern requirements

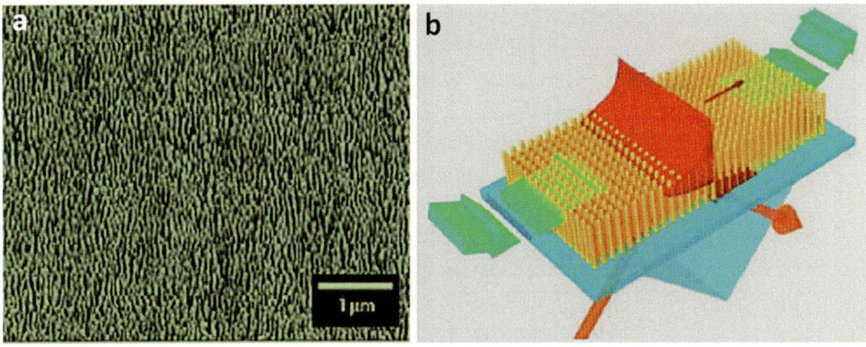

Fig. 11.12 Illustration of plasmonic metamaterial biosensor (Chen et al. 2012): (**a**) typical scanning electron micrograph of plasmonic nanorod metamaterial; and (**b**) schematic of the attenuated total internal reflection (ATR) measurements and flow cell

of biotechnology advancing towards new nanoscale designs and promising the manipulation on the nanoscale level, size-based selectivity, and selective chemical and biochemical nano-architectures. Localized surface plasmons (LSPs) of metallic nanostructures seem much more suitable to match these new trends, as well as to bring new functionalities, such as spectral tenability and strong enhancement of the local electric field. However, LSP-based sensors are known to provide sensing response to refractive-index change, which is at least an order of magnitude lower than SPPs with sensitivities not exceeding 100–300 nm per RIU in spectral interrogation schemes and probe depth as small as 10 times, making them applicable to only a very limited number of biological species. In order to improve the sensitivity, a metamaterial-based plasmonic biosensor by using an array of parallel gold nanorods oriented normally to a glass substrate has been developed. In this structure, the metamaterial consisted of an assembly of Au nanorods electrochemically growing into a substrate-supported, thin-film porous aluminum oxide template. The final structure represented an array of parallel nanorods occupying an area of up to 2 cm^2, as shown in Fig. 11.12a. Moreover, the structural parameters can be controlled by altering the fabrication conditions with typical dimensions being in the range of rod lengths between 20 and 700 nm, rod diameter of 10–50 nm, and separation from 40 to 70 nm, thus achieving a nanorod areal density of approximately $10^{10} \sim 10^{11}$ cm^{-2}. The lateral size and separations between the nanorods were much smaller than the wavelength of light used in the experiments, so only average values of nanorod assembly parameters were important, and individual nanorod size deviations had no influence on the optical properties that were well described by an effective medium model. For this reason, the optical properties of the nanorod arrays were very stable with respect to fabrication tolerances. The fabrication technology made it possible to fully or partially embed the nanorods into an alumina matrix. In the Fig. 11.12b, the illumination of the same nanorod structure in the ATR geometry revealed the new guided mode in the near-infrared spectral range (Kabashin et al. 2009). The metamaterial sensor worked similarly to a conventional SPP-based sensor, showing a redshift of the resonance in response to an increase in the refractive index. Furthermore, a change of the refractive index by 10^{-4} RIU caused a shift of the resonance by 3.2 nm even without any optimization of the structure. The corresponding minimum estimation of sensitivity of 32,000 nm per RIU exceeded the sensitivity of localized plasmon-based schemes by two orders of magnitude. The sensitivity of a guided wave offered by the nanorod metamaterial also exceeded the relevant parameter for commercial SPP-based sensors using spectral interrogation. Such a gain in sensitivity was due to a combination of the higher sensitivity of the metamaterial to bulk refractive-index change and a mass accumulation effect as a result of the large surface area of the nanoporous matrix. Therefore, the origin of such a high sensitivity gain for this biosensor was twofold. In nanorod arrays, the sensed substance was incorporated between the initially bare rods, and so the waveguide mode provided a better overlap between the sensing field and the sensed substance than SPR sensors. Furthermore, the effective dielectric constant $\varepsilon_{\mathrm{eff}}$ of the metamaterial strongly depended on the dielectric constant of the tested medium ε_{d} as a result of the modification of the plasmon–plasmon interaction in the nanorod array, thus leading to modification of resonant conditions of the guided-mode excitation caused by the sensed analyte. Another key advantage of the metamaterial-based sensor consisted of the essentially discontinuous porous nanotexture of the nanorod matrix. This enabled the implementation of new sensing geometries and strategies, not feasible with conventional film-based SPR. Indeed, by functionalizing the nanorods and immobilizing a receptor on their surface, one can follow the binding of a selective analyte with the receptor inside the nanorod matrix. The considerably increased surface area given by the nanoporous texture of the metamaterial significantly increased the amount of biomaterial that can be incorporated into the matrix within the probe depth available, maximizing the biological sensitivity of the system. Furthermore, the distance between the nanorods can be selected to match the size of biological species of interest, giving access to a further size selectivity option that was important for many tasks in the detection of immunoassays, virus and protein (Chen et al. 2012).

11.7.2 Thin-Film Sensors

Thin-film sensing utilizing interaction between electromagnetic waves and unidentified thin-film sample substance can provide important information for many applications throughout chemistry and biology. For example, the FSS consisting of periodic two-dimensional arrays of identical resonators have arisen as candidates for highly sensitive chemical or biological thin-film detection because it can be small and show a resonant frequency response that is tunable by design. However, these ideas mainly capitalized the structure of SRRs, whose natural oscillation frequencies depended critically on the permittivity of the boarding dielectrics. To achieve efficient thin-film sensing, some thin-film sensors based on metamaterials are proposed and fabricated. According to operating frequency of thin-film sensing, the metamaterial-based thin-film sensors are classified into three parts (Chen et al. 2012): microwave thin-film sensor, terahertz thin-film sensor, and plasmonic thin-film sensor.

1. Microwave thin-film sensor
   Despite the availability of resonators with weak free-space coupling, thin-film sensing with FSS devices is still a challenge because the sample substance has to be applied either to a specific portion of each resonator in the array or to the complete array area. Especially, when only small amounts of the substances are available, depositing the sample substances at several places introduces a high degree of inaccuracy and covering the whole area is not practical in most cases. In addition, the field confinement of the conventional thin-film sensor based on circular asymmetric double split resonator (aDSR) metamaterial is relatively weak, limiting its volumetric sensitivity. In order to sense minute amounts of sample substances, thin-film sensors have to feature a sharp transition in their frequency response. Moreover, the electric field must be confined to the portion of the sensor on which the sample substance is deposited. Based on such principles, the tip-shaped SRR metamaterial was developed as thin-film sensor to reduce device size and resonance frequency as well as to improve the Q-factor. In contrast to the traditional structures, the tip-shaped design exhibited a miniaturization and sharper dip on resonance in their transmission spectra. Furthermore, the sensor can deliver the sensitivity level of 16.2 MHz/μm and the error less than 2 μm nonlinearity when the uniform benezocyclobutene films with thickness between 100 nm and 50 μm were coated onto the fixed structure. To further improve the electric field distribution, a rectangular tip-shaped aDSRs with sharp tips was proposed based on the above ideas, which can offer a very high sensitivity at miniaturized scale. Figure 11.13a showed the schematic layout and electric distribution of the unit cells for both circular and rectangular resonators, compared with the traditional structure. In the case of the circular aDSR, the strongest field amplitude was located at the end pieces of the longer resonator arm with peak values of 13.7 V/m, while the field components inside the gap remained relatively small. In contrast to that behavior, the rectangular aDSR with tips strongly confined the field into the gap with peak values of 17.1 V/m so that this area became very sensitive to changes in the dielectric environment. Consequently, the rectangular design offered a miniaturization compared with circular structures. Furthermore, the tips at the end of the resonator arms concentrated the field components into a small area, increasing the volumetric sensitivity of the device. For example, the circular aDSR featured a resonant shift of 9, 24, and 48 MHz for the single covered square, the two covered squares, and the full coverage, respectively. This was considerably less than the values observed for the rectangular resonator with tips, where the corresponding shifts were 18, 36, and 78 MHz. Moreover, the resonant frequency of the rectangular aDSR without any overlayer lied at 5.993 GHz, which was roughly 77 % of the design frequency of the circular aDSR, located at

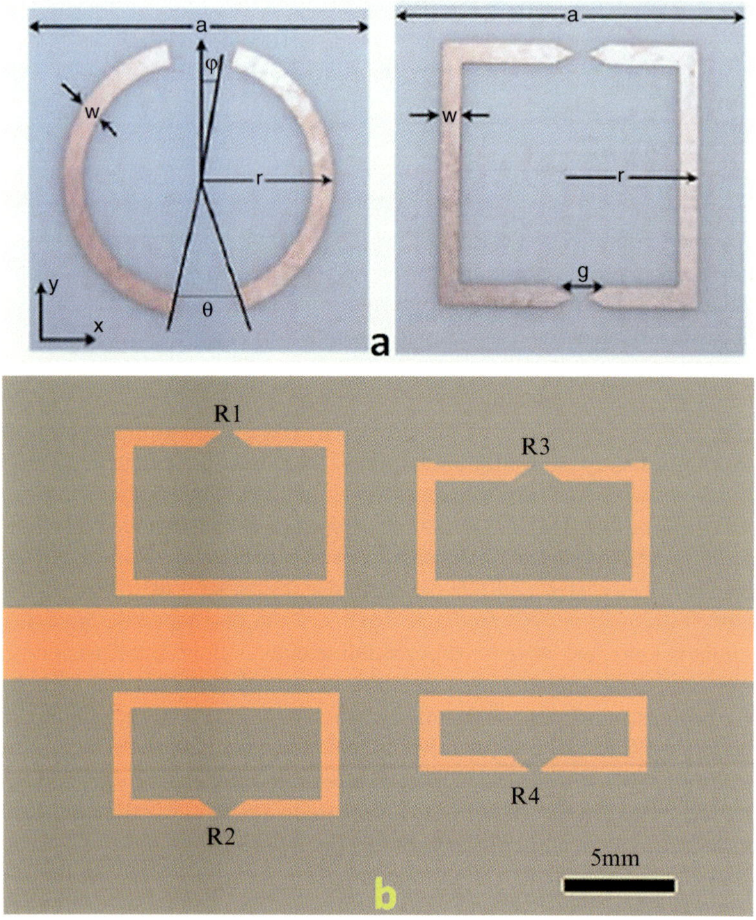

Fig. 11.13 Illustration of microwave thin-film sensor (Chen et al. 2012): (**a**) layout of (*left*) the circular aDSR and (*right*) the rectangular aDSR with field confining tips; and (**b**) diagram of four-channel sensor containing four SRR's structure

7.716 GHz. As both devices shared the same unit cell dimensions, a miniaturization by 23 % was achieved. To further improve the sensitivity of thin-film sensor, a nested SRR metamaterial-based microwave thin-film sensor which combined multiple SRRs in a compactly nested structure on a single chip was developed. Its advantage was the ability to obtain sharper and deeper dips in their transmission resonance, a lower operating resonant frequency per unit area, and more regions of high electric fields, compared with traditional SRRs previously used. This made the nested SRR structure very well suited for thin-film sensing applications at the microwave region. In addition, the nested SRR structure can achieve a higher resonant frequency shift, a higher sensitivity and a better linearity, compared with the traditional SRR structure. For example, an improved sensitivity of 28 MHz/μm or 0.41/μm was obtained by the nested SRR metamaterial-based thin-film sensor, while the traditional SRR metamaterial-based thin-film sensor demonstrated a sensitivity of 20.68 MHz/μm or 0.21/μm. Besides, the reduced nonlinear error of less than 182 nm (less than 7 %) in the nested SRR metamaterial-based thin-film sensor was obtained, compared with the nonlinear error less than 860 nm (72.6 %) of the traditional SRR metamaterial-based thin-film sensor data. Moreover, the miniaturization by 22.6 % was achieved by the proposed nested SRR structure, compared with the results of the traditional SRR structure (Al-Naib et al. 2008; Chen et al. 2012).

In addition, multichannel thin-film sensor was implemented by a set of microstrip-coupled SRRs with different dimensions, as shown in Figure 11.13b. Each SRR exhibited a unique high-Q resonance that was sensitive to the presence of a sample in a particular area. Hence, this SRR-based sensor can function (Withayachumnankul et al. 2011): (a) to detect different samples simultaneously to increase the throughput or (b) to characterize nominally identical samples at multiple frequencies to increase the sensor selectivity. Owing to the optimized design, sensing a low-permittivity film with a thickness as small as one thousandth of the operating wavelength was achievable.

2. Terahertz Thin-Film Sensor

To further improve sensitivity, metamaterials have arisen as candidates for highly sensitive chemical or biological thin-film detection since they can be small and show a resonant frequency response that is tunable by design. For example, small quantities of silicon (<1 ng), deposited as a film or overlayer on a planar THz metamaterial, can shift the resonant frequency by an easily measurable amount. Similarly, simulations of asymmetric SRRs indicate a possible scenario in which films as thin as 10 nm may be measured. To investigate the behavior of dielectric overlayers on metamaterials, the terahertz metamaterial-based thin-film sensors were fabricated by conventional photolithography techniques and consisted of square arrays of double SRRs, made from aluminum with 200 nm thickness on silicon substrates with 0.64 mm thickness. Uniform dielectric overlayers from 100 nm to 16 μm thick were deposited onto fixed SRR arrays in order to shift the resonant frequency of the electric response (O'Hara et al. 2008; Chen et al. 2012).

3. Plasmonic thin-film sensor

To overcome sensing limitations, a plasmonic thin-film biosensor was developed based on SRR array metamaterial as shown in Fig. 11.14 (Chang 2010). The plasmonic biosensor can not only substantially ease the aforementioned burdens (coupler-free, tunable operation frequencies and longer detection length), but also preserve the merits of the conventional SPR technique (excellent sensitivity, label-free, quick and real-time diagnose). In addition, the SRR structures can also exhibit multiple reflectance peaks so that the SRRs can be readily employed as biosensors, especially for real-time, label-free, and cell-level bimolecular thin-film detections by monitoring the shifts of reflectance peaks because of analytes binding to molecular receptors immobilized on the SRR surface. When applying thin dielectric layers with different thicknesses on the SRR array, the thin-film sensor showed the distinct sensing behaviors of each

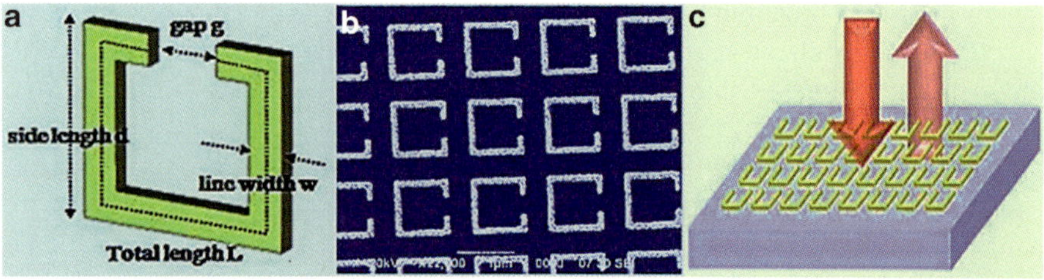

Fig. 11.14 Illustration of plasmonic thi-film sensor (Chang 2010): (**a**) the designed SRR unit cell; (**b**) SEM image of fabricated planar SRRs; and (**c**) schematic reflectance measurement upon the SRR-based plasmonic sensor. Here no optical coupler is required to excite plasmonic resonance

resonance mode in the multi-resonance reflectance spectra. The lower-order modes possessed greater sensitivity associated with stronger localized electromagnetic field leading to shorter detection lengths within five hundred nanometers, while the higher-order modes presented mediate sensitivity with micron-scale detection lengths to allow intracellular bio-events detection. As a result, the lower modes can be utilized to detect small targets and macromolecules including antibody–antigen interactions and the molecular recognition on the cell membrane, to gain the advantage of the excellent sensitivity and also to reduce noise from the dielectric environment, while the higher modes are facilitated to explore intracellular bio-events in live organelles and cells, due to their farther detection lengths in micron scale as well as label-free manner. Therefore, such multifunctional plasmonic biosensor can be readily employed to analyze activation-dependent cellular interactions and even a potential label-free bio-imaging device that other label-free techniques have not been achieved (Chen et al. 2012).

11.7.3 Wireless Strain Sensors

The ability to telemetrically measure strain is important to many aspects in daily life. Compared with traditional RF structures (e.g., rectangular and circular coils), the wireless strain sensors based on metamaterial can exhibit some advantages, including higher quality factors (Q factors) and sharper and deeper dips on resonance in their transmission. This made metamaterials very well suited for telemetric sensing applications. Moreover, metamaterial architectures can achieve higher resonance frequency shifts, leading to higher sensitivity and better linearity. In addition, by using SRR metamaterials, the operating resonance frequencies per unit area can be significantly reduced. This was especially critical for sensing applications that involve transmission through soft tissue (e.g., muscle) because such tissue strongly absorbed electromagnetic waves at very high operating frequencies (Melik et al. 2010; Chen et al. 2012).

In sum, applications of metamaterials in sensing provide novel opportunities for developing a new generation of sensing technologies. Metamaterials can improve the mechanical, optical, and electromagnetic properties of sensors, therefore, the metamaterial-based sensors are developing towards single molecular biosensor and high-throughput sensor arrays. However, like any emerging field, they also face many challenges. For example, the performance of sensors based on metamaterials is limited by fluctuation phenomena which may be caused by various external and internal mechanisms and result in noise appearing in the sensor readout. External noise sources include the quantum noise/shot noise, photodetector noise, and noise of processing circuitry, while intrinsic noise is mainly connected with the adsorption and desorption (AD) of analyte particles on the metamaterial. Moreover, the AD noise remains the main limiting factor of the ultimate performance of all such devices, because each metal–dielectric structure must be exposed to some kind of environment and thus AD noise will influence to a certain degree the performance of any metamaterial structure, as well as that of conventional SPR sensors and nanoplasmonic structures. To further avoid losses and scaling issues of metal-based structures, the purely dielectric configuration metamaterials instead of metal-structured metamaterials have been developed. In addition, metamaterial-based biosensors still need improvements in sensitivity and accuracy, mainly because of the difficulty in obtaining high-resolution, small feature sizes on substrates where conventional photolithography techniques are difficult to apply. This is mainly due to the inability to use conventional photolithography-based microfabrication techniques where chemical solutions are generally used. Thus, the advanced fabrication technology must be developed. The fulfillment of electromagnetic response to metamaterials in the visible and infrared areas can begin a new chapter of photonics related to such novel concepts and possible applications as safety imaging, remote sensing, and resonant devices. Moreover, with the arrival of micro- and nano-fabrication, new sensing possibilities are open for practical implementation of different metamaterials and the field is increasingly attracting more researches (Peng et al. 2007; Chen et al. 2012).

11.8 Future Prospects

Considering the fundamental research and application purposes, it is highly desirable to have truly bulk metamaterials, which are on the scale of tens of wavelengths, or even larger in all three dimensions. 3D metamaterials can be realized by focused iron beam milling through metal–dielectric multilayers (for example, the fishnet structure), or combining advanced planar micro-/nano-manufacturing techniques (such as optical lithography and EBL) for individual layers with sophisticated alignment procedures to stack different layers. Meanwhile, direct laser writing by multiphoton polymerization of a photoresist has emerged as a technique for the flexible fabrication of 3D metamaterials. This method utilizes a multiphoton absorption process in a photoresist that is transparent at the wavelength of the

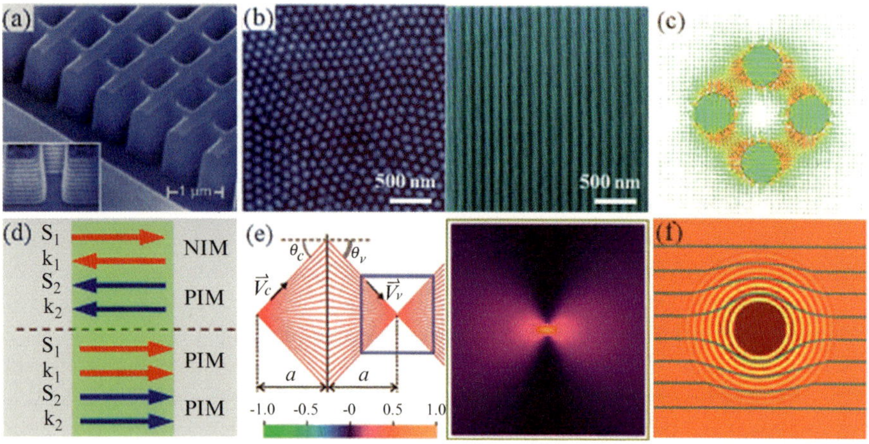

Fig. 11.15 Illustration of 3D metamaterial (Liu and Zhang 2011): (**a**) a 3D fishnet NIM structure fabricated by the top-down method (FIB), which consists of 21 total layers of alternating silver (30 nm) and MgF_2 (50 nm) films. The inset shows the cross-section view. Reprinted from ref. 106 with permission. (**b**) Silver nanowire metamaterials fabricated by the bottom-up technique. *Left* and *right* SEM images show the top and cross-section view, respectively. (**c**) Magnetic resonators composed of four metallic nanospheres. The structure may be synthesized from the self-assembly method. (**d**) Schematic of SHG when NIMs are involved, in comparison with normal SHG in only PIMs. (**e**) Focusing of electrons by a grapheme *p–n* junction. The *left panel* plots the trajectories of electrons, diverging from a source at the distance from the junction and becoming convergent after negative refraction. The *right panel* shows the interference-induced pattern in the charge current near the focal image. (**f**) A matter-wave cloaking system involves cold atoms within a three-dimensional laser-based optical lattice. Matter waves traveling through the medium bend around the atoms inside as though they are not there

laser used for creating the pattern. If the laser beam is tightly focused into the photoresist, the light intensity becomes sufficiently high to induce the multiphoton absorption and hence the polymerization process. By properly scanning and modulating the laser, an arbitrary three-dimensional connected pattern, either periodic or nonperiodic, can be created, as shown in Fig. 11.15a. However, using the aforementioned top-down techniques, it is extremely difficult and time-consuming to realize bulk optical metamaterials, because the feature size and spacing of metamaterials need to be precisely controlled at the nanoscale over a large domain, and the fabrication process is usually nonparallel. The bottom-up and self-assembly techniques may promise solutions to these fabrication challenges. For example, all-angle optical negative refraction, arising from the hyperbolic constant frequency contour of one unique indefinite metamaterial, has been experimentally demonstrated in metallic nanowire metamaterials with thickness about 10 times of the wavelength (Fig. 11.15b). This structure is fabricated by electrochemically growing metallic nanowires in a porous alumina template, which is prepared by the anodization method in a self-organized way. Such a method has proved to be a low-cost and high-yield technique for fabricating different kinds of nanostructures including nanowires, nanodots, and nanotubes. In addition, it has been shown that clusters of metallic nanoparticles exhibit pronounced electric and magnetic responses, which may be good candidates of building blocks for constructing optical NIMs (Fig. 11.15c). With advances in particle self-assembly techniques, it is possible to form the prescribed particle colloids, by tailoring the shape-dependent capillary interactions and surface energies to control the particle orientation and spatial arrangement in three dimensions. Ultimately, some atomic, molecular, and condensed matter systems may be used to realize metamaterials (Fan et al. 2010; Liu and Zhang 2011).

It has been shown that quantum interference effects similar to EIT can lead to a large induced chirality under realistic conditions, which enable negative refraction with minimal absorption. In addition, based on a two-component system in which one atomic or molecular component may be inverted. The existence of nonmagnetic amorphous solids and liquids can generate a NRI. Nonlinear metamaterials, whose properties depend on the intensity of EM waves, is an emerging area worth extensive investigation especially in the optical regime. The enhanced nonlinear optical properties by inserting nonlinear elements into the gap of SRRs, arises from the giant local-field amplification. Subsequently, theoretical work has predicted that nonlinear metamaterials could offer novel phenomena such as hysteretic transition, unusual wave mixing, and solitary wave propagation. One particular interesting phenomenon is the reversed Manley–Rowe relation and backward phase-matching condition for second-harmonic generation (SHG) or optical parametric amplification (OPA). SHG in normal dielectric materials is shown in Fig. 11.15d. Such extraordinary phenomenon could be demonstrated in metamaterials, such as the 3D fishnet structures shown in Fig. 11.15a, at proper frequencies. Moreover, artificial magnetic metamaterials could provide additional routes to boost the nonlinear process. In terms of applications, tunable metamaterials and memory devices have been experimentally demonstrated based on nonlinear metamaterials (Wang et al. 2008; Liu and Zhang 2011).

In addition, metamaterials may manifest fascinating phenomenon in the quantum world. In principle, the concept of metamaterials could be applied to any wave at any scale, including the matter wave which is the wave description of particles, such as electrons and neutrons, in quantum mechanics. For example, negative refraction and focusing of electrons can be achieved in grapheme (Fig. 11.15e), a monolayer of graphite. From solid state physics, electrons in the valence (conduction) band have a group velocity antiparallel (parallel) to the wave vector. This provides a delicate connection with optical negative refraction, which requires group and phase velocities of light opposite in signs. By examining a grapheme p–n junction, a point source of electron current in the n-type region radiates electrons to the interface, where they are negatively refracted into the p-type region and brought to a focus. This is just in analogy to light focused by an NIM slab. Moving forward, cloaks for matter waves have been designed with a three-dimensional optical lattice of laser beams, consisting of optical standing waves with gradually varying amplitude along the radial direction. The varying amplitude of the optical lattice changes the effective mass and band energy of the hidden object as another atom or wave moves near it, causing the stream of particles or incident matter waves to bend and come out on the other side of the object as if it was not there (Fig. 11.15f). Other proposals bridging metamaterials with quantum mechanics include modifying the emission rate of an emitter adjacent to carefully designed metamaterials with large photon densities of state, as well as exploring the possibility of repulsive Casimir forces that originate from the quantum zero-point oscillation of EM waves (Maslovski and Silveirinha 2010; Liu and Zhang 2011).

11.9 Summary

Optical metamaterials have attracted considerable attention due to the exciting potential applications ranging from perfect lenses and cloaking devices to subwavelength optical waveguides and the enhancement of MRI. Metamaterials are a class of composite materials artificially fabricated to achieve unusual electromagnetic properties not available in nature. Their unusual properties are not determined by their constituent materials but by man-made structures that are built from periodically arranged unit cells—resonators smaller than the electromagnetic wavelength involved. These unit cells act like artificial atoms and can be coupled to both the electric and magnetic field components of the electromagnetic waves, leading novel optical properties, such as a negative index of refraction.

The unique electromagnetic properties provided by metamaterials attract researchers from multiple disciplines. In turn, this will spark the merging of knowledge and expertise across different areas, further driving the astounding advance of metamaterials research. With the complete degree of freedom to control over material properties, what could be done is only limited by one's imagination.

References

Al-Naib IAI et al (2008) Thin-film sensing with planar asymmetric metamaterial resonators. Appl Phys Lett 93:083507–0835010

Bayatpur F (2009) Metamaterial-inspired frequency-selective surfaces. PhD dissertation, University of Michigan, Ann Arbor

Boltasseva A, Shalaev VM (2008) Fabrication of optical negative-index metamaterials: recent advances and outlook. Metamaterials 2:1–17

Chang YT (2010) A multi-functional plasmonic biosensor. Opt Express 18:9561–9569

Chaturvedi P (2009) Optical metamaterials: design, characterization and applications. PhD dissertation, University of Illinois at Urbana-Champaign

Chen Z et al (2004) Two- and three dimensional ordered structures of hollow silver spheres prepared by colloidal crystal templating. Adv Mater 16:417

Chen T, Li S, Sun H (2012) Metamaterials application in sensing. Sensors 12:2742–2765

Christian D et al (2008) Frequency selective surfaces for high sensitivity terahertz sensing. Appl Phys Lett 91:184102–184104

Drezet A et al (2007) Far-field optical microscopy with a nanometer-scale resolution based on the in-plane image magnification by surface plasmon polaritons. Phys Rev Lett 98:209730

Fan et al (2010) Self-assembled plasmonic nanoparticle clusters. Science 328:1135–1138

Formanek F et al (2006) Selective electroless plating to fabricate complex three-dimentional metallic micro/nanostructures. Appl Phys Lett 88:083110
Galisteo JF et al (2005) Self-assembly approach to optical metamaterials. J Opt A Pure Appl Opt 7: S244–S254
Houck AA et al (2003) Experimental observations of a left-handed material that obeys Snell's law. Phys Rev Lett 90:137401
Kabashin AV et al (2009) Plasmonic nanorod metamaterials for biosensing. Nat Mater 8:867–871
Kehagias N et al (2007) Reverse-contact UV nanoimprint lithography for multilayered structure fabrication. Nanotechnology 18:175303–175304
Kim KY (2004) Guided and leaky modes of circular open electromagnetic waveguides: dielectric, plasma, and metamaterial columns. PhD dissertation, Kyungpook national University
Lavoie BR et al (2012) Low-loss surface modes and lossy hybrid modes in metamaterial waveguides. Photonics Nanostruct 10(4):602–614
Liu Y, Zhang X (2011) Metamaterials: a new frontier of science and technologies. Chem Soc Rev 40:2494–2507
Liu Z et al (2007a) Far-field optical superlens. Nano Lett 7:403–408
Liu Z et al (2007b) Far-field optical hyperlens magnifying sub-diffraction-limited objects. Science 315:1686
Logeeswaran VJ et al (2006) Self-assembled microfabrication technology for 3D isotropic negative index materials. Proc SPIE 6393:639305–639310
Martin F et al (2003) Miniaturized coplanar waveguide stop band filters based on multiple tuned split ring resonators. IEEE Microw Wirel Compon Lett 13:511–513
Maslovski SI, Silveirinha MG (2010) Ultralong-range Casimir-Lifshitz forces mediated by nanowire materials. Phys Rev A 82:022511
Melik R et al (2010) Nested metamaterials for wireless strain sensing. IEEE J Sel Top Quantum Electron 16:450–458
Melville D, Blaikie R (2005) Super-resolution imaging through a planar silver layer. Opt Express 13:2127–2134
Moiseev SA, Kamli AA, Sanders BC (2010) Low-loss nonlinear polaritonics. Phys Rev A 81:033839(1–5)
O'Hara JF et al (2008) Thin-film sensing with planar terahertz metamaterials: sensitivity and limitations. Opt Express 16:1786–1795
Pendry JB (2000) Negtive refraction makes a perfect lens. Phys Rev Lett 85:3966–3969
Peng L et al (2007) Experimental observation of left-handed behavior in an array of standard dielectric resonators. Phys Rev Lett 98:157403–157406
Ponizovskaya EV, Bratkovsky AM (2009) Negative index materials with gain media for fast modulation. Appl Phys A Mater Sci Process 95:1137–1142
Reza A (2008) The optical properties of metamaterial waveguide structures. Master degree thesis, Queen's University, Kingston, Ontario, Canada
Smith DR et al (2003) Limitations on subdiffraction imaging with a negative refractive index slab. Appl Phys Lett 82:1506–1508
Subramania G, Lin SY (2004) Fabrication of three-dimensional photonic crystal with alignment based on electron beam lithography. Appl Phys Lett 85: 5037–5039
Tao H et al (2011) Metamaterials on paper as a sensing platform. Adv Mater 23:3197–3201
Wang BN et al (2008) Nonlinear properties of split-ring resonators. Opt Express 16:16058–16063
Weiland T et al (2001) Ab initio numerical simulation of left-handed metamaterials: comparison of calculations and experiments. J Appl Phys 90:5419–5424
Withayachumnankul W et al (2011) Metamaterial inspired multichannel thin-film sensor. IEEE Sens J 99:1–7
Wu W et al (2007) Midinfrared metamaterials fabricated by nanoimprint lithography. Appl Phys Lett 90:063107
Xiong Y et al (2007) Tuning the far-field superlens: from UV to visible. Opt Express 15:7095–7102
Yeh C, Shimabukuro F (2008) The essence of dielectric waveguides. Springer, New York
Zhang X, Liu Z (2008) Superlenses to overcome the difrrraction limit. Nat Mater 7:435–440

Chapter 12
Perspectives and Future Trends

Abstract A rebuilding of the information infrastructure is taking place to give instantaneous availability of data, voice, and video. This revolution of the Information Age is being gated more by the introduction of new materials and components, than by the design of systems, software, and networks. Electrons transmitted through metal wires have an information carrying capacity limited by the resistance and capacitance of the cable and the terminating electronic circuits. Photons transmitted through optical waveguides are capacity limited only by the dispersion of the medium. Each network node that requires transduction from photonics to electronics limits the performance and affordability of the network. The key frontier is the large scale integration and manufacturing of photonic components to enable the distribution of high bit rate optical streams to the individual information appliance. Microphotonics is the platform for large scale, planar integration of optical signal processing capability. The push towards smarter, ever-denser on-chip photonic networks drives a convergence of micro-, nano-, and plasmo-photonic techniques for progressively smaller devices and for improved functionality of modulators, switches, emitters, detectors, waveguides, resonators, tapers, and filters. This convergence includes composite components: monolithic integration of microstrip waveguides, 2D and 3D photonic-crystal elements, and metal/Si plasmon-optics that utilize buried or surface-mounted 2D arrays of metal stripes or nanodots. The development of optical interconnect devices including waveguides will proceed through a synergetic collaboration among material and processing technologies, design and fabrication of integrated optoelectronics, and optoelectronic packaging technology. This chapter will provide a brief review on perspectives and future trends of optical waveguides and their applications.

X.C. Tong, *Advanced Materials for Integrated Optical Waveguides*, Springer Series in Advanced Microelectronics 46, DOI 10.1007/978-3-319-01550-7_12,

12.1 Optical Waveguide Devices and Materials

Many optical devices including various waveguides have been focused on the applications in the wavelength range of several 100 μm to near 200 nm. Table 12.1 shows some examples classified by wavelength axis (OITDA 2009). The terahertz band is the extended wavelength range between radio and optical waves, and development is progressing for higher frequency radio waves and longer optical waves (lower frequency waves). Since the applications contain properties of both radio and optical waves, terahertz wireless, for example, was examined in terms of increasing the speed (higher frequency waves) of millimeter wave wireless. The wavelength range has been determined due to dispersions or losses of fiber in the near-infrared range for optical communication, and component development in each wavelength band is in progress depending on factors such as transmission distance, speed, and purpose such as multilevel photonic devices that contribute to update of communication records for the trunk line system; modules associated with standardization and speedup of Ethernet systems; optical sources and Si photonics; photoreceptors and modulators; light-emitting devices; and integration devices. In the visible and ultraviolet range, there has been progress in achieving high-output in the deep ultraviolet range and lengthening of long waves in nitride-based semiconductor devices, such as green color in a nitride-based semiconductor laser; blue violet–blue semiconductor lasers, ultraviolet LEDs, and GaN-based surface-emitting lasers; and white LED lighting and organic EL lighting. In addition to basic technologies and devices for new applications, a lot of device

Table 12.1 Applications in the wavelength range of several 100 μm to near 200 nm if some optical devices including various waveguides (OITDA 2009)

Wavelength		Module / Application	Device
THz	100 μm	·Wireless	·THz generator using electronic device ·Quantum cascade laser
	1.6 μm	·Compact wavelength tunable laser ~ Module	
Infra Red for fifer communication	1.3 μm	·10G-PON Module ·Laser for 100Gb Ethernet	·Laser for 100Gb Ethernet ·Si Photonica -Integrated device -Light emitting device
	1.0 μm		
	600nm	·Lighting -White LED -Organic LED	·Green laser using semipolar GaN ·Blue –ultraviolet ~blue diode laser ·VCSEL laser
Visible to Ultraviolet	400nm		
	200nm		·Ultraviolet LED

development and module development work for standardization and energy conservation have been performed from terahertz and mid-infrared bands, near-infrared (for optical communications), and visible and ultraviolet range (OITDA 2009).

12.1.1 Terahertz and Mid-Infrared Bands

Since terahertz band is located between light and radio waves and has physical properties of both waves, many possibilities have been investigated. For instance, high-speed wireless applications have been investigated for extension of the millimeter wave. In the existing millimeter wave (60–80 GHz), the maximum transmission speed is several Gb/s, whereas in terahertz waves, a transmission speed of 10–100 Gb/s is theoretically possible. However, due to increased atmospheric attenuation, the transmission distance is dramatically shortened. As for terahertz wireless generators, several methods are being investigated from radio aspects and optics aspects, and actual cases of transmissions have been reported in 1,000 Hz to several THz at approximately 1–10Gb/s for several km to ~10 km. With the use of electronic devices in radio wave generators, there has been progress in the achievement of higher frequency and higher output of oscillatory frequency, with reports of fundamental wave oscillation being up to 1 THz in resonant tunneling diodes, and current gain cutoff frequency and maximum oscillatory frequency being approximately 1 THz in HBT and HEMT. Investigations have also begun on Si-CMOS, possibly due to the large existing market for millimeter wave. On the other hand, THz wave oscillation using quantum cascade laser is being investigated in optic wave generators. In the existing quantum cascade lasers, room temperature oscillation is possible at 3.1–10.6 μm; however, room temperature oscillation in terahertz band has not yet been achieved. In optic waves, longer wavelengths (lower frequencies) are more difficult, and up to now the maximum oscillation temperature is 186 K at 3 THz. Investigations in active layer structure and factor analysis for room temperature oscillation are in progress (OITDA 2009).

Moreover, many applications within terahertz and mid-infrared bands can be considered. In addition to the quest for killer applications in the terahertz band, studies have been conducted to see if it is possible to add the features of terahertz band to existing applications in the wavelength range from millimeter wave to short wave and from infrared to long wave. For example, portable terahertz testing devices using time-domain spectroscopy have been made; an uncooled bolometer which was used for detecting infrared rays near 10 μm could detect up to the terahertz range and is expected to be applied in terahertz imaging and a wide range of other fields. With regard to the mid-infrared range (especially near 4–10 μm), quantum-cascade lasers that operate in room temperature are already being manufactured. A variable wavelength light source using an external resonator and watt class high output were achieved. It is expected to be applied in new fields such as medical care, processing and biology, not just for environmental measurement as done until now. Room temperature continuous operation and easy light detection are difficult using semiconductor lasers near 3 μm, which is on the border of near-infrared to mid-infrared region. However, infrared sensors using InGaAs/GaAsSb Type 2 quantum wells were operated around room temperature or Peltier cooling to achieve low-dark current and longer wavelengths with max cutoff wavelength of about 2.5 μm (OITDA 2011).

12.1.2 Near-Infrared Range

There have been advances in wavelength range of optical communications systems in terms of increase in communication capacity, spread and higher speeds of Ethernet systems in networks, and smaller size and energy conservation in optic transceivers used for interface. In the Ethernet systems, 10 Gb Ethernet (10 GbE) was introduced in the year of 2002, specifications for 40/100 GbE have been decided. The main difference between the systems in terms of transmission is that up to 10 GbE was mainly serial transmission, whereas in 40/100 GbE the transmission would be parallel (4, 10 lanes). Therefore, in the optical source, speedup of 25 Gbit/s in 1.3-μm band for 100 GbE (4 lanes) is being aimed for along with lower cost. There is progress in power conservation, along with investigations on speedup of direct modulation laser (10 → 25 Gbit/s) for the 10 km version, and changes in wavelength band of external modulation integration laser of trunk line system (1.55 → 1.3 μm band) in the 40 km version. In optical transceivers, power conservation and miniaturization of common industrial specifications is being aimed for, in addition to efforts to increase in number of ports for transmission devices. For this, it was planned to achieve miniaturization and power conservation, moving from 300 pin → *X2* → XFP → SFP+ by removing from the module the CDR which contains serial/parallel conversion IC and 3R functions. Furthermore, low-power integrated two-way IC in the SFP+ for 10 km and power conservation using uncooled operations and high temperature operations of EA/DFB optical sources in SFP+ for 40 km are being developed. Modulator integration and miniaturization in terms of mounting to modules are under progress for variable wavelength optical sources, and advances are being made in mounting to 38 pin butterfly, TOSA, and XFP using MZ model of modulator that is advantageous in a wide wavelength range. Moreover, monolithic integration of InP type ring resonator with external resonator that uses silicon waveguide is being investigated to miniaturize the variable wavelength optical source itself. Many material systems and structures are being explored for Si-based light emitters.

In an attempt to emit light in shorter wavelengths than the Si bandgap wavelength, EL luminescence using nanostructures or Si light-emitting diodes with ultrathin membrane that uses quantum effects have been performed. Other attempts include growth of Si/GaPN/Si types being achieved by using thin nitrogen series in an attempt to integrate Si and III–V group light emitters using epitaxial and light-emitting diodes using GaPN/GaAsN-SQW on Si. On the other hand, investigations are being done to make the band structure similar to that of a direct transition model, by applying stretching strain to Ge from the wavelength length side.

Improvements in luminescence efficiency brought about with the use of SiGe/ $FeSi_2$/SiGe are being investigated.

As for integration, investigations are being done on monolithic integration mainly of materials, such as photoreceptors, modulators, wavelength filters, and more, which can coexist with the Si-CMOS process (OITDA 2009).

With the increasing communication capacity in the wavelength range of optical communication systems, there has been progress in higher speeds, smaller size, and lower energy consumption for optical devices and optical transceivers. Ethernet was formally standardized to 40 and 100G in 2010. A 40GBASE-FR standard was added as IEEE802.3bg for 40 Gbit/s serial optical transmission. Whether the effective resonator length can be reduced by integrating passive waveguides for better efficiency of current injection in direct modulation lasers used as light sources is also being considered. As for semiconductor lasers integrated with a modulator, studies are being conducted to see if carrier discharge can be faster when the light absorbing layer has a tensile strained quantum well structure, instead of the conventional compressively strained quantum well structure, and it was reported to operate at 50 Gbit/s in the 1.3-μm band. In Si photonics, light-emitting and light-receiving elements on Si have been investigated. Experiments have been reported on optical interconnect link by integrating the light source, modulator, optical receiver, and optical waveguide together with the electronic circuitry for driving them, on a chip or within the module, including SOI or Si used in waveguides, Si or SiGe in modulators, Ge-on-Si in light receivers, InP system which is actually used for optical transmission in light source, and hybrid integrated optical and electronic circuits. Quantum cryptographic network is being considered, principally for security in optical communications. There is progress in development of a source of quantum entangled photon pairs using pseudo phasing in wavelength conversion, in addition to the conventionally used weak single photon, and there are hopes for investigations towards its practical use (OITDA 2011).

12.1.3 Visible and Ultraviolet Ranges

In visible and ultraviolet ranges, progress has been made in achievement of output-power improvement of light-emitting diodes and expansion of wavelength range of nitride-based semiconductor diodes. The previous wavelength of nitride-based semiconductor diode was less than 500-nm blue–green color fixing; however, a semi-polar GaN substrate (20–21) surface was used to achieve 531 nm at room temperature pulse and 520 nm at room temperature continuous operation. Also, room temperature continuous oscillation of 515 nm was achieved using the existing (0001) c-plane surface, and semiconductor diodes achieved green, a range previously not entered. In pure blue color, horizontal single-mode nitride-based semiconductor lasers (pulse 120 mW, continuous 60 mW) were also developed, in addition to the existing horizontal multimode. A shift from the short wavelength of 660 nm for DVDs to 638 nm with improved visual sensitivity had already been achieved in red waves, and the three primary colors of light (blue, green, and red) have been achieved by semiconductor laser. On the other hand, long-term reliability can be gained in the existing blue violet nitride-based semiconductor lasers, even during high-power output of 500 mW, because the speed of optical disk for recording has been increased. Towards new applications, investigations on surface luminescence laser are in progress, where photonic crystal-type and inclined mirror-type and vertical resonator surface-emitting lasers that use the top and bottom DBR mirror have been realized. There has been a growing interest in illumination in terms of energy and environmental conservation. In white LED lighting that uses a combination of nitride-based light-emitting diode (LED) and luminescent materials, properties (general color rendering index: 92, luminescence efficiency: 62 lm/W, total luminous flux: 500 lm) that are superior to light bulb-type fluorescent lamps have been developed. On the other hand, an LED is a point light source, whereas an organic EL is a 2D light source. During initial development, efficiency was low and there were issues with reliability, but with the development of phosphorescence luminescence materials and peripheral technologies, properties similar to those of fluorescent lamps have been obtained. Though there are some differences in LED lighting and organic EL, their practical implementation is expected to progress in the future in terms cost, purpose, design and so on, along with improved elemental characteristics (OITDA 2009).

Furthermore, development for new applications in addition to wavelength range expansion in semiconductor lasers or light-emitting diodes has progressed. White LED and photocatalyst in visible and ultraviolet bands were investigated for energy saving and energy creation. Luminance efficiency of white LED which is in the laboratory stage is 249 lm/W, which by far surpasses high-intensity discharge lamps (HID) and small fluorescent lamps (at that time, luminance efficiency of blue LED was 47 lm/W at 20 mA, 2.89 V). General lighting products are expected to be more efficient and will also have enhanced features with improvement in the color rendering properties and high output in the long run because of this development. Electrons and holes can be used as photocatalysts for generating hydrogen from reduced water and methane or formic acid from reduced carbon dioxide (artificial photosynthesis), and photocatalysts have been proved to be highly efficient and safe in GaN systems, which are eco-friendly and robust. With regard to the environment, deep ultraviolet rays (UV-C) with the shortest wavelengths in near-ultraviolet light have been investigated. Sterilization, breakdown of harmful substances, and water purification are possible in this wavelength range, but it cannot penetrate the surface of the earth, as it gets absorbed in the ozone layer. Therefore, from the viewpoint of environmental measurement, along with sensitivity to light, sensors are required to be solar-blind so they do not react to visible light and have long-term stability against ultraviolet rays. Diamond ultraviolet sensors were developed to meet these requirements (bandgap 5.5 eV, wavelength 225 nm). Moreover, agriphotonics, a new application of the visible range, is a promising food production technology. In addition to developing plant factories which can run on solar power with low power costs, development of enclosed type plant factories where rigorous environmental control is possible is underway. As a latest form of light source, LEDs which have higher wavelength use efficiency are now being used for plants in addition to conventional lamps. LED illumination costs more than other light sources, but LED has a good balance of irradiation of red light used for photosynthesis reaction and blue light used for phototropism and morphosis, and since LED lamps can be designed into any shape, they can be used in multistage cultivation devices. It is hoped that LED will come into common use once low cost and high efficiency are achieved (OITDA 2011).

12.1.4 Optical Interconnects

Emerging optical interconnection technologies offer both new capabilities and packaging challenges for the system designer. These technologies include parallel optics, high bandwidth plastic optical waveguides, and VCSELs, which cover an increasing range of wavelengths. Parallel optics offers low cost interconnection with perhaps the best use of backplane space.

Systems designers of computing and telecommunication systems have had so many choices for optical interconnection and different packaging approaches. Goals for packaging include the ability to make very small systems, the ability to distribute systems to solve thermal problems, achieving good waveguide management, getting the most bandwidth on and off and printed wiring board at the least use of space and the least disruption of backplane routine area, being able to insert boards having both optical and electrical interconnects without manual intervention with connectors, and the ability to flexibly manage heat dissipation in a system and on circuit boards. The use of optical interconnects between processors, boards, chips, and even gates (devices) can increase drastically the interconnection speeds. Optical interconnects will be vital to the development of high performance digital systems. The large bandwidth of optical waveguides provides much higher speed and large transmission distance than conventional electrical interconnects. Optical interconnects can provide also high parallelism because they are immune to EMI. Highly parallel board-to-board and chip-to-chip interconnects using 2D array devices and free space optical coupling are particularly appropriate for dealing with the massive amounts of data encountered in images (displays). With the boards spaced about an inch apart, for example, light could travel through free space from lasers on one board to detectors on another to carry information between the boards. In the high-speed, large capacity digital systems of the future, optical signals going in and from integrated optical functional devices should permit logic processing unlimited by transmission. Combining these devices with free-space parallel interconnects will provide powerful functions for processing extremely large amount of data. This is expected to lead to the development of optical computers and photonic switching systems, for instance (Suhir 2000). The development of optical interconnect devices have been proceeded through a synergetic collaboration among material and processing technologies, design and fabrication of

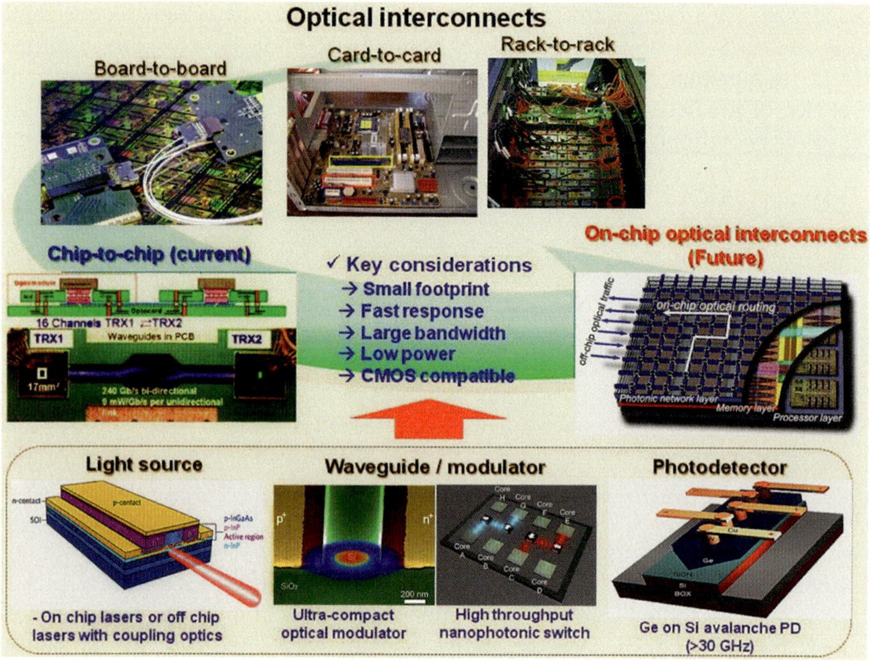

Fig. 12.1 Nanophotonics in integrated information system (Hu et al 2011)

integrated optoelectronics, and optoelectronic packaging technology. Figure 12.1 shows examples of optical interconnects and nanophotonics in integrated information system (Hu et al. 2011).

12.2 Advances of Micro-optics and Nanophotonics

Photonics is the science of generating, controlling, and detecting photons, and therefore nanophotonics studies the unique behavior of light at a nanometer scale. Correspondingly, microphotonics or micro-optics deals with the behavior of light at a micrometer scale. Emerging applications of micro-optics and nanophotonics are envisaged in diverse areas such as information processing, high-speed data communication systems, imaging, lighting, displays, manufacturing, life sciences and health care, and safety and security. The research aspects of micro- and nanophotonics have covered (Mihalache 2011): (a) silicon photonics, (b) spatial and spatiotemporal optical solitons (alias light bullets) in microwaveguide arrays and in arrays of evanescently coupled silicon-insulator nanowires, (c) spatial solitons in photorefractive materials, (d) nanoplasmonics, (e) photonic crystals, (f) metamaterials for micro- and nanophotonics including optical materials with negative refractive indices, (g) terahertz radiation and its applications, and

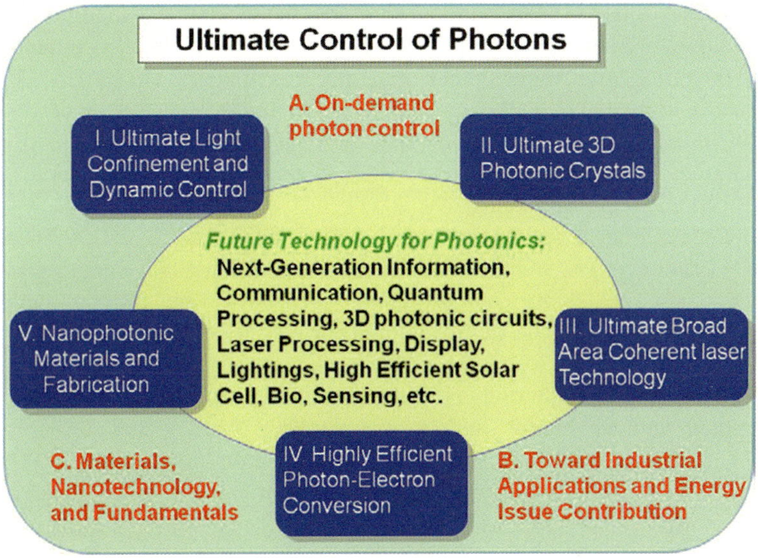

Fig. 12.2 Ultimate control of photons (Hu et al 2011)

(h) solid-state single-photon sources and nanometric size optical cavities for quantum information processing. For instance, the progress and vision in the ultimate control of photons for energy-efficient sensors, optical sources, and next-generation information processing can be represented by Fig. 12.2 (Hu et al. 2011).

12.2.1 Silicon Photonics

Silicon photonics is the photonic technology based on silicon chips, which can most applied in optical interconnections between digital electronic chips, such as all-optical high-speed signal processing with silicon-organic hybrid slot waveguides, periodically poled silicon, cladding-modulated Bragg gratings in silicon waveguides, light emission in silicon slot waveguides, enhancing nonlinearities in silicon photonic slot waveguides, and the possibility of light bullet formation in silicon-on-insulator (SOI) nanowires. Silicon (Si) has an indirect bandgap meaning that the upper and the lower electronic states (conduction and valence bands) do not occur at the same value of the crystal momentum. Light can penetrate much farther before being absorbed in an indirect-bandgap material than in a direct-bandgap one and this fact is crucial for photovoltaics (solar cells). Silicon is transparent in the fiber optic communication bands around 1,300 and 1,550 nm because the corresponding photon energies are less than the bandgap, which is about 1.1 eV for Si. That silicon performs various photonic functions can bring optical communication to the world of chip interconnects. Silicon chips communicate internally, or with other chips, using photons, in order to avoid the bandwidth limitation imposed by commonly used metallic interconnects. Other emerging applications include, e.g., low cost transceivers for 10–100 Gbit/s Ethernet (a transceiver is a device that has both a transmitter and a receiver which are combined and share common circuitry or a single housing), a platform for mid-infrared photonics, optically assisted analog-to-digital conversion, optical amplification, and lasing (Mihalache 2011).

Silicon's nonlinear optical properties, enhanced by tight optical confinement in Si/SiO_2 structures, i.e., SOI nanostructures, are producing wavelength generation and conversion, which are central functions in multiwavelength communications and signal processing. The SOI technology, which refers to the use of a layered silicon insulator-silicon substrate in place of conventional silicon substrates in semiconductor manufacturing, is a key technology in both micro- and nano-electronics and in micro- and nanophotonics. Another key technology in nanophotonics is the scanning transmission electron microscopy (STEM) for nanoscale visualization. STEM is a special technique in which an electron transparent sample is bombarded with a finely focused electron beam (with a diameter of less than 10 nm) that can scan across the sample. This technique provides high-resolution imaging of both inner structure and surface of a thin sample, as well as the possibility of chemical and structural characterization of both micrometer and nanometer sample domains through the evaluation of the X-ray spectra and the electron diffraction pattern. In a scanning transmission electron microscope, the size of the electron probe that is focused onto the specimen ultimately limits the spatial resolution. The resolution of a scanning transmission electron microscope is limited by the electron's de Broglie wavelength, e.g., for 300 kV electrons that limit is about 2 pm. Typically, the resolution is about 100 pm (about half the distance between atoms in some crystals) due to spherical aberration of electromagnetic lenses and finite size of electron source. By using a highly coherent focused electron probe in a fifth-order aberration-corrected transmission electron microscope, a crystal spacing of 50 pm has been resolved (Erni 2009).

Silicon photonics will play an important role in high-speed data communications, including sources of photons, electro-optic modulators, and photodetectors. A fundamental problem with Si is its lack of a static dipole moment, a consequence of its centrosymmetric crystal structure. This means that the linear electro-optic (Pockels) effect, that unique phenomenon that makes lithium niobate ($LiNbO_3$) and III–V semiconductors good electro-optic materials, is absent in Si. However, Si is not able to detect signals at standard communication wavelengths of 1.300 nm and 1,550 nm because such photon energies are less than its bandgap. Yet, these wavelengths represent standard telecommunication bands because optical fibers, to which devices must eventually interface, have low propagation losses in these frequency bands (about 0.2 dB/Km). The silicon's inability to absorb these wavelengths has been overcome by taking advantage of the small bandgap of germanium (Ge) grown on silicon. Raman scattering (a phenomenon that describes interaction of light with atomic vibrations of the crystal) and Kerr effect are two examples of nonlinear optical phenomena that only appear when silicon is pumped with high-intensity laser light (intensities $I > 10$ MW/cm^2). To avoid generating electrons and to prevent free carrier absorption, these devices use infrared light (at 1,500 nm), so that the photon energy is less than the bandgap of silicon. However, because of high intensities involved, two-photon absorption occurs. Important progress in the search for light amplification in silicon has been made. Stimulated Raman scattering (SRS) was also reported in SOI waveguides. Amplification of the Stokes signal, at 1,542.3 nm, of up to 0.25 dB has been observed in such SOI waveguides, using a 1,427-nm pump laser with a continuous-wave power of 1.6 W. It was also shown that the two-photon-absorption effects were found to be negligible at the pump power where SRS was observed (Claps 2003; Mihalache 2011).

Semiconductor nanostructures in the form of both quantum dots (QDs) and nanowires for emission, detection, guiding, and control of light have been widely studied. The large optical nonlinearities and the possibility of control of both linear and nonlinear optical properties by the size of the quantum dots are of special interest for emerging applications in integrated nanophotonic devices. The quantum efficiency of the absorption on quantum confinement levels in spherical QDs has been investigated. The size of QDs (with radius of about 1–3 nm) leads to neglect many body effects. For silicon QDs with radius of about 2.5 nm, which are embedded in amorphous silica, the absorption threshold shifts towards the far-infrared limit and that the spectral internal quantum efficiency reaches 4–5 % at the threshold (Iancu et al. 2011). Semiconductor QDs, which are robust and bright light emitters, have also captured special interest in life sciences, such as biology and medicine, namely, their capability for microorganism labeling. Silicon photonic nanowires can be used for on-chip communication and control and for chip-to-chip communications. These ultrasmall silicon light guides can be used as miniature chip-scale waveguides for transmitting data signals at very high bit rates (greater than 1 Tb/s). Diverse applications of semiconductor nanowires in the areas of optical waveguides, reconfigurable optical add-drop multiplexers, lasers (both optically pumped and electrically pumped), all-silicon lasers, photodetectors, antireflection coatings, sensors (biosensors), photovoltaics, quantum optics (single-photon sources), and for creating efficient tunable nanoantennas (optical antennas) are emerging. All-optical high-speed signal processing with silicon-organic hybrid slot waveguides has been performed. A 4-mm-long silicon organic hybrid nano-waveguide was fabricated with a record Kerr nonlinearity parameter of 10^5 W/km, which performed all-optical demultiplexing of 170.8–42.7 Gb/s. The performance of both planar and wire-like silicon slot waveguides (Si/SiO_2/Si nanometric size slot waveguides) as key components of on-chip light sources has been investigated; the spontaneous emission enhancement and waveguide coupling ratio for typical optical dopants such as Er in the low refractive index slot region containing SiO_2 were obtained. The enhanced luminescence efficiency and the strong coupling into a limited set of well-defined waveguide modes enable a new class of power-efficient waveguide-based light sources, which are compatible with complementary metal-oxide semiconductor (CMOS) technology (Jun et al. 2009). The electrical behavior of a multi-walled carbon nanotube network embedded in amorphous silicon nitride was studied by measuring the voltage and temperature dependences of the current. The current voltage curves present oscillations that are interpreted as due to percolation processes. The voltage percolation thresholds were related to the conductance minima. Surface acoustic wave sensors with carbon nanotubes and SiO_2/Si nanoparticles-based nanocomposites for volatile organic compounds detection were fabricated and experimentally studied. Also, the fabrication on thin aluminum nitride layers deposited by magnetron sputtering on high-resistivity (100)-oriented silicon substrates of surface acoustic wave resonators operating at 5 GHz was achieved (Stavarache et al. 2011; Mihalache 2011).

As a centrosymmetric crystal, bulk silicon lacks second-order optical nonlinearity, which is the key element in nonlinear optics applications. Hence, its lowest-order optical nonlinearity originates from the third-order susceptibility, which gives rise to both Raman and Kerr effects. Conventional poling processes, such as those used for $LiNbO_3$ and nonlinear polymers do not apply to silicon because as it lacks second-order optical nonlinearity in its native form. However, a new class of photonic devices based on periodic stress fields in silicon that enable second-order nonlinearity as well as quasi-phase matching has been developed. The periodically poled silicon waveguide is covered by a SiN cladding causing compressive stress and a SiN film is deposited astride the waveguide causing tensile stress. Mid-wave infrared radiation can be efficiently generated through difference frequency generation from near-infrared radiation with a conversion efficiency of up to 50 %. This can be achieved with a 20-ps pump pulse with wavelength $\lambda = 1{,}300$ nm and peak intensity of 1 GW/cm^2 (Hon et al. 2009). A cladding-modulated Bragg grating using a periodic placement of silicon cylinders along a waveguide was designed, fabricated, and experimentally validated using a SOI material platform. The coupling strength of the cladding-modulated distributed resonant structure was varied by changing the distance between the silicon cylinders and the waveguide. The period of the cladding modulation was chosen to satisfy the Bragg condition at a wavelength of about 1,550 nm for the TE-polarized mode and the waveguide cross section was chosen to be single mode at this wavelength. By exploiting the advantages of the horizontal silicon slot waveguide structure, the nonlinear interaction can be significantly increased compared to vertical slot waveguides. Thus, it has been shown that an optimized horizontal slot waveguide structure containing a 20-nm-thin optically nonlinear layer with low refractive index sandwiched between two silicon photonic wires of 220 nm wide and 205 nm high could enable a large Kerr nonlinearity coefficient of 2×10^7 W/km (Muellner et al. 2009; Mihalache 2011).

From the point of view of practical applications, lithium niobate is one of the key materials for the fabrication of both micro- and nano-optical waveguides, optical modulators, and various other linear and nonlinear optical devices. Photorefractive screening-photovoltaic solitons were observed in lithium niobate. Thus, 2D bright circular spatial optical solitons (self-confined optical beams) form due to a strong static bias field, externally applied, opposite to the photovoltaic internal field. Moreover, the formation of self-confined beams in erbium-doped lithium niobate was observed. Also, a very general dependence of the self-trapped beam waist on the refractive index change was experimentally derived, which is valid for every lithium niobate crystal, independent on its growing procedures or doping. Arrays of soliton waveguides in lithium niobate for parallel coupling were also investigated.

In micro- and nanophotonics, spatiotemporal optical solitons, alias "light bullets," are spatially confined pulses of light, i.e., electromagnetic wave packets self-trapped in both space and time (Malomed et al. 2005). These nondispersing and nondiffracting light structures could be used as natural information carriers. The "light bullet" arises because the spatiotemporal optical soliton can be thought of as a tiny bead of light propagating long distances without changing its shape. It is believed that the "light bullets" are the ideal information units in both serial and parallel transmission and processing information systems. Three-dimensional light bullets in 2D arrays of coupled waveguides was reported, in which these genuine 3D spatiotemporal solitons were excited by femtosecond pulses in a system featuring quasi-instantaneous cubic nonlinearity and a periodic, transversally modulated refractive index. Arrays of evanescently coupled waveguides support stable light bullets in a limited domain of their existence range. Another light bullet was set based on silicon photonic nanowires. The conditions for low-power spatiotemporal soliton formation in arrays of evanescently coupled SOI photonic nanowires have been thoroughly analyzed. Pronounced soliton effects can be observed even in the presence of realistic loss, two-photon absorption, and higher-order group-velocity dispersion (GVD). The well-established SOI technology offers an exciting opportunity in the area of spatiotemporal optical solitons because a strong anomalous GVD can be achieved with nanoscaled transverse dimensions, and moreover, the enhanced nonlinear response resulting from this tight transverse spatial confinement of the electromagnetic field leads to soliton peak powers of only a few watts for 100-fs pulse widths (the corresponding soliton energy being only a few hundreds fJ). The arrays of SOI photonic nanowires seem to be suitable for the observation of both bulk and surface light bullets because an adequate design of nanowires can provide dispersion lengths in the range of 1 mm and coupling lengths of a few millimeters (for 100-fs pulse durations). Spatiotemporal nonlinear optical effects would lead to spectral broadening in arrays of subwavelength silicon waveguides pumped with infrared femtosecond pulses (Benton et al. 2008; Gorbach et al. 2010).

Silicon-based plasmonic waveguides have been used as a means to confine and manipulate photonic signals. The high refractive index of silicon at telecommunication wavelengths (about 1,550 nm) assures strong confinement and a very high level of photonic integration with achievable waveguide separations of the order of 10 nm. The aluminum and copper plasmonic material platforms make silicon-based plasmonic waveguides fully compatible with the existing mature CMOS fabrication technology. It is also possible to compensate the intrinsic surface plasmon propagation losses and to achieve up to 10 Tb/s signal transfer rates (Krasavin and Zayats 2010; Mihalache 2011).

12.2.2 Nanoplasmonics

Nanoplasmonics aims to mold light flow at the nanoscale using metallic nanostructures (usually nanostructured noble metals such as Au and Ag) for a perspective on low-loss plasmonic metamaterials, mainly covering lasing in metal-insulator-metal subwavelength plasmonic waveguides, plasmonic slot waveguides, ultrafast active plasmonics, plasmonic solar cells, surface plasmon resonance sensors for biosensing and chemical sensing applications, and subwavelength plasmonic lattice solitons (PLSs) in arrays of metallic nanowires. It has been found that arrays of subwavelength holes in metallic films display unusual zero-order transmission spectra at wavelengths larger than the array period, beyond which no diffraction occurs. Light transmission through subwavelength metallic channels has been investigated, and a resonant cavity-enhanced light transmission mechanism in metallic gratings with subwavelength apertures was put forward for operation with light in the visible spectral range. Small variations of aperture shape or its spatial dimensions have a huge effect on the transmission properties of subwavelength metallic gratings. Metals structured at a nanometer scale can lead to improved and sometimes surprising properties; e.g., metals can display colors which vary with their size. These colors result from the coupling of light with the free electrons of nanostructured metal particles embedded in a surrounding dielectric or semiconductor matrix or nanometer size metal films deposited on a dielectric substrate to form surface plasmons. Surface plasmons are a specific kind of electromagnetic surface waves that occur at the interface between a noble metal (such as gold or silver) and a dielectric medium (such as glass or air). They can concentrate a huge electromagnetic field at the interface and, therefore, they strongly enhance both linear and nonlinear interaction between laser light and matter. Surface plasmons are considered a means to bridge nanoelectronics and nanophotonics. Photonic components are superior to electronic ones in terms of operational bandwidth, but the diffraction limit of light poses a significant challenge to the miniaturization and high-density integration of optical circuits. A possibility to avoid this problem is to exploit the hybrid nature of surface plasmons, which are light waves coupled to free electron oscillations in a metal that can be confined below the diffraction limit using subwavelength (nanometer size) noble metal structures. The simultaneous realization of strong light confinement and low propagation losses for practical applications proved to be a very difficult task. However metal/dielectric-based plasmonic waveguides constitute the key elements in developing ultra-compact integrated planar lightwave circuits in addition to other waveguide structures, such as silicon photonic crystal waveguides and metallic or dielectric-based slot waveguides, as well as silicon slot waveguides (Joannopoulos et al. 2008; Mihalache 2011).

Lasing in metal-insulator-metal waveguides filled with electrically pumped semiconductor cores, with core width sizes below the diffraction limit has been developed. These structures confine light (a TM mode) to waveguide core regions about half the diffraction limit in width, that is, about $\lambda/(2n)$, where n is the refractive index of the core material. The inherent losses in such subwavelength metal-insulator-metal waveguides can be overcome to create small plasmon mode lasers at wavelengths near 1,500 nm. 44-nm-diameter nanoparticles with a gold core and dye-doped silica shell has been used to make the nanometer-scale plasmonic laser to completely overcome the loss of localized surface plasmons by gain and achieve a surface plasmon amplification by stimulated emission of radiation (SPASER). SPASER radiation consists of surface plasmons that obey Bose-Einstein statistics and undergo stimulated emission but in contrast to photons can be localized at the nanoscale. SPASER operates on a dark mode; this means that enhanced local electromagnetic fields are generated but nothing is necessarily emitted in the far field. Such nanolasers can be achieved using a hybrid plasmonic waveguide consisting of a high-gain cadmium sulfide semiconductor nanowire, separated from a silver surface by a 5-nm-thick insulating gap (Noginov et al. 2009). Though the surface plasmons are intrinsically transverse magnetic (TM) polarized surface waves, the existence of transverse electric (TE) plasmons in bilayer graphene has been predicted. The plasmonic properties are much more pronounced in bilayer than in monolayer graphene. Nanoplasmonics enables a convergence of semiconductor electronics and nonlinear optics at nanoscale lengths and at femtosecond timescales. The ultimate goal of nanoplasmonics is to provide a new class of ultracompact (at nanometer scales) and ultrafast optical devices for all-optical information processing. Femtosecond optical frequency surface plasmon pulses propagating along a metal–dielectric waveguide (an Al/silica interface) were modulated on the femtosecond timescale by direct ultrafast optical excitation of the metal, thereby offering unprecedented terahertz modulation bandwidth. Therefore, femtosecond plasmon pulses can be generated, transmitted, modulated, and decoupled for detection in a single optical device. The resonance wavelength strongly depends on the metal nanoparticle size, shape, and on local dielectric environment. The sharp surface plasmon resonance can be used either in biosensing or in chemical sensing applications. These sensors use the absorbing light property of a nanometric thin noble metal layer (such as Au) deposited on a high refractive index glass substrate, which produces electron waves (surface plasmons) on the metal surface. This sharp resonance occurs only at a specific incidence angle (for a fixed value of the wavelength of the incident laser radiation) and is highly dependent on the metal surface, such that binding of a target analyte to a receptor on the metal surface produces a measurable optical signal. Surface plasmon resonance sensing combined with surface enhanced Raman scattering detection on periodic arrays of subwavelength metallic nanoholes opens a route for molecular detection to be integrated in lab-on-chip systems in order to increase the reliability of biosensing. For photodetectors in superconducting box-shaped plasmon–polariton optical waveguides, the light confinement regimes and the power absorption efficiency in the superconducting layer can be maximized by optimizing only the plasmon–polariton waveguide geometry. Thus, nanoplasmonics provides novel nanotools for both biosciences and medicine and promises integrated functionalities for future point-of-care diagnostics and cancer therapies (Baffou et al. 2010; Mihalache 2011).

Stable subwavelength PLSs can form in both 1D and 2D arrays of metallic nanowires embedded in nonlinear Kerr media. The main physical mechanisms for balancing the beam diffraction and the formation of subwavelength PLSs is provided by the tight confinement of the guiding modes of the metallic nanowires, combined with the strong nonlinearity induced by the enhanced field at the noble metal (e.g., Ag) surface. Moreover, vortical PLSs, which form in 2D arrays of metallic nanowires embedded into nonlinear media with both focusing and defocusing Kerr nonlinearities have been introduced. These subwavelength PLSs can be used to optically manipulate with nanometer accuracy the power flow in ultracompact photonic devices and systems (Ye et al. 2011; Mihalache 2011).

12.2.3 Photonic Crystals and Metamaterials for Micro-optics and Nanophotonics

The photonic crystals, specially engineered materials in which the atoms and molecules of a common crystal are replaced by macroscopic media with different dielectric permittivity so that the periodic potential is replaced by a periodic refractive index, allow a complete control over light propagation in such an artificial material. The photonic crystals display photonic bandgaps, i.e., light cannot propagate in certain directions with specific frequencies. There have been great progresses in this area, such as high-quality factor photonic crystal nanocavities and slow-light photonic crystal waveguides (Joannopoulos et al. 2008; Mihalache 2011).

Silicon photonic crystals allow, for instance, the fabrication of nanometric size waveguides and sharp waveguide bends, an active control of slow light on a chip, as well as the achievement of high-quality factor nanocavities. Photonic crystal nanocavities with a photon lifetime of 2.1 ns, which correspond to a quality factor of 2.5×10^6, have been fabricated by using photonic crystals with a triangular lattice of circular air holes with radii of 115 nm in a 250-nm-thick Si slab. The nanocavity itself consists of a line defect with the lattice constant increasing every two periods as it approaches the cavity center (Takahashi et al. 2007). Green light emission through third harmonic generation in a slow-light photonic-crystal waveguide has been achieved. Visible third harmonic-generation at a wavelength of 520 nm with a peak pump power of a few watts only was observed, and it has been demonstrated a strong third-harmonic-generation enhancement due to the reduced group velocity of the near-infrared pump signal. The photonic device consists of an 80-mm-long photonic-crystal waveguide in a 220-nm-thick air-suspended silicon slab, coupled to two tapered ridge waveguides. It was observed visible green light emission for only 10-W peak pump power due to both the tight light confinement within the photonic-crystal waveguide (the effective mode area was about 0.4 μm^2) and the energy density enhancement provided by the slow light mode (the group velocity was about $v_g = c/40$) (Corcoran et al. 2009; Mihalache 2011).

Metamaterials are artificially engineered structures that have unique properties, such as a negative refractive index, not attainable with naturally occurring materials; they can be artificial media structured on a size scale smaller than the wavelength of external stimuli. Developments in micro- and nanostructured metamaterials have given rise to negative refractive index media which have both negative dielectric permittivity and negative magnetic permeability in some frequency ranges. Negative refractive index metamaterials were first demonstrated at microwave frequencies. Metal-based negative refractive index metamaterials have been extensively studied in that spectral region. Negative-refractive-index metamaterials are more difficult to obtain at near-infrared or visible wavelengths due to both difficulties of fabrication and poor optical properties of metals at these frequencies. However, the first fabrication and experimental verification of transversely structured metal-dielectric-metal *multilayers* exhibiting negative refractive indices in the near-infrared spectral region (around 2 μm) were reported (Zhang et al. 2005). It was later experimentally demonstrated a 3D optical metamaterial having negative refractive index with a very high figure of merit of 3.5 (i.e., with a low loss) (Valentine et al. 2008). This metamaterial is made of cascaded fishnet structures, with a negative refractive index existing over a broad spectral range. The fishnet structure consisted of alternating nanometer-thick layers of Ag and magnesium fluoride (MgF_2) with thicknesses of 30 nm and 50 nm, respectively. Furthermore, an engineered metamaterial made of alternating layers of negatively refracting (a silicon photonic crystal) and positively refracting (air) materials strongly collimates a beam of near-infrared light at 1,550 nm. This result can be regarded as a first explicit experimental verification of the concept of "optical antimatter" (a slab of metamaterial appears to "annihilate" a slab of air of equal thickness) (Mihalache 2011).

Zero-average refractive-index bandgaps in photonic crystal superlattices consisting of alternating stacks of negative-refractive-index photonic crystals and positive refractive-index dielectric materials in the near-infrared range were reported (Kocaman et al. 2009). The fabricated nanostructured superlattices demonstrated the presence of zeroth-order bandgaps for a wide range of superlattice periods and unit cell variations. The use of photonic metamaterials as anisotropic media, in particular, nanostructured materials exhibiting tunable form birefringence, is of particular interest in the area of the so-called Dyakonov–Tamm lossless surface waves. These electromagnetic surface waves exist at interfaces between two homogeneous dielectric materials, one of them is isotropic and the other one is uniaxial with its optical axis aligned parallel to the interface. Propagation of both Dyakonov–Tamm waves and waveguide modes should occur in practice with negligible attenuation, in contrast to that of surface plasmon–polariton waves that are guided when the dielectric slab is replaced by an ultrathin noble metal film (Faryad and Lakhtakia 2011). Furthermore, the significance of Dyakonov–Tamm surface waves for optical sensing applications is quite obvious. A method for building different photonic structures in As_2S_3 amorphous chalcogenide thin films has been developed by using femtosecond pulses. Two-dimensional photonic configurations consisting of a regular assembly of rods (arranged in a triangular lattice) or micrometer period gratings have been fabricated. These chalcogenide photonic structures could be efficient for all optical processing in the infrared spectral range (wavelength of several micrometers). Two-dimensional photonic structures imprinted on the surface of arsenic sulfide glasses have been created by using femtosecond laser pulses. Laser-induced modification of materials (via techniques such as laser ablation, two-photon photopolymerization, and near-field laser lithography) is currently used to get micro- and nanostructured materials. Thus, laser-induced periodic surface structures were obtained on ZnO thin films deposited on sapphire substrate. Ripples with a period of about 150 nm, spaced by grooves with about 50 nm width and about 100 nm depth, have been obtained by scanning the sample surface with femtosecond laser beams. Photonic crystal cavities coupled with optical waveguides is able to transfer electromagnetic energy only in the forward direction. This simple photonic device has potential applications in photonic crystal microcircuits, especially in the waveguide intersections. The possibility of achieving a single-mode vertically coupled microring resonator that can be obtained by wafer bonding. In addition, nanoparticles of interest in both nanotechnology and in medicine (e.g., in the controlled release of drugs) can be produced by both anorganic and organic syntheses. Complex conjugate materials have been developed, where the ratio of the complex electrical permittivity and the complex conjugate magnetic permeability is real. They are characterized by a real refractive index and thus they allow non-attenuated propagation of electromagnetic waves. Such metamaterials could have important applications in miniaturized optical amplifiers and lasers (Dragoman 2011; Mihalache 2011).

Graphene is a low-dimensional material (only one atom thick) with an unusual electronic spectrum. It adds electro-optical capability to metamaterials in the infrared and terahertz spectral domains. This unique capability of graphene exploits the spectral shift of electromagnetic response driven by the applied gate voltage. A graphene-based ballistic diode is able to rectify an incident signal due to an oblique gate positioned between the two terminals of the device. The rectifying properties of the graphene-based ultrafast diode are thus tunable, in deep contrast with usual semiconductor-based diodes. Optical transparent and electrical conductive polymer/nanocarbon composite films have been synthesized by infiltration method (Sandu et al. 2011).

The conventional noble metals (Au, Ag) as plasmonic materials in the near-infrared and visible spectral ranges suffer from drawbacks such as large losses and incompatibility with the current semiconductor technology. These problems can be overcome by using an all-semiconductor-based approach to plasmonics. For instance, InAs-heterostructures are superior to other common semiconductors and metals for applications in plasmonic structures and metamaterials in mid-infrared frequency range (Mihalache 2011).

12.2.4 Terahertz Radiation and Its Applications

The terahertz (THz) radiation, i.e., the electromagnetic radiation with the wavelengths in the range 0.1–1 mm (3 THz–300 GHz) is a non-ionizing submillimeter microwave radiation. It cannot penetrate metal or water; however, it can pass through paper, clothing, wood, masonry, plastic, and ceramics. The envisaged applications comprise (Mihalache 2011) (a) medical imaging and clinical diagnostic, because the THz radiation is not expected to damage tissues and DNA, unlike X-rays; (b) security checks (THz radiation can penetrate fabrics and plastic, so it can be used in security screening, to uncover, e.g., concealed weapons on a person, remotely); (c) quality control of pharmaceutical and polymeric goods; (d) detection of contamination in food products; and (e) "Indoor" wireless communication. Various systems based on non-ionizing THz electromagnetic waves will widen mankind's scientific and technical potential to a similar extend the X-rays scanners did over the past century, such as generation of terahertz radiation on metallic nanostructured surfaces and nanoparticle-enabled terahertz imaging of biological tissues.

The radiation pattern and the efficiency of terahertz antennas based on graphene can be changed via the gate voltage applied on graphene. The phenomenon of optical rectification has been widely used to rectify ultrafast (picosecond or femtosecond) laser pulses from the visible to the terahertz frequency range. A resonant "incoherent" rectification process relies on the excitation of surface plasmons on nanostructured noble metal surfaces. Thus, the excitation of nanostructured gold and silver films with 800-nm femtosecond laser pulses resulted in the emission of terahertz radiation with an angle-dependent efficiency (Welsh and Wynne 2009).

The nanoparticle contrast agent-enabled terahertz imaging technique yields an enhanced sensitivity of the differential signal from cancer cells with nanoparticle contrast agents (e.g., gold nanorods). The THz reflection signal from the cancer cells increases by 20 % upon their irradiation with infrared light compared to cancer cells without gold nanorods. This enhanced sensitivity was due to the temperature rise of water in cancer cells by the excitation of surface plasmons. Therefore, THz cancer imaging can be realized with a micron resolution, which would facilitate the diagnosis of cancers at a very early stage (Oh et al. 2009; Mihalache 2011).

12.2.5 Nanophotonics and Quantum Information Processing

One of the main applications of nanophotonics to quantum physics and quantum information processing is to design single-photon sources based on the emission of cavities in photonic crystals or on the emission of quantum dots embedded in semiconductor nanowires which can be engineered to reduce the divergence of the far-field radiation. The semiconductor nanocrystals (comprising a few hundred to a few thousand atoms) constitute the ideal single-photon sources for quantum information applications. Several single-photon sources based on the emission of a quantum dot embedded in a semiconductor (GaAs) nanowire have been designed with the optical nanoantenna volume of the order of $0.05\ \lambda^3$. In contrast to other optical nanoantennas based on surface plasmons, the approach cannot rely on any resonance effect and the funneling that was actually achieved over a very broad spectral range, $\Delta\lambda = 70$ nm at $\lambda = 950$ nm. The slot-waveguide geometry for nanophotonic applications in the near-infrared range is also promising for quantum optical applications in the visible spectrum. To this aim, diamond- and GaP-based slot-waveguide cavities compatible with diamond color centers have been introduced with single-photon Rabi frequency on the order of 10^{11} rad/s (Hiscocks et al. 2009; Mihalache 2011).

Given the rapid growth of studies in micro- and nanophotonics, one can expect many new and exciting developments over the next decade. No doubt, soon one can expect a maturity of these fast growing research fields, leading to new and interesting physical phenomena and to the utilization of their huge technological potential in various areas of human activity, e.g., in reducing energy consumption in lighting systems, developing novel biomedical and chemical sensors, fabrication of optical sensors able to detect even single molecules, designing of new solar cells with enhanced efficiency, and implementation of ultrahigh-speed telecommunication systems (Mihalache 2011).

12.3 Trends in Applications

12.3.1 Optical Communication Networks

In access networks, there has been a major shift to optical fiber communications, from copper line communications such as DSL (digital subscriber line) and CATV (cable television).

Starting with conversions to fiber optics in long distance mainline and trunk lines in the 1980s and introduction of metro 10 Gb/s optical ring nets in the 1990s and 2000s, the system has evolved to fiber optic broadband network configurations such as core/metro/access. The following systems have been introduced in each of the networks: dense wavelength division multiplexing (DWDM) system that links two ultra-long distance points using repeated optical amplification relays, DWDM ring network system constructed with OADM nodes (Optical Add/Drop Multiplexer) that carries out optical branching and insertion and amplification, and

passive optical network (PON) system that carries out optical branching (down) and optical time division multiplexing (up). There have been dramatic cost reductions with the introduction of such systems that are based on technological innovations such as multiwavelength optical batch amplification and optical switches. For instance, the main advantage of the existing practical use DWDM system is that it bundles and sends 10 Gb/s signals at 40–80 wavelength, and the system capacity of each fiber in the core net and metro net is between 0.4 and 0.8 Tb/s.

With the introduction of optics technology, there has been a rapid speedup of Ethernet LAN (local area network). LAN optical technology would foray into the 100G era, ahead of WANs (wide area networks). In other words, the role of accelerating the speedup process of optics technology is shifting from WAN to LAN. Furthermore, the focus has begun to shift towards internal parts of equipment (chip-, board-, or system-level optical interconnection) including trunk optical transmission systems, photonic nodes, optical networking, access network, optic LAN/interconnect, and optical fibers (OITDA 2009).

1. Long-Haul Optical Transmission Systems
   There was remarkable progress in development of DSP-LSI for commercialization of 100 Gb/s/ch transmission systems. On the other hand, the world's first 100 Tb/s transmission experiment with ultrahigh spectral efficiency (11 b/s/Hz) was reported. In addition, physical limitations of optical fibers have become apparent, and there is a focus on the great increase in experiments on space multiplexed optical transmission using multicore fiber and multimode fiber which could possibly overcome these limitations (OITDA 2011).

   (a) Photonic Nodes
   In photonic nodes, Reconfigurable OADM (ROADM), which is an optical branching insertion node, is being introduced in the ring section where the traffic is concentrated. Multi-degree ROADM technology that enables optical interconnection between multiple rings is being developed. This technology is expected to control electricity consumption by reducing the installation surface area and the number of network devices. The development of element technology as a future technology for optical burst/optical packet switches is also advancing (OITDA 2009).

   Steady progress in the ROADM technology of optical switches which are key devices has led to the development of colorless, directionless, contentionless, and gridless types of switches. Regarding standardization, the new technology of ODUflex (GFP), an interface for transporting packet signals through generic framing procedure (GFP) in containers, for variable bit rate without disconnection of the operating service, was reported (OITDA 2011).

3. Optical Networking.
   With regard to optical networking, technology is being developed that integrates and controls optical path conversion having long distance/high capacity, with packet networking that efficiently multiplexes packet traffic coming from users. Control models such as power conservation network design theory, methods of fault recovery between layers, impairment routing that secures reachability of optical signals (method of establishing paths after considering transmission properties of individual optical paths), and so on have been investigated (OITDA 2009).

   Progress was made in standardization of flexible grids, and research and development on adaptive optical network technology using these grids has greatly advanced. Additionally, the new concept of network technology called software-defined network (SDN) that is capable of programming and controlling networks without regard to the networking devices was advocated and promoted (OITDA 2011).

4. Access Network.
   As for optical access, 1 Gb/s class FTTH has been launched, and R&D on next-generation optical access systems is being actively pursued in terms of high capacity and higher speed for achieving high-definition image distribution, etc. A move to full-scale technology development for 10G-EPON (10G Ethernet PON) has been seen with its standardization completed in September 2009, and technological developments in next-generation optical access systems applied in multilevel modulation technology and signal processing technology (OITDA 2009).

   Technological developments in 10G-EPON have been completed to the optical transceiver level and moved to the device development phase. Efforts for standardization are underway, such as Next Generation Passive Optical Network Phase 2 (NG-PON2) aimed at 40–100 Gb/s operations. Also, a series of reports on OFDMA-PON, Coherent-PON, WDM-PON, Hybrid-PON, OCDMA-PON, and so on for NG-PON2 have been released by academic societies. Especially, there is active development in next-generation optical access technology applying multi-level modulation and signal processing technologies (OITDA 2011).
5. Optical LAN/Interconnect.
   With regard to the optical LAN/interconnect field, technological developments in 40 and 100 Gb/s Ethernet were standardized in 2010. Moreover, maximum capacity of backbone routers/switches has already crossed 1 Tb/s per unit, and 4–10 Tb/s router/switches would be introduced. For this, the focus is on ultrahigh speedup with channel bit rate of $\geq$20 Gb/s and technology related to energy-saving optical devices/circuits/assemblies with 10 mW/Gb/s (OITDA 2009).

   Second-generation modules have begun to evolve in the electrical interface of the 100 GbE optical transceiver from the existing 10×10 to 4×25 Gb/s, and development has started on standardization of 100 Gb/s backplane technology. Additionally, development has begun on optical backplane with onboard transceiver for high-performance computers (HPC) and high-end routers which require high-capacity interconnects (OITDA 2011).
6. Optical Fibers.
   In the optical fiber field, as FTTH is being used increasingly, there are advances in technology development of low-bending-loss optical fibers used in premise/access systems that help in achieving improved degree of freedom for wiring and equipment miniaturization and skill-less line opening construction. Investigations on multicore fibers have also been initiated, aiming for breakthroughs in the physical limits of fiber optics capacity in trunk line systems (OITDA 2009).

   Concerning trunk line optical transmission systems, there has been remarkable progress in R&D related to speedup of transmission speed to 100 Gb/s per wavelength. Coherent reception based on digital signal processing has brought the long distance transmission system, having up to100 Gb/s $\times$ 100 waves, closer to reality. Also, mounted components are being standardized in accordance with the industry group OIF (Optical Internetworking Forum).

Investigations on further speedup and increased capacity of wireless communication systems and similar complex multi-valued modulation methods are also being initiated (OITDA 2009).

Moreover, development of space multiplexing multicore fibers and mode multiplexing few-mode fibers is progressing, and experiments on transmission by using them along with optical fiber amplifiers are being conducted, as their study became more active. Practical aspects of pulling them into homes and laying the low-bending-loss-type fibers supporting the wiring in homes is being studied for FTTH. In specialty fibers, steady progress has been seen in microstructured optical fibers, rare earth-doped optical fibers, and highly non-linear optical fibers (OITDA 2011).

In particular, free space optical communication technology has been investigated on satellite space optics, aircraft space optics, and building-to-building optical communication. It is hoped that this technology will complement the existing wireless communication infrastructure (OITDA 2011).

12.3.2 *Optical Memory and Information Processing*

Coupled with the spread of network distribution of contents and achievement of high capacity low-cost semiconductor memory, the marketability of optical memory came under threat. On the other hand, 3D displays, higher-definition cameras, and contents needing large capacity optical memory are being launched. Moreover, construction of archive systems using optical memory is being investigated, with the purpose of energy conservation. Technological developments in optical information processing technology are not restricted to the existing memory system or search systems, but have spread to other fields such as information security, measurement, and manufacturing (OITDA 2009).

Investigations on optical memory are classified into two fields: optical disk media for investigating the trends in recording and media manufacturing technologies and optical disk drives for investigating peripheral technologies including signal processing technology, servo technology, light sources, and light-receiving elements. Regarding BD standard BDXL format which enabled improvement in recording capacity through in-plane higher density and multiplexing, 100 GB rewritable disks were released in the industry in 2011. Recording capacity has thereby steadily improved. Also, next-generation archival system products with 12 optical disks in one large storage of 300 GB to 1.5 TB were launched (OITDA 2011).

1. Optical Disk Media and Drives.
   The specifications for Blu-ray Disc (BD) standards for 3D content were released in 2009, and the demand for optical memory with larger capacity has been growing. In response to this situation, technology developments in multilayer technology that further improves the capacity of the existing system (BD) were investigated. There have been advances in multilayer technology, particularly in rewritable optical disks (3 layer 100 GB) and recordable disks (10 layer

320 GB). In ROM, a 16 layer 400 GB disk has already been announced, and a capacity of over 100 GB in all existing optical memory shapes using multilayer technology has been demonstrated. Optical disks are strong in features such as long life, portability, and energy conservation, but increasing their capacity and transfer speeds are major issues for their adoption in archival uses and so on compared with magnetic disks and magnetic tapes. Next-generation large capacity technology has been explored, covering near-field optical disks, ultra-resolution disks, hologram, 3D bit by bit recording, and elemental technologies, such as signal processing technology, server technology, and components, used for achieving large capacity in the above mentioned fields. For instance, reading of contents using super resolution disks (50 GB (BD $\times$ 2), twice the speed), and 34 layer/204 GB using micro-reflector having just one layer of bit by bit volumetric records, have been achieved, thereby confirming steps forward in the appearance of technologies moving towards practical applications (OITDA 2009).

In the optical disk media field, the existing optical disks have been further developed and four fields of near-field disk and memory, hologram (page), microhologram, and microreflectors have been investigated. With regard to BDXL in BD format with room for expansion to three or four layers that was launched in June 2010, the recording capacity of each layer was increased from 25 to 33.4 GB max. In 2011, 100 GB rewritable disks were launched. Implementation of 51.2 GB per layer using a BD optical system was reported. In-plane capacity had reached its limits, but there is steady progress in technology for upgrading 5 in. disks from 512 GB to 1 TB class, by combining with multilayer technology. Reports were also presented on data storage life of optical disks, which are expected to be used for archiving. There are hopes for research on securing storage life, for utilizing the removability and drive compatibility of optical disks, and quantitatively evaluate the strong points of optical discs: less vulnerability to water damage, and so on. On the other hand, research continues on new methods of read/write technologies, aiming for a dramatic increase in the storage capacity of optical memory. Efforts for research are also being pursued on page holographic memory and bit recording microholograms, for achieving higher capacity by making use of phase information. It is possible to write phase information with a high signal to noise ratio despite weak light intensity, so technology for upgrading capacity by making use of optical properties attracts attention (OITDA 2011).

In optical disk drives, signal processing technology, servomechanism technology, and components have been investigated, such as 3D recording media by void records with guide grooves and tracking servos for multilayer optical disk media in servo technology. Looking at disk making technology, there are hopes for development of 3D recording media which has recording layers with no track grooves and super-multilayer optical disks with guide grooves. Consequently, in addition to development of media making technology, research on data address technology and servo technology is likely to gain prominence to achieve this. Also, research is progressing on technology analysis for stable rotation of 1.2-mm optical disks at 15,000 rpm or above, for achieving high-speed reading (OITDA 2011).

2. Optical Information Processing.
   Optical information processing has started to spread across various fields, information security technology, 3D information technology, and their respective technology. Their representative examples were photonic smart media and hierarchical memory based on near-field optical interaction, superhigh-speed 2D optical information retrieval system that uses hologram, and technology trends in digital optics (OITDA 2009).

   Optical information processing has been investigated by integrating optical technology and optoelectronic devices or high-speed computing technology, mainly including optical information processing technology, digital optics, optical correlation system, and hierarchical system based on near-field optical interaction. In particular, a measurement system has been developed by combining with high-speed computing technology based on optical technology and graphic processing unit (GPU) to surpass past performances. Moreover, research is underway on digital holography as ultra-deep 3D measurement technology. Quantitative high-speed 3D measurement is being implemented using interference measurements and reverse light wave propagation calculations by computer. By using a high speed camera, measurement of 150,000 frames/s ultra-high speed 3D is being done. Also, computing methods are being developed for faster calculation of light wave propagation by making use of GPU. 8,192 $\times$ 8,192 pixel image data conversion was possible at 56 ms by high-speed Fourier transformation. In scalar diffraction calculation, it is possible to use dual high-speed Fourier transformations to calculate light wave distribution after it has propagated some distance, therefore it is hoped to lead to development of a system that can reproduce a 3D domain in real time by using interferometric image acquired by digital holography (OITDA 2011).

12.3.3 Displays

There has been rapid growth in demand for low-cost displays and a switchover to a low-energy society, with a top priority being reduction in electricity consumed by displays. Regulations are being reinforced with specific energy conservation targets, such as Energy Star, EuP, and amendment of the Act on the Rational Use of Energy. There is an increasing need for cost reduction technology, power consumption reduction technology, and resource saving technology.

1. Electronic Display Devices
   3D images and flexibility have opened up new application fields for LCDs (liquid crystal displays). Since liquid crystal is a polarization control device and not a light-emitting element, 3D display is a perfectly suitable field of application. With regard to flexibility, there is progress in properties of organic and oxide semiconductors; and flexible displays are being developed using

plastic board. Having liquidity properties and voltage driving force, liquid crystals are a suitable display element for flexible displays. Moreover, display mode and structure that further improves the optical transmittance of backlight was developed as a step for prevention of global warming (OITDA 2009).

Recently, development of technology has driven towards new liquid crystal material, PDP energy saving, shutter data sharing and display systems not requiring glasses to promote 3D, flexible displays, commercialization of small and medium high-definition organic LED panels, and so on. Following the growth of tablet PCs, there has been rapid development of flexible displays with improved portability (OITDA 2011).

Organic electroluminescence (EL) television has been developed, although proper solutions need to be found for issues such as increasing the size, backplane, lower cost compared with liquid crystal displays, and PDP. However, in small size displays of up to 3 in., its market share is gradually growing in fields such as mobile phones and MP3 players. As for the backplane issue, the solution may lie in oxide semiconductors. Moreover, in spectacles-based 3D television, a refresh rate of 240 Hz or more may be required in the future, and the winds may be favorable for organic EL displays due to their fast response speed (OITDA 2009). In the near future, organic EL had no chance against LCD in large displays, but they became an equal competitor in the small to medium field (OITDA 2011).

Active research and development is being done in field emission display (FED) which is equipped with microelectromechanical system (MEMS) technology that uses field emitter array as the electron source. As for the light source, backlight with low environmental burden is being sought for liquid crystal display, and there is much more research on field emission lamps, in which it is possible to manufacture level surface/a typical light source without using mercury that causes environmental pollution. Active R&D has also begun on displays using quantum dots and on MEMS displays that display images by using MEMS micromirrors for optical reflection, diffraction, and interference phenomena or by using MEMS optical shutters for optical modulation (OITDA 2009).

In the large field, new large 3D TVs which could be watched without glasses or with passive glasses were launched in 2011. In the small to medium field, smart phones requiring no glasses, mobile phones, 3D portable game machines, 3D compatible head-mounted displays (HMD), and other various 3D products have been launched (OITDA 2011).

2. Electronic Paper
   In electronic paper, electrophoresis is the most popular method when it comes to digital book readers. However, there are active technological developments in other methods such as powder movement method, twist ball method, bistable liquid crystal method, electrochromic method, electrowetting method, and more. A developmental trend is also seen in applied surfaces in terms of all-purpose paper readers, price tags, POP advertisements, and so on (OITDA 2009).

3. Projection
   Investigations have been made on screen technology, new method optical system technology, light bubble technology, light source technology for projectors, and on the latest situation on all types of projectors (microprojectors, ultrashort focus projectors, 3D projectors, and so on).
4. 3D Display
   A full-blown market for 3D has been expected to be established although common measurement concepts for 3D display need properly formed, and there is a pressing global need to establish the measurement methods. Investigations were made on measurement methods for 3D displays, the status of its international standardization, and the latest situation of various 3D display devices (OITDA 2009).

12.3.4 Human Interface Technology

1. Image Sensors
   With regard to CMOS sensors, backside-illumination (BSI) technology has been widely used, and particularly with enhanced sensitivity and other features, its performance has improved. Progress in miniaturization of pixel size and development of large-size screens is drawing attention to 3D distance measurement, fast imaging, and also deployment in nanophotonics and application in the biomedical field. The applications of CMOS image sensors are expected to increasingly shift from in-vehicle to the biomedical field (OITDA 2011).
2. Earthquake Disaster and the Information and Communication System
   Internet media shall be the main source of information at the time of disasters from now on, and thus, there is a need to develop effective services. In doing so, if one considers the mental state of afflicted people, it is important to develop emergency support systems as an extension of familiar services people use every day (OITDA 2011).
3. Health Care and Optical Technology
   The social system and medical technology supporting health care is comprised of (a) health care as a social system, (b) remote medical service and new health care, and (c) medical information integrated system and cloud services; the most widely used optical interface technologies with health care are display devices, sensor devices, and high-speed wireless technology. Applications are smart phones, mobile terminals, cloud health services, medical information network of ambulances and mobile clinics doing the rounds in disaster areas, home health-care monitoring system using the mobile phone network, video surveillance system at the health-care site in conjunction with tablet terminals and wireless LAN, and so on (OITDA 2011).
4. Image Input Devices
   There have been improvements in the performance and functions of digital cameras, which account for the major share of the image input devices market.

This was brought about by devising a new layout of the microlens added to the photo element of CMOS image sensors, thereby improving the reduction in light quantity (peripheral light quantity) for images in high positions, and by upgrading the view finder for functional advancement. 3D input technology added functions to digital and video cameras. Also, an image input module based on a new concept was introduced as an ultrathin-image input module (OIDTA 2011).

5. Neurorehabilitation Support from Brain–Machine Interface
   Advancements in brain function measurement technology have brought progress in understanding of brain functions and research and development of technology called brain–machine interface (BMI). Potential applications of BMI in neurorehabilitation have been demonstrated, the supporting technology was developed, and some kind of feedback was given to the user's (patient's) brain. This kind of neurorehabilitation with the help of BMI is aimed at helping to rebuild neural circuits in the damaged area or the area around it in the brain (OITDA 2011).

12.3.5 Laser Processing and Optical Measurement

Laser processing and measurement generally cover fiber lasers, hybrid lasers, and chirped pulse oscillators as optical technology for manufacturing; scanners as processing-related technology; bonding technology using various lasers as processing technology; and digital holography and superhigh-speed image measurement as measurement technologies (OITDA 2009).

1. Light Source Technology for Manufacturing.
   The usefulness of fiber laser as the light source in manufacturing has been gaining prominence. Deep ultraviolet pulse light source is already used in factories in inspection equipment. The ultrashort pulse laser has already been used commercially for substrate splitting in the semiconductor field and as LASIK in the medical field, and its usage is expected to spread to other fields. The hybrid laser is a kind of laser combining the advantages of solid-state laser, semiconductor laser diode, and fiber laser, such as a type in which solid-state laser to amplify seed light from stable fiber laser, a type that excites solid state laser with semiconductor laser diode excitation fiber laser, and a type that serially couples multimode fiber and single-mode fiber in a semiconductor laser. In a chirping pulse oscillator, a new scheme is being researched where chirping pulse is directly generated inside the oscillator to replace the existing scheme where the ultrashort pulse from the laser oscillator is stretched outside to obtain the chirping pulse. With this method, amplification of chirping pulse can be done by attaching it as is to the amplifier, and it can be applied in simple laser processing also, as large pulse output can be obtained from the oscillator unit (OITDA 2009).

Among fiber lasers, there was competition between higher brightness and higher output using CW oscillation, but recently there is progress in development of pulsed fiber lasers. Pulsed fiber lasers were also developed in the past, but the pulse range was between femtosecond and several picoseconds. Now there is active development of pulsed fiber lasers in nanoseconds and subnanoseconds. By adopting a master oscillator power amplifier (MOPA) system in fiber lasers, it is possible to amplify seed light which has the desired pulse waveform, and use it in processing. Compared to solid state lasers, a much more stable system can be built with fiber lasers, as they do not need to draw light from free space even if amplified in multistages. Although it depends on the seed light control method, there is progress in development of technology for free selection of pulse width between 5 and 100 ns and for stable supply of picosecond pulses. Also, the development of technology for short pulsed and shorter wavelength lasers is progressing simultaneously. Technologies related to asymmetric polarization reversal structure are able to efficiently convert wavelength into UV light from a low-peak power-infrared CW light source. Moreover, longer life CLBO crystals useful in ultraviolet wavelength conversion have been introduced. High-output CW fiber lasers are widely used in the microprocessing field. However, with the emergence of shorter pulsed and shorter wavelength pulse fiber lasers, their use in the microprocessing field is expected to become widespread. But when using pulse fiber lasers, it is necessary to pay attention to the peak power of pulse rather than the average output (in W). The average output is derived from pulse energy × pulse repetition frequency, and if the pulse repetition frequency is high, the average output will also be high even when pulse energy is small. If pulse energy is small, the peak power is also small, resulting in a small pulse processing variation. If the repetition frequency is too high, the laser will constantly irradiate the material, and the processing phenomenon will become similar to thermal processing using CW lasers. Even with a pulse fiber laser, if processing does not reflect the pulse characteristics, there will be no great difference between pulse fiber laser processing and CW laser processing (OITDA 2011).

2. Scanner Technology.
   Scanner is an essential technology for improving the throughput of laser processing, and high output of lasers for processing is making it indispensable to have high-speed scanning of laser beams. Investigations were made on technologies that use optical encoders in location sensors by digitally controlling the Galvano scanner, which is a digital high-speed scanner that scans mechanically at a high speed. Moreover, the multi-prism scanner and KTN scanner that apply electro-optic effects, in which prism is electrically scanned, were investigated under electro-optic effect scanners (OITDA 2009).
3. Processing Technology.
   Until now, welding, the technique for joining metals, was in the mainstream as a technology for joining similar materials, but technologies are being developed that join different materials such as metals with plastics and others. Investigations have been made on junction technology using ultrashort pulses,

junction technology using CW laser, and junction technology using insert materials. It is now possible to input heat only around the junction part with precise heat input control of the laser. Now two pieces of glass can be joined, and different materials can be joined, for example, glass with metal, metal with plastic (resin), and so on (OITDA 2009).

In addition, laser light can be used to transmit energy through fibers and can achieve contactless material processing. Thus, there are hopes for its use in high-radiation environments by combining with remote control technology. Uses of lasers in pipe repair, thick plate cutting for reactor core disposal, and so on were introduced. The idea of silicon laser slicing has also been presented (OITDA 2011).

4. Measurement Technologies.
   Digital holography is a technology that reproduces numerical values in a computer and uses imaging elements such as CCD to digitize recorded images, instead of using conventional photographic plates. It is expected to be used in various applications since its high-speed processing can provide digitally obtained information to distant places, and the image developing process is not required. A superhigh-speed photography camera was developed in which the photographing speed can be one million frames/second or more. Color video camera broadcasting at two million frames/second is already being used at up to 410×720 (295,200) image pixels, and it is thought that its role will steadily grow with more improvements underway in ultrahigh speed and ultrahigh sensitivity (OITDA 2009).

   Moreover, lidar has been used for observation of vegetation on earth from the International Space Station. In addition, forestry biomass can be monitored by measuring the extent of life or height of trees from above (OITDA 2011).

12.3.6 Medical Technology in the Optical Industry

The trends in applications of optical technology to the medical industry and specifying the problems inherent to medical applications have been investigated. Optical coherence tomography (OCT) is a noninvasive tomographic imaging diagnostic technology. Near-infrared light topography uses brain blood flow information to obtain new knowledge on brain functions.

1. Optomedical Technology
   OCT is a technology that gives a tomographic image of a living body to a subepidermal depth of 1–2 mm at a spatial resolution of 10 μm. It is a biometry technology with simple equipment configuration that cleverly uses characteristics of optical waves. Based on low coherence optical interference, OCT selectively detects the direct optic elements being reflected from the interstitial part when light is exposed to the body tissue, and based on this, it constructs a 2D tomographic image. This technology was proposed in the beginning of 1990 and was actually implemented only 5 years later in ophthalmology for retinal diagnostics.

Now it is an indispensable method of examination in clinical ophthalmology and is expected to see further progress in the future. Moreover, with progress in achievement of high-resolution and high-speed examinations, starting with dermatology, its scope of clinical applications is spreading to digestive system surgery, obstetrics and gynecology, urology, etc. (OITDA 2009).

The near-infrared topography method was developed in 1995 based on the principles of brain blood flow measurement using a near-infrared spectroscopy method proposed. It images intracerebral hemodynamic variations in brain activity. By developing this from spot measurement to surface measurement, it has been implemented in clinical procedures such as determination of the language dominant hemisphere and examination of neurological seizure position. Moreover, in the fundamental neuroscience field, previous measurement technology had a difficult approach in measuring the language functions in newborn babies immediately after birth. With this technology, there is a newfound knowledge in this area, and it is becoming an important methodology in development science. Further, in near-infrared light topography, many parts can be configured in the semiconductor, and a noteworthy advantage is that it can be miniaturized in principle. Its expansion to versatile brain function imaging technology is expected, with progress in research and development (OITDA 2009).

2. Laser Medical Treatment Approved as Advanced Medical Treatment
   Advanced medical care is a health-care system and recognized as a system that uses ultramodern diagnostic technologies and health insurance treatment meant to ensure convenience and safety of the citizens, prevent increased burden on patients, and provide a wider range of options for citizens. It is important to keep track of optical technology approved as advanced medical care, because insurance is expected to be provided for it. In the future, a top priority would be to first understand the indications and their mechanisms in a scientific manner and then execute the treatment with laser (OITDA 2009). For instance, the investigation of laser treatment of varicose veins in the legs is attracting attention as a highly efficient method of treatment, as the rate of treatment success after 5 years is 95.4 % (OITDA 2011).
3. Photoacoustic Diagnosis
   Photoacoustic diagnosis technology is used for diagnosis after forming images from the propagation time of acoustic waves generated by light absorption in organisms or from the signal intensity. If a wavelength which can be easily absorbed by blood is used, a 3D image of the vascular network can be obtained. It is hoped this will be useful in diagnosis of cancer, brain functions, and arteriosclerosis (OITDA 2011).

12.3.7 Organic Devices and Materials

In addition to optical active and passive devices such as organic electroluminescent (EL) devices and polymer optical waveguides, the investigation of this area has covered organic device-related materials technologies and device processing and manufacturing technologies.

1. White Organic EL Illumination Technology
   The development of an organic EL device and its practical application has progressed since the base technology was developed. A high efficiency, longer life white organic EL element has been achieved. With further improvements in EL efficiency and progress in manufacturing technology, it will reach the same level as conventional LEDs. Basically, organic EL devices are considered to have the role of surface light sources and LEDs as point light sources. That is why a combination of both devices is expected to achieve a high-quality next-generation illuminated space which is both energy saving and comfortable (OITDA 2011).
2. Organic Transistor
   The development of new organic semiconductor molecules is very actively underway. In p-type materials, DNTT-based and C8-BTBT-based organic transistors with high mobility and stability can be obtained. On the other hand, in n-type materials also, high mobility has been obtained for PDIF-CN2 single-crystals. Processing using the printing method excels in productivity, and progress of research on this is directly linked to the development of low priced organic semiconductor devices, so it is gaining increasing attention in a wide range of industries other than the electronics industry, such as the printing industry (OITDA 2011).
3. Polymer Optical Waveguide
   Optical waveguides can realize complexity and high-density wiring and are also suitable for optoelectronic hybrid integration. As a result, it is being considered to use optical waveguides as short-distance optical interconnects at the board-to-board, the chip-to-chip, or on-chip level.

   It is hoped that polymer optical waveguides in particular will be widely used in board-level optical interconnection. Studies of materials which have low transmission loss, greater heat stability, and moisture resistance are being conducted. Also, modules using polymer optical waveguides are under development (OITDA 2011).
4. Organic Energy Conversion Devices.
   There is a focus on using organic energy conversion devices for an energy harvesting system which harvests energy from the surrounding environment (light, heat, oscillation, air, and so on.) and converts it into electricity. At present, the research is in the early stage, and various issues remain problems before it is put to practical use. However, these are promising devices for harnessing electrical energy in the future (OITDA 2011).
5. Organic Conductive Materials
   Research and development on organic conductive materials has been more active. This was triggered by the problem of depletion of resources of indium, a main component of tin-doped indium oxide (ITO) commonly used in

electrodes of FPDs such as LCD panels. Polyaniline, polypyrrole, and poly(3,4-ethylenedioxythiophene) are some of the leading candidate organic conductive materials, and studies focused on improving their conductivity are being conducted (OITDA 2011).

6. Nanoimprint Lithography
   There is progress in research and development on nanoimprint lithography using thermoplastic resin or light curing resin, a processing technology for forming nanoscale microstructures easily at low cost. The field of application has widened, and nanoimprint lithography is now reportedly being used in organic transistors and optical device patterning, in addition to optical elements. It is hoped that nanoimprint lithography will advance as a new micro-processing technology in the future, in addition to the conventional photolithography and electron beam lithography (OITDA 2011).
7. Inkjet Technology
   The application of inkjet technology in electronic device manufacturing is gaining attention as a technology which will bring a revolutionary change in conventional manufacturing processes, because it does not require vacuum technology, requires less amount of energy, and enables environmentally friendly manufacturing. There is active research and development on ink materials and ink surface energy controlling technology for use in an organic transistor manufacturing process and various other processes (OITDA 2011).

12.4 Summary

Many optical devices including various waveguides have been focused on the applications in the wavelength range of several 100 μm to near 200 nm. The terahertz band is the extended wavelength range between radio and optical waves, and development is progressing for higher-frequency radio waves and longer optical waves (lower-frequency waves). Since the applications contain properties of both radio and optical waves, terahertz wireless, for example, was examined in terms of increasing the speed (higher frequency waves) of millimeter wave wireless. The wavelength range has been determined due to dispersions or losses of fiber in the near-infrared range for optical communication, and component development in each wavelength band is in progress depending on factors such as transmission distance, speed, and purpose, such as multilevel photonic devices that contribute to update of communication records for the trunk line system; modules associated with standardization and speedup of Ethernet systems; optical sources and Si photonics; photoreceptors and modulators; light-emitting devices; and integration devices. In the visible and ultraviolet range, there has been progress in achieving high output in the deep ultraviolet range and lengthening of long waves in nitride-based semiconductor devices, such as green color in a nitride-based semiconductor laser; blue violet–blue semiconductor lasers, ultraviolet LEDs, and GaN-based surface-emitting lasers; and white LED lighting and organic EL lighting. In addition to basic technologies and devices for new applications, a lot of device development and module development work for standardization and energy conservation have been performed from terahertz and mid-infrared bands, near-infrared (for optical communications), and visible and ultraviolet range.

Emerging applications of micro-optics and nanophotonics are envisaged in diverse areas such as information processing, high-speed data communication systems, imaging, lighting, displays, manufacturing, life sciences and health care, safety and security, and more. The research aspects of micro-optics and nanophotonics have covered silicon photonics, spatial and spatiotemporal optical solitons (alias light bullets) in microwaveguide arrays and in arrays of evanescently coupled silicon-on-insulator nanowires, spatial solitons in photorefractive materials, nanoplasmonics, photonic crystals, metamaterials for micro- and nanophotonics including optical materials with negative refractive indices, terahertz radiation and its applications, and solid-state single-photon sources and nanometric size optical cavities for quantum information processing.

The future development of optical technology may cover or focus on information-processing photonics, safety and security photonics, optical user interface, optical communications, and optical processing. Given the rapid growth of studies in micro-optics and nanophotonics, one can expect many new and exciting developments over the next decade. No doubt, soon one can expect a maturity of these fast growing research fields, leading to new and interesting physical phenomena and to the utilization of their huge technological potential in various areas of human activity, such as in reducing energy consumption in lighting systems, developing novel biomedical and chemical sensors, fabrication of optical sensors able to detect even single molecules, designing of new solar cells with enhanced efficiency, and implementation of ultrahigh-speed telecommunication systems.

References

Baffou G et al (2010) Mapping heat origin in plasmonic structures. Phys Rev Lett 104:136805

Benton CJ et al (2008) Spatiotemporal quasisolitons and resonant radiation in arrays of silicon-on-insulator photonic wires. Phys Rev A 78:033818–033826

Claps R (2003) Observation of stimulated Raman scattering in silicon waveguides. Opt Express 11:1731–1739

Corcoran B et al (2009) Green light emission in silicon through slow-light enhanced third-harmonic generation in photonic-crystal waveguides. Nat Photon 3:206–210

Dragoman D (2011) Complex conjugate media: alternative configurations for miniaturized lasers. Opt Commun 284:2095–2098

Erni R (2009) Atomic-resolution imaging with a sub-50-pm electron probe. Phys Rev Lett 102:096101

Faryad M, Lakhtakia A (2011) Propagation of surface waves and waveguide modes guided by a dielectric slab inserted in a sculptured nematic thin film. Phys Rev A 83:013814

Gorbach AV et al (2010) Spatiotemporal nonlinear optics in arrays of subwavelength waveguides. Phys Rev A 82:041802

Hiscocks MP et al (2009) Slot-waveguide cavities for optical quantum information applications. Opt Express 17(9):7295–7303
Hon NK et al (2009) Periodically poled silicon. Appl Phys Lett 94:091116–091118
Hu EL, Brongersma M, Baca A (2011) Applications: nanophotonics and plasmonics. In: Roco MC, Mirkin CA, Hersam MC (eds) Nanotechnology research directions for societal needs in 2020: retrospective and outlook. Science Policy Reports. Springer, Dordrecht, pp 417–444
Iancu V et al (2011) Calculation of the quantum efficiency for the absorption on confinement levels in quantum dots. J. Nanoparticle Res 13:1605–1612
Joannopoulos JD et al (2008) Photonic crystals: molding the flow of light, 2nd edn. Princeton University Press, Princeton
Jun YC et al (2009) Broadband enhancement of light emission in silicon slot waveguides. Opt Express 17(9):7479–7490
Kocaman S et al (2009) Observations of zero-order bandgaps in negative-index photonic crystal superlattices at the near-infrared. Phys Rev Lett 102:203905
Krasavin AV, Zayats AV (2010) Silicon-based plasmonic waveguides. Opt Express 18(11):11791–11799
Malomed BA et al (2005) Spatiotemporal optical solitons. J Opt B: Quantum Semiclass Opt 7:R53–R72
Mihalache D (2011) Recent trends in micro- and nanophotonics: a personal selection. J Optoelectron Adv Mater 13(9):1055–1066
Muellner P et al (2009) Nonlinearity of optimized silicon photonic slot waveguides. Opt Express 17(11):9282–9287
Noginov MA et al (2009) Demonstration of a spaser-based nanolaser. Nature 460:1110–1112
Oh SJ et al (2009) Nanoparticle-enabled terahertz imaging for cancer diagnosis. Opt Express 17(5):3469–3475
OITDA (2009) Optoelectronic technology trends. Annual Technical Report. pp 17–25
OITDA (2011) Optoelectronic technology trends. Annual Technical Report. pp 18–26.
Sandu I et al (2011) Synthesis of optical transparent and electrical conductive polymer/nanocarbon composite films by infiltration method. Thin Solid Films 519:4128–4131
Stavarache I et al (2011) Electrical behavior of multi-walled carbon nanotube network embedded in amorphous silicon nitride. Nanoscale Res Lett 6:88
Suhir E (2000) Microelectronics and photonics—the future. Microelectron J 31:839–851
Takahashi Y et al (2007) High-Q nanocavity with a 2-ns photon lifetime. Opt Express 15:17206
Valentine J et al (2008) Three-dimensional optical metamaterial with a negative refractive index. Nature 455:376–379
Welsh GH, Wynne K (2009) Generation of ultrafast terahertz radiation pulses on metallic nanostructured surfaces. Opt Express 17(4):2470–2480
Ye F et al (2011) Subwavelength vortical plasmonic lattice solitons. Opt Lett 36:1179–1181
Zhang S et al (2005) Experimental demonstration of near-infrared negative-index metamaterials. Phys Rev Lett 95:137404

Index

X.C. Tong, *Advanced Materials for Integrated Optical Waveguides*, Springer Series in Advanced Microelectronics 46, DOI 10.1007/978-3-319-01550-7,

T

Printed by Printforce, the Netherlands